# Fundamentals of mold growth in indoor environments and strategies for healthy living

# Fundamentals of mold growth

## in indoor environments and strategies for healthy living

### edited by:

**Olaf C.G. Adan**
**Robert A. Samson**

*Wageningen Academic*
P u b l i s h e r s

ISBN: 978-90-8686-135-4
e-ISBN: 978-90-8686-722-6
DOI: 10.3921/978-90-8686-722-6

**Photo cover:**
Robert A. Samson
CBS Fungal Biodiversity Centre,
Utrecht, The Netherlands

**First published, 2011**

© **Wageningen Academic Publishers**
**The Netherlands, 2011**

# Preface

Molds play a crucial role in our daily life, affecting our well-being both in a positive and negative way. Molds can be found nearly anywhere, being present outdoors and indoors. They are spoiling food as well as giving it specific flavors, they are digesting plant and animal matter, they cause defacement of interior surfaces and they may release lightweight spores and fragments that travel through the air.

In our built environment, molds are often associated with dampness, introducing moisture control as the key to mold control. And for many years, molds and their implications to human health have been prominently on the agenda, referring to increased risk of respiratory symptoms, respiratory infections and exacerbations of asthma.

Today, indoor mold and moisture, and their associated health effects, are a society-wide problem. The economic consequences of indoor mold and moisture are enormous. Their global dimension has been emphasized in the 2009 Guidelines on Indoor Air Quality: dampness and mould of the World Health Organization, stating that: "The most important means for avoiding adverse health effects is the prevention (or minimization) of persistent dampness and microbial growth on interior surfaces and in building structures". Similarly, Krieger *et al.* (2010) concluded in their review for the US Centers for Disease Control that one of three interventions ready for implementation in houses was: "Combined elimination of moisture intrusion and leaks and removal of moldy items". These conclusions are in line with the leading principle of this book.

In this era of progress and prosperity, scarcity is now becoming a key problem globally. Scarcity in natural resources, scarcity in energy. In the next decennia, energy efficiency will dominate residential building and construction. First steps in response to the energy crises in the 20<sup>th</sup> century – i.e. thermal insulation and increased air tightness of the building envelope – obviously led to dampness problems. The question rises if next steps in energy efficiency will introduce new risks for adverse health effects of indoor molds.

This book aims to describe the fundamentals of indoor mold growth as a prerequisite to tackle mold growth in present building as well as in future energy efficient building. Without doubt, water is the key factor. A profound understanding of the mold-water relation lays the foundation for control strategies in any building, present and future.

The common approach to control mold growth risks on the basis of the *ambient* air humidity alone is no guarantee at all for a "mold-free" environment. Short humidity peaks may result in mold growth. This relation between mold growth and climate dynamics is particularly addressed in this book.

The book brings together different disciplinary points of view on indoor mold, ranging from physics and material science to microbiology and health sciences. The contents have been outlined according to three main issues: (1) *fundamentals*, particularly addressing the crucial roles of water and materials, (2) *health*, including a state-of-the-art description of the health-related effects of indoor molds, and (3) *strategies*, integrating remediation, prevention and policies. The latter has been added, as information dissemination among all relevant stakeholders is an essential step towards strategies for achieving healthy indoor environments.

The editors like to thank the authors for their excellent contributions and particularly David Miller, who also has continuously supported our research. We greatly acknowledge Jeannette Schouw of TNO for her indispensable support in (cross) checking references and editing figures.

*Olaf Adan and Rob Samson*

## References

Krieger J, Jacobs DE, Ashley PJ, Baeder A, Chew GL, Dearborn D, Hynes HP, Miller JD, Morley R, Rabito F and Zeldin DC (2010) Housing interventions and control of asthma-related indoor biologic agents: a review of the evidence. J Public Health Manag Pract 16(5) E-Supp: S11-S20.
World Health Organization (2009) WHO guidelines for indoor air quality: dampness and mould. WHO Regional Office for Europe, Copenhagen, Denmark, 228 pp.

# Table of contents

Preface                                                                                    7

1   Introduction                                                                          15
    *Olaf C.G. Adan and Robert A. Samson*

    Introduction                                                                          15
    Why is this book different?                                                           16
    The societal context of indoor fungal growth                                         18
    Introduction to indoor fungi                                                          21
    Water as the key factor                                                               22
    References                                                                            33
    Appendix 1. Thermodynamic definition of water activity and relative
               humidity                                                                  36

## Fundamentals

2   Water relations of fungi in indoor environments                                      41
    *Olaf C.G. Adan, Henk P. Huinink and Mirjam Bekker*

    Introduction                                                                          41
    Fungal response to long term variations in indoor climate                            42
    Short term variations in indoor climate                                              54
    Conclusions                                                                           63
    References                                                                            64

3   Fungal growth and humidity fluctuations: a toy model                                 67
    *Henk P. Huinink and Olaf C.G. Adan*

    Introduction                                                                          67
    A toy model for growth                                                                68
    Growth scenarios                                                                      73
    Comparison with experiments                                                           75
    Signposts for a better model                                                          78
    Conclusion and outlook                                                                80
    References                                                                            82

4   The fungal cell                                                                 83
    *Jan Dijksterhuis*

    Introduction                                                                    83
    Fungal growth                                                                   83
    The fungal colony as a unity                                                    87
    Considerations about aerial hyphae and fungal survival                         87
    References                                                                      98

5   Ecology and general characteristics of indoor fungi                           101
    *Robert A. Samson*

    Introduction                                                                   101
    Nomenclature                                                                   110
    References                                                                     114

6   Characteristics and identification of indoor wood-decaying
    basidiomycetes                                                                 117
    *Olaf Schmidt and Tobias Huckfeldt*

    Fungal species and significance                                                117
    Dry rot fungi: *Serpula* species, *Leucogyrophana* species, *Meruliporia*
    *incrassata*                                                                   123
    Cellar fungi ("wet-rot fungi"): *Coniophora* species                          135
    Indoor polypores: *Antrodia* species and *Oligoporus placenta*                139
    Diplomitoporus lindbladii                                                      151
    *Asterostroma* species                                                         152
    *Tapinella panuoides (Paxillus panuoides)*, stalkless paxillus, oyster rollrim 152
    *Trechispora* species                                                          154
    *Dacrymyces stillatus*, orange jelly                                           154
    Phellinus contiguus                                                            156
    Identification and characterization                                            156
    References                                                                     169

# Health

7   Health effects from mold and dampness in housing in western societies:
    early epidemiology studies and barriers to further progress          183
    *J. David Miller*

    Introduction                                                          183
    Early large-scale studies linking dampness and mold to health         187
    Biomarkers                                                            195
    Conclusions                                                           201
    Acknowledgements                                                      202
    References                                                            202

8   Aerosolized fungal fragments                                         211
    *Brett J. Green, Detlef Schmechel and Richard C. Summerbell*

    Introduction                                                          211
    The process of fungal fragmentation                                   215
    Contributions of hyphal fragments and particulates to the environment 218
    Collection, enumeration and quantification of hyphal and fungal fragments 224
    Implications for human health                                         231
    Conclusions and future perspectives                                   234
    Acknowledgements                                                      234
    References                                                            235

9   Mycotoxins on building materials                                     245
    *Kristian Fog Nielsen and Jens C. Frisvad*

    Introduction                                                          245
    Analytical methods                                                    247
    The fungi                                                             249
    Exposure                                                              264
    Conclusion                                                            265
    Acknowledgements                                                      265
    References                                                            266

10 WHO guidelines for indoor air quality: dampness and mold        277
   *Otto O. Hänninen*

   History of who guidelines for air quality                        277
   Focusing on indoor air                                           281
   State of the scientific evidence                                 286
   Potential global guideline targets                               292
   Indicators of dampness and microbial growth                     292
   Chemicals with specific role indoors                            294
   Relationship of guidelines to health impact assessment and management  296
   Interventions                                                    297
   Acknowledgements                                                 301
   References                                                       301

## Strategies – measuring

11 Moisture content measurement                                    305
   *Bart J.F. Erich and Leo Pel*

   Introduction                                                     305
   Basics of moisture measurements                                 305
   Measuring the moisture content in materials                     313
   Measuring moisture content in air                               324
   Measuring the moisture content on the scale level of the fungus 331
   References                                                       333

12 The fungal resistance of interior finishing materials          335
   *Olaf C.G. Adan*

   Introduction                                                     335
   Present standards to assess fungal resistance of materials for interior
   applications                                                     336
   A new concept to assess fungal growth                           341
   Next steps for improvement                                      344
   Pilot application on a wide range of materials                  346
   Towards performance requirements in building codes             349
   References                                                       350

13 Detection of indoor fungi bioaerosols                                    353
   *James A. Scott, Richard C. Summerbell and Brett J. Green*

   Introduction                                                             353
   Environmental sampling                                                   354
   Bioaerosols                                                              356
   Air sampling                                                             360
   References                                                               373

# Strategies – remediation

14 Mold remediation in North American buildings                            383
   *Philip R. Morey*

   Introduction                                                             383
   Examples of mold remediation strategies                                  386
   New York City guidelines                                                 395
   Containment and dust suppression during mold remediation                 397
   Quality assurance during mold remediation                                400
   North American guidelines on mold remediation                            404
   References                                                               409

15 Mold remediation in West-European buildings                             413
   *Thomas Warscheid*

   Introduction                                                             413
   Basic causes of indoor mold infestation                                  414
   Risk-assessment of mold growth                                           415
   Sanitation of mold infestations – goals and limits                       416
   Mold guidelines in Germany                                               417
   Limits and risks of mold guidelines                                      420
   Gaps and necessary improvements in mold guidelines and their analytical
   approach                                                                 422
   Challenges for future mold guidelines                                    424
   Health implications of indoor mold                                       425
   Intervention steps in sanitation                                         426
   Future prospects                                                         428
   References                                                               430

16 Protection of wood                                                    435
   *Michael F. Sailer and Waldemar J. Homan*

   Introduction                                                          435
   Wood protection by biocides                                           438
   Other ways of wood protection                                         446
   New developments                                                      458
   References                                                            459

17 Coating and surface treatment of wood                                 463
   *Hannu Viitanen and Anne-Christine Ritschkoff*

   Introduction                                                          463
   Microbial activity on coated surfaces                                 463
   Biocidal protection of paints on wood                                 468
   Novel trends in coating technology                                    473
   Testing the resistance and performance of coatings on wood            481
   References                                                            484

## Recommendations

18 Recommendations                                                       491
   *Olaf C.G. Adan and Robert A. Samson*

   Introduction                                                          491
   Fundamentals                                                          491
   Health                                                                492
   Strategies                                                            492
   References                                                            498

Contributors                                                             499

Index                                                                    505

# 1 Introduction

*Olaf C.G. Adan[1,2] and Robert A. Samson[3]*
*[1]Eindhoven University of Technology, Faculty of Applied Physics, Eindhoven; the Netherlands; [2]TNO, Delft, the Netherlands; [3]CBS-KNAW Fungal Biodiversity Centre, Utrecht, the Netherlands*

## Introduction

The mere facts that some 25% of social housing (i.e. more than 14 million units) in the 27 member states of the European Union is suffering from dampness and molds (Bonnefoy *et al.* 2003) and that approximately 4.6 million of current US asthma cases are estimated to be attributable to dampness and molds exposure (Mudarri and Fisk 2007) underline that indoor molds should be considered as a widespread and profound societal problem.

In the next decennia, energy efficiency will dominate residential building and construction. The European Commission has set a clear ambition in energy efficiency, i.e. a 20% reduction in greenhouse gas emission and energy consumption in new buildings in 2020. The Copenhagen Conference in 2009 underlined that climate change is one of the biggest challenges of our time and that deep cuts in the global emissions are needed to hold the increase in global temperature below 2 degrees Celsius at the end of this century (Anonymous 2009). It has been recognized internationally that clean production, increased green energy and energy efficient use will be key instruments to combat climate change.

The first real steps in energy conservation in the built environment were made in response to the energy crises of the 70s in the 20th century, as to reduce the dependency of the Western world on foreign energy supply and the economic consequences of rising energy prices. Since that time, molds and their implications on human health have been prominently on the agenda, as improved thermal insulation and increased air tightness of the building envelope obviously led to problems of dampness. The fear for mold problems – sometimes justified, mostly not – may have a restraining influence on the future upgrade of building energy performance.

Did we learn from our mistakes in the past? Do we really understand how to prevent mold growth in new and energy efficient buildings, and how to avoid molds and dampness after refurbishment? How to maintain a healthy indoor environment? Energy efficiency is also an important goal for health, but do new risks occur for adverse health effects of indoor mold?

*Olaf C.G. Adan and Robert A. Samson*

This books aims to describe the fundamentals of indoor mold growth, as a prerequisite to tackle mold growth in present building as well as in future energy efficient building. Like all living processes, mold growth requires the availability of water. Therefore, the interaction of molds and water forms a pivotal issue in this book. A sound understanding of this relationship lays the foundation for control strategies in any building, present and future. In that respect, the books gives follow-up to the recommendations of the 2005 International workshop "Fungi in indoor environments; towards strategies for living in healthy buildings" in the Netherlands that stressed the need to outline and disseminate the fundamental principles. The final Chapter integrates the main findings of this international forum of experts in the recommendations of this book.

## Scope

The book focuses on mold problems originating from airborne moisture in particular, being more complex than commonly assumed. It does not deal with the self-evident problems that are due to building defects – such as rain penetration or leakage.

## Why is this book different?

Numerous books exist with respect to indoor molds, addressing either the indoor mycobiota, their implications on human health or good practices of prevention and remediation. The understanding of causes of indoor mold growth is rooted in the building physics of the last century, and primarily refers to molds and surface condensation. In the late 80s, the causal relationship between surface condensation and mold growth received particular attention in the scientific community in response to the energy crises of the 70s – e.g. the IEA Program Annex XIV "Condensation and Energy" (Anonymous 1991) – as energy conservation measures obviously introduced a flood of indoor mold problems.

The interaction between thermal insulation and ventilation – key elements in the building energy balance – formed the heart of the scientific work at that time. On the one hand, thermal bridging, i.e. thermal short-circuiting in the building envelope, formed a pivotal issue, which resulted in many commercially available calculation tools and design aids for building practice. On the other, proper ventilation strategies, including end-user behavior, were studied profoundly. First, the main communication was to "close the windows" and seal air gaps and leaks, in order to reduce energy losses. Next, in many countries public campaigns were run spreading the opposite message to "open the windows" as to ensure a healthy indoor environment. For obvious reasons, these conflicting communications introduced much confusion in society. Nevertheless, the challenge to find a balance between energy conservation and good indoor environmental quality led to the introduction of more sophisticated ventilation systems and products (i.e. balanced ventilation, ventilation with heat recovery), which still is a main field of interest for research and development.

Nowadays, once again energy and energy efficiency have become major societal issues, accompanied by the introduction of greener systems and products. This time the drive behind it is more urgent in the context of sustainable development. Building physical principles for proper insulation are well known and thermal insulation products have become mature, even though a big challenge still exists to develop compact solutions for the existing stock (e.g. vacuum insulation panels). Huge progress has been made in ventilation systems, with many advanced systems on the market. However, despite these big steps forward, dampness and molds form a bottleneck in our current residential stock and their occurrence in future building is a question mark. It has become obvious that *neither thermal insulation nor ventilation forms a guarantee for a mold-free environment.*

This book differs from previous works. It brings together many disciplines, amongst which microbiology, materials science, physics and public health. The differences are found in a state-of-the-art discussion of three fundamental issues, which are not described in such comprehensive way in previous books:

### Difference 1. The response of molds to indoor climate dynamics

Worldwide observations point out that mold growth frequently occurs in indoor environments that are considered relatively dry – on average. In modern dwellings, about 60% of problem cases suffer from disfigurement in bathrooms, 40% in kitchens and only few % in other rooms. Mostly, the average air humidity is rather low, i.e. below 50% relative humidity, underlining the crucial role of dynamics.

Indoor climate dynamics are the rule rather than the exception. In this book, the present knowledge on fungal response to humidity dynamics is brought together, addressing both the scale levels of the biological cell and surface colony growth.

Commonly, fungal growth under transient humidity conditions is considered on the basis of proportional scaling of steady state growth data. This usual approach is oversimplified and incorrect as it ignores inertia effects of both the organism and the substrate. The living organism may influence the water balance at the substrate-air interface, either *passively* (e.g. as a diffusion resistance) or *actively* (e.g. through intra-cellular and extra-cellular mechanisms). The substrate may be porous, offering a large internal surface and volume that may act as a reservoir for moisture, leading to humid surface conditions even when the indoor air is dry. Both issues are discussed in this book (Chapters 2 and 3).

### Difference 2. The crucial role of materials in control strategies for indoor mold

For daily humidity fluctuations, i.e. particularly during showering and cooking, a material layer in the order of some tenths of millimeters to some millimeters at most fulfills the reservoir function, whereas the effective depth is normally much bigger in terms of thermal response. From the point of view of the fungus, such thin surface layer plays a crucial role to overcome periods of drought. *We propose an integral approach of thermal insulation and heating, ventilation and a proper choice of finishing materials as the foundation for any control strategy for molds in present and future indoor environments.* Especially in energy efficient building, surface materials form the missing link to effectively reduce indoor mold risks.

Consequently – considering fungal resistance a material property – assessment of interior finishes should take account of inertia effects. This principle should lead to better testing, which is outlined in a dedicated Chapter (i.e. Chapter 12) of the book.

Industrial trends towards eco-friendlier materials are often accompanied by an apparent increase in product susceptibility to molds, stressing the role of future materials in control strategies even more. Therefore, the book deals with recent (nano-based) advances in material technology to counteract this effect.

### Difference 3. Newest insights into adverse health effects

This book gives a state-of-the art description of the adverse health effects of indoor molds, showing that sensitization to mold is a significant (and possibly increasing) risk factor for asthma and that true markers of exposure should be identified (Chapter 7). Furthermore, it particularly addresses the recent findings with respect to aerosolized fungal fragments that highlight the importance of continued research to understand the potential health effects associated with fungal fragment exposure (Chapter 8).

## The societal context of indoor fungal growth

The problem of mold growth was familiar to our ancestors in the most ancient of times. Probably one of the oldest control strategies of fungal growth in houses is found in the Bible, Leviticus 14: 33-14: 57, where the removal of contaminated spots on walls and renders was prescribed as a first step. If not successful, the second step involved the house being torn down.

Molds still play a role in modern life and, understandably, are often associated with dampness (see Figure 1.1). In the Netherlands, a recent inventory of the housing stock shows that fungal problems occur in some 15% of Dutch dwellings. Belgium reported 20% in social housing, Germany about 30% and in the UK earlier indications were that 20-25% was severely affected. Dampness problems on this scale are not just typical of the temperate West-European climate, and may be found in warmer areas. In the coastal region of Israel, containing more than 50% of the country's population, some 45% of dwellings were found to suffer from condensation and mold problems.

World-wide observations pointed out that mold growth frequently occurs in indoor environments that are considered relatively dry – on average. In modern dwellings, about 60% of problem cases suffer from disfigurement in bathrooms, 40% in kitchens and only a few percent in other rooms. Mostly, the average air humidity is rather low, i.e. below 50% relative humidity, underlining the crucial role of dynamics. This value is far below the 80% RH, which is generally adopted by the scientific community as a threshold value to keep susceptible materials mold-free.

*Figure 1.1. A typical example of severe defacement of walls and ceiling due to mold growth in a bathroom.*

*Olaf C.G. Adan and Robert A. Samson*

Estimates of costs for repair, maintenance and redecoration are in the range of 1% of the annual turnover of the construction sector. For Europe, these costs will be in the order of 10 billion € per year, of which several billions are directly connected with mold.

### Implications for public health

Fungal growth does not only address esthetics and material degradation, but also health issues, like hypersensitivity, respiratory infections and allergic reactions such as rhinitis and asthma. It is only in recent years that the scientific community has become aware of the health effects of indoor fungi, in residential as well as in occupational buildings. Expert groups in both Europe (Bornehag *et al.* 2001, 2004) and the US (Institute of Medicine 2004) concluded that sufficient scientific evidence supports the association between indoor mold and adverse health effects for building occupants. The potential threat to human health as a result of indoor exposure to spores was brought to the attention by a report of deadly syndrome in infants in Cleveland, Ohio. Ever since, more than 40 similar cases were reported.

The World Health Organization (Bonefoy *et al.* 2003) recently showed in a pan-European study that residents in approximately 25% of the European social housing stock – i.e. more than 45 million people – are exposed to increased (i.e. 30-50%) health risks associated with exposure to indoor molds. First order approximations of the annual economic consequences, in terms of health care and worker loss, refer to tens of billion €. In the US, approximately 4.6 million cases of asthma (of a total of 22 million asthma cases) result from exposure to dampness and mold. The consequent economic cost is estimated to be 3.5 billion annually (Mudarri and Fisk 2007). For Europe, the sum of healthcare costs and lost work days as related to mold amounts to 5.8 billion (of a total of approximately 28 billion € asthma related costs).

### Legal context

Past years a rapid escalation of mold claims against builders and their insurance companies has occurred in the US, resulting in some thousands $ in the year 2000 to more than 3 billion $ in 2002. Melinda Ballard in Texas was the first house-owner in 2001 that received 32 million $ for health damage due to indoor mold. The rapid growth in mold litigation led to a growing tendency to drop mold coverage from insurance policies and to a shift from building insurance claims to liability or third party insurance claims.

Although until now mold claims in Europe are rather small and primarily focus on material damage, a member of The Supreme Court in the Netherlands early pointed out to be alert (Hoge Raad The Hague 2006).

## Introduction to indoor fungi

### The mycobiota

Research on the mycobiota in the indoor environment of several decades has resulted in a solid knowledge about the species composition. In the early studies on the fungi occurring indoors identification of the species was often incorrect mainly due to the lack of reference work and poor mycological skill of the researches. This was mainly reflected in using old and outdated nomenclature and not specifying species in common genera such as *Penicillium*, *Aspergillus* and *Cladosporium*. It is also important to note that much attention has been given to the ecological properties of the species and relating their occurrence with availability of water.

### Classification of indoor fungi

The classification of indoor fungi has undergone a radical chance in the last ten years. Formerly the classification and identification was many based on phenotypical characters and with the introduction of biochemical and molecular methods the genus and species concept has dramatically refined. In genera such as *Penicillium*, *Aspergillus* and *Fusarium* the species concept has been redefined by using a polyphasic approach, combing phenotypical, biochemical and molecular characters. The result is that the common indoor species are now better defined while in some cases this modern classification revealed undescribed taxa. For example in the Section *Usti* of *Aspergillus*, *Aspergillus ustus* used to be considered to be common, but polyphasic taxonomic studies showed that the new species *A. calidoustus* proved to be a common contaminant in houses.

### Surface growth versus degradation

Many fungi can penetrate and biodegrade materials. This mostly occurs in nature when certain species can grow inside plant or insect tissue with the help of special hyphae or organs and enzymes. The principal is valid for human pathogenic species. However, most indoor molds are restricted to surface growth with the exception of wood degrading species of basidiomycetes. Species, which occur on building materials, are saprophytes and their capability to grow on such substrates is the availability of nutrients and water. No detailed study of the surface growth by indoor mold has been carried out yet, but these observations are based on numerous microscopic analyses of many building samples.

### Health implications

The impact of the molds in indoor environments for society reached high levels, but the interest in these microorganisms often is not based on sound scientific data. Molds are always and everywhere present and their dominant presence should be avoided in any space where people or animals live. However, it is important to note that not all molds can evoke a health hazard or that molds all produce toxic substances which can be inhaled or exposed. The so-called "toxic black mold" is represented by species of *Stachybotrys* which in many cases is confused with other darkly pigmented species such as *Cladosporium* which do not produce any significant toxins. In Chapter 7 by Miller the health effects from mold and dampness in housing in western societies are described in great detail.

## Water as the key factor

There is no mold growth without water. Water can occur in different forms, it originates from different sources, and its availability can be expressed in many ways. Some parameters describing availability refer to the real state of water or water potential, i.e. how easily it can be taken up, and others refer to the amount of water that is available, i.e. are a capacitive quantity. This paragraph aims to give an overall introduction into water as the key factor for mold growth, and discusses the different quantities for water availability that are used throughout this book. In the appendix to this Chapter, the thermodynamic backgrounds are described.

### Water availability

#### The state of water in the air

The relative humidity RH is commonly used to describe the state of water in the environment of molds. RH is defined as the ratio between the actual water vapor content ($x$) per m$^3$ of air and the maximum amount of water vapor – so-called saturated content – that can be contained in the same volume at the given temperature. This is expressed as:

$$\text{RH} = \frac{x_{actual}}{x_{saturated}} \times 100\%,$$

ranging by definition between 0 (completely dry) and 100% (saturated, completely wet). Instead of the water vapor content ratio, the water vapor pressure ratio can also be used.

The relative humidity concerns water in the *vapor* phase, i.e. water in the air, which is normally invisible. It only becomes visible to the naked eye in the case of condensation or in the case of fog, i.e. when small droplets have formed. The saturated water content of the air is highly dependent on the air temperature (Figure 1.2). As temperature rises, the air is able to contain more water vapor. For example, air at 20 °C can contain 17.3 gram of water vapor per m³, whereas at 10 °C this is only 9.4 gram. Therefore, cooling the air leads to condensation of water vapor, i.e. the surplus of vapor turns into liquid water.

There are various equations available for approximation of the saturated water vapor content or pressure as a function of temperature, of which the Clausius-Clapeyron equation forms the basis (Chapter 11).

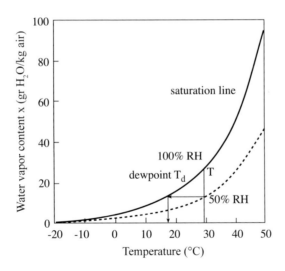

*Figure 1.2. The saturated water content as a function of the temperature. The ratio between the actual vapor content and the saturated vapor content at the same temperature represents the relative humidity RH at that temperature. Cooling the air does not alter the water vapor content until the so-called dew-point $T_d$ is reached. At this temperature, the actual vapor content at T equals the saturated vapor content at $T_d$. When lowering the temperature below the dew point $T_d$, condensation occurs and the surplus of water vapor turns in liquid water.*

*Olaf C.G. Adan and Robert A. Samson*

### The state of water in materials

The state of water in materials can be expressed in many different parameters, such as the water activity, osmotic pressure, fugacity, water potential and water content. The latter addresses the amount of water, and – as stated before – refers to a capacity, whereas the first three are based on the chemical potential, a thermodynamic quantity. The difference can be more or less understood on the basis of the following example: water tends to move from a high water potential to a low water potential, or a high water activity to a low water activity, even though the water content may be lowest at the highest potential or activity. As the water activity is a quantity that is commonly used in microbiology and food sciences, and the water content is widely applied in building physics and material science, both quantities and their relationship with the relative humidity will be discussed below.

### Water activity

The water activity includes qualitative information about the water availability in a material, either liquid or solid. It is defined on the basis of the chemical potential, as described in the appendix to this chapter. By definition, water activity ranges between 0 and 1, of which the latter value refers to pure water. Usually, in case of a solution, the water activity drops as a function of dissolved ingredients. The determination of the water activity, however, is complicated, as it *cannot be measured directly*. The actual water activity of an aqueous solution can only be measured, when the solution is in *equilibrium* with the air above it. The value of relative humidity of the air, expressed as a ratio, equals the water activity, when equilibrium between the liquid phase and the air has established and no net transfer of water between air and solution takes place (Figure 1.3). The same principle applies to the water activity of solid substrates,

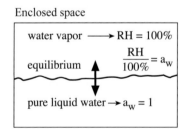
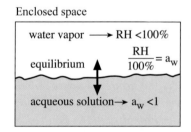

*Figure 1.3. The water activity $a_w$ of the solution equals the relative humidity RH (expressed as a ratio) only when hygric equilibrium has been established and no net transfer of water vapor takes place.*

either porous or non-porous (Figure 1.4). The water activity of the material can only be measured in case of equilibrium between substrate and air, i.e. by measuring the relative humidity. Chapter 11 discusses a variety of measuring principles.

Summarizing:

> Only in case of *equilibrium*, the water activity of the liquid or solid material equals the relative humidity of the air above it, expressed as a fraction.
>
> RH (%) = $a_w$ (–) × 100 or RH (–) = $a_w$ (–)

**Water content**

The water content of a solid material can be expressed as a volumetric fraction ($m^3/m^3$), a mass fraction (kg/kg) or a mass by volume ratio ($kg/m^3$). These quantities can be easily exchanged on the basis of the density of water and the dry material. Generally, the water content of a material is a function of the relative humidity under equilibrium conditions. This relationship can be graphically expressed in so-called sorption isotherms, showing the equilibrium water content as a function of RH for a specific temperature. For porous materials, the sorption isotherm is closely related to the physical properties and the geometry of the pore system. The binding of the water is due to different phenomena, amongst which surface adsorption (i.e. formation of molecular layers due to physical adsorption or chemisorption) and capillary condensation. The latter takes place at higher RHs, i.e. above 35%. Its principle has been established by Thomson – Lord Kelvin – in 1871 and shows that condensation takes place in a capillary at a RH below 100%, depending on the pore

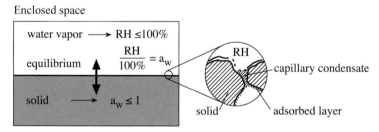

*Figure 1.4. The water activity $a_w$ of a substrate, either porous or non-porous, building material, food product or culture medium, can only be measured in case of equilibrium with the air above it. In that case, the actual moisture content of the material does not change.*

size. In smaller pores, capillary condensation occurs at lower RHs, for bigger pores higher RHs are required. For most materials, the water content increases with rising relative humidity, but in the high RH region, close to saturation (i.e. 100%), the shape of the curve becomes uncertain.

A more complicating factor, however, is due to *hysteresis*. In that case the moisture content is not a single valued function of the RH, but also depends on the history of the material (Figure 1.5). It is known that such hysteresis effects can be substantial for many common materials, especially organic materials. This implies that a water content measurement, even when it is 100% accurate, cannot be transferred into a water activity of the considered material or a corresponding relative humidity.

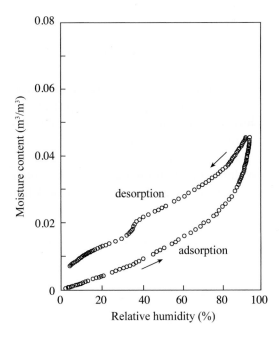

*Figure 1.5. A typical sorption-isotherm of a gypsum-based material, showing an adsorption and desorption branch in a hysteresis loop. The specific moisture content is not a single-valued function of the RH. Consequently, the water content is not an appropriate parameter to describe the material susceptibility to mold.*

### Transient conditions: the disequilibrium between air and material

The relationship between relative humidity, water activity and water content of the substrate is well defined under steady state equilibrium conditions, although previous history may introduce an ambiguous relationship between water content and RH. In case of *transient* conditions, this relation becomes much more complicated.

When moistening or drying of a material takes place, water is transported in the vapor and/or liquid phase, because of a difference in driving potentials. Reversely, as soon as the relative humidity of the air (expressed as a ratio) and the water activity of the material are not equal, transport of water will take place across the material-air interface when the surface layer is not an ideal vapor barrier. This disequilibrium causes a gradient of water potential and water content in the material, i.e. the water content and potential (or activity) vary as a function of position and time (Figure 1.6).

Generally, drying and moistening occur at different rates. The absorption of liquid water by a porous gypsum-based finishing material is usually very fast, whereas the

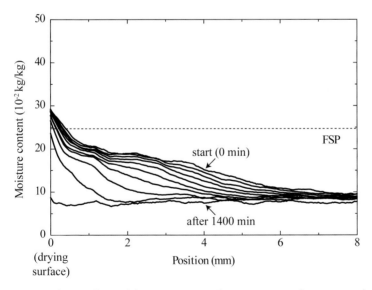

*Figure 1.6. Drying of coated wood (i.e. meranti with 100 mm acrylate coating), showing time dependent profiles of water content in a cross-section of the material (Van Meel 2009). Profiles are shown every 140 minutes. In this case, the material has been saturated with water and drying takes place at RH of about 50%. The dotted line refers to the so-called fiber saturation point FSP.*

drying can take a long period of time (Adan 1995). Opposite differences may be found e.g. in some coating systems, where uptake may be much slower than drying (Baukh *et al.* 2010).

From the mold point of view, the conditions at the material-air interface, i.e. the real *surface* conditions, play a pivotal role. These conditions are defined on the basis of the surface water activity, which is equal to the surface relative humidity assuming immediate thermodynamic equilibrium. Furthermore, the surface RH is linked to the moisture content via the sorption curve.

This relation between water activity and relative humidity point outs a crucial requirement for mold growth studies on culture media. In microbiology, culture media are used that have a well-defined nutrient composition and water activity, e.g. to characterize morphology or to examine growth in relation to water activity. An example of the latter is the work of Ayerst (1969), Smith and Hill (1982) and Magan and Lacey (1984), concerning germination and growth of *Cladosporium, Aspergillus, and Penicillium* species. It is stressed that

in any study of molds on culture media, the relative humidity of the air (expressed as a fraction) above it should be set at the same value of the defined water activity of the substrate.

Otherwise, the disequilibrium will introduce a change in the water activity, which may lead to misinterpretation of the water activity – growth relationship and uncertain reproducibility of the experimental data, since the precise conditions are unknown. This fact also raises doubts with respect to a sound interpretation of the aforementioned results of Ayerst, Smith and Hill and Magan and Lacey.

### Sources of water

As water is the key factor for mold growth, this book addresses the water relations of indoor fungi. We will deal with the fundamentals of the living cell response to humidity conditions, either steady-state or transient. In the indoor environment, water is available in different phases, i.e. in vapor or liquid form, originating from different sources.

### Sources of liquid water

Liquid water penetration may have many different causes, which are usually quite evident. Liquid water can originate from indoor sources, such as leakage of plumbing, or may come from the outdoor, such as rain, melting snow, and ground water. Exceptional cases are due to disasters like floods, hurricanes and firefighting. Many text books are available that deal with damages related to this kind of water sources. Generally, the building envelope is the most critical part of the building in terms of liquid water penetration. The building envelope forms the protection against the outdoor environment. It allows us to create an indoor environment, safe, maintaining the indoor temperature at a comfortable level and keeping wind and water out. The basic principles of good design of roof, façade, base floor and foundation are well known for many decades. The introduction of concrete foundation slabs and impermeable layers (e.g. plastics, foam glass) that separate the foundation and façade appeared very successful to tackle raising dampness, being most obvious in European areas below the sea level (such as the Netherlands). The cavity wall principle is another successful example, which is typically found in temperate and cold areas. This principle separates the basic functions of the wall i.e. air tightness (inner layer) and water tightness (outer layer). Despite this functional split, liquid water penetration may sometimes still occur. In such cases, causes are mostly found in bad construction or workmanship, inadequate design or lack of maintenance.

Liquid water penetration may occur at joints of constructional elements, e.g. around windows due to insufficient sealing. Or it may occur as a consequence of crack formation and lack of human interventions. Chapters 14 and 15 give a summary of typical cases and outlines the remediation strategies.

**The source is water vapor**

Water vapor in the indoor air is a much more complicated issue, since it involves differences in climate, in building and construction, and occupant's behavior (including heating, ventilation as pivotal issues). Although water vapor is invisible, the consequences can be serious. In overall, the most critical points are found at the building envelope, at the inner side or inside the structure, depending on the outdoor climate and the predominant vapor flow direction, and the thermal design of the envelope structure. (see Table 1.1).

Generally, the consequences of water vapor are shown as a result of surface condensation or interstitial condensation. Both phenomena are discussed hereafter.

**Interstitial condensation**

The condensation of water vapor inside a multilayered construction is a well know phenomenon for many decades. Much work has been done in West and North

*Olaf C.G. Adan and Robert A. Samson*

Table 1.1. *The consequences of water vapor as a function of climate zone and typical building characteristics.*

| Climate zone | Typical building characteristics | Predominant vapor flow direction | Critical points |
|---|---|---|---|
| Cold | high thermal insulation level, wood-frame construction | from indoor to outdoor | ventilated structures, interstitial condensation |
| Temperate and maritime | moderate thermal insulation level | from indoor to outdoor | thermal bridging, surface condensation |
| Warm and maritime | low thermal insulation level, air conditioning | from outdoor to indoor | interstitial condensation and air conditioning systems |

European countries and Canada, starting with the early work of Glaser (Glaser 1958) in the late 50s concerning vapor transport in walls of freezing chambers. The Glaser method combined steady-state vapor diffusion and steady-state heat conduction. Follow-up in physical modeling considered capillary water displacement, water flow by gravity, enthalpy transfer, transient effects and initial moisture content as a starting condition and driving rain as a boundary condition (De Vries 1958, Luikov 1966).

The scientific interest in interstitial condensation phenomena boosted during and after the 70s, when large scale application of thermal insulation took place. With the advent of personal computers and the progress of computer power, more advanced models were developed to predict the response of the envelope (Van der Kooi 1971, Sandberg 1973, Nielsen 1974, Kießl 1983, Kohonen 1984). Simultaneously, some researchers published methodologies that extended the usability of the Glaser method (Vos 1969, Hens 1978). At the end of the eighties the first computer codes to predict transient heat and moisture response of the envelope parts became commercially available (Pedersen 1990).

Normally, interstitial condensation is a slow process, with a typical time scale in the order of years, except for ventilated structures. Research in both North America and Europe learned that air displacement had an amazing impact on the hygro-thermal response of the building envelope. Crucial lessons were learnt in practice with the so called "cold" roofs. Such roofs were based on thermal insulation at the inner side, usually above the ceiling and below the watertight barrier, and removal of vapor

**30**                                    **Fundamentals of mold growth in indoor environments**

inflow from the inside by ventilation above the ceiling. Interstitial condensation caused many wood-based structures to collapse in a relatively short period of time due to biodegradation. In 1990, the International Energy Agency launched a program to include air a in the hygrothermal modeling of the envelope (Anonymous 1996). Beginning of the new millennium, a general methodology has been widely accepted, reflecting the international consensus on fundamental principles (Adan *et al.* 2004, Carmeliet *et al.* 2004, Hagentoft *et al.* 2004).

Nowadays, our knowledge of the physical backgrounds of interstitial condensation in envelope structures has become mature and adequate solutions for good design and construction have been transferred into building practice and standards. The basic design principles to avoid interstitial condensation are simple: thermal insulation at the cold side, a vapor tight barrier at the warm side, no ventilated structures. This book will not deal with these principles. We refer to the extensive amount of handbooks and tools that are available.

**Surface condensation**

Interstitial condensation is not visible, except for the case of drastic consequences to the structure. Surface condensation can be visible. When surface condensation occurs, liquid film formation on the surface takes place. When the surface is not porous, and does not absorb water, the liquid film will be visible to the naked eye, either in the form of a film of condensed water, or in the form of droplets that may be running of the surface. A clear example is the condensed mirror during showering, or the condensation on single glazing during cold winter days.

However, *surface condensation may not always be visible*. Porous materials may absorb water due to capillary forces, and a liquid water film at the surface may be absent due to a high absorption speed, even though surface condensation is a momentaneous phenomenon. Many plasters or renders shows such behavior.

Condensation of water vapor occurs when the temperature of the air reaches or drops below the so-called dew-point temperature (see previous section "Water availability"), or when the vapor content of the air reaches its maximum due to excessive vapor production. In the first case, surface condensation is associated with thermal bridging and heating, in the second it is usually linked with inefficient or insufficient ventilation.

In cold and temperate climate regions, thermal bridging is a phenomenon that is well understood. Thermal bridging implies that a higher heat flux occurs locally, resulting in a lower inner surface temperature. Numerous examples can be found with respect to this thermal short circuiting, often related to the local absence of the thermal insulating layer. Another thermal bridging effect is related to geometric effects, i.e. in corners, where a cooling effect occurs sine the outer surface is larger than the inner surface. Moreover, in case of non- or poorly insulated structures, the surface temperature can be decreased even more due to furniture that shields the surface from heating by warmer parts of the indoor environment.

Many calculation tools are commercially available to assess building structure design in terms of thermal performance. Many solutions have been developed to overcome thermal bridging and today many tools exist for quick and overall analysis of the actual performance of structures in practice (see Figure 1.7).

For a long time, surface condensation has been assumed a primary requirement for indoor mold growth. *Strictly speaking, saturated conditions will not favor mold growth.* Moreover, optimal conditions for growth of virtually all indoor molds take place at relative humidities below 100%, as will be shown in Chapter 2.

However, as a consequence of water take-up, surface condensation may lead to a moist surface material, which may induce molds to grow. In that case, the drying period becomes the key factor for molds. Such transient effects, i.e. the factor time, form the core issue of this book (Chapter 2 and 3).

A

B

C

*Figure 1.7. Thermal bridging effects in corners. (A) Typical pattern of mold growth on a wall paper in a corner. (B) Thermal picture based on thermography of an outer corner, showing the lowest temperature in the 2- and 3-dimensional edges. (C) Calculated temperature distribution in a 3-dimensional corner of a building structure that separates the indoor and outdoor climate.*

## References

Adan OCG (1995) Determination of moisture diffusivities in gypsum renders. HERON 40 (3): 201-215.

Adan OCG, Brocken H, Carmeliet J, Hagentoft C-E, Hens H and Roels G (2004) Determination of liquid water transfer properties of porous building material and development of numerical assessment methods: introduction to the EC HAMSTAD project. J Therm Env Build Sci 27: 253-260.

Anonymous (1991) Energy conservation in buildings and community systems programme: Annex XIV Condensation and energy, Vol. 1-4, International Energy Agency.

Anonymous (1996) Energy Conservation in buildings and community systems programme: Annex XXIV Heat, air and moisture transfer through new and retrofitted insulated envelope parts (Hamtie), Vol. 1-5, International Energy Agency.

Anonymous (2009) the Copenhagen Accord, Decision -/CP.15, 18 December 2009. Available at: http: //unfccc.int/resource/docs/2009/cop15/eng/l07.pdf.

Ayerst G (1969) The effects of moisture and temperature on growth and spore germination in some fungi. J Stored Products Res 5: 127-141.

Baukh V, Huinink HP, Erich SJF, Adan OCG and van der Ven LGJ (2010) NMR imaging of water uptake in multilayer polymeric films: stressing the role of mechanical stress. Macromolecules 43: 3882-3889.

Bonnefoy XR, Braubach M, Moissonnier B, Monolbaev K and Röbbel N (2003) Housing and health in Europe: preliminary results of a Pan-European study. Am J Public Health 93: 1559-1563.

Bornehag CG, Blomquist G, Gyntelberg F, Järvholm B, Malmberg P, Nordvall L, Nielsen A, Pershagen G and Sundell J (2001) Dampness in buildings and health. Nordic interdisciplinary review of the scientific evidence on associations between exposure to dampness and health effects. Indoor Air 11: 72-86.

Bornehag CG, Sundell J, Bonini S, Custovic A, Malmberg P, Skerfving S, Sigsgaard T, Verhoeff A and Euroexpo (2004) Dampness in buildings as a risk factor for health effects. A multidisciplinary review of the literature (1998-2000) on dampness and mite exposure in buildings and health effects, Indoor Air 14: 243-257.

Carmeliet J, Adan OCG, Brocken H, Cerny R, Hall Ch, Hens H, Kumaran K, Pavlik Z, Pel L and Roels G (2004) Determination of the liquid water diffusivity from transient moisture transfer experiments. J Therm Env Build Sci 27: 277-305.

De Vries DA (1958) Simultaneous transfer of heat and moisture in porous media. Trans Am Geophysical Union 39: 909-916.

Glaser H (1958) Wärmeleitung und Feuchtigkeitsduchgang durch Kühlraumisolierungen, Kältetechniek 3: 86-91.

Hagentoft C-E, Adan OCG, Adl-Zarrabi B, Becker R, Brocken H, Carmeliet J, Djebbar R, Funk M, Grunewald J, Hens H, Kumaran K, Roels G, Sasic Kalagasidis A and Shamir D (2004) Assessment method of numerical prediction models for combined heat, air and moisture transfer in building components: benchmarks for one-dimensional cases. J Therm Env Build Sci 27: 327-354.

Hens H (1978) Condensation in concrete flat roofs. Building Res Pract 6 (5): 292-309.

Hoge Raad (High Supreme Council) (2006) The Hague, 2 June 2006, nr. C05/164, LJN: AW 6167.

Hoppe W, Lohmann W, Markl H and Ziegler H (1983) Biophysik. Springer Verlag, Heidelberg, Germany.

Institute of Medicine (2004) Damp indoor spaces and health. Committee on Damp Indoor Spaces and Health, Board on Health Promotion and Disease Prevention. The National Academies Press, Washington, DC, USA. Available at: http://www.nap.edu.

Kießl K (1983) Kapillarer und dampfformiger Feuchtetransport in mehrschichtigen Bauteilen. PhD Thesis, Universität Essen, Essen, Germany.

Kohonen R (1984) A method to analyse the transient hygrothermal behaviour of building materials and components. DSc-Dissertation VTT, Publication 21, Espoo, Finland.

Luikov AV (1966) Heat and mass transfer in capillary porous bodies. Pergamon Press, Oxford, UK.

Magan N and Lacey J (1984) Effect of temperature and pH on water relations of field and storage fungi. Trans Br Mycol Soc 82: 71-81.

Moore WJ (1978) Physical Chemistry. Longman, London, UK.

Mudarri D and WJ Fisk (2007) Public health and economic impact of dampness and mold. Indoor Air 17: 226-235.

Nielsen A (1974) Fugtfordelinger i gasbeton under varme- og fugttransport. PhD thesis, Meddedelse 29, Danmarks Tekniske Hoejskole, Laboratoriet for Varmeisolering, Lyngby, Denmark.

Pedersen CR (1990) Combined heat and moisture transfer in building constructions. PhD thesis, Report 214, Technical University of Denmark, Copenhagen, Denmark.

Sandberg PI (1973) Byggnadsdelars fuktbalans i naturligt klimat, PhD-Dissertation, Report 43, Lund Institute of Technology, Lund, Sweden.

Smith SL and Hill ST (1982) Influence of temperature and water activity on germination and growth of *Aspergillus restrictus* and *A. versicolor*. Trans Br Mycol Soc 79: 558-559.

Thain JF (1967) Principles of osmotic phenomena. The Royal Institute of Chemistry, Heffer, Cambridge, UK.

Van der Kooi J (1971) Moisture transport in cellular concrete roofs. PhD thesis, Eindhoven University of Technology, Eindhoven, the Netherlands.

Van Meel PA (2009) Moisture transport in coated wood. MSc Thesis, report nr N/TPM 2009-04, Eindhoven University of Technology, Eindhoven, the Netherlands.

Vos BH (1969) Internal condensation in structures. Building Sci 3: 191-206.

*Olaf C.G. Adan and Robert A. Samson*

## Appendix 1. Thermodynamic definition of water activity and relative humidity

### The water activity

From the Gibbs-Duhem equation

$$\Sigma_i \, n_i \cdot \mathrm{d}\mu_i = -S \cdot \mathrm{d}T + V \cdot \mathrm{d}P \tag{1}$$

where $n_i$ is the number of moles of $i$ in the system, and $V$, $T$ and $S$ the volume, the temperature and the entropy of the system, respectively, it follows that the variation of the chemical potential $\mu_i$ of substance $i$ with pressure $P$ is given by

$$v_i = \left( \frac{\partial V}{\partial n_i} \right)_{P,T,\{n/ni\}} = \left( \frac{\partial \mu_i}{\partial P} \right)_{T,\{n\}} \tag{2}$$

If substance $i$ is an ideal gas then

$$v_i = \left( \frac{\partial \mu_i}{\partial P} \right)_{T,\{n\}} = \frac{R \cdot T}{P} \tag{3}$$

and hence

$$\mu_i = \mu_{i,0} + R \cdot T \cdot \ln(P) \tag{4}$$

If $i$ is a component of a mixture of gases, $P$ is replaced by $p_{i,v}$, the partial pressure of $i$. The reference state $\mu_{i,0}$ is the chemical potential of the gas under a partial pressure of $10^5$ Pa, and is a function of the temperature $T$ only.

The concept of an ideal gas is useful in discussions of thermodynamics of gases and vapors. Many cases of practical interest are treated adequately by ideal gas approximations. Description of the properties of solutions, however, is much more complicated. Since the vapor pressure of a component above the solution is a good measure of the tendency to escape from the solution, and therefore reflects the physical state of affairs within the solution, the activity $a_i$ of a component $i$ in the solution was introduced. When *equilibrium* between the component $i$ and its vapor exists, the chemical potential $\mu_{i,1}$ of $i$ in the solution equals the chemical potential $\mu_{i,v}$ of $i$ in the vapor phase.

$$\mu_{i,l} = \mu_{i,v} = \mu_{i,0} + R \cdot T \cdot \ln(p_{i,v}) \tag{5}$$

Comparison to the chemical potential of pure solvent $\mu_i^*$ in a reference state introduces the activity $a_i$ of $i$ in the solution. It is emphasized that the activity depends strongly on the reference state chosen.

In case of pure solvent $i$

$$\mu_{i,l}^* = \mu_{i,0} + R \cdot T \cdot \ln(p_{i,v}^*) \tag{6}$$

From Equation 5 and 6 follows

$$\mu_{i,l} - \mu_{i,l}^* = R \cdot T \cdot \ln(a_i) \tag{7}$$

Two definitions of the activity $a_i$ can be found in the literature (Thain 1967, Moore 1978, Hoppe *et al.* 1983)

$$a_i = \frac{p_{i,v}(P_H, T)}{p_{i,v}^*(P_H, T)} \tag{8}$$

and

$$a_i = \frac{p_{i,v}(P_H, T)}{p_{i,v}^*(P_{H,fixed}, T)} \tag{9}$$

with

$$P_{H,fixed} = 10^5 \text{ Pa} \tag{10}$$

As long as the vapor behaves as an ideal gas, the activity $a_i$ of a component $i$ in the solution equals the ratio of the partial pressure of $i$ above the solution to the vapor pressure of pure $i$ in the reference state. Again, it is emphasized that *equilibrium* between the liquid and vapor phase is required. The first definition (Equation 8) of the activity implies that the activity of the pure solvent always is 1 (Thain 1967). The consequence of the second definition (Equation 9 with Equation 10) is that the activity slightly depends on the total hydrostatic pressure $P_H$ (Moore 1978, Hoppe *et al.* 1983).

For non-ideality, a new function called fugacity $f$ was introduced (Moore 1978). In that case, the activity $a_i$ is calculated from the ratio of $f_i$ of $i$ in the solution and $f_i^*$ of pure $i$ in a reference state. However, since the vapor pressure in the indoor environment is sufficiently low ($<2.3\times10^3$ Pa), non-ideality may be neglected.

### The relative humidity

$$RH = \frac{p_{w,v}(P_B,T)}{p^{\bullet}_{w,v}(P_B,T)} \cdot 100\% = \frac{p_v(P_B,T)}{p_{vs}(P_B,T)} \cdot 100\% \tag{11}$$

Although Equation 11 resembles the definition of the activity in Equation 8, consideration of the assumptions underlying Equation 8 shows marked differences. The *RH* concerns water in the vapor phase, whereas $a_w$ refers to water in another state, either liquid or bound to the substrate. Furthermore, the *RH* is a measurable quantity, and $a_w$ can only be deduced from other quantities such as the *RH* under equilibrium conditions and the osmotic pressure (see Chapters 1 and 2).

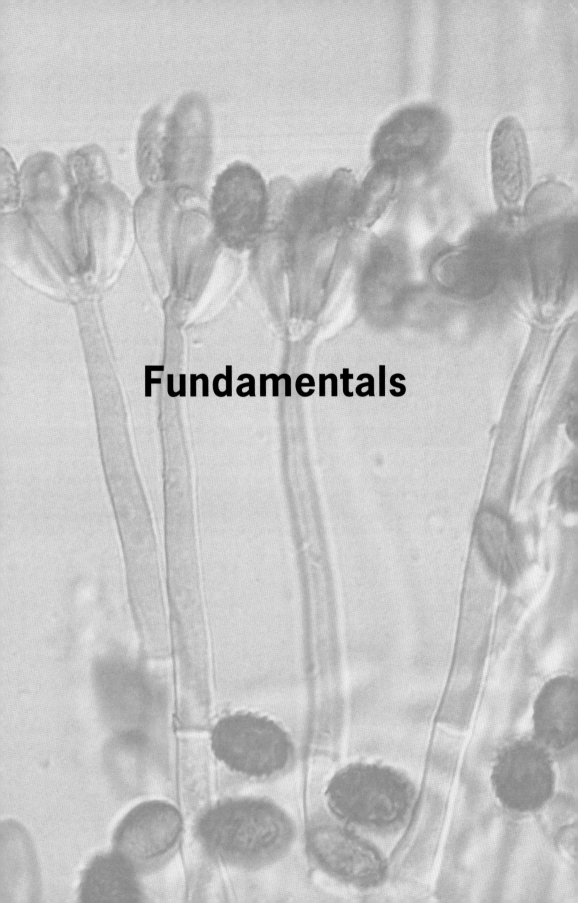

# Fundamentals

# 2 Water relations of fungi in indoor environments

*Olaf C.G. Adan[1,2], Henk P. Huinink[1] and Mirjam Bekker[1]*
*[1]Eindhoven University of Technology, Faculty of Applied Physics, Eindhoven, the Netherlands; [2]TNO, Delft, the Netherlands*

## Introduction

It is without doubt that there is no fungal growth without water. This relationship should underlie any control strategy for indoor fungal growth, but appears to be more complex than commonly assumed. In microbiology, the water relations of fungi are commonly studied under well-defined, stable conditions of water activity and temperature. The actual reality of the indoor environment, however, can be highly transient. Time dependent changes in humidity, temperature and air movements introduce inertia effects both in the biological system of the growing fungus and the physical system of materials, constructions and building. Typically, fluctuations in the environmental parameters can be grouped into two categories.

Firstly, long term variations on a yearly basis and a cyclic pattern that is related to the change of the seasons. It is obvious that the geographic location plays a pivotal role in the typical pattern of outdoor conditions, e.g. the continental, arctic and maritime climate show a highly different picture with respect to outdoor humidity and temperature. For a typical temperate climate, such as in Western Europe and the Eastern part of the US, the difference between the lowest and highest monthly temperatures are moderate (in the order of 15-20 °C), and the relative humidity is rather stable. For example, in the Netherlands, the monthly temperature varies between 2.8 and 17.4 °C, with a yearly average of 9.8 °C, and the relative humidity ranges between 73 and 89%, with a yearly average of about 80%. In tropical area's extremes are more pronounced, e.g. the monthly relative humidity usually is in the range of 60-90%.

Secondly, short term variations on a daily basis, determined mainly by the typical pattern of household activities such as cooking, showering, heating and cooling in the presence of people. It is self-evident that also outdoor conditions play a role and interact with human behavior. Generally, the extremes in humidity differ widely. During showering, the indoor humidity may rapidly change from a value below 50% to a peak value of 100%, leading instantaneously to surface condensation.

This chapter deals with the water relations of fungi in the indoor environment. Fungal growth will be discussed in relation to long term and short term fluctuations of humidity and temperature. With respect to the latter, the inertia effects of the material will be considered and the so-called "time-of-wetness" concept will be introduced.

The modern world is facing a crucial step with respect to sustainability of the built environment, including the need to increase energy conservation and resource efficiency. Direct actions in the near future will concern the increase of the thermal performance of the building envelope as well as the further promotion of eco-friendlier products. Both will impact the indoor environment and consequently may have a potential effect on indoor fungal growth. This chapter will shortly address such potential implications of the international $CO_2$ and energy agreements (Kyoto, Bali) on indoor mold growth.

## Fungal response to long term variations in indoor climate

### Fungal response to steady-state humidity conditions

The extreme situation that humidity conditions and temperature remain unchanged during time is studied well in microbiology. In microbiology, the water activity $a_w$ is commonly adopted as a measure for water availability instead of the relative humidity *RH* (under equilibrium conditions, both are directly related, see Chapter 1). Most studies in mycology focus on steady state conditions, i.e. fungi were mostly grown on agar media with fixed water activities and temperature. One should be cautioned that these do not reflect the circumstances for growth that actually occur in building practice.

Like all living processes, fungal growth requires the availability of water. Generally, a saturated environment of 1.0 $a_w$, is taken as the upper limit for fungal development. However, in the early work of Scott (1957) it was already pointed out that fungal growth is impossible in pure water, probably being a consequence of the absence of dissolved nutrients. A minimum and optimum water activity can be considered main features for growth of every type of fungus. The minima may differ widely and the differences may easily be in the range of 0.2-0.3 $a_w$, whereas the optimum water activity of most filamentous fungi are in a narrow range of 0.9-0.99, mostly considerably greater than 0.9 (Ayerst 1969).

Indoor fungi can be grouped according to their water tolerance. Lacey *et al.* (1980) introduced a scheme for categorization that has been widely accepted. They distinguished hydrophilic ("water-loving") and xerophilic ("dry-loving") species on the basis of the minimum water activity required for growth. Hydrophilic species are only suited to growing at $a_w$ above 0.9, whereas the latter category is able to grow at conditions of reduced $a_w$ below 0.9.

The $a_w$ influences each of the main phases of the growth cycle. In studies of *Aspergillus*, *Cladosporium* and *Penicillium* species (Ayerst 1969, Magan and Lacey 1984), the minimum value required for germination was lower than that to sustain linear growth, and likewise the minimum value for linear growth was usually lower than the minimum required for asexual sporulation. In the majority of studies, differences of 0.02 $a_w$ between minima for germination, linear growth and sporulation have been recorded.

The ultimate lower $a_w$ limit for fungal germination has been studied by several workers. The existence of this limit is often attributed to the distortion of the helical structure of the DNA molecule due to dehydration (Chen and Griffin 1966). From an early review on fungal water relations, Scott (1957) found that some fungi could grow at $a_w$'s as low as 0.64 after 1-2 years incubation, with occasionally conidia of *Eurotium echinulatum* germinating at 0.62. Andrews and Pitt (1987) showed that *Eurotium* sp. was capable of growing at 0.68 $a_w$, and Harrewijn (1979) mentioned 0.65 $a_w$ as the ultimate lower limit. At the opposite end for hydrophilic species, the highest value is found for the basidiomycete micro fungus *Sistotrema brinkmannii* with a minimum $a_w$ of 0.96-0.97. An extensive listing of the published minimum $a_w$ of filamentous fungi encountered in buildings is given by Flannigan and Miller (2001).

It is emphasized that the lower limit for germination may not always be sharply defined, since it depends to some extent on the length of the observation period. Furthermore, many reports of fungal growth at the lower limit refer to substrates of high nutrient status.

## Fungal growth on construction and finishing materials

### Construction materials

Fungal growth on constructional materials, other than wood and wood-based products, hardly received any attention, probably due to the fact that their biodeterioration is not common in the indoor environment. Decay of wood and wood-based products is considered separately in this book; see Chapters 6, 16 and 17. The early work of Coppock and Cookson (1951) concerned painted wood and various types of plain, white-washed and cement-rendered brick. Except for red clay brick, all types of material showed slight to moderate growth within a 4-week period at steady-state *RH*'s in the range of 80-95% and a temperature of 27 °C.

They suggested that mold growth should be related to geometric properties of the substrate, i.e. porosity and possibly pore-size distribution, although no proof could be deduced from their experiments.

## Paints

Fungal growth on finishing or decorating materials has received more attention, although it has been focused on exterior rather than on interior applications. Becker *et al.* (1986) and Becker and Puterman (1987) studied the effect of paint porosity, as related to the pigment volume concentration, on growth of *Aspergillus niger* and *Penicillium expansum* on polyvinyl acetate and acrylic emulsion paints at 97% *RH*. They concluded that the Critical Pigment Volume Concentration seems to be a transition point, as fungal growth increased significantly for *PVC*'s above this value. Grant *et al.* (1989) studied the minimal water requirement for fungal growth on woodchip paper (either plain or emulsion painted) patterned wallpaper, vinyl wall covering and emulsion painted gypsum plaster. On emulsion painted woodchip paper *Aspergillus versicolor* and *Penicillium chrysogenum* were able to grow after 3 weeks exposure to 25 °C and steady-state *RH*'s down to 79%. Furthermore, their study showed that the nature of the base substrate affected fungal growth on the coating, but there was no consistent pattern for all fungal species examined.

## Plasters and renders

Manufactured plasters often contain organic additives that cause susceptibility to fungal growth. Fungal growth on interior plasters has been studied by several researchers. Becker and Puterman (1987) found fungal growth on plain gypsum boards (within 5 days after inoculation) and on manufactured lime-cement plasters (within 9 days after inoculation), both at 97% *RH*. Francis (1987) tested gypsum plasters in a 10-week period at *RH*'s of 86%, 92% and 97%. Only at 86% plasters remained virtually unaffected. In experiments of Herback (1990) traces of fungal growth occurred on manufactured gypsum after a 5-week incubation at 97% *RH*. Our cryo-SEM observations of the common indoor mold *Penicilium chrysogenum* on gypsum revealed that all growth stages occur within 73 hours after inoculation at 97% *RH* and 21 °C temperature (Figure 2.1).

Morgenstern (1982) studied effects of polyvinylacetate (PVAC) additives of gypsum-lime plasters on the susceptibility to fungal growth. These additives are often applied to improve strength, elasticity and adhesion. Experiments under conditions near saturation at 23 °C clearly demonstrated an accelerated and intensified fungal development as a consequence of PVAC addition. Initially, the alkalinity of the

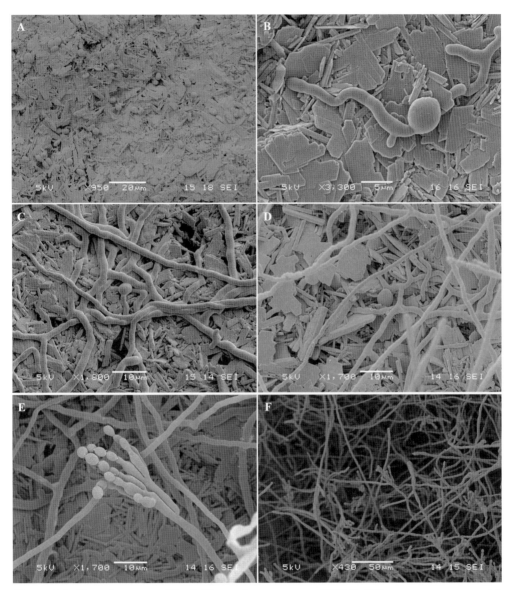

*Figure 2.1. Growth stages of* Penicillium chrysogenum *(CBS 52E8) on pure gypsum with additional nutrients, incubation at 21 °C and 97% RH in the dark. (A) Conidia on the gypsum surface, initial inoculum, 15 min. (B) Swollen conidium with germination tube, 23 h. (C) Extended hyphal growth, 45 h. (D) Formation of aerial hyphae, 51h. (E) Sporulating conidiophore, 69 h. (F) Mycelial mass, with sporulating conidiophores, covering the substrate, 73 h.*

plaster inhibits fungal growth, but owing to carbonation, pH alters towards more tolerable values. The experiments showed that PVAC additives nearly halved the time needed for entire carbonation. Besides, PVAC additives resulted in increased water content of the plaster. Both hydrolysis of PVAC and carbonation were related to this effect. Although polyvinylacetate itself cannot be used as a carbon source by the species of fungi tested, hydrolysis resulted in the formation of polyvinylalcohol and acetate, the latter being a possible nutrient.

**Wallpaper and boards**

Pasanen *et al.* (1992) investigated fungal growth on wallpaper, plywood, gypsum board and acoustical fiber board. Both the microflora on material sampled from damp buildings and fungal growth of a mixture of the same species in a range of steady-state *RH*'s in the laboratory was studied. *Aureobasidium pullulans, Cladosporium cladosporioides, Penicillium verrucosum, Stachybotrys chartarum* and *Trichoderma viride* were found. For all materials considered, the minimum *RH* required for fungal growth was 83-96% depending on the species involved. Furthermore, their study indicated that successive growth of the species used takes place in relation to the *RH* conditions.

Similarly, based on field observations and lab studies on wall paper, Grant *et al.* (1989) suggested primary colonizers growing at $a_w$ below 0.8 (e.g. *Penicillium* spp. and *Aspergillus versicolor*), secondary colonizers growing in the $a_w$ range 0.8-0.9 (e.g. *Cladosporium* spp.) and tertiary colonizers growing at $a_w$ above 0.9 (e.g. *Stachybotrys chartarum* and *Aureobasidium pullulans*).

**In summary**

Mold growth is dependent on water and proper nutrients. Usually, the nutritional requirements are minimal and satisfied either by the material constituents or by a minor contamination of the surface by dust or other deposits (Becker and Puterman 1987, Ginbergs *et al.* 1993).

There is no doubt that water usually is the only limiting factor for sustainable control. Although mold is often related to surface condensation (*RH*=100%), experimental evidence shows that most fungi readily germinate and grow on substrates in equilibrium with *RH*'s below saturation. Moreover, *virtually all indoor molds have optimum conditions for growth in a RH-range of 90-100%*.

> *Pragmatically*, there is consensus in the scientific community that surfaces can be kept free from mold growth if the *relative humidity* of the adjacent air is maintained *below 80%.*

### Fungal response to steady-state temperatures

Next to water activity $a_w$ or relative humidity $RH$, another physical parameter affects fungal growth substantially: temperature. Fungi posses no means of controlling internal temperature and the temperature within the cell is therefore determined by the external temperature.

Fungi have a wide tolerance of temperatures and depending on the temperature range for growth, they can be allocated to difference classes. The highest tolerance is found in the psychrophiles (being able to grow at or even below the freezing point, and a maximum at about 20 °C) and the thermophiles (being able to grow at temperatures above 20 °C and a maximum of 50-55 °C or higher). An example of the latter category is *Rhizomucor pusillus*. By far the majority of indoor fungi fall into the intermediate category of mesophiles. Some of the mesophiles (including some species of *Cladosporium* and *Penicillium*) can grow at freezing temperatures, but their optima are between 20-30 °C; others, such as *Aspergillus fumigatus*, are able to grow at temperatures of above 50 °C, but are not considered thermophile because of their minima below 20 °C. Although some species can grow or adapt to refrigerator temperatures, most indoor species have minimum temperatures for germination and growth between 5 and 10 °C. The optimum temperature for mycelial growth is generally between 22 and 35 °C (for a large number of mesophilic molds it is around 25 °C, i.e. above normal room temperature), whereas the maximum temperature ranges between 35 and 55 °C.

Growth and survival are not necessarily affected in the same way. Normally, temperatures below the minimum affect growth, but not viability. At these low temperatures fungi may remain dormant for considerable periods of time.

Generally spores are more resistant to high temperatures than the vegetative mycelium. The temperature threshold to kill spores within a 30 minutes period is 60-63 °C (Gaudy and Gaudy 1980, Bravery 1985).

### Water and temperature interactions

Temperature has an important bearing on the minimum water activities or relative humidities for growth. Actually, each developmental stage in growth is a function of both temperature and *RH*, which is even more complicated in practice as temperature variations also affect the *RH*.

In case of growth on culture media, the simultaneous effect of water activity $a_w$ and temperature T can be expressed in so-called isopleths, i.e. T/$a_w$-diagrams with lines connecting points of equal growth rate (for which usually linear extension of hyphae is taken as the starting point) or points of equal time required for germination. Ayerst (1969), Smith and Hill (1982), and Magan and Lacey (1984) found that *Alternaria, Cladosporium, Fusarium* and *Stachybotrys* species show shallow curves over a wide temperature but limited $a_w$ range, whereas *Penicillium* and *Aspergillus* species tend to grow within narrower temperature ranges and are more tolerant of low $a_w$'s. Generally, it can be said that:

1. When the temperature becomes more divergent from the optimum, molds become less tolerant for lower $a_w$ for growth. Grant *et al.* (1989) indicated similar response in experiments on fungi isolated from the indoor walls and air on emulsion painted wood-chip paper. A decreasing tolerance of lower *RH* occurred when the temperature deviated more from the optimum growth temperature.
2. The maximum tolerance of $a_w$ extremes is exhibited at approximately the optimum T, which is often near the upper end of the temperature range, and conversely the maximum tolerance of temperature extremes is near the optimum $a_w$.
3. Growth rate and the duration of the lag phase are increasingly sensitive to changes in $a_w$ and T near the limiting values.
4. Most xerophilic species can be identified with isopleths showing maximum growth at $a_w$ levels markedly below 1.0. In experiments of Magan and Lacey (1984) *Eurotium amstelodami, Eurotium repens* and *Aspergillus versicolor* grew fastest between 0.90 and 0.95 $a_w$. Observations of Smith and Hill (1982) showed a similar pattern for *Aspergillus versicolor* and *Aspergillus restrictus*.

### The effect of seasonal variations in climate conditions

The maximum tolerance of humidities is exhibited on materials of a high nutritional content under optimum temperature and other conditions such as alkalinity. In building practice, however, nutrient availability and temperature are seldom likely to be optimal, implying that fungi will almost invariably need higher water activities or relative humidities than the minimum values indicated in lab experiments on culture media. Nevertheless, Ayerst (1969) suggested that the isopleth diagram may be considered a stable feature of the fungal species, meaning that the type of substrate and its nutritional status only affect the actual growth rate of each isopleth proportionally, but not the shape.

Assuming that inertia effects of the organism and substrate during slow fluctuations of $RH$ and T may be neglected, the isopleth diagrams can be used to gain insight in the response of fungi to long term variations in the indoor climate.

In common with others, Ayers (1968) and Magan and Lacey (1984) identified two typical types of isopleths diagrams, i.e. for hydrophilic and xerophilic species, respectively. The latter showed a relatively wide range of $a_w$ and temperature, with an optimum that is clearly below the saturation level $a_w$=1. In sharp contrast to these xerophiles is the narrow and more restricted range of the hydrophiles, with optimum conditions close to $a_w$=1. Figure 2.2 gives two typical examples, which have been used by Adan (1994) in simulations of fungal growth on a building envelope surface.

As a typical case for such simulations, the temperate climate in Western Europe and the Eastern part of the US is taken, and the long term response is approached on a monthly basis. In these regions, the indoor surface temperature is usually lower, sometimes much lower (in the order of some degrees C) than the indoor air temperature. This implies that the $RH$ at the surface is higher – or much higher – than the indoor $RH$. Using monthly mean outdoor and indoor conditions, the surface temperature pattern of building envelope parts can be calculated as a function of its thermal characteristics. Next, assuming an ideal mixture of the water vapor in the indoor air, this temperature pattern can be translated into a surface $RH$ pattern.

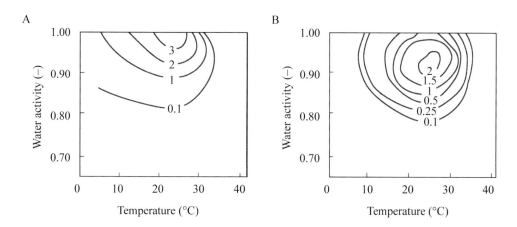

*Figure 2.2. Growth of (A)* Penicillium martensii *(after Ayerst 1969) and (B)* Aspergillus versicolor *(Magan and Lacey 1984) on agar as a function of water activity and temperature. The lines, or isopleths, connect points of equal growth rate (mm/day).*

Subsequently, the resulting set of surface T-*RH* combinations can be transferred into a relative growth pattern using isopleths and interpolation.

Although one should bear in mind the previously mentioned assumptions and cautions, isopleth-based simulations indicate some trends with respect to indoor fungal growth:

- In temperate and cold regions, fungal growth on thermal bridges is a well known phenomenon. Thermal bridges are thermal short-cuts in the envelope, showing an increased outflux of heat, leading to a decreased inner surface temperature and consequently an increased surface *RH*. Simulations indicate (Adan 1994), however, that the coldest spot on the thermal bridge does not always result in the most profuse growth, which is in line with observations in reality (e.g. in corners). Especially in case of xerophiles this is most obvious. In this case, *surface temperature* is dominating growth rates, and not the *RH*.

> When the surface *RH* is sufficiently high, the *surface temperature* becomes a crucial parameter for fungal growth rates.

- Weather conditions affect indoor temperatures and humidities and therefore should be considered in the overall risk analysis of fungal growth. As a first-order approach, risks can be related to the entire mold growth area at the surface or the integrated growth pattern. This also overcomes the fact that severest growth not necessarily occurs at the coldest spot. Monthly based simulations for a temperate climate and an insulated cavity wall with a solid thermal bridge are indicating the fungal response on this time scale. In general, simulations show for both the hydrophiles and xerophiles that the highest risk of growth appeared to be associated with the winter season, January being the most critical. Figure 2.3 gives some examples. The xerophilic species started deviating from this behavior under exceptional circumstances of low thermal performance (severe thermal bridging). In this case, most abundant growth occurs in autumn and spring. During winter months, the low temperatures at the coldest spot of the thermal bridge reduced growth rates strongly. These results are consistent with field observations of successive growth, starting with xerophiles in late autumn/early winter and more hydrophilic species as the winter progressed (Grant *et al.* 1989).

### Global $CO_2$/energy objectives and indoor mold: the temperature ratio criterion

Since the energy crises in the 70s of the $20^{th}$ century, the consequences of molds are prominently on the agenda, partly due to the relation of energy conservation with the indoor climate. Actually, the increase of thermal insulation and increased

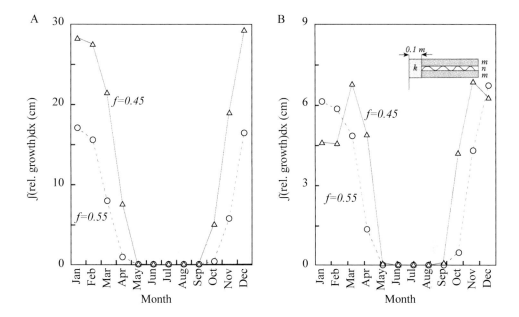

*Figure 2.3. Integrated growth as a function of the month for (A) hydrophilic and (B) xerophilic mold species. In both figures a thermal bridge is used (k) with a temperature ratio f=0.45 and f=0.55, connected to a well-insulated cavity wall with f=0.84. Monthly mean indoor RH are based on the upper limit for monthly averages in the Dutch dwelling stock.*

air tightness is often associated with an increased risk of dampness and mold. This fear – sometimes justified, mostly not – has a restraining influence on the upgrade of building energy performance, even though there is no consistent scientific evidence. This paragraph deals with the effects of increased thermal insulation in temperate and cold climates.

The first-order and quasi steady-state approach described in the previous paragraph brings together 3 different sets of influential factors of indoor fungal growth in the frame of long term climate variations:
1. The surface relative humidity, which is obviously linked to the average indoor and outdoor air humidity, as well as to the surface temperature.
2. The surface temperature, being a consequence of the local thermal resistance of the envelope and the indoor and outdoor temperatures.
3. The relation between fungal growth and surface temperature and relative humidity under steady state conditions, as expressed in the contour plots in the isopleth diagrams.

Since the thermal quality of the building envelope influences both the surface *RH* and temperature, a criterion for the thermal performance is an obvious step towards risk management of indoor fungal growth. The thermal quality of a cross-section of the building envelope can be expressed by the temperature ratio *f*:

$$f = \frac{T_{i,sur} - T_{e,air}}{T_{i,air} - T_{e,air}}$$

Where $T_{i,sur}$ is the interior surface temperature and $T_{i,air}$ and $T_{e,air}$ are the indoor and outdoor reference temperature, respectively. To measure *f*, $T_{i,air}$ is related to a reference position in the indoor air. The temperature ratio can be calculated from the heat transfer coefficients at the construction and air surfaces and the thermal properties and dimensions of the constituent materials. If this expression is used to characterize a thermal bridge, *f* is related to the lowest temperature of the inner surface under steady-state conditions. The temperature ratio ranges by definition between 0 and 1, wherein increasing values refer to an increasing thermal insulation level.

In the temperate European and Eastern US climates, the winter season is most critical in terms of fungal growth. Using typical isopleth diagrams of xerophiles and hydrophiles (Figure 2.2), growth rates can be expressed as a function of the temperature ratio. In Figure 2.4, growth of both is given for various indoor humidity conditions, corresponding to the lower and upper limit of the monthly mean climate conditions found in the West European housing stock, and an equally increased value that refers to extremely humid conditions (e.g. occurring in swimming pools). Both the hydrophilic and xerophilic curves show an initial increase in growth rate with increasing thermal insulation level. The actual shape of the curves in this range is uncertain, however, since condensation is occurring (indicated by the circle and arrow in the curve) and the isopleths are based on data for $a_w < 1$. Nevertheless, a decreasing growth rate with decreasing *T* (or *f*) may reasonably be expected.

It is obvious that – principally – a further increase of the thermal insulation level of the building envelope will reduce risks for fungal growth. In some cases, thermal insulation may go hand in hand with increased air tightness, leading to elevated humidity levels indoors. The upper curve (II) in Figure 2.4 represents the worst-case scenario based on large scale measurements in the housing stock, including such side effects.

More interesting is that the simulations indicate a temperature ratio in the range of 0.59-0.73 to prevent fungal growth. This implies that:

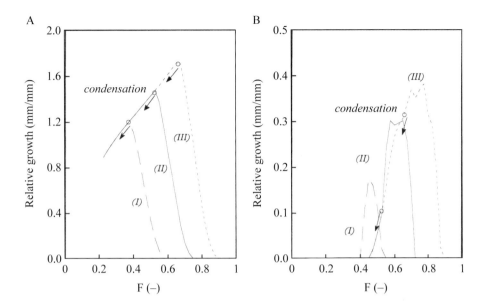

*Figure 2.4. Relative growth of (A) hydrophilic and (B) xerophilic species under January mean climate conditions, as a function of the temperature ratio f. Lines correspond to the lower (I) and upper (II) limit of monthly mean indoor conditions in West European dwellings, as well as to extreme indoor conditions such as in swimming pools. The circle and arrow denote the range of surface condensation.*

> The present temperature ratio criterion in Western European countries (typically around 0.65) may *not* be on the safe side to prevent mold growth on interior surfaces.

A safe temperature ratio criterion of 0.73 is consistent with the value that can be deduced on the basis of the 80% *RH* threshold, using the January conditions. Such an increase of the temperature ratio criterion to a value of or above 0.73 may face serious barriers for implementation and have drastic implications on the way of constructing, leading to "Scandinavian-like" designs in temperate climate zones. Especially in case of refurbishment of the building stock – which is the priority in view of the global energy targets in the built environment – , connections between building elements (e.g. at window frames) are a challenge. Thin insulation systems and compact solutions will be required to realize the energy objectives.

*Olaf C.G. Adan, Henk P. Huinink and Mirjam Bekker*

## Short term variations in indoor climate

Although requirements for the thermal performance of the building envelope or a *RH* threshold value can be used as a first step to reduce mold risks, it is *no guarantee* at all for a mold-free environment. It is the rule rather than the exception that the indoor climate is dynamic. Practically, it is often unavoidable that the *RH* exceeds 80%, if only bathrooms are considered. Despite this every day reality, knowledge of the effects of transient humidity conditions on fungal growth is only minor. This part deals with an overview of the state-of-the-art with respect to fungal response to indoor climate dynamics.

### The discrepancy between indoor climate and micro climate

Transient indoor humidities throw light on the difference between the *indoor climate* and the so-called *micro climate*. This micro-climate is defined on the basis of physical conditions at the material surface, i.e. in the immediate surroundings at the scale and dimensions of the mold. The surface conditions may highly differ from the ambient conditions due to 4 effects:

1. Temperature differences. The surface temperature may substantially differ from the indoor air temperature. The effect is most obvious at the building envelope, where heat transport through the construction may cause substantial temperature gradients between surface and air as well as along the surface. Thermal bridging has been treated in the first part of this chapter, concerning long term variations, and common daily temperature fluctuations hardly affect the conclusions in the previous paragraph (Adan 1994).
2. Airborne water vapor transport. When water vapor is emitted (e.g. during production processes or household activities), it usually has a higher temperature than the surrounding air. This induces an upstream of warm and moist air, creating a boundary flow along ceilings and walls that moistens surfaces *before* mixing with the indoor air and removal by ventilation. In terms of relative humidity, the differences between ambient air and surface may easily rise to 50% (see Figure 2.5). In that case surface condensation occurs as the air temperature usually is above the surface temperature, even with inner walls.
3. Moisture retention at the surface. The reservoir function of the substrate may create humid surface conditions during periods that the ambient air is dry again. For short-term, daily indoor humidity changes, a thin surface layer in the order of a few mms is actively involved in buffering effects (Van der Well and Adan 1999). This means that the water exchange with the outer finish, often an organic paint, plays a crucial role. The reservoir feature may lead to substantial *prolongation*

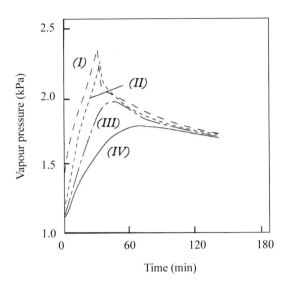

*Figure 2.5. The local water vapor pressure as a function of time, during and after a 30 minutes period of water vapor emission. The water vapor pressure is measured at several heights at 2 m distance from the water vapor source: (I) 2.9 m, close to the ceiling (II) 2.3 m, (III) 1.6 m and (IV) 0.8 m in the living zone. In this experiment, the water vapor source was at 0.8 m height and air and wall temperatures in the room were adjusted to 20 °C.*

of favorable humidity conditions at the surface, as will be shown in the next paragraph.

4. The fungal ability to influence its local environment. Experimental evidence indicates that the living organism itself may influence the water balance at the substrate-air interface, either *passively* (e.g. as a diffusion resistance) or *actively* (e.g. via exudates). This potential ability to adjust the immediate environment is hardly understood, and – consequently – its effect usually ignored. Fungi harness themselves against drought stress, both through intra-cellular and extra-cellular mechanisms:

   – *Intra-cellular amassing of so-called compatible solutes.* Nearly all fungi possess an internal (over)pressure, called turgor, a driving force for hyphal extension. The actual turgor pressure is related to the water in- and out-fluxes. A net out-flux of water during drying or increasing osmolarity of the environment causes dehydration and turgor reduction. The uptake and/or synthesis of so-called compatible solutes (e.g. organic osmolites like mannitol and trehalose) enable the cell to counteract these effects. For instance, the fungus *Serpula lacrymans*, which is able to grow over inert surfaces (e.g. plastics), accumulates high levels of trehalose.

– *Extra-cellular secretion of agents.* Fungi may affect the water availability in its immediate surroundings. Read *et al.* (1983) showed that fully frozen hydrated hyphae, germ tubes and spores cultured or grown in humid environments are frequently coated with water droplets or water films concealing surface textures. Similarly, Adan and Samson (1994) observed that rapid cryofixation of growing *P. chrysogenum* on gypsum and coatings revealed coherent structures of conidia, hyphae and water (see Figure 2.6). These observations were unexpected since no water in the liquid phase was added prior to incubation. Such agglomerates of aqueous residuals and fungal structures may contain hydrophilic exudates. A study of Van Wetter *et al.* (2000) indicated that fungi may secrete polysaccharides that effectively absorb water and thereby serve fungal re-growth. None of these mechanisms, however, have been studied yet in the context of micro-climate dynamics.

### The time-of-wetness concept

In order to deal with indoor climate dynamics, we introduced the so-called *time-of-wetness* (TOW) as a first order approximation of mold growth risks (Adan 1994). This time-of-wetness actually represents the fraction of time (ranging between 0 and 1) during which the relative humidity in the immediate vicinity (or micro-environment) of the fungus is above a threshold level, for which usually the 80% *RH* value is taken (Figure 2.7).

Next to time-related effects due to intermittent water vapor production and airborne vapor transport, the substrate itself may play a crucial role in defining the surface humidity conditions. In that case inertia effects become important. Generally, material dampness originating from the indoor air can be attributed to moistening by *capillary* condensation and adsorption (i.e. the binding of molecular layers of water to the surface) at *RH*'s below saturation or to absorption of liquid water due to saturated conditions near the surface (i.e. *surface* condensation). For porous finishing materials like gypsum, transient *RH*'s show no hygroscopic inertia effects (i.e. do not extent the TOW), owing to the low value of the moisture diffusivity for vapor transport during transient air humidities (Adan 1994). Surface condensation, however, may affect the TOW significantly, because of the fast water absorption by the gypsum.

Particularly on porous materials, the formation of liquid water during surface condensation is mostly not visible, because of very rapid take-up. For example, during

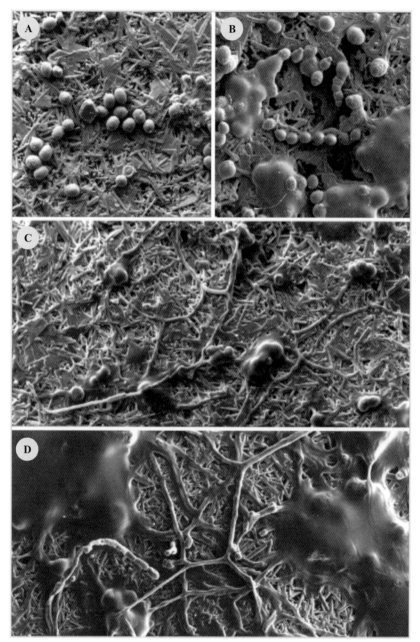

*Figure 2.6: Stages of growth of* P. chrysogenum *(CBS 401.92) on plain gypsum, incubated at 97% RH, cryo-preserved. (A) Conidia, initial inoculum, ×810; (B) swelling of conidia, 21 h, ×770; (C) initial germination and hyphae formation, 41 h, ×600; (D) hyphal growth over the surface and formation of conidiophores, 65 h, ×920. Note the amorphous agglomerates of fungal structures and aqueous residuals in (B), (C) and (D).*

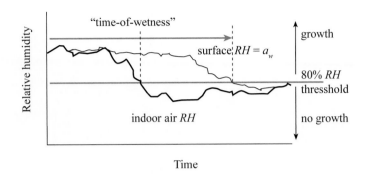

*Figure 2.7. Schematic presentation of the time-of-wetness.*

showering, condensation may be visible on mirror and tiles, but not on gypsum renders or on joints between tiles. The porosity not only introduces absorption, but also the ability of water retention. Such reservoir feature may lead to substantial prolongation of favorable humidity conditions at the surface, i.e. hugely extend the time-of-wetness.

The response of a common gypsum render is a typical example of this effect. A 10 minutes period of surface condensation (e.g. typical for showering) may easily lead to a *surface* RH above 80% during a period of time of more than 6 hours, even when the average indoor air humidity is below 60%. Only a layer of some tenths of mm appears to be effectively involved in storage of this moisture peak. In other words: the reservoir function of just a thin layer extends the time-of-wetness at the surface from less than 0.01 to 0.33 (see Figure 2.8).

### Fungal response to indoor climate dynamics

The most crucial question is: "How do molds respond to that?" Considering fungal growth dynamics at a phenomenological level, only few experimental data have become available in the past decades (Pasanen *et al.* 1992, Adan 1994, Viitanen 1997). Generally, fungal growth under changing humidities could not be explained by extrapolation of growth under constant humidity conditions. However, the TOW concept, proved to be of great value to reveal the growth response of the organism, as will be described in the next Chapter.

Extensive experiments of Adan for *Penicillium chrysogenum* on porous gypsum substrates and acrylic coatings indicated a typical *non-linear* relationship between

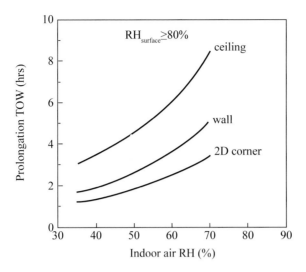

*Figure 2.8. Time-of-wetness at a gypsum surface as a function of the ambient air humidity after moistening through 10 minutes showering (Adan 1994).*

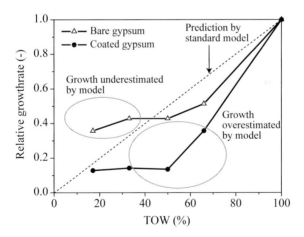

*Figure 2.9. Relative growth rate of* Penicillium chrysogenum *on bare and coated gypsum as a function of the time-of-wetness TOW. The doted line represents the growth prediction assuming steady-state water relations (Adan 1994). Obviously, at low TOW values, this model prediction may underestimate growth on plain gyspum, whereas at higher TOW values this may lead to overestimation for coated gypsum.*

growth and TOW. For both types of material, Figure 2.9 shows growth as a function of the TOW and also includes the predicted growth assuming a linear relationship. Obviously, compared to the steady-state situation, growth rates *may not be proportionally scaled* with the time fraction that conditions are favorable to growth. Two growth regimes are observed. First of all, in case of intermediate and high values of TOW (with fast fluctuations, i.e. typically in the order of hours) *deceleration* of growth occurs compared to the same period of time of favorable humidities in the steady-state situation. Viitanen (1997) found similar effects for a mixed flora of fungi, primarily *Aspergillus* and *Penicillium* spp.

Secondly and even more interesting, the experimental data at very low values of the TOW indicate that *acceleration* of growth takes place compared to the steady-state situation. The response in this low TOW region is of most interest, since it demonstrates that

> Very *short peaks in humidity* – even below saturation *RH* – may result in growth

and that reduction of the TOW in that range has a marked effect on fungal development. In this case, prediction of growth on the basis of steady-state data may result in (severe) underestimation.

Studies on the response of *P. chrysogenum* on a microscopic level showed a consistent picture with these findings. Cryo-SEM observations of Adan and Samson (1994) on *Penicillium chrysogenum* revealed that *all developmental stages in the fungal growth cycle* take place, no matter what TOW value was chosen. Under transient humidity conditions, the stages of germination, mycelium growth and sporulation were more than proportionally delayed compared to the steady-state exposure to high humidity (Figure 2.10), which is in line with their macroscopic findings on the propagation of biomass. The most explicit retardation obviously took place in the formation of spore-forming structures, or conidiophores.

Studies that relate fluctuating humidity conditions with separate stages of the fungal growth cycle are sparse. Park (1982) made one of the first attempts and studied the recovery of a variety of fungi on cellophane membranes after drying and rewetting. Surprisingly, *Cladosporium cladosporioides* and other phylloplane fungi that had been incubated under dry conditions for up to two weeks were able to restart growth within 1 hour. In fact, growth restarted from an intact and originally growing apex rather than from a newly formed branch. Other fungi, including *Penicillium* species, needed 50 hours to initiate re-growth of hyphae. Obviously, these fungi were able

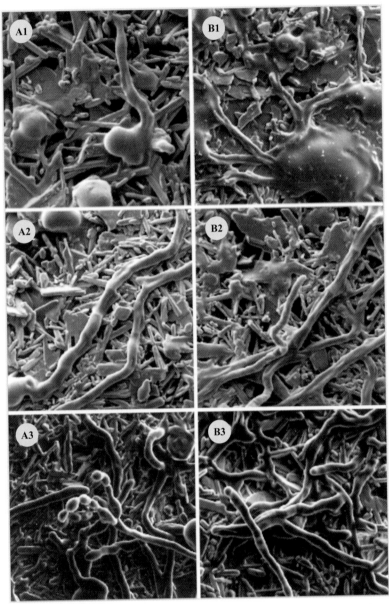

*Figure 2.10. Cryo-SEM observations of* Penicillium chrysogenum *on gypsum incubated at a TOW of 0.5. i.e. at an alternating pattern of 6 hours periods of 97% and 58% RH (left) and 6 hours periods of 97% and 33% RH (right) respectively. (1) swelling and germination after 8 days (2) hyphal growth after 2 weeks and (3) early conidiophore formation after 6 weeks. Note that the average RH is 78% and 65%, respectively. During steady state exposure at 97%, swelling of conidia occurred after 21 hours, germination and early hyphae formation after 41 hours and hyphal growth and conidiophore formation after 65 hours.*

**Fundamentals of mold growth in indoor environments**

to preserve a dynamic cellular structure through a period of drought. The cellular events during these stages, however, have never been studied in detail.

## Global $CO_2$/energy objectives and indoor mold (revisited)

Without doubt, the world-wide ambition to reduce $CO_2$ emissions will have drastic and inevitable implications for the building sector. In Europe, buildings are responsible for approximately one third of the $CO_2$ emission and 40% of total energy consumption, of which 80% is related to energy during the use or service life of the building. It is clear that in many countries, such as the EU East-European member states and many US states, increased *thermal insulation* levels will be required as a first step towards a greener environment. As has been discussed previously, increasing the thermal insulation basically will decrease the risk for indoor surface molds. In temperate regions, a major step in terms of minimized risk can only be made, if the lowest thermal performance of envelope parts corresponds to a temperature ratio criterion above 0.7. In other words: a Scandinavian or Canadian-like thermal performance will reduce surface mold growth risk to a minimum, no matter whether the indoor climate is steady-state or not.

A second crucial factor in the energy balance of a building concerns the intended and unintended air exchange of indoor and outdoor air, i.e. *ventilation and infiltration*, respectively. Ventilation is needed to provide an acceptable indoor air quality, to dilute odors and to limit the concentration of $CO_2$ and airborne pollutants. Many international standards exist with respect to minimum ventilation rates, such as the ASHRAE Standard 62, which will not be dealt with within the context of this book.

Generally, in (future) highly insulated indoor environments (i.e. at insulation levels as suggested above), ventilation rates and strategies will be determined by removal of $CO_2$ and airborne pollutants, since the high surface temperature will allow a rather high indoor humidity before the surface *RH* becomes critical for growth. At lower insulation levels, ventilation will become more and more an instrument to reduce risks for indoor mold growth, as indicated previously in Figure 2.4.

In highly insulated indoor environments or interior rooms, the surface temperature of walls, floors and ceiling will be closely to or equal to the indoor air temperature. However, in the case of water vapor production, for example during showering, ventilation *cannot* reasonably prevent that interior surfaces will be moistened. The produced water vapor is usually at a higher temperature and surface condensation occurs, before the humid air can be removed by ventilation. Although such moisture peak will usually be rather short, moisture retention in a thin material layer may cause a substantial prolongation of a high surface humidity, i.e. a prolongation of the time-of-wetness. The previous paragraphs showed that drying can be much slower than moistening. In that case, the surface material becomes the dominating factor for mold growth risks. A sustained high ventilation rate lowers the indoor humidity level, but may hardly affect the TOW in case of such retaining material, as the drying process of the material is internally limited, i.e. due to internal transport in the material (see Figure 2.8).

In conclusion, the effect of ventilation in highly insulated indoor environments is only minor in terms of its potential to reduce risks of mold growth. Real sustained control of indoor mold in such environments should take account of indoor humidity dynamics, i.e. short *RH* peaks, and should therefore consider the appropriate application of *finishing materials as the key instrument.* This pleads for performance requirements in building codes with respect to finishing materials. Chapter 12 deals with recent developments with respect to assessment of the material resistance in that respect.

## Conclusions

The water relations of fungi in the indoor environment differ for long term and short term fluctuations of humidity and temperature. Indoor climate dynamics are a reality, and both the fungal response and inertia effects in surface materials should be considered in risk control of indoor mold. In the case of climate dynamics, prediction of growth cannot be done on the basis of steady-state data, and controlling mold growth risks on the basis of the *ambient* relative humidity alone is no guarantee at all for a "mold-free" environment. Similarly, the use of *average* values of the air humidity to analyze the indoor climate with respect to mold growth is misleading.

In case of highly transient indoor humidity conditions, the properties of the finishing layer on walls, floor and ceiling, and especially their moisture reservoir function, play a pivotal role with respect to surface mold growth. Because of this, a targeted application of "mold resistant" finishes is recommended, particularly in rooms where intermittent vapor production takes place. Chapter 12 will deal with this more explicitly.

*Olaf C.G. Adan, Henk P. Huinink and Mirjam Bekker*

# References

Adan OCG (1994) On the fungal defacement of interior finishes. PhD thesis, Eindhoven University of Technology, Eindhoven, the Netherlands, 224 pp.

Adan OCG and Samson RA (1994) Fungal disfigurement of interior finishes. In: Singh J (ed.) Building mycology. Management of decay and health in buildings. Chapman and Hall, London, UK, pp. 130-158.

Andrews S and Pitt JI (1987) Further studies on the water relations of xerophilic fungi, including some halophiles. J Gen Microbiol 133: 233-238.

Ayerst G (1969) The effects of moisture and temperature on growth and spore germination in some fungi. J Stored Prod Res 5: 127-141.

Becker R and Puterman M (1987) Verhütung von Schimmelbildung in Gebäuden. Teil 2: Einfluß der Oberflächenmaterialien. Bauphysik 4: 107-110.

Becker R, Puterman M and Laks J (1986) The effect of porosity of emulsion paints on mould growth. Durability Build Mat 3: 369-380.

Bravery AF (1985) Mould and its control. Information Paper 11/85, Building Research Establishment, Garston, Watford, UK.

Chen AW and Griffin DM (1966) Soil physical factors and the ecology of fungi IV. Interaction between temperature and soil moisture. Trans Br Mycol Soc 49: 551-562.

Coppock JBM and Cookson ED (1951) The effect of humidity on mould growth on constructional materials. J Sci Food Agric 2: 534-537.

Flannigan B and Miller JD (2001) Microbial growth in indoor environments. In: Flannigan B, Samson RA and Miller JD (eds.) Microorganisms in home and indoor work environments. Diversity, health impacts, investigation and control. CRC Press, Boca Raton, FL, USA, pp. 35-68.

Francis A (1987) Schimmelproblemen in gebouwen. Determinatie, groei-omstandigheden, gevoeligheid van diverse afwerkingen, bestrijding. PhD Thesis, Katholieke Universiteit Leuven, Leuven, Belgium.

Gaudy AF and Gaudy ET (1980) Microbiology for environmental scientists and engineers. McGraw-Hill Book Company, London, UK.

Grant C, Hunter CA, Flannigan B and Bravery AF (1989) The moisture requirements of moulds isolated from domestic dwellings. Int Biodeterioration 25: 259-284.

Grinbergs L, Hyppel A, Höglund I and Ottoson G (1993) Wet-room wall systems – Mould resistance. In: Erhorn H, Reiß J and Szerman M (eds.), Proceedings of the International Symposium Energy Efficient Buildings (Design, Performance and Operation) of the CIB Working Commission W67 "Energy Conservation in the Built Environment" and IEA-SHC Working Task Group XIII "Low Energy Buildings", March 9-11, Leinfelden-Echterdingen, Germany, IRB Verlag, Stuttgart, Germany.

Harrewijn GA (1979) Elementaire microbiologie. Centraal Instituut voor Voedingsonderzoek TNO, Zeist, the Netherlands.

Lacey J, Hill ST and Edwards MA (1980) Micro-organisms in stored grains: their enumeration and significance. Trop Stored Prod Inform 39: 19-33.

Magan N and Lacey J (1984) Effect of temperature and pH on water relations of field and storage fungi. Trans Br Mycol Soc 82: 71-81.

Morgenstern J (1982) Einfluß von Polyvinylacetat-Zusätzen in Putzmortel auf die Schimmelbildung. Material und Organismen 17: 241-251.

Park D (1982) Phylloplane fungi: tolerance of hyphal tips to drying. Trans Br Mycol Soc 79: 174-178.

Pasanen A-L, Heinonen-Tanski H, Kalliokoski P and Jantunen MJ (1992) Fungal micro-colonies on indoor surfaces – an explanation for the base level fungal spore counts in indoor air. Atmos Environ 26: 121-124.

Read ND, Porter R and Beckett A (1983) A comparison of preparative techniques for the examination of the external morphology of fungal material with the scanning electron microscope. Can J Bot 61: 2059-2078.

Scott WJ (1957) Water relations of food spoilage microorganisms. In: Mark EM and Steward GF (eds.), Advances in food research. Academic Press, New York, NY, USA, pp. 83-127.

Smith SL and Hill ST (1982) Influence of temperature and water activity on germination and growth of *Aspergillus restrictus* and *A. versicolor*. Trans Br Mycol Soc 79: 558-559.

Van der Well GK and Adan OCG (1999) Moisture in organic coatings – a review. Progr in Organic Coatings 37: 1-14.

Van Wetter MA, Wösten HAB, Sietsma JH and Wessels JG (2000) Hydrophobin gene expression affects hyphal wall composition in *Schizophyllum commune*. Fungal Genet Biol 31: 99-104.

Viitanen HA (1997) Modelling the time factor in the development of mould fungi – the effect of critical humidity and temperature conditions on pine and spruce sapwood. Holzforschung 51: 6-14.

# 3 Fungal growth and humidity fluctuations: a toy model

Henk P. Huinink[1] and Olaf C.G. Adan[1,2]
[1]Eindhoven University of Technology, Faculty of Applied Physics, Eindhoven, the Netherlands; [2]TNO, Delft, the Netherlands

## Introduction

As all other organisms, fungi grow preferably under certain climatological conditions. All fungal species have a specific combination of relative humidity and temperature at which the growth is optimal. For a number of fungi the growth has been characterized as function of the relative humidity and temperature and plotted in so-called isopleths (Ayerst 1969, Smith and Hill 1982, Magan and Lacey 1984), which are contour plots of growth rates as a function of these two parameters. Such studies have provided much insight in the influence environmental conditions on the growth of fungi under steady-state conditions.

However, in practice the (micro-)climate of a fungus fluctuates in time with frequencies determined by the local surroundings of the organism. For example, in a bathroom a few times a day the walls of the room are subjected to warm humid air during a limited amount of time, while showers are taken. Generally, only during these short periods the relative humidity is sufficiently high to promote fungal growth. By combining isopleths with data on the micro-climate predictions can be made for growth under fluctuating conditions (Adan 1994, Clarke *et al.* 1999, Sedlbauer 2002). Although many of these studies differ in a number of details, they all have in common that steady-state data is extrapolated to predict growth under transient conditions. In fact these studies assume that organism responds infinitely fast on fluctuations in the local relative humidity and temperature. We will refer to this approach as the *standard model*.

The validity of this approach is subject to debate. In case that the climate fluctuations are limited to variations in the humidity (alternations of wet $RH>80\%$ and dry periods $RH<80\%$), the standard model simply predicts that growth rates are reduced by the ratio $\phi$ of the actual time that the organism has a wet environment and the total time (called the time-of-wetness TOW). Deviations of the standard model have been observed (Adan 1994, Viitanen 1996 1997), which have to be attributed to fact that the organism needs time to adapt its growth to changes in the micro-climate, given that the micro-climate is really the climate experienced by the organism. When the humidity fluctuations are characterized by a frequency $f$ (Hz) and the response of the organism by a time constant $\tau$ (s), then the standard model will hold as long as $f\tau \ll 1$ (note that $f = T^{-1}$ and $T$ (s) is the time constant of the humidity fluctuations). Therefore, deviations of the standard model indicate that this condition is no longer met.

Unfortunately there is a lack of experimental data on the growth response of fungi to changes in the humidity. As far as we know, there is only one study in the literature in which quantitative date is reported on response times of particular parts of a fungus (hyphae in this case) (Park 1982). Although a number of sophisticated models have been developed that relate fungal growth with processes like hyphal tip extension, branching processes, nutrient translocation, substrate heterogeneity, etc. (Prosser 1994a,b, Meškauskas 2004, Boswell 2007), application of these models to growth in fluctuating environments does make sense as long as there is no quantitative data on the response of the processes on fluctuations. A review of these models is therefore beyond the scope of this book, and the reader is referred to the aforementioned papers.

In this chapter a simple model is used in order to reconsider existing experimental data on fungal growth under fluctuating humidity conditions. It will be shown that this toy model, despite its simplicity and its numerous assumptions, is a powerful tool for pointing out the directions in which experimental research in this area has to move. In fact it is the simplicity of the model that enables to pin point which experimental observations have simple explanations and which not. It will be shown that the activation process of the fungus after a period of drought is the most important prerequisite for understanding the problem. In the preceding sections subsequently the model will be presented, growth scenarios will be outlined, the scenarios will be compared with the existing experimental data and conclusions will be drawn.

## A toy model for growth

### Philosophy

In this section a toy model is introduced for fungal growth in a fluctuating micro-climate. The model is not a detailed description of both the micro-structure and the processes on the micro-scale (on the level of single spores, hyphae, etc.). In contrary, as much processes and structural elements are lumped in a few parameters. The advantage of such a crude model is that its outcomes are easier to analyze. Due to its limited amount of parameters and variables, its simple mathematical structure and its clarity with respect to the assumptions, agreement with experimental data proves that there are simple explanations for the observations. Disagreement proves that essential parameters and variables have been overlooked and shows directions for modifications.

## Fluctuations of the micro-climate

A key ingredient of the model is its description of the micro-climate of the fungus: humidity and temperature. In order to keep the problem as simple as possible, the temperature is assumed to be constant and only the relative humidity varies. In Figure 3.1 it is shown how the humidity varies in the model. The relative humidity as experienced by the fungus cycles stepwise between a low value $RH_{min}$ and a high value $RH_{max}$. A complete cycle has a length $T$ and the duration of a period with high relative humidity is $T_w$. In the remainder of this chapter to the periods with $RH=RH_{max}$ and $RH=RH_{min}$ will be referred as wet and dry period respectively. It has to be stressed that the notions wet and dry respectively refer to steady-state circumstances at which there is significant growth of nearly no growth. The distinction between wet and dry is determined by the growth characteristics of a fungus. In practice for many organisms $RH=80\%$ seems to be reasonable choice for distinguishing wet and dry conditions (Adan 1994). Further, in this model the TOW equals $\phi \equiv T_w/T$.

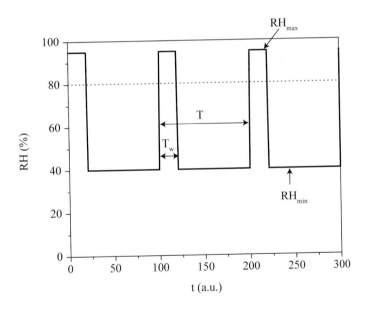

*Figure 3.1. Wet/dry cycles as experienced by the fungus. The solid curve represents relative humidity. The dotted curve marks the relative humidity, which is defined as the boundary between a wet and dry environment for the fungus. A complete cycle has a length T and the wet period with a duration $T_w$.*

**Fundamentals of mold growth in indoor environments**

## Fungal response on wet and dry periods

The fungal organism is assumed to exists in two different states: active (growing) biomass $A$ and passive (not growing) biomass $P$. Obviously, the total biomass is the sum of both $F=A+P$. Three processes occur. First of all, active biomass can multiply itself by growth (tip extension, branching, new germinations, etc.), which happens with a rate constant $k$. Under steady-state conditions with $RH=RH_{max}$ the growth rate equals $k$. Secondly, when the micro-climate changes from wet to dry, then biomass in the active state will be transferred into the passive state, $A{\rightarrow}P$, with a rate $\lambda$. Thirdly, as soon as the circumstances change from dry to, activation of the biomass will occur, $P{\rightarrow}A$, with a rate $\mu$.

During the wet period active biomass multiplies and passive biomass is activated. The evolution of the biomass can be described with the following set of differential equations,

$$\frac{dA}{dt} = kA + \mu P, \tag{1a}$$

$$\frac{dP}{dt} = - \mu P. \tag{1b}$$

In the dry period active biomass multiplies or is made passive. The biomass can be calculated with,

$$\frac{dA}{dt} = (k - \lambda) A, \tag{2a}$$

$$\frac{dP}{dt} = \lambda A. \tag{2b}$$

The model, as written down with the Equations 1a-2b, has three important features that deserve attention. First, self-interactions of the biomass are ignored, because the growth rate is assumed to be constant in the Equations 1a and 2a. Generally in later stages of the growth the effective growth rate drops due to depletion of nutrients or the productions of inhibitors leading to so-called sigmoid growth curves. Second, viability issues are ignored. In principle passive biomass can always be activated. In some sense, viability can be introduced in the model without changing it by setting $\mu=0$, but that introduces another assumption that all passive biomass is by definition not viable. Third, all details of the fungal life-cycle have been lumped into one single variable: biomass. By assuming constant rates for all the processes in the biomass, it is assumed that internal structure is always the same.

## Growth and wet/dry cycles

Analytical expressions can be obtained for the development of the biomass during either a wet or a dry period. It follows from solving the Equations 1a and 1b that in the wet period the active and passive biomass changes during an interval $\Delta t$ as

$$A(\Delta t) = A(0)\exp(k\Delta t) + P(0)\frac{\mu}{\mu + k}\left[\exp(k\Delta t) - \exp(-\mu\Delta t)\right], \tag{3a}$$

$$P(\Delta t) = P(0)\exp(-\mu\Delta t). \tag{3b}$$

For the dry period a similar set of equations can be obtained by solving the Equations 2a and 2b.

$$A(\Delta t) = A(0)\exp([k - \lambda]\Delta t) \tag{4a}$$

$$P(\Delta t) = A(0)\frac{\lambda}{k - \lambda}\left[\exp([k - \lambda]\Delta t)) - 1\right] + P(0) \tag{4b}$$

In Figure 3.2 an impression is given how the active and passive biomass could develop during a climate cycle according the Equations 3a-4b. In this particular example the active biomass (solid line) increases due to growth and activation of passive biomass (dotted line). After the wet-to-dry the active biomass deactivates quickly transition and nearly all biomass ends up in the passive state. The net increase of biomass is marked by thin horizontal dotted lines.

In order to predict the biomass evolution for $n$ wet/dry cycles an iterative procedure is used,

$$\begin{bmatrix} A \\ P \end{bmatrix}_n = \overline{\overline{D}} \cdot \overline{\overline{W}} \cdot \begin{bmatrix} A \\ P \end{bmatrix}_{n-1}, \tag{5}$$

where the active and passive biomass after cycle number $n$ is represented by the vector on the left hand side of the equation. The biomass obtained by multiplying the biomass after $n-1$ cycles with the matrices $\overline{\overline{W}}$ and $\overline{\overline{D}}$, which represent the processes during the wet and dry period respectively. The matrix $\overline{\overline{W}}$ is given by

$$\overline{\overline{W}} = \begin{bmatrix} \exp(k\phi T) & \frac{\mu}{\mu + k}\left[\exp(k\phi T) - \exp(-\mu\phi T)\right] \\ 0 & \exp(-\mu\phi T) \end{bmatrix}, \tag{6}$$

and is constructed by using the Equations 3a and 3b and $\Delta t = \phi T$. The matrix $\overline{\overline{D}}$ can be constructed from the Equations 4a and 4b with $\Delta t = [1-\phi]T$ and obeys the following expression,

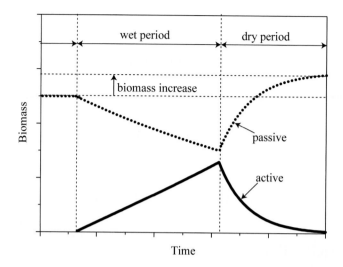

*Figure 3.2. The evolution of the active and passive biomass during a humidity cycle. The horizontal dotted lines mark the net increase of biomass. The vertical dotted lines mark the dry-to-wet and wet-to-dry transitions.*

$$\overline{\overline{D}} = \begin{bmatrix} \exp([k-\lambda][1-\phi]T) & 0 \\ \dfrac{\lambda}{k-\lambda}\left[\exp([k-\lambda][1-\phi]T)-1\right] & 1 \end{bmatrix}. \tag{7}$$

When exponential growth is assumed, then an effective growth rate $R$ can be calculated by using

$$F(nT) = F(0)\exp(RnT). \tag{8}$$

Note that $nT$ is the total time after $n$ cycles. The ratio $R/k \leq 1$, because the maximal growth rate equals $R=k$ in the absence of a dry period $\phi=1$.

### Dimensionless parameters

In the toy model the growth is determined by five parameters: three parameters defining the behavior of the organism ($k$, $\mu$ and $\lambda$) and two parameters characterizing the climate ($T$ and $\phi$). It has to be realized that the real number of independent parameters is only four, since the Equations 6 and 7 are fully characterized by $\tilde{k} \equiv kT$, $\tilde{\mu} \equiv \mu T$, $\tilde{\lambda} \equiv \lambda T$ and $\phi$. Studies on the influence of the frequency of the wet/dry events can be done as follows. In the model the frequency can be increased by decreasing the cycle time $T$. In order to study the frequency dependency of a particular system

($k$, $\mu$ and $\lambda$ fixed), the parameters $\tilde{k}$, $\tilde{\mu}$ and $\tilde{\lambda}$ have to be varied while the ratio $\tilde{k}{:}\tilde{\mu}{:}\tilde{\lambda}$ is kept constant.

These dimensionless parameters $\tilde{k}$, $\tilde{\mu}$ and $\tilde{\lambda}$ are useful, because they simplify the analysis. This can be illustrated with $\tilde{\lambda}$. When $\tilde{\lambda} \gg 1$, then the deactivation of the active biomass occurs quickly on the timescale of a humidity cycle. As a consequence the biomass follows the humidity cycles easily by switching from the active to the passive state and *vice versa*. The opposite is true when $\tilde{\lambda} \ll 1$. In this case the biomass will not switch from active to passive while the humidity drops during the cycle. Similar explanations hold for the parameters $\tilde{k}$ and $\tilde{\mu}$.

Furthermore, by using those dimensionless parameters a more universal picture of fungal growth is obtained, which can be tested experimentally. In fact, the toy model predicts that two organisms with completely different growth characteristics ($k$, $\mu$ and $\lambda$) behave similar given that the ratio $k{:}\mu{:}\lambda$ is constant and the cycle time of the wet/dry events is tuned such that the dimensionless parameters are similar for both organisms.

## Growth scenarios

Numerous calculations have been performed with the described toy model. Here, the discussion will be limited to $\tilde{k}/\tilde{\lambda} < 1$, which is according Equation 7 a prerequisite for a decrease of active biomass in a dry period. In Figure 3.3 the relative growth rate, $R/k$, is plotted as a function of the time-of-wetness $\phi$. Curves for two different sets of dimensionless parameters ($\tilde{k}$, $\tilde{\mu}$ and $\tilde{\lambda}$) are shown.

### Case I – Activation limited growth

For $\tilde{k}=0.1$, $\tilde{\mu}=1$ and $\tilde{\lambda}=10$ the relative growth rate $R/k$ is concave function of the time-of-wetness $\phi$. Detailed exploration of the parameter space has made clear that the following condition has to be met for this scenario: deactivation should be a faster process then activation $\tilde{\mu}/\tilde{\lambda} < 1$. Therefore this growth behavior will be called *activation limited growth*. An interesting feature of this growth regime is its dependency on the frequency of the wet events. With decreasing frequency, i.e. increasing cycle times $T$, the relative growth $R/k$ increases and its dependency on $\phi$ becomes more linear ($R/k \uparrow \phi$). It can be concluded that the standard model is a limiting case of activation limited growth when the cycle time is sufficiently long (fast switching between active and inactive state $\tilde{\mu} \gg 1$ and $\tilde{\lambda} \gg 1$).

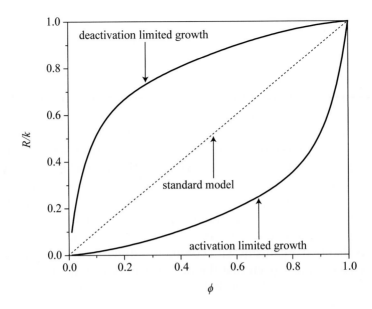

*Figure 3.3. Typical curves of the relative growth rate R/k as a function of the time-of-wetness φ (solid lines): activation limited growth (k̃=0.1, μ̃=1 and λ̃=10) and deactivation limited growth (k̃=0.1, μ̃=10 and λ̃=1). The dotted line represents the prediction of the standard model.*

### Case II – Deactivation limited growth

In case that $\tilde{k}$=0.1, $\tilde{\mu}$=10 and $\tilde{\lambda}$=1 the relative growth rate $R/k$ is convex function of the time-of-wetness $\phi$. Investigation of the parameter space has shown that activation should be a faster process then activation $\tilde{\mu}/\tilde{\lambda}$>1 for such growth behavior. Therefore this growth behavior will be called *deactivation limited growth*. With decreasing frequency, i.e. increasing cycle times $T$, $R/k$ decreases and becomes more linear dependent of $\phi(R/k \downarrow \phi)$. Again the standard model is a limiting case of deactivation limited growth when the cycle time is sufficiently long ($\tilde{\mu}\gg1$ and $\tilde{\lambda}\gg1$).

The model predicts that the transition point between activation limited growth and deactivation limited growth and *vice versa* occurs when activation and deactivation rates are in balance: $\tilde{\mu}/\tilde{\lambda}\approx1$. Interestingly, at this transition point the relative growth is a linear function of the time-of-wetness ($R/k\approx\phi$) even at high frequencies, where the assumptions of the standard model are violated. Therefore, experimental data that show a linear relation between the growth rate and the time-of-wetness, $R\propto\phi$, have to be interpreted with some care.

## Comparison with experiments

### Relative growth and time-of-wetness

Mathematical models are valuable as tools for either predictions of new phenomena or explaining the origins of known phenomena. The toy model, as discussed in this chapter, is used in the latter way. The best way to test the validity of the predicted growth scenarios is to compare with experiments in which the time-of-wetness is varied systematically. There is only one report in the literature in which such an investigation is presented (Adan 1994). In this study the growth of *P. chrysogenum* was monitored on bare gypsum and gypsum coated with an acrylic paint. The experiments were performed at constant temperature but under transient humidity conditions; i.e. alternating periods of high, $RH_{max}$=97%, and low relative humidity, $RH_{max}$=58%. In Figure 3.4 the obtained values for $R/k$ are plotted as a function of $\phi$ for a cycle time $T$=12 h.

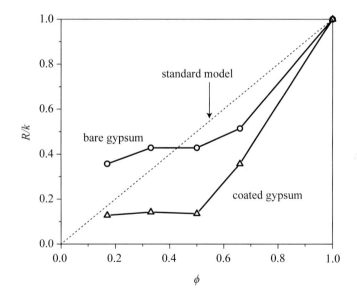

*Figure 3.4. The relative growth rate of* P. chrysogenum *as a function of the time-of-wetness. Both the bare and coated gypsum were exposed to an alternating pattern T=12 h) of high, RH$_{max}$=97%, and low relative humidity, RH$_{max}$=58%. The open spheres and triangles are the actual data points (Adan 1994) (the solid line only functions as a guide for the eye). The dotted line refers to the standard model.*

By comparing Figure 3.4 with the outcomes of the toy model two important features are observed, which can be explained with the presented toy model and will be discussed subsequently. First of all, the growth behavior of *P. chrysogenum* in this particular experiment deviates from the standard model. The observed growth rates are non-linear functions of the time-of-wetness, indicating that a growth model has to take into account the response of a fungus to a wet/dry or a dry/wet transition. Second, the growth curves are concave functions of $\phi$. This observation can be explained as a manifestation of activation limited growth; i.e. the activation process after a dry/wet transition is slow compared to the deactivation process after a wet/dry transition. From this it can be concluded that the activation of *P. chrysogenum* after rewetting the system takes a considerable amount of time (perhaps of the order of a few hours). The importance of the activation process is confirmed by other studies. Studies of the response of a variety of colony margins on rewetting have shown that the time for growth to retain can be very long (Park 1982).

Figure 3.4 contains also two features that cannot be explained by the presented toy model. First, despite the fact that the experimental data points support a concave shape of the growth curve, the growth curve should have a convex shape at low values of the time-of-wetness, $\phi<0.10$, in order to start at $R/k=0$ in the absence of wet periods $\phi=0$. Such a curve (convex for $\phi<0.10$ and concave for $\phi>0.10$) cannot be obtained with the presented toy model, see Figure 3.3. Second, the growth curve for *P. chrysogenum* crosses the curve of the standard model. As discussed in the previous section, the standard model is a limiting case of the toy model. Within the parameter space that has been explored, the toy model does not predict such phenomenon.

### Frequency dependency of growth

The frequency dependency of growth can be measured by changing the cycle time $T$ of the wet/dry events while keeping the time-of-wetness $\phi$ constant. The scarce experimental reports do not provide data for a simple picture (Adan 1994, Viitanen 1997). During experiments with mold fungi on pine and spruce sapwood it was observed that the relative growth rate $R/k$ increased with increasing $T$ (decreasing frequencies) (Viitanen 1997). In this report it is stated that for $T>24$ h only the total time that the system was subjected to a wet climate determined the degree of growth. Note that this is exactly the same as stating that the growth rate is proportional to the time-of-wetness ($R \propto \phi$), which indicates that the standard model holds for $T>24$ h. Summarizing, with decreasing frequency the growth rate increases and at low frequencies standard-model-behavior is observed. According to the toy model this experimental finding should be attributed to the fact that the organism is in the activation limited growth mode. As explained in the section on growth scenarios the toy model reproduces the standard model at low frequencies, long cycle times, which is observed in this experiment.

The outcomes of experiments with *P. chrysogenum* on coated and bare gypsum cannot be interpreted so easily with the toy model (Adan 1994). Whereas on coated gypsum the growth rate was not sensitive to the frequency, on bare gypsum it strongly increased with increasing frequency. The latter observation is the opposite of the experimental findings with mold fungi on wood (Viitanen 1997) and cannot be explained by the toy model with an activation limited growth scenario (see section on growth scenarios). The discrepancy between experiment and model are not conclusive, since the difference between the two different experiments itself (coated and bare gypsum) is not understood.

## Making up the balance

Now the predictions of the toy model have been confronted with the existing experimental data, it is time to make up the balance. First of all, it has to be concluded that quantitative experimental data is scarce. Obviously, this hampers progress with respect to the topic of interest. It is difficult to find suitable data to test even a model as simple as the toy model that is presented in this chapter. As a consequence, it is hard to decide what parts of the model have to be improved. Since the mathematical structure is only a translation of biological concepts, this means that the biological concepts cannot be tested. Progress in this field is only possible when systematic quantitative experimental data becomes available.

Second, the toy model seems to explain parts of the existing experimental data. With the so-called activation limited growth scenario the concave non-linear relationship between the relative growth rate $R/k$ and the time-of-wetness $\phi$ could be explained. Also part of the data on the influence the frequency could be explained with this scenario of the toy model (increasing growth rates with decreasing frequency). Therefore, it can be concluded that one of the merits of the toy model is that it draws the attention towards the response rates after wet-to-dry and dry-to-wet transitions ($\lambda$ and $\mu$). Therefore, future experimental studies should try to characterize these response rates as quantitative as possible for different organisms on different substrates. Future studies should also try to link these rates with processes on the micro-scale (on the level of a single spore or hypha).

Third, there are aspects of the experimental observations that cannot be explained with the toy model: the behavior $R/k$ at small values for $\phi$ and some of the experimental regarding the frequency influence. As the complete curves, as presented in Figure

3.4, cannot be reproduced by the toy model, it has to be concluded that the toy model is too simple and ignores biological processes that are important for understanding fungal growth under humidity fluctuations. In the next section directions for future modeling attempts will be pointed out.

## Signposts for a better model

### Introduction

Obviously, the presented toy model overlooks relevant biological processes. Improvement of the existing model or development of a completely new model is only possible when it is known what biological processes have to be included. As already mentioned only systematic experimental work of strong quantitative nature can lead to a scientific breakthrough at this point. In this section optional future directions for modeling will be pointed out. First, in next paragraph a minor modification of the toy model will be discussed to prove that the problem of progress is not a matter of mathematics. A simple mathematical adaptation leads to much better outcomes. However, such adaptation also leads to the introduction of new unknowns. Second, the succeeding paragraph will address a few issues that have to be incorporated in future modeling studies.

### Two activation regimes

It is the behavior of the toy model at small values of $\phi$ which indicates that the model is too simple. A few modifications of the model have been tried. Interestingly, changes in the mathematics of the deactivation process do not lead to improvements and sometimes make the discrepancy between experiment and theory more pronounced. This is caused by the fact, that the mathematical structure of the deactivation rate especially plays role at high values of $\phi$ (short dry periods), which are described rather well by the toy model. More promising results were obtained by modifying the mathematical structure of the activation process. The activation rate was made a function of time,

$$\mu(t) = \begin{cases} \mu_{max} & \wedge \ t \leq t^* \\ \mu_{min} & \wedge \ t > t^* \end{cases}, \tag{9}$$

where $\mu_{max} > \mu_{min}$ and $t^*$ is the time when the system switches from high to low activation rates ($t^*$ is measured from the beginning of the dry-to-wet transition). The biological meaning of Equation 9 is that the organism adapts itself to the length of the wet period. Initially after the start of the wet period activation of the passive biomass goes quickly. As the wet period continues the activation rate drops.

In Figure 3.5 $R/k$ is plotted as a function $\phi$ for a system characterized by $\tilde{k}=0.1$, $\tilde{\mu}_{max}=10$, $\tilde{\mu}_{min}=0$ and $\tilde{\lambda}=1$. The switch from high to low activation rates occurs at $t^*/T=0.01$. Note that as long as $\phi<t^*/T$ the wet periods are so short that the organism never switches to a low activation rate. As long as $\phi<t^*/T$ the system is characterized by $\tilde{\mu}_{max}/\tilde{\lambda}=10$ and follows a deactivation limited growth scenario; $R/k$ is convex function of $\phi$ in this regime (see section "growth scenarios"). With a longer time-of-wetness, $\phi>t^*/T$, the organism switches during the wet period to a state with a low activation rate and the system transfers to deactivation dominated growth.

Clearly, the discrepancy between the outcomes of the original toy model and the experimental data has been resolved by using Equation 9. It has introduced a convex shape at low values of $\phi$ and the curve of the standard model can be crossed. The price for this improvement is the introduction of two new unknowns: the validity of the biological concept behind Equation 9 and the parameterization. The good news is that such specific unknowns raise new scientific questions, which serve as signposts for both experimental work and modeling attempts. To be more specific, it

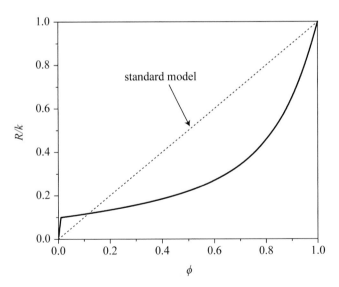

*Figure 3.5. The relative growth rate as predicted with a modified toy model (solid line): $\tilde{k}=0.1$, $\tilde{\mu}_{max}=10$, $\tilde{\mu}_{min}=0$ and $\tilde{\lambda}=1$. During a wet period the system switched from high to low activation rate at $t/T=t^*/T=0.01$. The dotted line represents the standard model.*

seems that future research should especially focus on the activation processes after a dry-to-wet transition.

### Pathways for future modeling

The introduction of Equation 9 in the previous paragraph has raised two questions. First, what experimental justification can be given for the added unknowns? Second, what biological concepts are not included in the present toy model? Especially the details of the activation processes have to be addressed. In this paragraph a basic assumption of the original toy model will be examined in order to clarify the issues that should be incorporated by a new model and deserve experimental validation.

The life-cycle of the fungal organism and the resulting differentiation of the biomass is completely ignored in the toy model. In the model the biomass is treated without accounting for its development in time. The composition (the structure of the mycelium, the number of fruit bodies, etc.) of the biomass is considered to be constant in time. Further, the rates biological processes are not influenced by time; the variables that describe the kinetics ($k$, $\mu$ and $\lambda$ are constants in the original toy model. The only temporal variation in the biomass is its state of activity; the biomass is present in two states active $A$ and passive $P$.

Therefore, a future model should include (parts of) the composition of the biomass and its temporal variations. First of all, variables for the density of the biomass, the number of fruit bodies and the spore density have to be included. Second, parameters for the rates of the different steps in the fungal life-cycle have to be introduced. These parameters have to be linked with the duration of the wet and dry periods in the climate cycle. Third, the activation of passive biomass after a dry-to-wet transition has to be specified. Activation via germination of spores has to be distinguished from activation of by resuming growth of the hyphae. Therefore, at least two rate parameters are needed for describing the activation process. An important consequence of these three modifications could be that the preferential pathways for activation (germination or resuming growth of the hyphae) become dependent on the time-of-wetness and the frequency of the wet/dry events.

## Conclusion and outlook

The growth of fungi under alternating wet and dry periods has been modeled with the aim of identifying the important issues for future research. To that end a toy model has been developed, which leaves out a lot of biological details. In the model the biomass can adopt two states: active and passive. Active biomass grows with a constant rate $k$. In a wet period passive biomass is activated with a constant rate $\mu$. During a dry period active biomass is deactivated rate $\lambda$. The simplicity of the model has proven to be its explanatory power; both agreement and discrepancy between model results and experimental data can easily be attributed to key parameters and assumptions of this toy model.

The toy model proved that the *standard model* as long as activation of biomass at the beginning of a wet period and deactivation of biomass in the dry period are fast processes on the time scale of the humidity cycles. Therefore, the standard model only works in case of a slow varying climate (weeks). Deviations from the standard model are predicted by the toy model in case of fast fluctuations (hours and days). These deviations have been observed in various experiments. This implies that predictions of fungal growth based on climate data in combination with isopleths (steady-state growth rates) can only be justified for long term climate variations.

Parts of the experimental data on fungal growth in the presence of alternating wet and dry periods could be explained with the model in terms of a so-called *activation limited growth* scenario. The rate of the activation of biomass after dry-to-wet transitions seems to be the key parameter for understanding the growth behavior. Agreement between experimental data and model outcomes could be obtained by assuming that activation of biomass is a slower process then deactivation.

Discrepancy between model predictions and experiments was observed for cycles with short wet periods (a small time-of-wetness). The interesting feature of this disagreement is that it again points towards the importance of the activation process. By making the activation rate time dependent the experimental trends could be reproduced. From this it can be concluded that future research should focus on the details of the response of a fungus on the transition from a dry to a wet period.

Experimental studies should focus on two aspects. First of all, the influence of rewetting on different stages of the *life-cycle* should be understood. Activation of biomass can occur via germination of spores or resuming growth by existing hyphae. The importance of both activation processes may depend on the frequency of the climate cycle. Maybe there are different preferential pathways for activation for different frequencies. Second, progress in this area is only possible with experimental data of a quantitative nature. Refining simple models, as presented in this chapter, does not make sense as long as there is nearly no experimental data for validation.

# References

Adan OCG (1994) On the fungal defacement of interior finishes. PhD thesis, Technische Universiteit Eindhoven, Eindhoven, the Netherlands, pp. 187-201.

Ayerst G (1969) The effects of moisture and temperature on growth and spore germination in some fungi. J Stored Prod Res 5: 127-141.

Boswell GP, Jacobs H, Ritz K, Gadd G and Davidson FA (2007) The development of fungal networks in complex environments. Bull Math Biol 69: 605-634.

Clarke JA, Johnstone CM, Kelly NJ, McLean RC, Anderson JA, Rowan NJ and Smith JE (1999) A technique for the prediction of the conditions leading to mould growth in buildings. Build Env 34: 515-521.

Magan N and Lacey J (1984) Effect of temperature and pH on water relations of field and storage fungi. Trans Br Mycol Soc 82: 71-81.

Meškauskas A, Fricker MD and Moore D (2004) Simulating colonial growth of fungi with the neighbouring-sensing model of hyphal growth. Myc Res 108: 1241-1256.

Park D (1982) Phylloplane fungi: tolerance of hyphal tips to drying. Trans Br Mycol Soc 79: 174-178.

Prosser JI (1994a) Kinetics of filamentous growth and branching. In: Gow NA and Gadd GM (eds.) The growing fungus, Chapman & Hall, London, UK, pp. 301-318.

Prosser JI (1994b) Mathematical modeling of fungal growth. In: Gow NA and Gadd GM (eds.) The growing fungus, Chapman & Hall, London, UK, pp. 319-335.

Sedlbauer K (2002) Prediction of mould growth by hygrothermal calculation. J Therm Env Build Sci 25: 321-336.

Smith SL and Hill ST (1982) Influence of temperature and water activity on germination and growth of *Aspergillus restrictus* and *Aspergillus versicolor.* Trans Br Mycol Soc 79: 558-560.

Viitanen HA (1996) Factors affecting the development of mould and brown rat decay in wooden material and wooden structures. PhD thesis, Uppsala, Sweden.

Viitanen HA (1997) Modeling the time factor in the development of mould fungi – the effect of critical humidity and temperature conditions on pine and spruce sapwood. Holzforschung 51: 6-14.

# 4   The fungal cell

*Jan Dijksterhuis*
*CBS-KNAW Fungal Biodiversity Centre, Utrecht, the Netherlands*

## Introduction

Fungal problems arise when there are water problems. This will be the case after calamities such as floods and also when condensation of water occurs inside the house due to a cold bridge. However, even when these calamities do not happen, it is still remarkably difficult to maintain an entirely fungus free house. Bathrooms, for example are notorious for the development of fungi to such an extent that their presence seems to be the rule and not the exception.

This presence of fungal development occurs while the *average* humidity in a house during the day is below the minimal value needed for fungal development (Adan 1994). In laboratory situations the great majority of the studies on fungi are performed in stationary conditions or after the variation of one parameter.

Adan introduced the concept of time-of-wetness (TOW) into the mycological area. The term hitherto only known from physics, expresses the fraction of time for which the relative humidity exceeds a threshold value. A TOW of 0.1 means that of every 24 hours, 2.4 hour are characterized by a humidity that allows fungal development. For the fungal colony this means that intermittent periods of development are followed by longer periods of humidity below the level that supports growth and as such an extension of the fungal hyphae. For the fungal cell this implies that development stops at regular time intervals and has to resume within the allowed time window, which may be a number of hours.

## Fungal growth

Some authors have characterized fungi as a kind of tube-dwelling amoeba. Inside a sturdy cell wall, which is variable in thickness, the living fungal cell or cells reside. The long tubes or filaments are called hyphae, which form branches at regular distances (Figure 4.1). At later stages of growth, these branches grow into the air to form the aerial hyphae (Figure 4.1).

The fungal cells are separated by cross-walls, a characteristic of the higher fungi; but the separation is not complete. There are pores inside the cross-walls, enabling the cells to communicate (cross-walls are nicely visible in Figure 4.1). As such the fungal

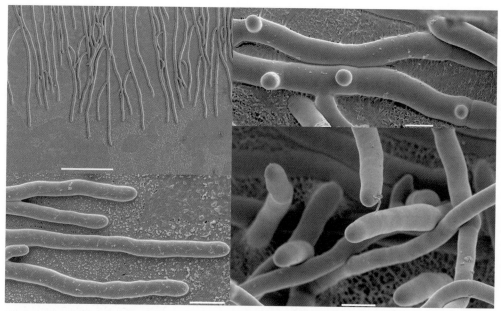

*Figure 4.1. Different stages of hyphal growth of a* Penicillium *species as illustrated by means of cryoSEM. (top left) An overview of the leading edge of a fungal culture where hyphae form lateral branches. (top right) Detail of leading hyphae. (bottom left) Initials of aerial hyphae that appear directly behind septa. Note that the fungal cell in the middle, which has formed a lateral branch, now forms an aerial hypha. (bottom right) A further stage of aerial growth. Bars are 100, 5, 5 and 10 μm (clockwise).*

mycelium can be regarded as a continuum, a so-called syncytium. This can be a risky situation when hyphal cells become damaged due to mechanical stress or as a result of the action of cell-wall degrading enzymes or cytotoxic compounds. Fungal hyphae often face these threatening conditions in soil where they thrive. "Small animals" such as mites and springtails can physically damage the fungal cell and many bacteria form antifungal compounds (see e.g. Dijksterhuis *et al.* 1999). If the integrity of a fungal cell were to be destroyed and the connection between the cells remained open, cytoplasm would flow freely from the mycelium. However, many fungi have developed an amazing device for dealing with these situations, namely the Woronin body. Small, spherical and optical dense bodies hover near the septal pore and may even be connected to them. When a sudden cytoplasmic flow indicates a breakage of the hyphal continuum, these bodies are pressed against or in the septal pore and block further devastating loss of cytoplasm. Mutants are prepared that could not functionally form Woronin bodies and these fungi literally bleed to death (Jedd and Chua 2000).

The apex (tip of the hyphae) is a very dynamic cell area, where different cell organelles cooperate to ensure extension. The most notable feature of this area is a structure called the Spitzenkörper (Spk), a conglomerate of different types of vesicles containing cell-wall constituents such as chitosomes, structures that enable the formation of the backbone of the fungal cell wall, chitin (see for instance Riquelme *et al.* 2007). Chitin is also the most important compound of the exoskeleton of crustaceans and one of the most abundant polymers in the world. In Figure 4.2, three images of a living hypha of the fungus *Rhizoctonia solani* are shown with a brightly stained Spk inside the hyphal tip. Here, a membrane-staining dye is taken up into the cell and stains the vesicles that accumulate into the Spk. In addition, the Spk also contains other cell components such as actin fibers, ribosomes, microtubules and proteins like γ-tubulin. The area directly behind the Spk contains a well-ordered array of organelles such as elongated mitochondria, the energy fabrics of the cells, or microbodies (Meijer 2007), organelles that play an important role in the synthesis of secondary metabolites, such as penicillin by the fungus *Penicillium chrysogenum*. Further apical of the apex, nuclei, the Golgi apparatus and tubular vacuoles complete the orchestration of the remarkable functional unity of this specialized area. The complexity of the fungal

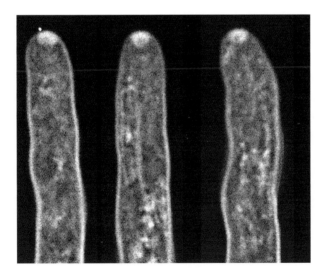

*Figure 4.2. Three different pictures of a growing hypha of the fungus* Rhizoctonia solani, *seen by means of confocal microscopy. The hypha is stained with a fluorescent dye, FM 4-64, that stains membrane structures without doing damage to the cell. At the top of the hyphae a bright staining structure is visible, which is the Spk. The different pictures are taken less than 5 min apart and the dynamics of growth is clearly illustrated. Note that the positioning of the Spk varies slightly.*

cytoplasm is illustrated in a picture of a hypha of the fungus *Talaromyces macrosporus* that expresses a fluorescent protein in the cytoplasm showing many different areas, i.e. not a uniform mass (Figure 4.3). The Spk is functionally synonymous with the VSC, the Vesicle Supply Center, the area that is the organization centre for the many vesicles that will fuse with the plasma membrane. The extension of the cell at the apex is correlated with a plastic and very thin cell wall that is still in the act of formation (Wessels 1990) due to the recent secretion (exocytosis) of cell-wall building blocks. Fungal hyphae produce many proteins and these have to pass through this cell wall prior to excretion. Enzymes are very important for the fungus while it digests often complex compounds such as plant cell walls outside the cell, and feeds itself with these products of enzymic degradation.

The VSC is strongly correlated to the direction of fungal growth. For example, when it moves to the right, the hyphae will bend to that direction while the cell wall is deposited mostly to that side of the apex. The Spk is a complex and sensitive structure. Upon sudden changes in, for example, temperature or after touch it simply disappears and fungal growth ceases, often accompanied by a swelling of the hyphal tip. The growing hyphal apex even has a characteristic tapering, which is visible in the aerial hyphae of Figure 4.1 and which has been described by mathematical modeling (Bartnicky-Garcia *et al.* 1989) as a cotangent function. One can literally recognize growing hyphae.

The unique organization and sensitivity of this area of the growing fungal cell is very important in light of the context of the previously posed questions. However, several

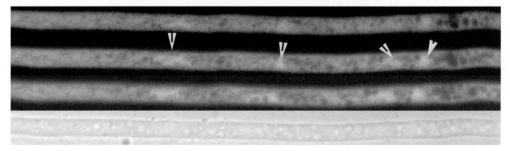

*Figure 4.3. A hypha of* Talaromyces macrosporus *that expresses a fluorescent protein, GFP, illustrates the complex structure of a living fungal hypha. The three top micrographs show three levels of focus (depth), taken by means of confocal microscopy. It shows the distribution of the fluorescent protein through the fungal cell. The arrow shows different zones of more intense fluorescence. There are also non-fluorescent areas, which are the nuclei. The bottom picture shows how the hypha appears with transmitted light.*

questions can be asked: (1) does it react to changing humidity; (2) is it robust enough to resume growth quickly enough after a period of low relative humidity?

## The fungal colony as a unity

As stated in the previous paragraph, fungal growth can start at subapical areas, further down the hyphal apex, but then as branches. Branch formation (Figure 4.1) begins with an area of weakening of the cell wall, followed by the establishment of a hyphal tip at that site. Some branches only go for a short while and fuse with another very short branch formed on another hypha. This process is called anastomosis and is important in that it enables the fungal colony to form a network, a true flock of interconnected hyphae, the mycelium.

That the fungal network can act as a unity is nicely illustrated in the case of the fungus *Rhizoctonia solani*. This fungus can form survival structures called sclerotia that appear as brownish structures on a colony. It can be infected by another fungus *Verticillium biguttatum,* which can penetrate through the cell wall into the cytoplasm of the broad hyphae of *Rhizoctonia,* forming a small club-like structure that withdraws nutrients from the fungal network. As a result the fungus *Rhizoctonia solani* is not able to form sclerotia even in remote parts of the colony when infection with *V. biguttatum* is only confined to a very small part of the colony (Van den Boogert and Deacon 1994). This indicates that different areas of a colony are able to communicate. However, *Rhizoctonia* has many different types of anastomosis; mycelia that do not belong to compatible groups refuse to anastomose. In fact, they do anastomose, but then the fused cells die and no connections are established. Remarkably, when two colonies belonging to different anastomosis types are inoculated, they grow over the surface of the culture, but refuse to become a unit. When one of the strains is infected with *V. biguttatum,* only that culture does not form the sclerotia; the other colony does, while growing on the same medium.

## Considerations about aerial hyphae and fungal survival

As mentioned in the previous paragraph, after a period of growth along the substrate, hyphae are formed that grow into the air. Aerial hyphae appear to be "normal hyphae", with an apex shape similar to that of the horizontally growing hyphae (see Figure 4.1). Is the aerial mycelium identical to the substrate mycelium? The next paragraphs give a survey of the different unique aspects of the aerial mycelium that ultimately brings us back to the area of fungal survival.

## Orientation

The orientation of these hyphae has changed from "horizontal" or "substrate bound" to "aerial". In *Penicillium* (Figure 4.1) I consider this as a certain acquired property of the hyphae and not as a result of a coincidental change in the axis of the horizontal hyphae. Aerial hyphae "choose" the air and are in that respect different from horizontally growing hyphae. In Figure 4.1 one can see that initially a horizontally growing cell forms lateral branches but, at a later stage, branch initials are invariably formed perpendicularly behind a septum and grow into the air. It is tantalizing to think that this occurs while there is no space for a lateral branch left. Recently, we observed hyphae of *Aspergillus niger* growing through a membrane with pores with a diameter of 10 µm. At the periphery of the colony, hyphae grew through the pores and resumed their horizontal growth. At another location where the membrane was above areas where aerial hyphal were more common, the hyphae that grew through the pores continued vertical growth upon leaving the pore. This illustrates that, in at least a number of fungi, aerial hyphae are differentiated fungal structures that perform a fixed axis of orientation away from the substrate. Figure 4.4 shows several hyphae of *Fusarium oxysporum* growing into the air amidst horizontally growing hyphae, but from this picture it is not clear whether the aerial hypha starts as an initial on a substrate hypha. Thus, variation of hyphae to reach the air might occur between fungal species

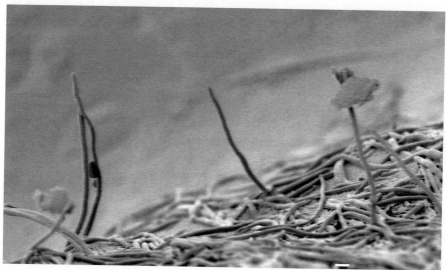

*Figure 4.4. Aerial growth of* Fusarium oxysporum *as seen by cryoSEM. Individual hyphae choose the air.*

## Colonization and breaking the water film

Aerial hyphae enable the fungus to cross (short) distances through the air to reach novel sites of nutrients. In soil, small air pockets between particles and hyphae enable the fungi to cross these, an opportunity the bacteria do not have. Crossing of small distances is seen in the case of *F. oxysporum* in a multititer plate (Figure 4.5). In the fungus *Schizophyllum commune* the ability of hyphae to grow into air was correlated with the production of proteins called hydrophobins. When present in an aqueous solution these proteins lower the water surface tension markedly. Furthermore, they self-assemble on the fungal cell wall as a very thin layer rendering the hyphae hydrophobic (Wösten *et al.* 1993, 1999). These aspects mean that (1) fungal hyphae are able to penetrate a thin film of water, which on a microscale is a formidable force for keeping structures flattened, and (2) fungal hyphae do not fall back to the "substrate" as a result of wetting of the mycelium. A related protein, SC4, is present inside air channels in the macroscopic fruit bodies of *S. commune* and may prevent "drowning" of the fungus as a result of water-influx through capillary forces (Lugones *et al.* 1999, Van Wetter *et al.* 2000). There are other cell wall proteins present in *S. commune*, for instance SC 15 has functions similar to that of SC3 (Lugones *et al.* 2004). SC15 is thought to have a more prominent function when the fungus grows on birch wood, the natural substratum for the fungus (De Jong *et al.* 2006). The presence of uncolonized wood near the fungal colony provokes aerial hyphae formation towards this attractive food source. Even mutants that do not form SC3 or SC15 or both, form aerial mycelium (De Jong *et al.* 2006). This indicates that other factors are involved and that aerial hyphae can perform chemotaxis (growth against a gradient of a certain attractive compound). Remarkably, colonized wood does not attract hyphae.

## Respiration and secretion

The fungus *Aspergillus oryzae* forms abundant hyphal mycelium on a wheat flour model substrate (Rahardjo *et al.* 2004). The authors used a reaction diffusion model to describe the different gradients that evolve in growth situations of the fungal colony on a solid substrate. During growth on a medium the substrate hyphae become interconnected and form a dense biofilm layer while the aerial mycelium originates from this layer into the air. Hyphae can also grow into the substrate (penetrative hyphae). Inside the substrate enzymes are released, which degrade polymers (e.g. starch) into monomers that are taken up into the cytoplasm. Oxygen levels drop very quickly in this layer, when pores between the hyphae of the substrate biofilm are filled with water. Measurements with micro-electrodes revealed that the oxygen concentration dropped below 2.5% within 50 µm depth and to 0% within approx. 80 μm (Rahardjo *et al.* 2002). Microelectrodes also showed that the oxygen concentration remained 21% at a depth of 4-5 mm inside the aerial mycelium.

Cultures that were covered with a gas-permeable polycarbonate membrane did not form aerial hyphae. The $O_2$ flux in the aerial mycelium was ten-fold higher compared to the overlay culture. Up to 90% of the respiration was confined to the aerial mycelium after 70 hours of growth. The biomass in cultures after this cultivation time was 75% higher compared to membrane-covered cultures. In addition, secretion of the starch-degrading enzyme α-amylase was also approximately twice the level of cultures without aerial mycelium (Rahardjo *et al.* 2005a). The question remains whether the aerial hyphae themselves secrete enzymes; a situation which would not be very logical, since there is nothing to degrade in mid-air. The alternative explanation would be that enzyme production is accelerated inside the substrate hyphae as a result of the presence of the aerial hyphae. Amylase production in this study was proportional to the oxygen consumed. Can the aerial hyphae be regarded as "the lungs" of the fungal culture?

Te Biesebeke *et al.* (2006) identified a putative haemoglobin-like protein in *A. oryzae* and also overproduced these domains, which resulted in higher oxygen uptake of the fungal filaments. The presence of oxygen-binding proteins in fungi is highly suggestive of possible oxygen transport through the mycelium towards areas of low oxygen levels (inside the biofilm of the substrate hyphae). Oxygen transport would lead to higher levels of metabolism including enzyme secretion (see above).

Vinck *et al.* (2005) studied the production of the enzyme glycoamylase by the fungus *Aspergillus niger* by means of the expression of GFP-binding protein and observed that only part of the substrate hyphae could be regarded as an enzyme-producer. Hitherto, no observations have yet been made to confirm that these aerial hyphae do not secrete. There are clues that conidiophores, the structures that produce spores, do not show expression of GFP under the glycoamylase promoter, when they are growing into the air. However, while conidiophores are different structures from aerial hyphae, they can be formed from them.

### Nutrients, metabolites and water

One should not assume that aerial hyphae secrete enzymes intended for nutrient degradation, and that expression of the genes responsible for these enzymes would be absent. The enzymes needed for cell wall polymerization and structural proteins (including the hydrophobins) still need to be secreted, but will reside on or inside the cell wall.

It has to be concluded that all the nutrients needed for growth of these hyphae must originate from the substrate mycelium. So, transport of nutrients is vital for these hyphae, which can reach out for many millimeters into the air. I have observed hyphae of the fungus *Arthrobotrys oligosporus* growing inside Petri dishes towards the upper lid of the dish, growing over the inner surface and forming spore-forming structures after having grown for several centimeters. Figure 4.5 shows hyphae of *Fusarium oxysporum* crossing air spaces and growing over a polystyrene surface. *Aspergillus oryzae* can form aerial hyphae that measure up to 4-5 mm on millet wheat grain (Rahardjo *et al.* 2005b) and *Schizophyllum commune* grows through the air towards uncolonized birch wood (De Jong *et al.* 2006).

Aerial hyphae might have a very active metabolism and may produce secondary metabolites residing both on or inside the cell wall, and may even affect the structure. For instance, deposits or crystals are formed on aerial hyphae. Furthermore, secretion of water containing (secondary) metabolites may also occur as the formation of exudate in many fungi. These are droplets which evolve on aerial mycelium during certain growth stages of the fungal colony. They contain different compounds (J.C. Frisvad, DTU, Denmark, personal communication). For instance, exudates can be yellow or even dark-brown indicating the presence of pigments. Inherent to the formation of exudates is transport of water through the cytoplasm and out of the hyphae. Coggins *et al.* (1980) observed hyphae of the fungus *Serpula lacrymans* that grew from agar on a coverslip over the surface of a Petri dish at high humidity. They observed that droplets were formed from individual hyphae, which was dependent on intact cells; severed hyphae did not show an increase in the diameter of the formed droplets. This was also achieved by adding 0.1 mM sodium azide to the centre of the delivering culture. When 30 µl 1M KCl was added, the increase of the diameter of droplets was halted, but lower amounts of solutes such as 100 mM KCl or 200 mM sucrose had some puzzling effects, including an initial decrease in droplet formation and a subsequent increase above the level of the controls. Brownlee and

*Figure 4.5.* Fusarium oxysporum *crosses the air and grows over the polystyrene surface.*

Jennings (1981a) inoculated the fungus in a reservoir of fluid medium on Perspex Leucocyte migration plates, which colonized from that, the plastic surface (kept on glass rods inside a Petri dish). They studied the formation of the droplets when different solutions were added to the fluid medium. High concentrations of glucose or KCl added to the food solution resulted in a reduction of droplet volume and thus diminished droplet formation and fungal extension. The authors concluded that both the vapor pressure of the atmosphere above the droplet as well as the water and solute uptake at the food source had a bearing on droplet formation. Oligomycin, a blocker of proton channels, markedly reduced the droplet formation. All these data suggest that water is actively driven out of the hyphae. Olaf Adan (personal communication) observed exudate formation in agar cultures of *Penicillium chrysogenum* and found that the characteristic formation of droplets by this fungus varies with the water activity of the air, with the growing temperature and with the growing stage of the colony. Most droplets appeared at the optimum growth temperature and at a high relative humidity. In some cases droplets appeared and disappeared during a certain "window of time".

### Survival

Hyphae that grow into the air or on surfaces are different from substrate attached hyphae, with respect to the availability of nutrients and need the internal transport of nutrients from the substrate hyphae. Aerial hyphae also need extra requirements for survival in dry conditions and to break the water surface. They can also change to become substrate attached hyphae again (see Figure 4.5 and De Jong *et al.* 2006). Hyphae possibly influence inert surfaces such as plastic and rock particles and might even "etch" or damage the surfaces (Gusse *et al.* 2006, Milstein *et al.* 1992) due, for instance, to the action of externally produced hydrogen peroxide or acids.

Do aerial hyphae have a higher stress-tolerance than substrate hyphae? This is relevant for fungal colonies which have to survive transient periods of relative humidity. It is clear that aerial hyphae do encounter dryer conditions than the substrate hyphae. They leave the safety of the hyphal mat for the colonization of new substrate that is physically abroad. Given the time they need to complete this quest, they may die after having transferred the living cytoplasm to new horizons. Collapsed aerial hyphae can be seen on colonies that have been faced with dry air after opening of the Petri dishes. Microscopic techniques for observing the physical state of aerial hyphae are sparse. Cryo-electron microscopy freezes aerial hyphae instantly, but could as artifact lead to collapsed cells. On the other hand, these cells may have collapsed as a result of the transfer of the fungal cell through ambient air, outside the relatively high humidity of the culturing state. Finally, fungal hyphae might collapse but remain living and will resume growth after rewetting of the cells. Park (1982) studied hyphae of 15 different species and their ability to re-grow after drying periods. Fungal colonies were grown on cellophane over malt agar for 3-7 days and the membrane was transferred to Petri dishes in ambient laboratory air. Fungi were from the "phylloplane" type (*Alternaria alternata, Aureobasidium pullulans, Botrytis cinerea, Cladosporium cladosporioides* and *Epicoccum purpurascens*) that grow naturally on leaf surfaces; the "storage type" (*Penicillium brevicompactum*; *P. canescens, P. chrysogenum, P. herquei* and *P. nigricans*); and the "soil" type (*Fusarium solani, Gliocladium roseum, Humicola grisea, Trichoderma harzianum* and *Verticillium lateritium*).

The author stated that: "In the colonies stored dry for 1 and 2 weeks the phylloplane fungi all recommenced growth from the apical apices that were present in the colony margin at the time of drying, sometimes with some sub-apical branching, but also by simple re-growth of the same apex. Moreover this re-growth was detectable in less than 1 h in all the isolates, and in two (*A. alternata* and *B. cinerea*) in less than 25 min". The biological consequences of these observations are huge for fungal biology. It means that the dynamic, complex interactions inside the apex of the fungal hyphae can be "frozen" in time up to 14 days and "simply" resume growth within 30 min to 1 hour. The soil and storage fungi showed growth not from the margin but from residual fungal compartments (possibly conidia or chlamydospores) and reached the colony margin after long lag periods (5 to 50 hours). With respect to growth of indoor fungi that deal with changing humidities, this would give an ecological benefit for the leaf-dwelling fungi over the normal abundant indoor fungi, namely *Penicillium* species and *Aspergillus* species. The biology of this phenomenon has not been addressed since Park's observations.

In our current research a *Cladosporium* cf *cladosporioides* from paint in a building was isolated, which was visible as a small dried-out colony (Figure 4.6). The small piece of paint and wood was then kept for a minimum of 6 weeks packed in paper and under ambient dry conditions. Then the piece was transferred to an agar surface and within 20 hours numerous hyphae had been formed. When the same fungus was grown on polycarbonate membranes for 24 hours and the membranes were taken from the agar and dried in ambient air no re-growth was observed from hyphal apexes upon re-wetting. There is probably a requisite minimal drying time so that the cell can react to the changes; the membranes were dried in minutes.

The hyphae of *Serpula lacrymans* cross inert surfaces for considerable distances to reach new areas of nutrients (in the case of this fungus, wood to be degraded). Growth over a surface of aged lime-sand plaster containing 1.5% (w/w) water was more or less uniform. When no water was added to the plaster, growth was approximately

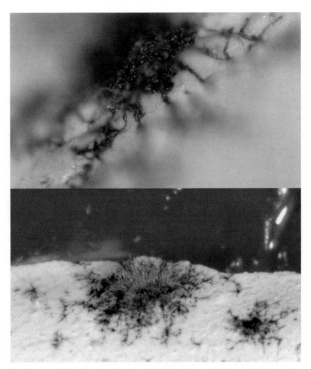

*Figure 4.6. A dried colony of* Cladosporium *cf* cladosporioides *on paint (top) that had formed numerous hyphae 20 hours after transfer of the piece of wood and paint to an agar surface (bottom).*

6 times slower and occurred in a series of fan-like waves (Clarke *et al.* 1981). The authors state that the fungus does change the water content of the plaster when growing over it. Mycelium will lose water to the plaster when the water potential there is lower. The diffusion of water through the plaster may determine how fast the water potential increases close to the mycelium of the fungus. *S. lacrymans* does transport water through the mycelium from the inoculum site towards the hyphae that grow over the inert medium, but it has to be remembered that water can be formed as a result of the metabolism of translocated carbohydrates. Hyphae of the fungus do accumulate large amounts of compatible solutes such as trehalose, arabitol and to a lesser extent mannitol, which may render the hyphae more drought-resistant and protect the cell constituents against damage (Brownlee and Jennings 1981b). However, there are discrepancies in the study and the authors conclude that: "The most probable reason for the discrepancy lies in the fact that the measurements of relative humidity refer to the bulk of the plaster. When water is lost from the

mycelium, it is not unreasonable to suppose that any dramatic change of ψ of the plaster only occurs close to the hyphae".

The question remains whether aerial hyphae are important for the indoor situation. A fungal colony growing on a wall in a bathroom will face periods of low water activity. Aerial hyphae can cross millimeters or more, when a water source is available elsewhere, but also need a minimal water activity to be formed. The fungus *Aspergillus oryzae* forms abundant aerial mycelium on a wheat grain medium and measurements indicate growth up to 5 mm length under very humid conditions (Rahardjo *et al.* 2002). Aerial mycelium formation is dependent on the external situation and aerial hyphae will dry out if their relatively large surface areas are exposed to dry air. Microscopic investigation of building material, however, indicates that indoor fungi produce little aerial mycelium but abundant sporulating structures (R.A. Samson, unpublished data).

### Survival structures and their formation

As described, the fungal mycelium consists of hyphae that grow on and within a substrate or hyphae that stick into the air. The latter may differ with respect to growth and secretion parameters. In the colony there may be areas of increased stress resistance that may improve survival during variable indoor conditions. The colony is also able to differentiate into other structures that form cells with notable stress resistance and that are vehicles for distribution to other places by means of air or water transport. These are the spores, that can have many different appearances, a fact reflected in the variety of their names; conidia, ascospores, sporangiospores, etc. The spores are often produced on spore-forming structures that also show marked variation among the fungal species. These structures are called conidiophores and contain different types of cells. Vinck (2007) indicated that (aerial) conidiophores do not tend to express secreted enzymes, which is expected as in mid-air normally no digestable substrates are found. Figure 4.7 illustrates that conidiophores can be very different from other aerial hyphae. In this picture the size of the stipe of the conidiophore of *Aspergillus niger* is huge compared to the "tiny" aerial hyphae. The function of stipes and related structures is entirely different from hyphae. They grow for a relatively fixed distance and then start the process of spore formation. In case of the fungal genera *Aspergillus*, *Penicillium* and *Cladosporium*, dominant indoor fungi, spore formation occurs in massive numbers (see Figure 4.8, for *Aspergillus*) and the spore-forming structures all show "multiplication of growing ends" to ensure simultaneous formation of spores. This is illustrated in Figure 4.9, where the spore-forming apparatus of *Cladosporium* sp. is shown. The conidiophores might be relatively vulnerable structures, while many delicate processes such as mitosis, cell

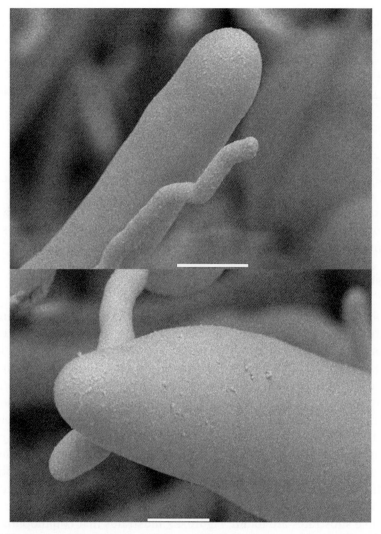

*Figure 4.7. Two stages of development of the so-called stipe in case of* Aspergillus niger. *This is the structure that leads to the formation of conidia, spores that will be distributed into the air. A "normal" aerial hypha is present near the stipe, which illustrates the differences between the two types of cells. Bars are 10 and 5 μm.*

differentiation and organelle transport take place on a large scale. For instance, when nuclear division fails, the survival compartment (the spore) will never germinate. After maturation of the spore, its stress resistance however can be higher than any part of the fungal culture. Remember the work of Park (1982) where either growth

*Figure 4.8. Many spore-forming structures of* Aspergillus niger *as observed by cryoSEM. The scale bar is 100 μm large and illustrates that these structures are just visible with the naked eye. The difference in size means that the conidiophores are at different stages of development. The number of spores (conidia) that can be formed is enormous. Bar is 100μm.*

of the hyphae could resume in a short time or with the phylloplane fungi such as *Cladosporium* or much later from spore structures, as was the case with *Penicillium* and *Aspergillus*. Remarkably, the spores of *Cladosporium* are more sensitive than those of the other two genera, as if this survival habit of the hyphae were balanced by that of the spore.

All in all, this chapter has highlighted the different cell types that are present in the fungal colony. Some cells actively grow on substrate or into the air. Conidiophores also grow out into the air or develop in the substrate to form many survival cells. The fungal colony consists of different types of cells that are very likely to respond with varying sensitivity to the variable conditions that occur in the indoor environment. The study of the biology of fungi in these man-made conditions is completely novel and would open a whole new area of research.

*Figure 4.9. The branched appearance of these spore-forming structures of the fungal genus* Cladosporium *is clearly visible here; they enable the conidiophore to produce many spores simultaneously.*

## References

Adan O (1994) On the fungal defacement of interior finishes. PhD Thesis, University of Eindhoven, Eindhoven, the Netherlands.

Bartnicki-Garcia S, Hergert F and Giertz G (1989) Computer simulation of fungal morphogenesis and the mathematical basis for hyphal(tip) growth. Protoplasma 153: 46-57.

Brownlee C and Jennings DH (1981a) Further observations on tear or drop formation by mycelium of *Serpula lacrymans.* Trans Br Mycol Soc 77: 33-40.

Brownlee C and Jennings DH (1981b) The content of soluble carbohydrates and their translocation in mycelium of *Serpula lacrymans*. Trans Br Mycol Soc 77: 615-619.

Clarke RW, Jennings DW and Coggins CR (1980) Growth of *Serpula lacrymans* in relation to water potential of substrate. Trans Br Mycol Soc 75: 271-280.

Coggins CR, Jennings DH and Clarke RW (1980) Tear or drop formation by mycelium of *Serpula lacrymans*. Trans Br Mycol Soc 75: 63-67.

De Jong J (2006) Aerial hyphae of *Schizophyllum commune*: their function and formation. PhD Thesis, University Utrecht, Utrecht, the Netherlands.

Dijksterhuis J, Sanders M, Gorris LGM and Smid EJ (1999) Antibiosis plays a role in the context of direct interaction during antagonism of *Paenibacillus polymyxa* towards *Fusarium oxysporum*. J Appl Microbiol 89: 13-21.

Gusse AC, Miller PD and Volk TJ (2006) White-rot fungi demonstrate first biodegradation of phenolic resin. Environ Sci Tech 40: 4196-4199.

Jedd G and Chua N-H (2000) A new self-assembled peroxisomal vesicle required for efficient resealing of the plasma membrane. Nat Cell Biol 2: 226-231.

Lugones LG, De Jong JF, De Vries OMH, Jalving R, Dijksterhuis J and Wösten HAB (2004) The SC15 protein of *Schizophyllum commune* mediates formation of aerial hyphae and attachment in the absence of the SC3 hydrophobin. Mol Microbiol 53: 707-716.

Lugones LG, Wösten HAB, Birkenkamp KU, Sjollema KA, Zagers J and Wessels JGH (1999) Hydrophobins line air channels in fruiting bodies of *Schizophyllum commune* and *Agaricus bisporus*. Mycol Res 103: 635-640.

Meijer W (2007) Microbody dynamics in *Penicillium chrysogenum*. Autumn Meeting, Section Mycology of the Dutch Society of Microbiology, Utrecht, the Netherlands, 16 November 2007.

Milstein O, Gersonde R, Huttermann A, Chen MJ and Meister J (1992) Fungal Biodegradation of ligopolystyrene graft copolymers. Appl Environ Microbiol 58: 3225-3232.

Park D (1982) Phylloplane fungi: tolerance of hyphal tips to drying. Trans Br Mycol Soc 79: 174-178.

Rahardjo YSP, Korona D, Haemers S, Weber FJ, Tramper J and Rinzema A (2004) Limitations of membrane cultures as a model solid-state fermentation system. Lett Appl Microbiol 39: 504-508.

Rahardjo YSP, Sie S, Weber FJ, Tramper J and Rinzema A. (2005a) Effect of low oxygen concentrations on growth and α-amylase production of *Aspergillus oryzae* in model solid-state fermentation systems. Biomol Eng 21: 163-172.

Rahardjo YSP, Weber FJ, Haemers S, Tramper J and Rinzema A (2005b) Aerial mycelia of *Aspergillus oryzae* accelerate α-amylase production in a model solid-state fermentation system. Enzym Microb Tech 36: 900-902.

Rahardjo YSP, Weber FJ, Le Comte EP, Tramper J and Rinzema A (2002) Contribution of aerial hyphae of *Aspergillus oryzae* to respiration in a model solid-state fermentation system. Biotechnol Bioeng 8: 539-544.

Riquelme M, Bartnicki-García S, González-Prieto JM, Sánchez-León E, Verdín-Ramos JA, Beltrán-Aguilar J and Freitag M (2007) Spitzenkörper localization and intracellular traffic of green fluorescent protein-labeled CHS-3 and CHS-6 chitin synthases in living hyphae of *Neurospora crassa*. Eukaryot Cell 6: 1853-1864.

Te Biesebeke R, Boussier A, Van Biezen N, Braaksma M, Van den Hondel CAMJJ, De Vos WM and Punt PJ (2006) Expression of *Aspergillus* hemoglobin domain activities in *Aspergillus oryzae* grown on solid substrates improves growth rate and enzyme production. Biotechnol J 1: 822-827.

Van den Boogert PHJF and Deacon JW (1994) Biotrophic mycoparasitism by *Verticillium biguttatum* on *Rhizoctonia solani*. Eur J Plant Pathol 100: 137-156.

Van Wetter M-A, Wösten HA, Sietsma HJ and Wessels JGH (2000) Hydrophobin gene expression affects hyphal wall composition in *Schyzophyllum commune*. Fungal Genet Biol 31: 99-104.

Vinck A (2007) Hyphal differentiation in the fungal mycelium. PhD Thesis, University Utrecht, Utrecht, the Netherlands.

Vinck A, Terlou M, Pastman WR, Martens EP, Ram AF, Van den Hondel CAMJJ and Wösten HAB (2005) Hyphal differentiation in the exploring mycelium of *Aspergillus niger*. Mol Microbiol 58: 693-699.

Wessels JGH (1990) Role of cell wall architecture in fungal tip growth generation. In: Heath IB (ed.) Tip growth in plant and fungal walls, Academic Press, San Diego, CA, USA, pp. 1-12.

Wösten HA, Van Wetter MA, Lugones LG, Van der Mei HC, Busscher HJ and Wessels JG (1999) How a fungus escapes the water to grow into the air. Curr Biol 9: 85-88.

# 5 Ecology and general characteristics of indoor fungi

*Robert A. Samson*
*CBS-KNAW Fungal Biodiversity Centre, Utrecht, the Netherlands*

## Introduction

The mycobiota of indoor environments contains about 100-150 species and this is only a fraction of the more than 100,000 species of described fungi (Samson *et al.* 2010, Miller 2011). Most species belong to the so-called anamorphic fungi which have been known as Deuteromycetes, Hyphomycetes or Fungi Imperfecti. They can produce high concentrations of spores and compounds which can affect our health. In addition to these micro fungi, a number of Basidiomycetes can be found indoors, growing on wood in buildings and are considered important degraders of wooden buildings material (see also Chapter 6, 16 and 17).

In this chapter the ecological aspects of mold growth indoors is discussed. Furthermore the important growth characteristics of the various species are described to provide a better understanding of how molds can live indoors and on building materials and how they can multiply and distribute. Some important indoor fungal genera are discussed.

## What are fungi?

Fungi are often called molds and belong to a separate Kingdom: the Fungi (Crous *et al.* 2009). These organisms grow in the form of multicellular filaments, called hyphae. A connected network of these tubular branching hyphae has multiple, genetically identical nuclei and is considered a single organism, referred to as a colony or in more technical terms a mycelium. Most fungi multiply by a sexual life cycle and the characteristic formation of sexual spores (zygospores, ascospores and basidiospores) and according to their life cycle these organisms can be classified in the main order Zygomycetes (or Zygomycota), Ascomycetes (Ascomycota) or Basidiomycetes (Basidiomycota). Also the yeasts, which do not produce true mycelium, are regarded conventionally as one-celled structures, belong either to the Ascomycetes or Basidiomycetes.

Most species found indoors are lacking a sexual stage and are often referred as Fungi Imperfecti, Deuteromycetes or Hyphomycetes. However most of these fungi belong phylogenetically to the Ascomycetes. Some genera as *Wallemia, Schizophyllum* and *Sistotrema* and the yeasts *Sporobolomyces* are basidiomycetes. In current mycological terminology a sexual or perfect state is called teleomorph while the asexual (or imperfect stage) is the anamorph. For most asexual indoor molds therefore the correct terminology is anamorphic fungi.

The formation of the sexual life cycle can be important for some important indoor species. For example, in *Eurotium* the anamorph (asexual) state belongs to *Aspergillus* and this stage is formed immediately after the spore has germinated and occurs when sufficient moisture is present. As soon as the fungus has formed its mycelium and conidiophores, it will continue growing. At high water activity, it will continue to produce mycelium and the anamorph stage, but as soon as the water activity will decrease, the fungus will start to form fruiting bodies and ascospores (of *Eurotium*). The thick-walled ascospores will allow the fungus to survive and can produce a new generation as soon as water will become available.

### Growth of a filamentous fungus

After germination, the spores of indoor molds produce a mycelium and colonize a substrate. In nature many fungi can penetrate plants, insects or human tissue by special organs and the production of enzymes. However, indoor molds which have building materials, textiles or other matrices as substrates, restrict their growth to the surface as seen by numerous light microscopic examination of building material samples. Cryo-scanning electron microscopy as described in Chapter 2 also indicates that fungal growth is restricted to the surface. The limitations of growth are determined by the availability of water and nutrients. On various substrates (e.g. gypsum board, wall paper) nutrients are readily available, whereas on other substrates (e.g. tiles, glass) a minimum of carbohydrates still allows for growth.

The spore of anamorphic fungi is called a conidium. The structure bearing conidia is a conidiophore. The formation of conidia varies between the different genera and their efficiency to produce air borne propagules mainly is determined by the mode of conidium formation (Cole and Samson 1979). For example species of *Aspergillus* and *Penicillium* produce numerous dry conidia which easily become airborne and this explains the dominance of these fungi in indoor environment. In other species sporulation occurs in slimy and wet heads or in fruiting bodies. These various modes of sporulation largely depend on the special cells that produce the conidia and the ecological conditions. Table 5.1 list the common indoor fungi and their conidiogenesis (mode of conidium formation).

*Table 5.1. Conidiogenesis of the common indoor molds.*

| Conidiogenesis | Genera |
|---|---|
| Phialides with dry conidia | *Aspergillus, Paecilomyces, Penicillium, Stachybotrys (=Memnoniella), Wallemia* |
| Phialides with slimy conidia | *Acremonium, Clonostachys, Exophiala, Fusarium, Gliomastix, Phialophora, Stachybotrys, Trichoderma, Verticillium* |
| Phialides in pycnidia | *Phoma* |
| Annellides | *Scopulariopsis* |
| Thallic | *Geotrichum* |
| Blastic arthric | *Aureobasidium, Chrysonilia, Cladosporium, Geomyces, Oidiodendron* |
| Blastic single | *Botrytis, Epicoccum, Tritirachium* |
| Poroconidia | *Alternaria, Curvularia, Ulocladium* |

According to the formation of conidia and spores the indoor mycobiota can be classified as follows:

- *Conidia produced as dry single spores or in chains.* These fungi are common in air because they produce numerous conidia which easily become airborne. They can grow in wet environments but are typical xerotolerant and xerophilic species. To this group belong the genera *Aspergillus, Penicillium, Paecilomyces, Scopulariopsis, Wallemia, Alternaria, Curvularia, Ulocladium, Botrytis, Epicoccum, Tritirachium, Chrysonilia, Cladosporium, Geomyces* and *Oidiodendron*. Species of *Aspergillus* and *Penicillium* produce numerous conidia in long chains (Figure 5.1A-B), whereas *Cladosporium* (Figure 5.1C), *Alternaria* (Figure 5.1D) and *Ulocladium* are typical fungi that produce regular branched but fragile conidiophores.
- *Conidia produced in wet or slimy heads.* In general, these fungi are not easily found in air samples, because the conidia stick together in a wet or slimy head. Typical representatives are *Stachybotrys* (Figure 5.2A-B), *Acremonium, Trichoderma, Aureobasidium* and *Fusarium*. The presence of these fungi indicates that water is readily available, such as in bath rooms, etc.
- *Conidia or spores produced in fruiting bodies.* There are only a few indoor molds such as *Phoma* (Figure 5.3A), where conidia formed inside a fruiting body. At maturity the conidia are released via an apical opening. In this group also the indoor Ascomycetes can be grouped. In *Eurotium* (Figure 5.3B) the sexual ascospores are produced in a closed fruiting body, while in *Chaetomium* ascospores are formed in a perithecium, a fruiting body with an apical opening. The species belonging to

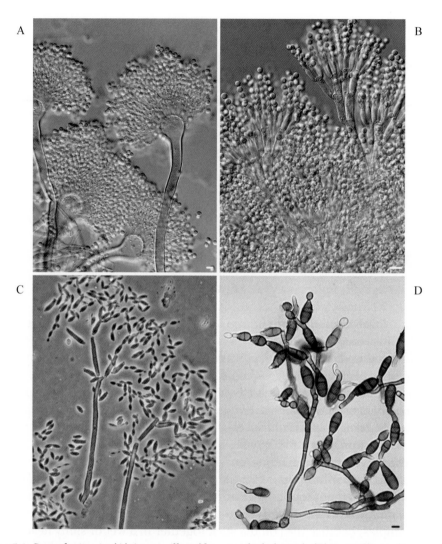

*Figure 5.1. Sporulation in (A)* Aspergillus *(dry conidial chains), (B)* Penicillium *(dry conidial chains), (C)* Cladosporium *(branched conidiophores fragmenting in conidia) and (D)* Alternaria *(conidia in branched chains). Scale bar=10 µm.*

the Zygomycetes have a different mode of sporulation, where spores are produced in a sporangium (Figure 5.3C-D). These fungi can only grow at high water activity.

All species in the indoor environment have their own individual growth conditions and their presence gives a good indication of the condition of the building. Samson

*Figure 5.2. Sporulation in* Stachybotrys. *(A) conidiophores with clustered conidia in slimy heads, (B) conidiophores showing the cells producing the conidia. Scale bar=10 μm.*

*et al.* (1994) suggested that several species can be considered as indicator organisms to signal moisture and/or health problems. For example the many species in the genus *Aspergillus* differ from each other in their growth requirements. Consequently, identification only to the generic level is not sufficient, and further identification to species level is essential.

## The mycobiota of indoor enviroments

Many fungi can produce numerous spores or other propagules and this explains that there can be high concentrations in the air. In nature fungi occur on various decaying substrates and also grow on plants and animals. Many spores are deposited in soil and this is a major source for contamination. For the indoor air the outdoor air is an important source for fungal spores and other propagules. The spore concentration, both outdoors and indoors, is depending on the presence of vegetation and decaying substrates and on climatological influences. Shelton *et al.* (2002) compared the concentrations of indoor and outdoor air and found that both correlated well. Concentrations were high in summer and autumn, while they were low in winter and spring, but there was no significant variation in ratio of indoor and outdoor air between the seasons. However, in arctic climates Reponen *et al.* (1992) found that concentrations of indoor molds were higher that outdoors, which could be explained by lower fungal growth and sporulation due to the lower outdoor

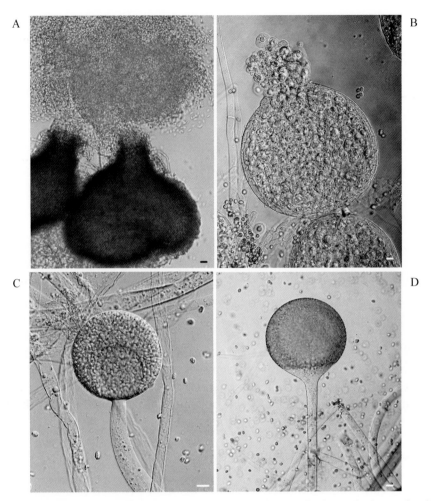

*Figure 5.3. Spore formation in (A)* Phoma *(pycnidium or fruiting body in which conidia develop and are released from an apical opening), (B)* Eurotium *(ascospore formation in a closed fruiting body), and Zygomycete sporangia of (C)* Mucor *and (D)* Rhizopus. *Scale bar (A-B)=10 µm and (C-D)=20 µm.*

temperatures. On the other hand the air spora in subtropical and tropical regions is different than in the temperate climates. A comparison of data collected in Brazil by Oliveira *et al.* (1993) and those obtained from Denmark by Larsen and Gravesen (1991) showed a higher incidence of *Penicillium*, *Aspergillus* and *Fusarium*. Little data is available on the indoor air spora in (sub) tropical regions.

Although there is a correlation between outdoor and indoor air it is important to note that the composition of the species can be different. Horner *et al.* (2004) who studied outdoor and indoor air of 50 houses in Atlanta observed that the concentrations were similar but that *Penicillium* species were more dominant outdoors than indoors. Since every fungus is characterized by its physiological, biochemical and pathogenic properties it is therefore important that the detected molds are identified up to species level.

Table 5.2 lists the most common molds found indoors and is compiled by our own research. In general these species are found more or less world-wide, although recent studies have shown that the mycobiota indoor in subtropical and tropical regions contain more thermotolerant species (unpublished data). Although some species can also be found outdoors, the origin of the indoor mycobiota is often questioned. If we compare the composition of the indoor mycobiota as listed in Table 5.1, a striking similarity is seen with the mycobiota in food (Pitt and Hocking 2009; Samson *et al.* 2010) and this source is often seen as the source of fungi in indoor environments.

McGregor *et al.* (2008) provided lists of fungi from certain moldy building materials, including insulation material, gypsum wall board, wood, textiles and ceiling tiles. Common species on these most wetted substrates are species of *Acremonium, Aspergillus, Cladosporium, Penicillium, Paecilomyces* and *Stachybotrys.* On these buildings materials the typical xerophilic species of *Aspergillus, Eurotium* and *Wallemia* are absent, whereas their presence on food is rather common. These fungi are also common in house dust and this is also seen as an important source.

In most surveys the indoor mycobiota is determined by detection of viable colony forming units (CFU), while non-viable spores, conidia and fungal fragments are of course also present in indoor air. Amend *et al.* (2010) collected dust samples from six continents from 61 buildings and offices, shops and a church. DNA was extracted from these samples and PCR amplified from the loci of ITS and D1 and D2 regions of the large subunit gene (LSU). These were sequenced in multiplex using 454 GS FLX titanium technology. Contrary to common ecological patterns, it was shown that fungal diversity was significantly higher in temperate zones than in the tropics. Remarkably, building function has no significant effect on indoor fungal composition, despite stark contrasts between architecture and materials of some buildings in close proximity. This molecular study does not allow species identification but the diversity was expressed as operational taxonomic units (OTUs). This showed however that most OTU's represented members of the Dothideomycetes, which include species of *Cladosporium, Alternaria* and *Epicoccum.* Although one would expect a common

*Robert A. Samson*

*Table 5.2. Common molds in indoor environments.*

| | | |
|---|---|---|
| Absidia corymbifera | Eurotium amstelodami | Penicillium rugulosum |
| Acremonium murorum | Eurotium chevalieri | Penicillium simplicissimum |
| Acremonium strictum | Eurotium herbariorum | Penicillium spinulosum |
| Alternaria tenuissima | Eurotium rubrum | Penicillium variabile |
| Aspergillus calidoustus | Exophiala dermatitidis | Phialophora fastigiata |
| Aspergillus candidus | Fusarium culmorum | Phialophora verrucosa |
| Aspergillus clavatus | Fusarium solani | Phoma glomerata |
| Aspergillus flavus | Fusarium verticillioides | Phoma macrostoma |
| Aspergillus fumigatus | Geomyces pannorum | Pyronema domesticum |
| Aspergillus niger | Geotrichum candidum | Rhizopus stolonifer |
| Aspergillus penicillioides | Mucor circinnolides | Rhodotorula mucilaginosa |
| Aspergillus restrictus | Mucor plumbeus | Schizophyllum commune |
| Aspergillus sydowii | Mucor racemosus | Scopulariopsis brevicaulis |
| Aspergillus terreus | Oidiodendron griseum | Scopulariopsis candida |
| Aspergillus versicolor | Oidiodendron rhodogenum | Scopulariopsis fusca |
| Aspergillus westerdijkiae | Paecilomyces lilacinus | Serpula lacrymans |
| Aureobasidium pullulans | Paecilomyces variotii | Sistotrema brinkmannii |
| Botrytis cinerea | Penicillium brevicompactum | Sporobolomyces roseus |
| Candida peltata | Penicillium chrysogenum | Stachybotrys chartarum |
| Chaetomium aureum | Penicillium citreonigrum | Stachybotrys (Memnoniella) |
| Chaetomium globosum | Penicillium citrinum | echinata |
| Chrysonillia sitophila | Penicillium commune | Syncephalastrum racemosum |
| Cladosporium cladosporioides | Penicillium corylophilum | Trichoderma harzianum |
| Cladosporium herbarum | Penicillium crustosum | Trichoderma longibrachiatum |
| Cladosporium macrocarpum | Penicillium decumbens | Trichoderma viride |
| Cladosporium sphaerospermum | Penicillium expansum | Tritirachium oryzae |
| Cryptococcus laurentii | Penicillium funiculosum | Ulocladium alternariae |
| Curvularia lunata | Penicillium glabrum | Ulocladium atrum |
| Emericella nidulans | Penicillium olsonii | Ulocladium chartarum |
| Epicoccum nigrum | Penicillium palitans | Wallemia sebi |

occurrence of xerophilic species such as *Wallemia* and *Eurotium,* their absence in this study is explained by the restriction of the methodology used.

## Fungal products

Molds can produce several extra-cellular material. For the exposure to occupants of moldy buildings the production of mycotoxins has attracted a lot of interest. Unfortunately the data found on the internet are mostly incorrect. Many indoor molds are incorrectly claimed to produce mycotoxins although it is not proven whether these species can produces these toxic compounds on building material or other substrates found indoors. The exposure to mycotoxins in the indoor environment has been discussed in detail by Ammann (2002, 2003) and this issue has been controversial. Reports on mycotoxins detected indoors are relatively rare. For example Englehart *et al.* (2002) detected sterigmatocystin, a toxin known from the common indoor species *Aspergillus versicolor*, in carpet dust. Bloom *et al.* (2009) studied the prevalence of selected, potent mycotoxins and levels of fungal biomass in samples collected from water-damaged indoor environments in Sweden during a 1-year period. One hundred samples of building materials, 18 samples of settled dust, and 37 samples of cultured dust were analyzed for sterigmatocystin, gliotoxin, aflatoxin $B_1$, and satratoxin G and H. Sixty-six percent of the analyzed building materials samples, 11% of the settled dust samples, and 51% of the cultured dust samples were positive for at least one of the studied mycotoxins.

Mycotoxins can be found in spores, but also in and on hyphae and in the substrate where fungal growth occurred. Brasel *et al.* (2005) found macrocyclic trichothecenes in small particles from *Stachybotrys* contaminated ceiling tiles, demonstrating that toxins can also be inhaled via small fungal fragments. The significance of small fungal fragments is also an important issue as an allergen, as is described in Chapter 8. In Chapter 9 Nielsen and Frisvad give a more detailed account of the species which are relevant and can produce mycotoxins.

Many volatiles organic compounds are found in indoor air and a group of compounds is from microbial origin (MVOCs). These compounds, which are produced by fungi as well as by bacteria, can have an impact on indoor air quality. MVOC's are produced during active fungal growth and some species, such as *Aspergillus versicolor* and *Penicillium commune* are known to produce strongly smelling compounds. Methods for fast, easy and inexpensive detection of MVOCs are not currently available and this has hampered the research on the significance of these fungal compounds, particularly explaining health complaints. Ryan (2011) has discussed the future of MVOC's and the problems of approach to determine its role in air quality.

*Robert A. Samson*

# Nomenclature

For a non-mycologist the naming of fungi is often unclear and confusing. Fungal taxonomy is a rapidly changing field and consequently names which have been used for many decades sometimes change. The correct nomenclature name of a taxon reflects the current state of taxonomic knowledge. Another important problem which is associated with the identification of molds is that much confusion exists about the naming of species. Recent molecular taxonomic approaches also have changed the old taxonomies which were based on morphology. For example, the taxonomic status of *Memnoniella echinata* was not discussed for many years, although this fungus which produces dry chains of conidia closely resembles species of *Stachybotrys* on ecological and biochemical criteria. Recent phylogenetic work has demonstrated that *Memnoniella* cannot be separated from *Stachybotrys* and therefore the current and correct name is *Stachybotrys echinata* (see also Samson *et al.* 2010). Table 5.3 lists some common species with their incorrect names.

## Identification

To understand the impact of fungal growth indoors, it is always important to carry of out a correct identification. This is particularly true for investigating the correlation of the fungi found in a particular building and the reported health hazards. Fungal identification is still based on morphological criteria and Samson *et al.* (2010) provide keys, description of illustrations to assist the identification. In the last years morphological criteria for common fungi have shown many restrictions, because molecular taxonomic tools have changed the species concept. There are many examples where the species identification cannot be based anymore solely

*Table 5.3. List of common species with the old or incorrect name.*

| Species | Incorrect name |
| --- | --- |
| *Epicoccum nigrum* | *Epicoccum purpurascens* |
| *Aspergillus calidoustus* | *Aspergillus ustus* |
| *Aspergillus westerdijkiae* | *Aspergillus ochraceus* |
| *Penicillium glabrum* | *Penicillium frequentans* |
| *Penicillium glaucum* | Refer to various *Penicillium* species |
| *Penicillium chrysogenum* | *Penicillium notatum* |
| *Penicillium verrucosum* | Refer to species of subgenus *Penicillium* |
| *Verticillium lecanii* | *Lecanicillium lecanii* |

on morphology and that a molecular diagnosis using analyzing a particular gene sequence is necessary. For example in *Aspergillus fumigatus* a recent taxonomic study based on a polyphasic approach showed that *Aspergillus lentulus* is also rather common, but that a distinction between the two species is difficult. Other examples are *Aspergillus westerdijkiae/A. ochraceus*, and species of *Trichoderma, Fusarium* and *Cladosporium* (see also Samson *et al.* 2010).

## Important fungal genera of the indoor mycobiota

In the following notes a description of the most common and relevant fungal genera is given. For a more detailed account of the specific species we refer to the references provided in this section.

### Aspergillus

*Aspergillus* species are common contaminants on various substrates. In subtropical and tropical regions they occur more commonly than *Penicillium* spp. Identification of aspergilli is crucial because many species are human pathogens (e.g. *A. fumigatus, A. terreus*) or notorious mycotoxin producers (e.g. *A. flavus, A. versicolor*). They are further important for their allergenic properties.

Colonies are usually growing rapidly, colored white, yellow, yellow-brown, brown to black or shades of green, mostly consisting of a dense felt of erect conidiophores. Conidiophores are unbranched with a swollen apex (vesicle). Phialides are borne directly on the vesicle (uniseriate) or on metulae (biseriate). Conidia in dry chains form compact columns (columnar) or diverging (radiate). Conidia are one-celled, smooth or ornamented, and hyaline or pigmented. Species may produce Hülle cells (large, thick- and smooth-walled cells) or sclerotia (firm, usually globose, masses of hyphae). Teleomorphs: *Eurotium, Emericella, Neosartorya* and other genera.

The classification was mainly based on morphological characters. Raper and Fennell (1965) divided the genus into 18 groups and accepted 132 species with 18 varieties. The genus now contains more than 250 species, with about 70 named teleomorphs. For recent overviews about the taxonomy and accepted list of species see Samson and Varga (2007) and Varga and Samson (2008).

### Cladosporium

This genus has a world-wide distribution. Several species are plant pathogens or are saprophytic and more or less host-specific on old or dead plant material. Cladosporia are common in indoor environments and play an important role for their allergenic properties.

Colonies are mostly olivaceous-brown to blackish-brown or with a greyish-olive appearance, velvety or floccose becoming powdery due to abundant conidia, and rather slow-growing. Conidiophores are very fragile, erect, straight or flexuous, unbranched or branched only in the apical region, with geniculate sympodial elongation in some species. Conidia occur in branched chains and are one-celled, ellipsoidal, fusiform, ovoid or (sub)globose, often with distinct scars, pale to dark olivaceous-brown, smooth-walled, verrucose or echinulate. (Blasto)conidia are mostly formed on denticles in groups of 1-3 at the apex of the conidiophore, subapically below a septum or on the tip of previously formed conidia.

In most handbooks a few indoor species are listed as *Cladosporium cladosporioides, C. sphaerospermum, C. herbarum* and *C. macrocarpum*. However recent phylogenetic studies have shown that most of these taxa represent so-called cryptic species, representing many new species (Schubert *et al.* 2007, Zalar *et al.* 2007, Dugan *et al.* 2008).

## *Fusarium*

Most *Fusarium* species occur in soil or on plants and have a world-wide distribution. The genus includes plant pathogens of crops widely grown around the globe, e.g. wheat and maize. A few fusaria can be found in indoor environments and are mostly associated with wet niches, including cooling units of air-conditioning systems. The current taxonomy of *Fusarium* is not clear, and molecular taxonomic data has changed the classification and number of species radically. For a simplified list of fusaria occurring in food and indoor the reader is referred to Samson *et al.* (2010).

## *Penicillium*

The genus *Penicillium* is a large genus with more than 400 species. Many species are common contaminants on a wide variety of substrates such as soil. However, a great number are found on food where they spoil the products and produce mycotoxins. Indoors, several species can occur of which *Penicillium chrysogenum, P. brevicompactum, P. olsonii* and *P. glabrum* are the most common ones. Because of the widespread occurrence of penicillia and their potential for producing mycotoxins, correct identification is important when studying possible *Penicillium* contamination. However, the phenotypic characters of many species are very similar and identification is not very simple. Purely morphological characteristics can be used for identification, but the isolates should be grown under standard conditions using malt extract agar and Czapek-yeast extract agar at 25 °C for 7 days.

The species concept in *Penicillium* is currently based on a polyphasic approach, using a combination of the morphology of the conidiophores and conidia, colony characteristics on various media and temperatures, extrolite profiles and molecular data (mainly partial β-tubulin sequences) (Frisvad and Samson 2004, Samson and Frisvad 2004). Samson *et al.* (2010) keyed out the most commonly occurring penicillia in an dichotomous key. Furthermore an electronic key for species of *Penicillium* subgenus *Penicillium* is provided at http://www.cbs.knaw.nl/penicillium.htm. This key is based on a database with phenotypical and molecular (partial s-tubulin sequences) characters. A database for identifying indoor penicillia is also available at http://www.cbs.knaw.nl/indoor/.

## Scopulariopsis

Colonies of this genus vary from white, creamish, grey or buff to brown or even blackish, often darkening with age, but never green like *Penicillium*. The conidiogenous cells are cylindrical or with a slightly swollen base, annellate (rings formed after conidium formation), single or borne on whorls. Conidia are in chains with a broadly truncate base. *Scopulariopsis brevicaulis* is rather common in indoor air and often found on cellulose material such as wall paper. For a more detailed account of descriptions and key, see Samson *et al.* (2010).

## Stachybotrys

Species of *Stachybotrys* have a world-wide distribution and occur on paper, wallpaper, seeds (e.g. wheat, oats), soil, textiles and dead plant material. There are several species of *Stachybotrys*, but the most well-known taxon is *S. chartarum*. *S. chlorohalonata* has a similar morphology as *S. chartarum* but produces a green extracellular pigment and has broadly ellipsoidal conidia. *Memnoniella echinata* is now accommodated in *Stachybotrys* on basis of the close genetic relationship, although the species produces conidia in chains rather than conidia in slimy heads.

## Yeasts

Indoor yeasts are relatively rare and mostly species of *Rhodotorula*, *Sporobolomyces* and *Cryptococcus* are detected. Yeasts mostly reproduce vegetatively by budding. Exceptions are species of the genus *Schizosaccharomyces*, which reproduces by a fission process, and of the genera *Sporobolomyces*, *Sterigmatomyces* and *Fellomyces*, which form buds on short stalks. The morphology of the unicellular yeast cell indicates the genus, but further identification should be performed using a number of physiological tests, including assimilation of a number of sugars. Commercial identification kits provide computer programs for processing the results obtained with them. These include the API 20C and API YEAST-IDENTTM kits in which growth tests are carried out in small wells on plastic strips. Biolog Inc. markets the Microlog™ system for yeasts which, by using the redox dye tetrazolium violet, detects whether a microorganism can utilize a given carbon source. For a recent review the reader is referred to Barnet *et al.* (2000a,b). Polyphasic identification systems for yeasts have been published by Robert and Skoke (2006) and Robert *et al.* (2008).

**Basidiomycetes**

The Basidiomycetes are a large group mostly comprising species with typical fruiting bodies or basidiocarps, that are mostly known as toadstools and bracket fungi. In the indoor environment, some species which do not possess macroscopic fruiting bodies, such as *Sistotrema*, may occur in air samples. Typical basidiomycete species in buildings are the wood-decaying fungi, among which *Serpula lacrymans* is the most notorious species, causing dry rot. Airborne spores of *S. lacrymans* can be present in high concentrations and allergenic reactions are reported in the literature (Murphy *et al.* 1995). Most basidiomycetes are difficult to grow on artificial media and they often do not form sporulating fruiting bodies, but only sterile mycelium.

Chapter 6 describes the characteristics and identification of wood-decaying basidiomycetes. Other relevant literature is published by Wagenfuhr and Steiger (1966), Coggings (1980), Bravery *et al.* (1987), Hennebert and Balon (1996) and Schmidt (2006).

## References

Amend AS, Seifert KA, Samson, RA and Bruns, TD (2010) Indoor fungal composition is geographically patterned and more diverse in temperate zones than in the tropics. Proc Natl Acad Sci USA 107: 13748-13753.

Ammann HM (2002) Indoor mold contamination – a threat to health. Part I. J. Environ Health 64: 43-44.

Ammann HM (2003) Indoor mold contamination – a threat to health. Part II. J. Environ Health 66: 47-49.

Barnett JA, Payne RW and Yarrow D (2000a) The yeasts: characteristics and identification, 2nd edition. Cambridge University Press, Cambridge, UK.

Barnett JA, Payne RW and Yarrow D (2000b) Yeast identification PC program. J.A. Barnett, Norwich, UK.

Bloom E, Nyman E, Must A, Pehrson C and Larsson L (2009) Molds and mycotoxins in indoor environments – a survey in water-damaged buildings. J Occup Environ Hyg 6: 671-678.

Brasel TL, Douglas DR, Wilson SC and Strauss DC (2005) Detection of airborne *Stachybotrys chartarum* macrocyclic trichothecene mycotoxins on particulates smaller than conidia. Appl Environ Microbiol 71: 114-122.

Bravery AF, Berry RW, Carey JK and Cooper DE (1987) Recognising wood rot and insect damage in buildings. Building Research Establishment, Garston, Watford, UK.

Coggins CR (1980) Decay of timber in buildings. Dry rot, wet rot and other fungi. Rentokil Ltd., East Grinstead, UK.

Cole GT and Samson RA (1979) Patterns of development in conidial fungi. Pitman, London, UK.

Crous PW, Verkley GJM, Groenewald JZ and Samson RA (eds.) (2009). Fungal biodiversity. CBS laboratory manual series 1. CBS-KNAW Fungal Biodiversity Centre, Utrecht, the Netherlands.

Dugan FM, Braun U, Groenewald JZ and Crous PW (2008) Morphological plasticity in *Cladosporium sphaerospermum*. Persoonia 21: 9-16.

Engelhart S, Loock A, Skutlarek D, Sagunski H, Lommel A, Färber H and Exner M (2002) Occurrence of toxigenic *Aspergillus versicolor* isolates and sterigmatocystin in carpet dust from damp indoor environments. Appl Environ Microbiol 68: 3886-3890.

Frisvad JC and Samson RA (2004) Polyphasic taxonomy of *Penicillium* subgenus *Penicillium*. A guide to identification of the food and airborne terverticillate penicillia and their mycotoxins. Stud Mycol 49: 1-173.

Hennebert GL and Balon F (1996) Les mérules des maisons. Artel, Namur, Belgium.

Horner WE, Worthan AG and Morey PR. (2004) Air- and dustborne mycoflora in houses free of water damage and fungal growth. Appl Environ Microbiol 70: 6394-6400.

Larsen LS and Gravesen S (1991) Seasonal variation of outdoor viable microfungi in Copenhagen, Denmark. Grana 30: 467-471.

McGregor HJ, Miller JD, Rand T and Solomon J (2008) Mold ecology: recovery of fungi from certain moldy building materials. In: Prezant B, Weekes DM and Miller JD (eds.) Recognition, evaluation, and control of indoor mold. AIHA, Fairfax, VA, USA.

Miller, JD (2011) Mycological investigations of indoor environments. In: Flannigan B, Samson RA, Miller JD (eds.). Microorganisms in home and indoor work environments: diversity, health impacts, investigations and control. 2$^{nd}$ edition. CRC Press, Boca Raton, FL, USA, in press.

Murphy, DMF, Morgan, WKC and Seaton A (1995) Hypersensitivity pneumonitis. In: Morgan WKC and Seaton A (eds.) Occupational lung diseases. W.B. Saunders, Philadelphia, PA, USA, pp. 525-567.

Oliveira MTB, Santos Braz RF and Ribeiro MAG (1993) Airborne fungi isolated from Natal, State of Rio Grande do Norte – Brazil. Rev. Microbiol., São Paulo 24: 198-202.

Pitt JI and Hocking AD (2009) Fungi and food spoilage. 3$^{rd}$ edition. Springer, Dordrecht, the Netherlands.

Raper KB and Fennell DI (1965) The genus *Aspergillus*. Williams and Wilkins, Baltimore, MD, USA.

Reponen T, Nevalainen A, Jantunen M, Pellikka M and Kalliokoski P (1992) Normal range criteria for indoor air bacteria and fungal spores in a subarctic climate. Indoor Air 2: 26-31.

Robert V and Szoke S (2006) BioloMICS Software.

Robert V, Groenewald M, Epping W, Boekhout T, Smith M and Stalpers J (2008) CBS Yeasts Database. Centraalbureau voor Schimmelcultures, Utrecht, the Netherlands.

Ryan TJ (2011) Microbial volatile organic compounds. In: Flannigan B, Samson RA, Miller JD (eds.). Microorganisms in home and indoor work environments: diversity, health impacts, investigations and control. 2nd edition. CRC Press, Boca Raton, FL, USA, in press.

Samson RA and Frisvad JC (2004) *Penicillium* subgenus *Penicillium*: new taxonomic schemes, mycotoxins and other extrolites. Stud Mycol 49: 1-257.

Samson RA and Varga J (2007) *Aspergillus* systematics in the genomic era. Stud Mycol 59: 1-206.

Samson RA, Flannigan B, Flannigan M, Verhoeff AP, Adan OCG, Hoekstra ES (eds.) (1994) Health implications of fungi in indoor environments. Elsevier, Amsterdam, the Netherlands.

Samson RA, Houbraken J, Thrane U, Frisvad JC and Andersen B (2010) Food and indoor fungi. CBS laboratory manual series 2. Centraalbureau voor Schimmelcultures, Utrecht, the Netherlands.

Schmidt O (2006) Wood and tree fungi. Biology, damage, protection, and use. Springer, Berlin, Germany, 334 pp.

Schubert K, Groenewald JZ, Braun U, Dijksterhuis J, Starink M, Hill CF, Zalar P, De Hoog GS and Crous PW (2007) Biodiversity in the *Cladosporium herbarum* complex (*Davidiellaceae, Capnodiales*), with standardisation of methods for *Cladosporium* taxonomy and diagnostics. Stud Mycol 58: 105-156.

Shelton BG, Kirkland KH, Flanders WD and Morris GK (2002) Profiles of airborne fungi in buildings and outdoor environments in the United States. Appl Environ Microbiol 68: 1743-1753.

Varga J and Samson RA (2008) *Aspergillus* in the genomic era. Wageningen Academic Publishers, Wageningen, the Netherlands.

Wagenführ R and Steiger A (1966) Pilze auf Baumholz. A. Ziemsen Verlag, Wittenberg-Lutherstadt, Germany.

Zalar P, De Hoog GS, Schroers H-J, Crous PW, Groenewald JZ and Gunde-Cimerman N (2007) Phylogeny and ecology of the ubiquitous saprobe *Cladosporium sphaerospermum*, with descriptions of seven new species from hypersaline environments. Stud Mycol 58: 157-183.

# 6 Characteristics and identification of indoor wood-decaying basidiomycetes

*Olaf Schmidt and Tobias Huckfeldt*
*University of Hamburg, Wood Biology, Hamburg, Germany*

## Fungal species and significance

The indoor wood-decaying basidiomycetes cause considerable economical damage. Approximately 80 species have been found in North German buildings. This chapter describes their significance as well as classical morphological and molecular techniques for identification.

Indoor wood-decaying fungi may be considered to be the most important "wood fungi" as they deteriorate wood at the end of the economical series: forestry – timber harvest – storage – woodworking – indoor use. For Britain, it has been estimated that the cost of repairing fungal damage of timber in construction in 1977 amounted to £3 million per week (Rayner and Boddy 1988). In the northern hemisphere, mainly coniferous wood is used as interior structural timber. The most important wood-degrading fungi within buildings in Europe and North America are therefore fungi that cause brown rot (see Table 6.1) in conifers.

There have been several evaluations of the frequency of the various species involved in house-rot. The most prevalent wood-destroying fungi found in a survey of 1,500 buildings in New York State from 1947 to 1951 were *Hyphodontia spathulata* (Schrad.) Parmasto, *Gloeophyllum sepiarium*, *G. trabeum* and *Antrodia xantha* (Silverborg 1953). An investigation of 3,050 buildings in Poland revealed the wood-decaying fungi to consist of *Serpula lacrymans* (54%), *Coniophora puteana* (22%) and *Antrodia vaillantii* (11%) (Ważny and Czajnik 1963). The fungal population based on a 21-year survey of 1,200 biotic damages in buildings of the former East Germany comprised 35% *S. lacrymans*, 15% *Coniophora* spp. and 9% *"Poria"* (Schultze-Dewitz 1985) and that of an evaluation of 749 damaged buildings in Belgium between 1985 and 1991 comprised 59% *S. lacrymans*, 10% *C. puteana, C. marmorata*, 10% *Donkioporia expansa* and 2% *A. vaillantii, A. sinuosa* and *A. xantha* (Guillitte 1992). Of a total number of 3,434 decay fungi in Norwegian houses from 2001 to 2003, the most frequent were *Antrodia* species (18%), *C. puteana* and *S. lacrymans* (each 16%) and *G. sepiarium* (3%) (Alfredsen *et al.* 2005).

*Table 6.1. Species, type of rot, abundance and significance of basidiomycetes found in 2,000 German buildings during a nine-year investigation.*

| Species | Rot | Abundance | Significance |
|---|---|---|---|
| *Serpula lacrymans* (Wulfen) J. Schroet. | b | 691 | from cellar to basis of roof |
| *Coniophora puteana* (Schumach.) P. Karst. | b | 446 | from cellar to roof |
| *Donkioporia expansa* (Desm.) Kotl. & Pouzar | w | 252 | from cellar to roof |
| *Antrodia* spp. | b | 270 | from cellar to roof |
| *Coprinus* spp. | w | 102 | cellar, on walls, half-timbering, ceilings with reed-layer, oak rafter |
| *Oligoporus* spp. | b | 82 | from cellar to roof |
| *Tapinella panuoides* (Batsch) E.-J. Gilbert | b | 76 | cellar, below bathrooms, windows, roof, half-timbering |
| *Serpula himantioides* (Fr.) P. Karst. | b | 74 | from cellar to roof |
| *Asterostroma cervicolor* (Berk. & Curtis) Massee | w | 69 | from cellar to roof |
| *Coniophora marmorata* Desm. | b | 69 | from cellar to roof |
| *Leucogyrophana pinastri* (Fr.) Ginns & Weresub | b | 66 | cellar, subfloor, beam-ends, bathrooms |
| *Antrodia xantha* (Fr.) Ryvarden | b | 57 | roof, windows |
| *Gloeophyllum abietinum* (Bull.) P. Karst. | b | 47 | windows, doors, bathrooms, roof |
| *Antrodia vaillantii* (DC.) Ryvarden | b | 41 | from cellar to roof |
| *Trechispora* spp. | w | 40 | windows, cellar, below bathrooms, roof, half-timbering, door frame |
| *Antrodia sinuosa* (Fr.) P. Karst. | b | 30 | from cellar to roof |
| *Gloeophyllum sepiarium* (Wulfen) P. Karst. | b | 30 | windows, doors, roof, glue laminated timber beam |
| *Gloeophyllum trabeum* (Pers.) Murrill | b | 28 | windows, doors, roof, laminated timber |
| *Phellinus contiguus* (Pers.) Pat. | w | 27 | windows, half-timbering, shingle roofs |
| *Phanerochaete* spp. | w | 23 | moist wood constructions, windows |
| *Grandinia* spp., *Hyphoderma* spp., *Hyphodontia* spp. | w | 22 | cellar, roof, purlin, beams |
| *Gloeophyllum* spp. | b | 20 | windows, doors, bathrooms, roof, half-timbering |
| *Resinicium bicolor* (Alb. & Schwein.) Parmasto | w | 16 | indoor chipboard, windows, beam ends |
| *Oligoporus placenta* (Fr.) Gilb. & Ryvarden | b | 15 | flooring, staircases, windows, half-timbering, insulation layer |
| *Leucogyrophana pulverulenta* (Sowerby) Ginns | b | 14 | subfloor, beam-ends, bathrooms, kitchens |

| Species | | | Occurrence |
|---|---|---|---|
| Lentinus lepideus (Fr.) Fr. | b | 13 | cellar, subfloor, beam-ends, windows |
| Trametes spp. | w | 12 | half-timbering, windows, purlin at overhang |
| Dacrymyces stillatus Nees | b | 11 | windows, doors, facades, half-timbering, bedroom floor |
| Leucogyrophana spp. | b | 11 | cellar, bath-rooms, half-timbering, beam-ends, |
| Trechispora farinacea (Pers.) Liberta | w | 11 | roof, timber below defect sanitary facilities |
| Boletus spp., Xerocomus spp. | ? | 10 | cellar, basis of roofs |
| Hypochniciellum molle (Fr.) Hjortstam | w | 10 | window, beam-end, roof ceiling, cellar floorboards |
| Radulomyces confluens (Fr.) M.P. Christ. | w | 10 | church door, windows, cornice timber |
| Diplomitoporus lindbladii (Berk.) Gilb. & Ryvarden | w | 9 | subfloor, roof, half-timbering |
| Fomitopsis rosea (Alb.: Schwein.) P. Karst. | b | 9 | beams under floor, ceiling beams, cellar flooring |
| Schizopora paradoxa (Schrad.) Donk | w | 9 | rafter, window, ceiling |
| Coniophora arida (Fr.) P. Karst. | b | 7 | floorboard, beams |
| Tomentella spp. | ? | 5 | cellar, subfloor |
| Cylindrobasidium laeve (Pers.) Chamuris | w | 6 | roof, half-timbering in a mill |
| Fomitopsis pinicola (Sw.) P. Karst. | b | 6 | roof, insulation layer |
| Oligoporus rennyi (Berk. & Broome) Donk | b | 6 | cellar ceiling |
| Trametes hirsuta (Wulfen) Pilát | w | 6 | windows |
| Trechispora mollusca (Pers.) Liberta | b | 6 | roof insulation |
| Antrodia serialis (Fr.) Donk | b | 5 | beam-ends in upper floor, windows |
| Asterostroma laxum Bres. | w | 5 | staircase, bath-room, cellar, brickwork |
| Pleurotus ostreatus (Jacq.) P. Kumm. | w | 5 | rubbish below living space, ship, chipboard inside wall, fitted carpet, mobile home |
| Bjerkandera adusta (Willd.) P. Karst. | w | 4 | beam, window |
| Hyphoderma puberum (Fr.) Wallr. | w | 4 | oakwood rafter |
| Leucogyrophana mollusca (Fr.) Pouzar | b | 4 | cellar |
| Phlebiopsis gigantea (Fr.) Jülich | w | 4 | windows |
| Pleurotus dryinus (Pers.) P. Kumm. | w | 4 | chipboards in outer walls and flooring, roof, ship |
| Fibulomyces mutabilis (Bres.) Jülich | w | 3 | sommer house door, rafter in a barn |
| Hyphodontia microspora J. Erikss. & Hjortstam | w | 3 | roof with roofing felt, half-timbering, window |
| Lycoperdon pyriforme Schaeff. | n | 3 | ground under cellar |
| Perenniporia tenuis (Schwein.) Ryvarden | w | 3 | windows, doors, sill |
| Radulomyces confluens (Fr.) M.P. Christ. | w | 3 | windows, timber in cornice |

*Table 6.1. Continued.*

| Species | Rot | Abundance | Significance |
|---|---|---|---|
| Trechispora microspora (P. Karst.) Liberta | w | 3 | softwood rafter |
| Antrodia gossypium (Speg.) Ryvarden | b | 2 | wall |
| Antrodia crassa (P. Karst.) Ryvarden | b | 2 | engine room |
| Botryobasidium spp. | w | 2 | cellar |
| Cerinomyces pallidus G.W. Martin | b | 2 | windows outside |
| Crustoderma dryinum (Berk. & M.A. Curtis) Parmasto | w | 2 | oak half-timbering |
| Dacrymyces capitatus Schwein. | b | 2 | facades |
| Dacrymyces tortus (Willd.) Fr. | b | 2 | church door-frame and sill |
| Grifola frondosa (Dicks.) Gray | w | 2 | oak half-timbering with soil contact, engine room |
| Heterobasidion annosum (Fr.) Bref. | w | 2 | introduced with moist timber |
| Hyphodontia nespori (Bres.) J. Erikss. & Hjortstam | w | 2 | gable roof, half-timbering |
| Hyphoderma praetermissum (P. Karst.) J. Erikss. & Å. Strid | w | 2 | flooring, half-timbering, windows outside |
| Perenniporia medulla-panis (Jacq.) Donk | w | 2 | sill |
| Pleurotus cornucopiae (Paulet) Rolland | w | 2 | wooden ship |
| Pleurotus pulmonarius (Fr.) Quél. | w | 2 | oriented strand boards in a wall |
| Pluteus cervinus P. Kumm. | w | 2 | timber within wall, monastic ceiling |
| Stereum rugosum Pers. | w | 2 | oak half-timbering with soil contact |
| Trametes versicolor (L.) Lloyd | w | 2 | window, kitchen door |
| Agrocybe praecox (Pers.: Fr.) Fay. | ? | 1 | moist flooring after leakage |
| Amylocorticiellum cremeoisabellinum (Litsch.) Zmitr. | b | 1 | roof extension |
| Amylocorticiellum subillaqueatum (Litsch.) Spirin & Zmitr. | b | 1 | ceiling |
| Amylostereum areolatum (Chaillet ex Fr.) Boidin | w | 1 | upper floor sill |
| Amyloxenasma allantosporum (Oberw.) Hjortstam & Ryvarden | w | 1 | vestibule window |
| Antrodia malicola (Berk. & M.A. Curtis) Donk | b | 1 | beam inside wall |
| Antrodia sordida Ryvarden & Gilb. | b | 1 | moist roof beam |
| Athelia fibulata M.P. Christ. | w | 1 | window outside |
| Basidioradulum radula (Fr.) Nobles | w | 1 | flooring with wall contact |
| Basidioradulum crustosum (Pers.) Zmitr, Malysheva & Spirin | w | 1 | unknown |
| Ceraceomyces sublaevis (Bres.) Jülich | w | 1 | timber below shower-tub |

| Species | Rot | | Location |
|---|---|---|---|
| Clitopilus hobsonii (Berk. & Broome) P.D. Orton | ? | 1 | roofing felt with root contact |
| Coniophora fusispora (Cooke & Ellis) Cooke | b | 1 | cellar |
| Crepidotus sp. | w | 1 | kitchen flooring |
| Dacrymyces punctiformis Neuhoff | b | 1 | facades |
| Hyphodontia alutaria (Burt) J. Erikss. | w | 1 | purlin |
| Hyphodontia arguta (Fr.) J. Erikss. | w | 1 | roof flooring |
| Hyphodontia aspera (Fr.) J. Erikss. | w | 1 | upper floor |
| Hyphodontia breviseta (P. Karst.) J. Erikss. | w | 1 | roof of a barn |
| Hyphodontia floccosa (Bourdot & Galzin) J. Erikss. | w | 1 | cellar |
| Hyphodontia juniperi (Bourdot & Galzin) J. Erikss. & Hjortstam | w | 1 | staircase |
| Hyphodontia pruniacea J. Erikss. & Hjortstam | w | 1 | rafter |
| Hyphodontia radula (Pers.) Langer & Vesterh. | w | 1 | old oak rafter |
| Hypochnicium geogenium (Bres.) J. Erikss. | w | 1 | bath |
| Hypoxylon sp. | w | 1 | roof |
| Laetiporus sulphureus (Bull.) Murrill | w | 1 | engine room |
| Leccinum sp. | ? | 1 | wet cellar brickwork with root contact |
| Leucogyrophana romellii Ginns | b | 1 | cellar flooring |
| Melanogaster broomeanus Berk. | ? | 1 | cellar brickwork |
| Merulius tremellosus Schrad. | w | 1 | cellar |
| Peniophora pithya (Pers.) J. Erikss. | w | 1 | roof |
| Phellinus pini (Brot.) Bondartsev & Singer | w | 1 | flooring timber |
| Pycnoporus cinnabarinus (Jacq.) P. Karst. | b | 1 | window |
| Pycnoporellus fulgens (Fr.) Donk | b | 1 | laminated board stacks |
| Schizophyllum commune Fr. | w | 1 | roof |
| Sistotrema brinkmannii (Bres.) J. Erikss. | w | 1 | ceiling under concrete |
| Tomentella crinalis (Fr.) M.J. Larsen | ? | 1 | cellar brickwork |
| Trametes ochracea (Pers.) Gilb. & Ryvarden | w | 1 | window |
| Trametes pubescens (Schumach.) Pilat | w | 1 | unknown |
| Trechispora invisitata (H.J. Jacks.) Liberta | w | 1 | half-timbering, |
| Trichaptum abietinum (Dicks.) Ryvarden | w | 1 | roof, introduced with moist timber |
| Volvariella bombycina (Schaeff.) Singer | w | 1 | flooring |

b=brown rot, w=white rot, ? ability to rot or type of rot not known.

A 9-year investigation in 2,000 German buildings revealed the presence of over 100 basidiomycetes (Table 6.1). Several of them, e.g. *Cerinomyces pallidus, Grifola frondosa, Pluteus cervinus* and *Radulomyces confluens*, were demonstrated for the first time in houses (Huckfeldt and Schmidt 2006a). Some of the rare indoor species (Table 6.1) normally occur on trees or on timber in outdoor use. The anamorphic *Ptychogaster rubescens* (brown rot) was detected twenty-one times. Soft-rot decay was found in 280 buildings. Additional fungal indoor wood-damage is due to discoloration by blue-stain fungi and molding by Deuteromycetes.

Fungal names listed in the older literature must not be taken as absolutes as sometimes misleading synonyms were used formerly, like *Poria vaporaria* for different *Antrodia* species and *Oligoporus placenta*, and species were often only roughly identified in situ based on the morphology of their fruit bodies and strands and only rarely after isolation by their mycelial characteristics. Precise molecular methods were not available for the identification of indoor basidiomycetes until the 1980s. For example based on molecular methods, 20% of cultures of *Antrodia* species and *Oligoporus placenta* that were obtained from various culture collections and which had been determined by classical methods were shown to be misnamed (Schmidt and Moreth 2003).

Most indoor basidiomycetes (Table 6.1) are distributed across several families within the Aphyllophorales. *Coprinus* species from the Agaricales family have been found, The Boletales are represented by common fungi like *Serpula, Meruliporia, Coniophora* and *Tapinella*. Among the Dacrymycetales within the Heterobasidiomycetes are *Cerinomyces pallidus* and *Dacrymyces stillatus*.

The sexuality of all species is not yet known (Ryvarden and Gilbertson 1993/1994). Most basidiomycetes, including *Serpula lacrymans* (Harmsen *et al.* 1958), have a bifactorial (tetrapolar) incompatibility system. The sexual cycle of *S. lacrymans* could be completed in culture (Schmidt and Moreth-Kebernik 1991a), permitting monokaryon production for classical intra- and interstock breeding (Schmidt and Moreth-Kebernik 1991b). Only dikaryons show clamps, while only monokaryons form an abundance of arthrospores. A small sample of ten dikaryotic isolates from Germany, UK and Australia showed only four and five different A and B factors, respectively, probably indicating that the fungus has a low outbreeding efficiency. Parental strains (and $F_1$-dikaryons) grew faster than their derived monokaryons, the latter have been found tolerate slightly higher temperature (growth at 28 °C) and a higher boron concentration in wood decay tests (Schmidt and Moreth-Kebernik 1990). Seventy-five world-wide isolates have been grouped into eight vegetative compatibility groups (VCGs), some of which are distributed on different continents (Kauserud *et al.* 2006a).

In the following, indoor basidiomycetes that had been isolated several times from German buildings are described. Lesser frequent isolations are only listed in Table 6.1. Fungal names and authors (Table 6.1) are mainly according to Index Fungorum. Emphasis is placed on *S. lacrymans*, which is the most common indoor basidiomycete in several central European countries. *Meruliporia incrassata* (Berk. & M.A. Curtis) Murrill, the North American pendant to *S. lacrymans*, is also described. The following morphological descriptions are mainly based on observations and measurements in damaged buildings and on laboratory incubations on wood samples (Huckfeldt and Schmidt 2006a,b). The fungi for the latter pure culture studies had been previously re-identified by rDNA-ITS sequencing (Schmidt and Moreth 2002/2003). For some genera like *Coniophora* and *Leucogyrophana*, the important indoor species are shown for the first time as comparative pictures.

## Dry rot fungi: *Serpula* species, *Leucogyrophana* species, *Meruliporia incrassata*

Brown-rot causing dry-rot fungi comprise *Serpula lacrymans, S. himantioides, Leucogyrophana mollusca, L. pinastri, L. pulverulenta* and *Meruliporia incrassata*. The fungi belong to the Coniophoraceae in the Boletales.

### Serpula lacrymans, *true dry rot fungus*

### General

*Serpula lacrymans* is the most dangerous house-rot fungus in central, eastern, and northern Europe. It grows however also in cooler areas of Japan (Doi 1991), Korea, India, Pakistan and Siberia (Krieglsteiner 2000), in New Zealand and southern Australia (Thornton 1991), Oceania and North America (Kauserud *et al.* 2007a). The data concerning its involvement in fungal indoor damage reach from 16% in Norway (Alfredsen *et al.* 2005) to 59% in Sweden (Viitanen and Ritschkoff 1991). For example, the annual repair costs of dry rot damage amount to at least £150 million in Great Britain (Jennings and Bravery 1991).

Since the fundamental work by Hartig (1885), Mez (1908) and Falck (1912), *S. lacrymans* belongs to the best-investigated fungi. The older observations and results are described by Liese (1950), Bavendamm (1951a, 1969), Lohwag (1954), Cartwright and Findlay (1958), Harmsen (1960), Savory (1964), Findlay (1967), Coggins (1980),

Segmüller and Wälchli (1981) and Grosser (1985). A literature search lists 1,200 publications (Seehann and Hegarty 1988). Younger reviews and laboratory results to the biology and physiology are by Jennings and Bravery (1991), Viitanen and Ritschkoff (1991), Eaton and Hale (1993), Huckfeldt *et al.* (2005), Schmidt and Huckfeldt (2005), Huckfeldt and Schmidt (2006a), and Schmidt (2006, 2007).

As cause of the special danger of *S. lacrymans* the following features were specified in the literature: Its omnipresent spores germinate on damp wood or other cellulosic materials (paper, cardboard), and the mycelium can reach wood by growing over and through substrates that do not serve as a nutrient. For initial colonization, it only needs low wood moisture content. Wood moisture %u refers to the mass of the dry state: u (%)=[(mass of wet wood − mass of dry wood) : mass of dry wood] × 100. The conventional wisdom is that it is the only fungus that can infect so-called dry timber (at least 21%u) and brickwork (at least 0.6% water content) and widely spread by mycelium and its highly developed strands (name: small serpent), thereby growing over and through wood and several other materials, like porous or ruptured brickwork or its wall joints, and supplying channels for electricity (Coggins 1991, Jennings 1991). However, recent laboratory experiments showed that *S. lacrymans* is not unequalled as also other indoor fungi were able to colonize dry wood (Tables 6.2 and 6.3). Coggins (1991) stressed that the initial colonization of a substrate, as for example the growth through wall joints, occurs by the youngest hyphae of the vegetative mycelium, in contrast to the infection way of *Armillaria* species that do this by means of rhizomorphs. The strands develop as a secondary mycelium behind the growth front and serve rather to transport nutrients to the hyphal margin than water. An acute infection is often for a longer time not recognized due to the hidden way of life. Spores and still alive mycelia can lead to re-infections in the case of inappropriate remedial treatments (Bravery *et al.* 2003).

Based on comprehensive investigations in infected buildings, the danger of *S. lacrymans* seems to be due to the fact that it is the only fungus which has all important abilities to colonize a building: namely, to grow through inorganic materials, to infect wood below fiber saturation point, to form dense surface mycelium preventing the substrate from drying, and to survive in dry timber (Table 6.2). However, considering separate capabilities, other fungi are more powerful.

*Serpula lacrymans* occurs in buildings from cellar to the roof basis (Schultze-Dewitz 1985, 1990). Poorly ventilated houses and all buildings with high relative humidity in connection with damages to the structural fabric are particularly endangered. Growing surface mycelium and strands are often found on wood in an atmosphere of constantly over 95% air humidity. There was no growth on agar at 95% humidity.

*Table 6.2. Important characteristics of indoor wood-decaying basidiomycetes for spreading and damage in buildings.*

| | Strand formation | Growth through brickwork | Spreading from brickwork | Colonization of dry wood[1] | Retarding substrate drying by surface mycelium |
|---|---|---|---|---|---|
| *Serpula lacrymans* | yes | yes | yes | yes | well |
| *Coniophora puteana* | yes | yes | unknown | yes | little |
| *Donkioporia expansa* | no | no | no | yes | very well |
| *Antrodia vaillantii* | yes | rarely | unknown | yes | moderately |
| *Serpula himantioides* | yes | yes | unknown | yes | little |
| *Leucogyrophana pinastri* | yes | yes | doubtful | no | little |

[1] Colonization of dry wood if a near moisture source is available.

*Table 6.3. Cardinal points of wood moisture content (%u) of some indoor wood-decaying basidiomycetes for colonization and decay of wood (after Huckfeldt and Schmidt 2006a).*

| Species | Minimum for colonization | Minimum for decay | Optimum for decay | Maximum for decay |
|---|---|---|---|---|
| *Serpula lacrymans* | 20 | 26 | 45-140 | 240 |
| *Coniophora puteana* | 17.5 | 22 | 36-210 | 262 |
| *Donkioporia expansa* | 21 | 26 | 34-126 | 256 |
| *Antrodia vaillantii* | 19-22 | 29 | 52-150 | 209 |
| *Gloeophyllum abietinum* | 20 | 22 | 40-208 | 256 |
| *Gloeophyllum sepiarium* | 28 | 30 | 46-207 | 225 |
| *Gloeophyllum trabeum* | 25 | 31 | 46-179 | 191 |
| *Leucogyrophana pinastri* | 30 | 37 | 44-151 | 184 |

Important causes of dry rot infections are building defects that affect increased wood moisture content (Paajanen and Viitanen 1989). The mycelium is sensitive to draught and humidity removal, generally to climatic changes, so that it often develops in false ceilings and false soil areas under floors and behind wall coverings, from where it spreads. Because of this hidden way of life, often only fruit bodies on brickwork, baseboards, doorframes or stairway steps show that the higher floors are already infected. In extreme cases all timbers as well as large parts of the brickwork have to be removed. Old buildings, which had insulating windows as the only measure of heat insulation, are particularly at risk: in that case, the moisture in the building condenses on other weak spots like empty spaces of the brickwork at the back of heaters (Huckfeldt *et al.* 2005).

Except in homes, *S. lacrymans* occurs on mine timber and also in the open. It has been only found in a few natural environments, namely in the Himalayas (Bagchee 1954, Bech-Andersen 1995, White *et al.* 2001), Northern California (Cooke 1955, Harmsen 1960), in the Czech Republic (Kotlaba 1992) and in East Asia (Kauserud *et al.* 2004a). Molecular analyses showed that there are two varieties of the species, one residing naturally in North America and Asia (var. *shastensis*), and another lineage including specimens from all continents, both from natural environments and buildings (var. *lacrymans*). From mainland Asia as the likely origin of the latter form, a few genotypes have migrated worldwide to Europe, North and South America, Japan and Oceania (Kauserud *et al.* 2007a).

## Morphology

The rust-brown fruit bodies (Figure 6.1) are mostly 1-5 mm thick (maximum 3 cm) and to 2 m wide, resupinate to effused-reflexed and imbricate, and sometimes stalactite-like. From shakes and vertical planes grow pad and bracket-like basidiomes. The gyroso-reticulate hymenophore is named merulioid (former generic name: *Merulius*). The margin is white-yellowish, often bulging and always with a sharply limited front. The fruit body is first monomitic, later dimitic with fibers. Particularly at the margin, as also with the mycelium, arise liquid drops of neutral pH value due to guttation, which led to the naming *lacrymans* (watering). Fresh fruit bodies have a pleasant smell, but putrefy after sporulation and then easily stink from ammonia. The old, dry, then black-brown fruit bodies hardly show the merulioid structure. Fruit bodies develop over the whole year, with an amassment in the late summer until winter (Falck 1912). Affected areas in a building are often widely covered with yellow-brown, elliptical, thick-walled spores (9-11 × 4.5-5.5 μm) with small, pointed extension at an end and partly with up to five intracellular oil droplets (Hegarty and Schmitt 1988, Nuß *et al.* 1991). First fructification in laboratory culture was obtained by Falck (1912). Cymorek and Hegarty (1986a) stimulated fructification by 12 °C incubation and by natural temperature change in the open (cool) (Hegarty and Seehann 1987; Hegarty 1991). Fruit bodies relatively often developed in pure

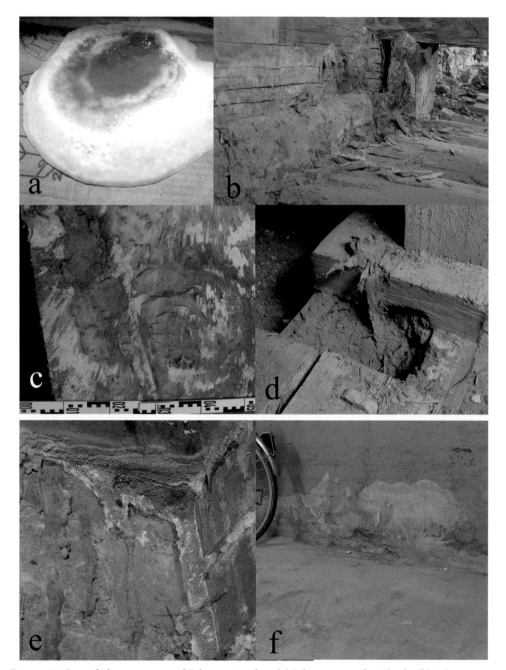

*Figure 6.1.* Serpula lacrymans *and* S. himantioides. *(a)* S. lacrymans *fruit body, (b)* S. lacrymans *mycelium and strands, (c)* S. lacrymans *mycelium on insulating material, (d)* S. lacrymans *strands inside decayed wood, (e)* S. himantioides *fruit body, (f)* S. himantioides *mycelium and strands.*

cultures, if the mycelium was first incubated for 4 weeks at 25 °C on malt agar and then at 20 °C and natural daylight (Schmidt and Moreth-Kebernik 1991a).

During initial growth, with sufficient humidity and standing air, often a white, woolly thick surface mycelium develops, which is rapidly interspersed by the typical strands. Yellow to wine-red to violet discolorations (inhibition colors) by restraining influences [light, accumulation of toxic metabolites, increased temperature: Zoberst (1952), Cartwright and Findlay (1958)] are characteristic and led to the former generic name *Merulius*, going back to the yellow beak of the male blackbird *Turdus merula* (Coggins 1980). Older mycelium collapses to removable, dirty grey to silvery skins, in which the branched strand system is embedded.

The silver-grey, grey to brown (young: white), to 3 cm thick, to 2 m long, and on their surface fibrously roughened strands (Figure 6.1) break when being dry with audible cracking (Falck 1912). Strands contain hyaline, partly yellowish (old: brown) vegetative hyphae (2-4 μm in diameter) with large clamps, straight, refractive and hardly septate, clampless fibers [(3-)4.5-(5) μm thick, within strands near fruit body to 12 μm thick], and not or rarely branched vessels (5-60 μm thick) with bar-like or warty wall thickenings (to 13 μm high). Strands are formed only in surface mycelium, and there as well by dikaryotic as by monokaryotic mycelium, and not in substrate mycelium and reach (at 20 °C) 5 mm length increase per day (Nuß *et al.* 1991).

**Physiology**

Conifers are preferred. Hardwoods with dark heart like oak and chestnut are more resistant than light species (Wälchli 1973). Beside wood and brickwork, composite woods (chipboards, fiberboards), carpets and textiles are attacked and insulating materials (Grinda and Kerner-Gang 1982) like mineral wool and are through-grown and damaged (Bech-Andersen 1987a).

*Serpula lacrymans* can be differentiated from other dry rot fungi (*S. himantioides*, *Leucogyrophana* species, *M. incrassata*) by its relatively low temperature maximum of 26-27 °C (Table 6.4), although monokaryons (Falck 1912, Schmidt and Moreth-Kebernik 1990) and some wild Himalayan isolates tolerated slightly higher temperature (Palfreyman and Low 2002). Whereas a two-week incubation of *S. lacrymans* on agar showed that the fungus was able to survive 27.5 °C, short-time tests on agar revealed a temperature tolerance to 50 °C (also Mirič and Willeitner 1984). In wood samples, some isolates survived 65 °C for 4 hours (Table 6.4). Since the infected wood had been slowly dried prior to the heat treatment, the mycelium was able to develop heat-tolerant arthrospores (Huckfeldt and Schmidt 2006a). The

Table 6.4. Influence of temperature (°C) on mycelial growth and survival of indoor wood-decaying basidiomycetes (after Schmidt and Huckfeldt 2005).

| Species | Optimum on agar | Maximum on agar | Lethal on agar after 2 weeks | Lethal on agar after hours | Lethal in wood after 4 hours |
|---|---|---|---|---|---|
| Serpula lacrymans | 20 | 26-27 | 30 | 55 (3 hours) | 50-70 |
| Serpula himantioides | 25-27.5 | 32.5 | >35 | | 65 |
| Leucogyrophana mollusca | 25-27.5 | 32.5 | 30-≥35 | | 75 |
| Leucogyrophana pinastri | 20-27.5 | 32.5 | >35 | | |
| Coniophora puteana | 22.5-25 | 27.5-≥37.5 | 32.5-≥37.5 | 60 (3 hours) | 70-75 |
| Coniophora marmorata | 20-27.5 | 25-≥37.5 | 35-≥37.5 | | |
| Coniophora arida | 25 | 27.5 | 35 | | |
| Coniophora olivacea | 22.5-25 | 32.5-35 | 35-≥37.5 | | |
| Antrodia vaillantii | 27.5-31 | 35 | 37-40 | 65 (24 hours) | >80 |
| Antrodia sinuosa | 25-30 | 35 | 37-42.5 | 65 (3 hours) | |
| Antrodia xantha | 27.5-30 | 35 | 40-42.5 | | |
| Antrodia serialis | 22.5-25 | 32.5-35 | 37.5-42.5 | | |
| Oligoporus placenta | 25 | 35 | 40-45 | 65 (24 hours) | >80 |
| Gloeophyllum abietinum | 25-27.5 | 37.5-42.5 | 40-42.5 | | >95 |
| Gloeophyllum sepiarium | 27.5-32.5 | ≥45 | ≥45 | 60 (3 hours) | >95 |
| Gloeophyllum trabeum | 30-37.5 | ≥45 | ≥45 | 80 (1 hour) | >95 |
| Donkioporia expansa | 28 | 34 | >40 | 65 (24 hours) | >95 |

supposed lethal values of the fungus around 55 °C are, however, the basis for the eradication with hot air in the roof space of buildings, performed in some European countries (Koch 1991, Steinfurth 2007). Spores of *S. lacrymans* did not survive 1 h at 100 °C (Hegarty *et al.* 1986).

Wood moisture is the most important factor influencing wood decay by fungi and consequently also the most important factor to be taken into consideration in terms of wood protection. The conventional opinion is that *S. lacrymans* can decay so-called dry timber (below the fiber saturation point of approximately 30%u wood moisture) (name: dry rot fungus) because the species is particularly effective to transport water by means of mycelium and strands from a moist source to the infestation of dry wood (Wälchli 1980, Coggins 1991). However, strands develop behind the mycelial growth front (Nuss *et al.* 1991). Their main significance lies in the nutrient transport (translocation) (Savory 1964, Watkinson *et al.* 1981, Bravery and Grant 1985, Jennings 1987), and they do not act as a conduit for water transportation (Ridout 2000, 2007). Nitrogen was also transported from the soil under houses to wood decay in the interior (Doi 1989, Doi and Togashi 1989). Water can be transferred through the mycelium to change the water potential of the substratum. Copious droplets of liquid (guttation) occur at the hyphal apices at the mycelial front (Jennings 1991) and on the fruit bodies. The fungus only attacks timbers in a building if those become wet by water penetration or water formation by condensation, but it can only continue to decay the wood it the timbers remain wet. If the water source is removed, then all fungal wood-decaying acitivity ceases (Bravery *et al.* 2003). Resistance to dryness has already been shown by Theden (1972) to vary at 20 °C from 1 year for *S. lacrymans* to more than 11 years for *Oligoporus placenta.* In laboratory tests, the minimum wood moisture for initial wood colonization by *S. lacrymans* was 20% (Huckfeldt and Schmidt 2006a). Experiments have shown that the fungus is not alone in decaying dry wood. Several fungi colonized wood of around 20%, if the samples were 20-30 cm away from the agar that provided water (Tables 6.2 and 6.3). The optimum wood moisture content for decay was high and varied among the species up to 210%. Older literature data quote e.g. for *S. lacrymans* 30 to 60% (Cartwright and Findlay 1958, Wälchli 1980). The maximum moisture of *S. lacrymans* for decay of 240% was higher than the 90-180% maxima found by Theden (1941), Wälchli (1980) and Viitanen and Ritschkoff (1991).

The erroneous opinion that the enzymatic decomposition of wood alone can produce water for *S. lacrymans* to survive is found even in recent publications. However, wood carbohydrates are not completely degraded to ATP, $CO_2$ and 56% water, but approximately 40% of the consumed cellulose is used via metabolites for fungal biomass (Weigl and Ziegler 1960). Furthermore, water production from carbohydrates is the rule for all breathing organisms.

Infected timber parts can exhibit just so much moisture to enable a slight growth for a long period of time (Grosser 1985). The danger of re-infection may derive from the dryness-resistant spores. The duration of the germ ability was said for *S. lacrymans* to amount to 20 years (Grosser 19085). In infected buildings, *S. lacrymans* produces an abundance of basidiospores, and they were assumed to be the main agent of dispersal (Falck 1912, Schultze-Dewitz 1985). Falck (1912) calculated the spore release by a 1 $m^2$ fruit body to $3\times10^9$ spores per hour. However, according to Wälchli (1980) the infection occurs by mycelium that is brought in with timber from other remedial treatments and via wooden boxes or shoes.

House-rot fungi causing brown-rot accumulate oxalic acid (oxalate) in rather large quantities and acidify their environments (Rypáček 1966). The preferential indoor occurrence of *S. lacrymans* was assumed to be caused by the intensive synthesis and secretion of oxalic acid (Bech-Andersen 1985): excess acid was neutralized to Ca-oxalate by calcium from brickwork or by chelating with iron from metals (Bech-Andersen 1987a,b, Paajanen 1993, Palfreyman *et al.* 1996). Considering the features listed in Table 6.2, fungi like *S. lacrymans* and *C. puteana* have an advantage because they can acidify alkaline brickwork for subsequent growth through walls. However, oxalic acid production is not assumed to be the main reason for indoor occurrence. Fungi like *A. vaillantii*, which also produce much oxalic acid, often grow outdoors: in woodlands, the large amount of produced oxalic acid complexes calcium in the production of calcium sediments (Jennings and Lysek 1999). The implication of calcium in oxalate precipitation was also shown for *M. incrassata* (Jellison *et al.* 2004). Dry rot attack in buildings is often found in the mostly moist ends of wooden beams, which are not separated from the brickwork. Such timber has contact to brickwork, which favors oxalic acid neutralization and also an appropriate wood moisture for decay. Dry wood in central parts of a room is rarely attacked.

The reduction in the pH value of the environment by fungi is thought to favour the activity of non-enzymatic systems and cellulolytic activity (Goodell *et al.* 2003). Oxalate serves as an acid catalyst for the hydrolytic breakdown of wood polysaccharides, attacking hemicelluloses and amorphous cellulose and thus increasing the porosity of the wood structure for hyphae, enzymes and low-molecular agents. Oxalic acid is also implicated in copper tolerance of fungi (see *Antrodia vaillantii*). Hastrup *et al.* (2005) showed 11 out of 12 isolates of *S. lacrymans* to be tolerant against copper citrate.

During controversies, for example in the context of house buying, frequently the question of the infection date by *S. lacrymans* plays a role, for whose determination the daily average mycelial growth is often used. According to Jennings (1991), the linear mycelial extension of *S. lacrymans* on wood, brickwork and insulants ranges from 0.65 to 9 mm per day. Assuming a five-mm radial increase per day on malt agar at optimal temperature, 15 cm growth result per month. However, due to the changing and not always optimal conditions in buildings and because different isolates of the fungus exhibited considerable differences in growth rate [1.5-7 mm/day: Cymorek and Hegarty (1986b), Seehann and Von Riebesell (1988)], an exact age determination on the basis of the mycelial extension is impossible. Similar isolate variation has been found for the decay of pine sapwood samples, ranging among 25 isolates from 12 to 56% in 6 weeks (Cymorek and Hegarty 1986b, Thornton 1991). Different isolates differed likewise in their sensitivity to wood preservatives (Abou

Heilah and Hutchinson 1977, Cymorek and Hegarty 1986b, Ważny and Thornton 1989, Ważny *et al.* 1992).

Important is also the judgment that the mycelium in a building is alive or dead. Subculturing on malt agar is possible, but isolations from mycelium are often contaminated by molds. Vital staining with fluorescein diacetate revealed to be suitable (Koch *et al.* 1989, Bjurman 1994, Huckfeldt *et al.* 2000).

## Serpula himantioides, *wild merulius*

*Serpula himantioides* is common in the open, in Europe frequently on spruce wood, stumps, structural timber in outdoor use, and rarely on living trees. It is also found in buildings (Table 6.1, Falck 1927, Harmsen 1978, Grosser 1985, Seehann 1986). The rust-brown, resupinate, sometimes membrane-like, smooth to merulioid, annual fruit bodies (Figure 6.1e) are to 2.6 mm thick, produce yellow-brown, thick-walled spores (9-11 × 5-5.5 µm). The white to grey-brown, root-like strands (to 2 mm in diameter; Figure 6.1f) are not surrounded by thick mycelium as with *S. lacrymans* and contain fibers [(2-)2.5(-4.5) µm in diameter, sometimes not clearly distinguishable from *S. lacrymans*]. The species is tetrapolar. Three genetic lineages within the morpho-species were shown by multiple gene genealogies and AFLP analysis (Kauserud *et al.* 2006b).

## Leucogyrophana *species*

Three *Leucogyrophana* species (Figure 6.2) grow in the forest on fallen stems and branches, but also on wood in indoor use: *L. mollusca, L. pinastri* (Schulze and Theden 1948, Siepmann 1970) and *L. pulverulenta* (Harmsen 1953). They differ from *Serpula* by smaller spores (Ginns 1978, Pegler 1991, Breitenbach and Kränzlin 1986). *Leucogyrophana pulverulenta* is rather common in Denmark and Germany. At least, *L. pinastri* needs a higher wood moisture content than *S. lacrymans* (Tables 6.2 and 6.3).

## Leucogyrophana mollusca, *soft dry rot fungus*

The fungus occurs only on softwoods. The resupinate, orange to yellow-brown (old: grey-blackish), easily separable fruit bodies (1-2 mm thick, to a few decimeters wide; Figure 6.2a) with white, cottony-frayed margin have a merulioid hymenophore with tooth-like elevations, producing yellowish-brown spores (6-7.5 × 4-6 µm). Round, often somewhat irregular, brown-violet to grey-black sclerotia (1-6 mm; Figure 6.2b), often in groups, are found in the mycelium. The strands (Figure 6.2b) inhabit also

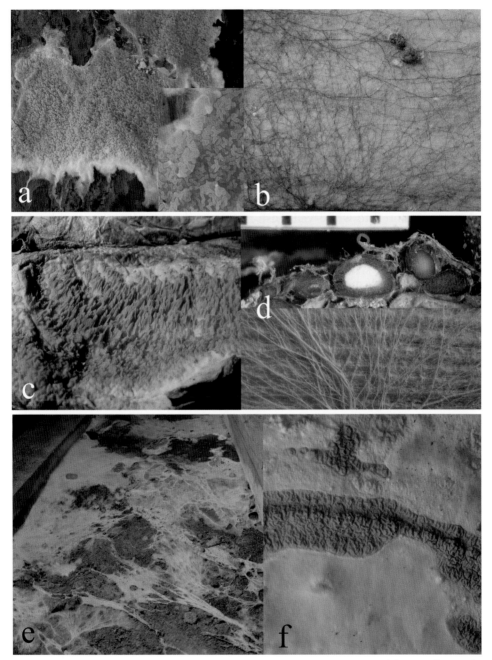

*Figure 6.2.* Leucogyrophana mollusca, L. pinastri *and* L. pulverulenta. *(a)* L. mollusca *fruit body, (b)* L. mollusca *strands and sclerotia, (c)* L. pinastri *fruit body, (d) top:* L. pinastri *cut sclerotia, bottom:* L. pinastri *young strands, (e)* L. pulverulenta *fruit body within mycelium, (f)* L. pulverulenta *fruit body detail.*

brickwork and are hair-like, first cream-yellow, soon red-brown to black, below 0.5 mm thick, separated from mycelium ("barked"), somewhat flexible when dry and fragile when old. There are no fibers. Vessels (to 25 µm thick) are numerous, in groups and show bar-thickenings. Some vegetative hyphae are bubble-like swollen to 10-25 µm in diameter and always have clamps.

### Leucogyrophana pinastri, *mine dry rot fungus, yellow-margin dry rot fungus*

The species grows probably only on softwoods. The resupinate, first yellow-orange, then olive-yellow to brown (old: grey-black) fruit bodies (to 1 m wide; Figure 6.2c) with the merulioid (at margin) to irpicoid to hydnoid (centre), to 1 mm thick hymenophore with cream-colored to white-yellowish margin and pale-yellow lower side, which contains brown to black strands (below 0.1 mm thick) produce yellow, thick-walled spores (4.5-6 × 3.5-4.5 µm). The first yellowish, then grey-brown, somewhat flexible when dry, hair-thin strands (under 0.5 mm in diameter; Figure 6.2d, bottom) are sometimes covered by lighter mycelium, do not contain fibers and are also found on brickwork (Table 6.2). The vessels are to 25 µm in diameter, often partly thickened, occur in bundles and have bars (mostly indistinct). Vegetative hyphae (1.5-6 µm thick) have large clamps. Small and oblong (to 2.5 mm long) sclerotia (sometimes absent; Figure 6.2d, top) are brown to grey. Sclerotia on the lower side of fruit bodies are brown to violet-black, 1-3 mm long, to 1.5 mm thick, and have a white core and a dark, 0.1-0.2 mm thick outer layer.

### Leucogyrophana pulverulenta, *small dry rot fungus*

The resupinate, first sulphur-canary yellow, then olive-yellow to cinnamon-brown (grey-black when old), thin (to 1 mm thick) fruit bodies (to 20 cm wide; Figure 6.2e,f) with crème-white and indistinct margin and smooth to merulioid hymenophore produce hyaline to yellow, iodine-negative, light-yellow in KOH, asymmetrically elliptical, thick-walled spores (5-6 × 3.5-4.5 µm). Sclerotia are absent. The white, crème-yellow to grey strands (to 2 mm thick) are indistinct, just as embedded as those of *S. lacrymans*, somewhat flexible, when dry always brittle, not clearly separated, do not contain fibers and also occur on brickwork. The septate vessels (to 20 µm thick) are numerous and often in bundles. Bar-thickenings are indistinct or absent. Vegetative hyphae (1-4 µm thick; 6 µm thick in the trama; small hyphae partly with thickened cell wall) have large clamps.

## Meruliporia incrassata, *American dry rot fungus*

Whereas *S. lacrymans* is restricted in North America to the northern parts of the USA and Canada, *Meruliporia incrassata,* being a warm-temperature fungus, occurs particularly in the southern states and the Pacific northwest of the USA (Verrall 1968, Palmer and Eslyn 1980, Gilbertson and Ryvarden 1987, Burdsall 1991, Zabel and Morrell 1992, Eaton and Hale 1993, Jellison *et al.* 2004). Two isolates from the USA and Canada grew best between 22.5 and 25 °C (Schmidt 2003). Burdsall (1991) named 24-30 °C as the optimal temperature range for growth and above 36 °C as the lethal temperature. Jellison *et al.* (2004) quoted 28-30 °C as the optimal range for growth and 40 °C for lethal temperature. Sapwood and heartwood of many gymnosperms and angiosperms are attacked. The fungus was rarely found on standing trees, infrequently on felled logs and stumps, on structural timber outdoors such as in mills, lumber yards, on shingles, on bridge timber and posts. It inhabits moist wood or wood located near a permanent or intermittent water supply if the wood is untreated (Palmer and Eslyn 1980). Some characteristics of wood decay are similar to those of *S. lacrymans*: The fungus is sensitive to drying. At 90% relative humidity, artificial inoculations were dead in 10 days. Burdsall (1991) emphasized its water-conducting strands.

The monomitic fruit bodies are similar to those of *S. lacrymans,* annual, resupinate to effused, 20 cm or more in size, thin (1-12 mm in diameter), fleshy and easily separable with a whitish to buff margin, brittle when dried, first appearing as a felted pad of mycelium with formation of pores beginning at the centre and subsequent fertile to the margin. The hymenophore is poroid (1-3 unequally circular to angular pores per mm), occasionally merulioid, whitish to buff or ochre-grey when fresh and grey-brown to black when dry. The thick-walled oblong to ellipsoid spores are variable in size (8-16 × 4-8 µm). The strands occur first as vein-like structures in the mycelium, often extending into soil or brickwork, appear whitish when young and browny-black with age (Eaton and Hale 1993), are 0.3-5.1 cm in diameter and up to 9 m long (Palmer and Eslyn 1980). Illustrations of mycelium and strands are by Zabel and Morrell (1992).

## Cellar fungi ("wet-rot fungi"): *Coniophora* species

The genus *Coniophora* (Coniophoraceae, Boletales) comprises about 20 species occurring worldwide with a broad host range primarily on conifers (Ginns 1982). Seven species occur in Europe (Jülich 1984) and five in Germany (Krieglsteiner 1991). *Coniophora puteana* is frequently associated with brown-rot decay in European buildings (Table 6.1). The older European literature summarized several species to

*C. puteana.* This fungus was said to be the most common species in new buildings. It however occurs also in old buildings, on stored wood, timber in soil contact like poles, piles, sleepers and on bridge timber as well as rarely on stumps and as wound or a weakness parasite on living trees (Bavendamm 1951b, Grosser 1985, Breitenbach and Kränzlin 1986). Of 177 basidiomycetes on American mine timbers, 83 isolates were *C. puteana* (Eslyn and Lombard 1983). The fungus was estimated to be twice as common as *S. lacrymans* in the UK (Eaton and Hale 1993). It comprised over 50% of the inquiries at the Danish Technological Institute (Koch 1985), 16% in Norway (Alfredsen *et al.* 2005), and 13% at the Finnish Forest Products Laboratory (Viitanen and Ritschkoff 1991). The fungus has been used for approximately 70 years as a test fungus for wood preservatives in Europe. It also occurs in the USA, Canada, South America, Africa, India, Japan, Australia, and New Zealand. Multilocus genealogies of three DNA regions revealed the occurrence of three cryptic species in the morpho-taxon *C. puteana*, with an indication that the species complex originated in North America (Kauserud *et al.* 2007b).

Further cellar fungi that attack indoor timber in Europe are *C. marmorata* and rarely also *C. arida*, *C. fusispora* and *C. olivacea* (Table 6.1). In Europe, the cellar fungi cause with about 10% frequency the two to third most common fungal indoor wood decay after *S. lacrymans*. In buildings, the fungi do not occur, like the name misleadingly suggests, only in cellars, but they can ascend everywhere on damp timber up to the roof (Schultze-Dewitz 1985, 1990). Beside softwoods, *C. puteana* attacks also several hardwoods (Wälchli 1976). As a so-called wet rot fungus (Bravery *et al.* 2003) with relatively high requirement for moisture from 30 to about 70%u and the optimum around 50%, all timber in the area of damp walls (beam ends and wall slats), damp floors and ceilings in kitchens, bathrooms and toilets as well as all timber in areas with water vapour development (swimming pools, launderettes) is endangered. *In vitro*, minimum moisture of *C. puteana* for colonization of sprucewood was 17.5% and for decay 22%. The optimum moisture content was broad, from 36 to 210% (Table 6.3). Damage by the cellar fungi is quite comparable with that one of *S. lacrymans* and can even exceed it. A fresh floorboard can be completely destroyed in one year, so the danger exists that furniture or persons can fall through. These types of damages occurred in Germany frequently during the building boom in the postwar years, if insufficiently dried wood were used, or the homes had not sufficiently dried before they were moved into and drying was retarded by humidity-impermeable painting, linoleum, or carpet. The cellar fungi belong to the fast-growing house-rot fungi and reached on agar at 23 °C up to 11 mm radial increase per day. The optimum temperature (Table 6.4) was between 20 and 27.5 °C and the maximum was between 25 and about 37.5 °C. Isolate Ebw. 1 of *C. puteana* survived on agar 15 min. at 60 °C (Mirič and Willeitner 1984) and 3 h at 55 °C. In slowly dried wood samples, even 4 hours at about 70 °C were withstood. The data concerning a possible dryness resistance of the fungus vary: according to observations in practice, it did not survive when drying. Up to 7 years were however survived in dry wood in the laboratory (Theden 1972). There was isolate variation with regard to the sensitivity to wood preservatives (Gersonde 1958).

The diagnosis of damage by cellar fungi is not always easy, since fruit bodies are rare and colonized wood shows frequently no or only meagre surface mycelium (Käärik 1981). The fruit bodies resemble those of the *S. lacrymans*, are however thinner. The species *C. puteana* is easy to recognize of the warty knots on the hymenophore (name: carrying cones). Characteristic on agar are double and multiple clamps. The initial stages of the rot are frequently ignored, since hardly infection signs become visible on exposed wood exterior surfaces, e.g. on baseboards, while the wood at the backside is already completely rotten and overgrown by thread-thin, radiate to root-like, brown to black strands. Early signs of rot are often dark discolorations under paint.

The species can be differentiated on the basis of their fruit bodies (Ginns 1982, Jülich and Stalpers 1980). With regard to isolates in culture, *Coniophora* cannot be differentiated at the species level by morphological and cultural characteristics (Stalpers 1978). Thus, isolations from buildings were summarized as *C. puteana/C. marmorata* (Guillitte 1992). Sequencing of the rDNA-ITS separated four species (Schmidt *et al.* 2002/2003).

## Coniophora puteana, *(brown) cellar fungus*

The monomitic, annual, resupinate, first white-yellow, then light to dark brown fruit bodies (to 4 mm thick, to 3 meters wide; Figure 6.3a) with warty knots (to 5 mm thick) have an indistinct and fibrous margin, are firmly attached, fragile when dry and produce yellow-brown spores (9-16 × 6-9 µm). The strands (Figure 6.3b) are first white, soon brown to black, to 2 mm wide, to 1 mm thick, root-like, hardly removable and fragile, partly with brighter centre and occur often also on brickwork (Table 6.2) and other inorganic substrates. The pale to brown, branched fibers (2-4 µm thick) are somewhat thick-walled, however with relatively broad and usually visible lumen. The septate vessels (10-30 µm thick) are often deformed, without bars and surrounded and interwoven by fine hyphae (0.5-1.5 µm thick). Vegetative hyphae (2-8 µm thick) without or rarely with clamps (also multiple clamps) produce hyaline to brown drop-shaped secretions (1-5 µm) holding the hyphal net together (sclerotization). Blackish discolored wood is seen beneath the strands.

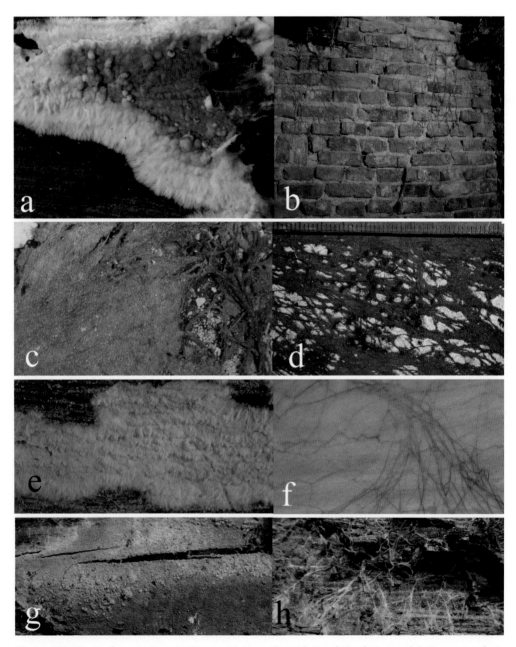

*Figure 6.3.* Coniophora puteana, C. marmorata, C. arida *and* C. olivacea. *(a)* C. putena *fruit body, (b)* C. putena *strands, (c)* C. marmorata *fruit body, (d)* C. marmorata *strands, (e)* C. arida *fruit body, (f)* C. arida *strands, (g)* C. olivacea *fruit body, (h)* C. olivacea *strands.*

### Coniophora marmorata, *marmoreus cellar fungus*

The dimitic, annual, resupinate, pale to olive-brown, felty and separable fruit bodies (to 0.4 mm thick, to 15 cm wide; Figure 6.3c) with a grey margin produce light-yellow to brownish, smooth, thick-walled spores (7-7 × 6-7 μm ) with apiculus and oil drop. The flabby, light to dark-grey mycelium is interwoven by strands and does not show drops. The brownish to black, to 2 mm thick net-shaped strands (Figure 6.3d) are easily separable. Hyaline to brown, branched, septate, thin-walled vegetative hyphae [2-5(-6) μm in diameter] without bars are rarely clamped. Multiple clamps are more common than in *C. puteana*. Hyaline to brown, unbranched, septate vessels (to 31 μm diameter) are in close contact with surrounding hyphae. Long, wiry, brown fibers have 2.5-4 μm diameter.

### Coniophora arida, *arid cellar fungus*

The monomitic, annual, resupinate, white-ochre to yellow-brown, smooth to felty and firmly attached fruit bodies (to 1 mm thick, to 10 cm wide; Figure 6.3e) with a small, cream-colored, fine-frayed margin produce elliptical (5.5-8 × 9-13 μm), smooth, thick-walled spores with apiculus. Strands (Figure 6.3f) are rare, white to brown and approximately 0.1 mm thin. Hyaline to brown, branched, thin-walled, vegetative hyphae (1-3 μm diameter) are septate and rarely show multiple clamps. Fibers are indistinct, brown and thick-walled.

### Coniophora olivacea, *olive cellar fungus*

The monomitic, annual, resupinate, olive-brown, firmly attached, smooth to warty fruit bodies (to 1 mm thick, to 10 cm wide; Figure 6.3g) have a cream-colored, small (1-2 mm broad) margin, fraying with strands. The white, cream-colored to brown, thin (to 1.1 mm in diameter) strands (Figure 6.3h) have a lighter core with a darker outer layer. Hyaline to brownish, septate, clampless, thin-walled vegetative hyphae have 3-7 μm diameter. Vessels are numerous, hyaline to light-brown, to 26 μm thick and thin- to somewhat thick-walled (wall-thickness to 1 μm). Fibers occur only in the outer area of strands and are brown and thick-walled (3-4 μm diameter, partly thickened, then to 4 μm diameter). Thick-walled, septate cystidia (6-12 × 85-188 μm) without basal clamp are encrusted with crystals.

## Indoor polypores: *Antrodia* species and *Oligoporus placenta*

Four *Antrodia* species (Coriolaceae, Aphyllophorales) and *O. placenta* (Bjerkanderaceae, Aphyllophorales) may be assigned to the common indoor polypore fungi. The species occur circumglobal in the coniferous forest zone, mostly on softwoods (Findlay 1967, Domański 1972, Coggins 1980, Lombard and Chamuris 1990, Grosser 1985, Lombard 1990, Ryvarden and Gilbertson 1993/1994, Krieglsteiner 2000, Bernicchia 2005). The indoor polypores form a group of brown-rot fungi that are associated with the decay of softwoods in buildings. In Central Europe, these fungi belong after *S. lacrymans*, and together with the cellar fungi to the most common indoor decay fungi. They accounted for 14% of indoor decay fungi in Denmark (Koch 1985) and Finland (Viitanen and Ritschkoff 1991). A survey in California ranked *A. vaillantii*, *A. sinuosa*, *A. xantha* and *O. placenta* with 29% occurrence as the main group (Wilcox and Dietz 1997). The indoor polypores have similar biology and distribution (Lombard and Gilbertson 1965, Donk 1974, Breitenbach and Kränzlin 1986, Lombard and Chamuris 1990, Bech-Andersen 1995, Schmidt and Moreth 1996, 2003, Schmidt 2006).

*Antrodia vaillantii* is widely distributed in Europe, but rather rare in Fennoscandia. It is the most frequent fungus in British mines (Coggins 1980). *Antrodia sinuosa* is circumpolar in the boreal conifer zone, widespread in Europe, North America, East Asia, North Africa, and Australia (Domański 1972). The species was in Sweden with 1,045 damages between 1978 and 1988 with 13% portion the most common indoor polypore (Viitanen and Ritschkoff 1991). *Antrodia serialis* attacks logs and piles, causes heart rot in standing trees and occurs widespread, also in Himalaya and Africa (Seehann 1984, Breitenbach and Kränzlin 1986), rarely (1.4%) in buildings (Viitanen and Ritschkoff 1991, Coggins 1980), within the roof area, in cellars and under corridors (Domański 1972). *Antrodia xantha* (Domański 1972) occurs in Europe and North America on fallen stems, branches, stumps, in greenhouses (Findlay 1967), on windows (Thörnqvist *et al.* 1987), on timber in swimming pools and in flat roofs (Coggins 1980). *Oligoporus placenta* is rare, but widespread in Europe except for the Mediterranean. In North America, the fungus is the most common wood decayer in ships (Findlay 1967) and was exported to Great Britain (Coggins 1980). In North America, *O. placenta* and *A. serialis* are common on mine timber and poles (Gilbertson and Ryvarden 1986).

The indoor polypores differ in their fruit body, spore morphology (Jülich 1984, Ryvarden and Gilbertson 1993/1994) and sexuality. Some species also fruit in laboratory culture, which supports identification of mycelia and tests for sexuality. *Antrodia vaillantii* is tetrapolar heterothallic (Lombard 1990), *A. serialis*, *A. sinuosa* and *O. placenta* are bipolar (Domański 1972, Stalpers 1978). Three *Antrodia* species develop strands (Falck 1912, Stalpers 1978, Jülich 1984), *O. placenta* only *in vitro*. However, the vegetative mycelium that has been isolated from decayed wood is hardly distinguishable (Nobles 1965). Due to the limited differentiating features, misinterpretations occur. Furthermore, the nomenclature has a confusing history and is still not always uniform (Cockcroft 1981). Fungi have been variously classified as *Polyporus, Poria, Amyloporia, Fibroporia* (Domański 1972). Misleading synonyms in the older literature such as *Polyporus vaporarius* and *Poria vaporaria* have been used for different species, viz. *A. vaillantii* (Bavendamm 1952a), *A. sinuosa*, and *O. placenta*. According to Ryvarden and Gilbertson (1994), the Reddish sap polypore, formerly *Tyromyces placenta* (Fr.) Ryvarden, was placed in *Oligoporus*, since the genus *Tyromyces* is restricted to fungi causing a white rot. The epithet *placenta* must remain in the female gender. Older synonyms are *Postia placenta* (Fr.) M.J. Larsen & Lomb., *Poria placenta* (Fr.) Cooke sensu J. Eriksson, *Poria monticola* Murr., and the haploid standard strain *Poria vaporaria* (Pers.) Fr. sensu J. Liese (Domański 1972). *Postia* is a nomen provisorium/nudum in the sense of Fries and illegitimate in the sense of Karsten. Isolate MAD 698 of `*Postia placenta*´ was thoroughly investigated in view of brown-rot decay mechanisms (e.g. Clausen *et al.* 1993, Highley and Dashek 1998). Difficulties may increase because *O. placenta* separates into the forms *placenta* with salmon-pink fruit bodies (Reddish sap polypore) and *monticola*, never with reddish stain (Domański 1972). Monokaryotic isolates of *O. placenta* were used for testing wood preservatives in Germany (*Poria vaporaria* standard strain II) and are obligatory in the European standard EN 113 (named *Poria placenta* FPRL 280). Even current literature uses the names *Postia placenta* and *Poria placenta*.

For species identification in the case that only mycelium is present, rDNA-ITS sequencing separated the five fungi (Schmidt and Moreth 2003).

For an easier understanding during a practical valuation of a fungal damage, the different fungi are often summarized as "indoor polypores" or as "*Vaillantii* group", particularly because they differ from the cellar fungi and *S. lacrymans* by their mycelia, strands, and fruit bodies. The polypores, particularly *A. vaillantii*, form a well-developed white and cottony surface mycelium without inhibition colors, which, thus, can be confused with the young mycelium of *S. lacrymans*. The mycelium may spread ice flower-like over the substrate, that of *S. lacrymans* is converted with ageing into silvery-grey skins, and that of the cellar fungi is dominated by fine black strands. White (*A. vaillantii*), to string-thick, smooth and flexible strands develop within the mycelium and rarely grow over non-woody substrates and through porous brickwork (Table 2; Grosser 1985), the latter, however, less intensive than by *S. lacrymans*. The white to yellow (*A. xantha*) or reddish (*O. placenta* f. *placenta*) fruit bodies show pores that are visible with the naked eye. The dry wood has the typical brown-cubical rot. The size of the cubes caused by the polypores is often smaller than those by the cellar fungi and *S. lacrymans*. The cube size varies however also as a function of the wood moisture content and may also depend on the size of the infected timber. The polypores attack predominantly coniferous woods in damp new and old buildings, from cellar to roof, particularly in the upper floor, furthermore mine timber, stored timber as well as timber in outside use, particularly in the soil/air zone, such as poles and sleepers. They also attack trees as wound parasites and live on stumps and fallen trees (Krieglsteiner 2000). *Antrodia serialis* was found in over-mature Sitka spruce trees (Seehann 1984).

In the laboratory, wood samples of about 20% moisture content were colonized (Table 6.3). As so-called wet-rot fungi (Coggins 1980, Bravery *et al.* 2003), they need wet timber with moisture contents from 30 to 90% for a long time. According to literature, the optimum is around 45%. Laboratory experiments revealed that minimum moisture for wood decay by *A. vaillantii* was 29% and the optimum 52 to 150% (Table 6.3). With timber drying, *Antrodia* species were supposed to die (Bavendamm 1952a, Coggins 1980) or only to stop growth (Grosser 1985). In the laboratory, up to 11 years were survived by dryness resistance (Theden 1972), so that fungi may come to life again. There is also resistance to high temperature: *Antrodia vaillantii* and *O. placenta* survived on agar 3 hours at 65 °C. *Antrodia vaillantii* and *O. placenta* withstood heat of 80 °C for 4 hours in slowly dried wood samples (Table 6.4), which has to be considered in view of a possible treatment of infected homes with hot air.

The indoor polypores, especially *Antrodia vaillantii*, are resistant to copper (Da Costa and Kerruish 1964, Schmidt and Moreth 1996, Green and Clausen 2003) mainly due to the excretion of oxalic acid (Rabanus 1939, Da Costa 1959, Sutter *et al.* 1983, 1984, Jordan *et al.* 1996) and subsequent formation of Cu-oxalate (e.g. Collett 1992a,b, also Humar *et al.* 2001). *Antrodia vaillantii* decreased the life-time of timber impregnated with chromated copper arsenate and borate, respectively. Chromium, which plays a role in the fixation reactions of the elements in wood, and arsenate as well as borate were solubilized by oxalic acid and subsequently washed out by rain. Copper was precipitated into the insoluble form of the oxalate, rendering the copper inert (Stephan *et al.* 1996). The fungal leaching effect of elements like chromium and arsenate has been used for biological remediation of chromium-copper-treated wood waste by *A. vaillantii* and further fungi (Leithoff *et al.* 1995, Kartal *et al.* 2004). Isolate variation occurred (Da Costa and Kerruish 1964, Collett 1992a,b), and monokaryons were more tolerant than their parental strains (Da Costa and Kerruish 1965). *In vitro*, *A. vaillantii* was the most copper-tolerant fungus among the five species and produced most oxalic acid (Schmidt and Moreth 2003). *Antrodia vaillantii* is also tolerant to arsenic (Göttsche and Borck 1990, Stephan and Peek 1992).

## Antrodia vaillantii, *mine polypore, broad-spored white polypore*

The dimitic, annual, resupinate, first white (Figure 6.4a), then light yellow to grey fruit bodies (to 1 cm thick; Figure 6.4c) grow on the wood underside or above as pads (when dry as corky layer) and produce hyaline spores (5-7 × 3-4 µm) in 2-4 circular-angular pores (to 12 mm long) per mm hymenium. The strands (Figure 6.4b) are pure white, felty, 0.2-7 mm in diameter, ice flower-like and flexible also when dry. Fibers are numerous, white, flexible, 2-4 µm thick, mostly unbranched, insoluble in 5% KOH and sometimes with "blown up" segments. Vessels are not rare but in old strands difficult to isolate, to 25 µm in diameter, partly with thick walls and reduced lumen and without bars. The vegetative hyphae (2-6 µm diameter) with rare clamps (sometimes medallion clamps) are also somewhat thick-walled. Surface mycelium (Figure 6.4d) is white to cream-colored. It reaches up to some square meters size in no-draught or under-floor areas, later also stalactite-like growth from occurs above.

## Antrodia sinuosa, *white polypore, small-spored white polypore*

The dimitic fruit bodies (Figure 6.4e) are similar to those of *A. vaillantii*, annual, resupinate, to 5 mm thick with 1-3 circular-sinuous pores (to 3 mm long) per mm hymenium and produce hyaline spores (4-6 × 1-2 µm). The strands (Figure 6.4f) are similar to those of *A. vaillantii*. Thin- to thick-walled, hyaline, septate, clamped, branched vegetative hyphae have 1-4 µm diameter. Hyaline thick-walled fibers (2-3 µm diameter) are wriggled, iodine-negative, without KOH reaction, and are rarely branched.

## Antrodia xantha, *yellow polypore*

The dimitic, annual, resupinate, first yellowish, then pale, crème-white, crusty to bracket-shaped, to 10 mm thick and 1 m wide fruit bodies (Figures 6.5a and 6.5b) have 3-7 circular-angular pores (to 5 mm long) per mm hymenium. Pores are absent at the margin. Small knots (to 8 mm size) occur on vertical substrates, partly growing together. The size of the hyaline, thin-walled, iodine-negative allantoid spores with apiculus is 4-5 × 1-1.5 µm. The strands are similar to those of *A. vaillantii*, but partly yellow discolored, later often pale and then undistinguishable from *A. vaillantii*. Hyaline, thin- to somewhat tick-walled, septate vegetative hyphae (2-3 µm diameter) are common. Fibers are frequent, hyaline, 2-4 µm in diameter, straight-wiry, somewhat thick-walled to thick-walled, mostly unbranched, iodine-negative, and swell in 5% KOH.

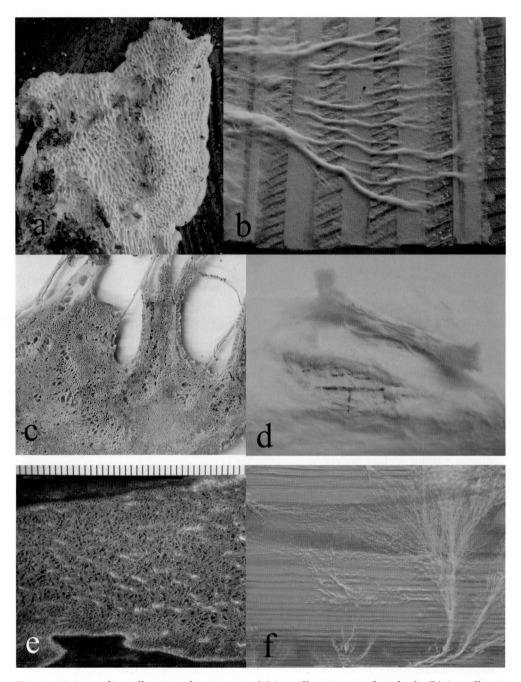

*Figure 6.4.* Antrodia vaillantii *and* A. sinuosa. *(a)* A. vaillantii *young fruit body, (b)* A. vaillantii *strands, (c)* A. vaillantii *old fruit body and strands, (d)* A. vaillantii *mycelium, (e)* A. sinuosa *fruit body, (f)* A. sinuosa *strands.*

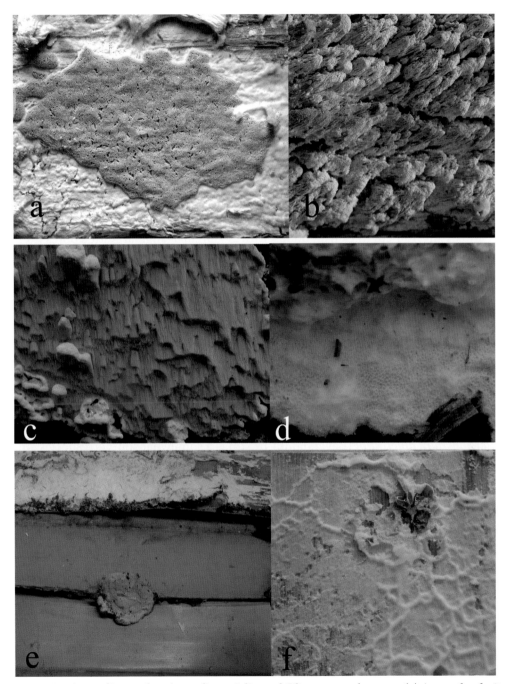

*Figure 6.5.* Antrodia xantha, Antrodia serialis *and* Oligoporus placenta. *(a)* A. xantha *fruit body, (b)* A. xantha *old fruit body, (c)* A. serialis *fruit body, (d)* A. serialis *fruit body detail, (e)* O. placenta *fruit body (form:* placenta*), (f)* O. placenta *strands in laboratory culture.*

## Antrodia serialis, *effused tramete, row polypore*

The dimitic, annual to biennial, resupinate to pileate, single and in rows, first white to crème-ochre, then pink-spotted, to 6 mm thick and to a few decimeters wide fruit bodies (Figures 6.5c and 6.5d) with 2-4 circular, partly slitted pores (to 5 mm long) per mm hymenium, with distinct and wavy margin produce hyaline spores (4-7 × 3-5 µm). Strands have not yet been observed.

## Oligoporus placenta, *(reddish) sap polypore*

The monomitic, annual, resupinate, either white to grey-brown (form *monticola*) or later pink to salmon-violet (reddish form *placenta*; Figure 6.5e) (Domański 1972), easily passing, to 1 cm thick fruit bodies with 2-4 circular-angular-slitted pores (to 15 mm long) per mm hymenium produce hyaline spores (4-6 × 2-2.5 µm). The strands observed on wood samples in laboratory culture (Figure 6.5f) were white, partly yellowing, easily refractable and to 1 mm in diameter. Fibers and vessels were rare or absent.

## *Gill polypores:* Gloeophyllum *species*

Three *Gloeophyllum* species (Gloeophyllaceae, Aphyllophorales) are relevant to indoor wood. The fungi have similar fruit bodies and life conditions (Hof 1981a,b,c, Grosser 1985), and are thus united as "wood gill polypores". They are widespread in Europe, North America, North Africa, and Asia on conifers and hardwoods. *Gloeophyllum abietinum* (Bavendamm 1952b) is a somewhat southern species, and *G. trabeum* a southern species. The three species are bipolar. Monstrous (tap-, pin-, antlers- or cloud-like) fruit bodies grow in the dark and in wood cavities and shakes.

The tree fungi are predominantly saprobic, *G. sepiarium* and *G. trabeum* occur exceptionally on living trees. The species belong to the strongest brown-rot fungi of stored and finished coniferous (*G. trabeum* also on hardwoods) structural timber that is again moistened, like poles, posts, fences, sleepers and mining timber. The gill polypores occur in buildings after moisture damages or incorrect structure on roofing timbers, on façades, outside doors, balconies, and on timber in saunas and mines. The fungi also belong to the most important destroyers of conifer windows (e.g. Alfredsen *et al.* 2005) that had accumulated moisture (moisture optimum 40-210%, Table 6.3) due to inappropriate window construction and handling faults by the user. For example, 3.5 million (7%) of wooden windows were destroyed by fungi, predominantly by *G. abietinum*, in Germany between 1955 and 1965. Fungi survived in the sun-warmed and dry window timber due to their heat and dryness resistance

[*G. abietinum*: 5-7 years survival in dry timber: Theden (1972)]. Fungi cause decay first only in the wood interior. The serious brown rot under the varnish layer is often only recognized if fruit bodies develop.

### Gloeophyllum abietinum, *fir gill polypore*

The trimitic, perennial, pileate (2-8 cm wide) and broadly attached fruit bodies (Figure 6.6a) grow often in rows, tile-like, and resupinate on timber lower side. The upper surface is hirsute to velutinate, in age zonate, scrupose to warted or smooth, when young whitish-yellow-brown, then rusty yellow, reddish-brown to dark grey and black when old with a wavy and sharp margin and black discoloration in KOH. The hymenophore with wavy lamellae (8-13 per cm, behind margin) and anastomosing mixed with poroid areas produces hyaline, smooth, thin-walled, cylindrical, iodine-negative spores (3.5-4.5 × 9-11 µm) with apiculus. The surface mycelium (Figure 6.6b) is cream-colored, ochre to dark-brown and underneath crème-white. First bright reddish, then red-brown to grey strands of few centimeters of length with dark-brown fibers (1.5-3 µm thick) and clamped vegetative hyphae (2-4.5 µm thick), but without vessels, have been rarely found in laboratory culture.

### Gloeophyllum sepiarium, *yellow-red gill polypore, conifer mazegill*

The trimitic, annual to perennial, pileate, broadly sessile, dimidiate, rosette-shaped, often imbricate in clusters from a common base or fused laterally, to 7 cm wide, 12 cm long and 6-8 mm thick fruit bodies (Figure 6.6c) have a slightly wavy margin. The upper surface when young is yellowish-brown, then reddish brown and grey to black when old, scrupose, warted to hispid, finally zonate and often differently colored. The hymenophore has straight lamellae (15-20 per cm, behind margin) with the edges of lamellae golden brown in active growth and later umber brown and the side surface of lamellae ochre-brown, and is usually mixed with daedaleoid to sinuous pore areas (1-2 per mm). The surface mycelium (Figure 6.6d) is white, cream-colored to light-brown and rarely contains cylindrical arthrospores (3-4 × 10-15 µm). First bright, then yellowish to ochre-brown strands of few centimeters of length, usually covered by mycelium and containing yellow to brown fibers (2-4.5 µm thick), hyaline, clamped vegetative hyphae (2-4 µm thick), but no vessels, have been rarely found in laboratory culture.

### Gloeophyllum trabeum, *timber gill polypore*

The dimitic, annual to perennial, pileate, sessile, imbricate fruit bodies (Figure 6.6e) with several basidiomes from a common base or elongated and fused along wood

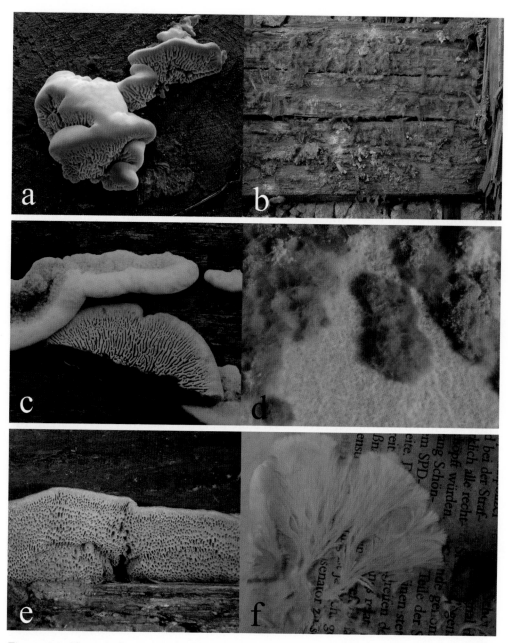

*Figure 6.6.* Gloeophyllum abietinum, G. sepiarium *and* G. trabeum. *(a)* G. abietinum *fruit body, (b)* G. abietinum *mycelium, (c)* G. sepiarium *fruit body, (d)* G. sepiarium *upper brown mycelium on lower white mycelium, (e)* G. trabeum *fruit body, (f)* G. trabeum *mycelium.*

cracks are to 3 cm wide, 8 cm long and 8 mm thick. The upper surface is soft and smooth, hazelnut to umber-brown to grayish when old, weakly zonate to almost azonate and has a lighter, ochre to umber-brown margin. The hymenophore is semi-lamellate or labyrinthine to partly poroid (2-4 per mm). There are rarely lamellate specimens with up to four lamellae per mm along the margin. The surface mycelium (Figure 6.6f) is white, beige, yellow-orange to grey-brown. White-beige to yellow-orange-grey-brown, below 1 mm thick, not clearly defined strands, usually covered by mycelium, with yellow to brown, septate fibers (1-4 µm thick), and hyaline, clamped vegetative hyphae (2-4 µm thick), but without vessels, have been observed in laboratory culture.

## Donkioporia expansa, *oak polypore*

*Donkioporia expansa* (Coriolaceae, Aphyllophorales) is only recognized since the 1910s as relevant for practice and since about 1985 as important decay fungus in buildings (Mez 1908, Kleist and Seehann 1999) an mines (Harz 1888). Assumable, the species was often overlooked despite the less common decay type of a white rot in buildings and the large size of its fruit bodies (Figure 6.7a). Further reasons it was overlooked may be that damage is often restricted to wood interior and not noticed until fruit bodies appear and that the fruit bodies are inconspicuously embedded in plentiful surface mycelium.

The fungus is reported for Central Europe and North America, for Germany preferentially in the south. At least in Europe it is almost exclusively restricted to structural timber, such as from *Quercus, Castanea, Fraxinus, Populus* and *Prunus*. In spite of the name Oak polypore, *D. expansa* is common in Germany on indoor timber of *Picea* and *Pinus*.

The Oak polypore inhabits damp areas in kitchens, bathrooms, WC, cellars, occurs on beams, under floors, cow-sheds, in mines, on bridge timber and cooling tower wood [Azobé, Bangkirai: Von Acker *et al.* (1995), Von Acker and Stevens (1996)]. Continuous high wood moisture (e.g. defective sanitary facilities) promotes growth. The fungus is often found at beam-ends that are enclosed in damp walls. At initial attack, the timber surface remains often nearly intact. In laboratory culture, minimum wood moisture for wood colonization was 21% and for wood decay 26%. Greatest wood mass losses occurred between 34 and 126%. Moisture maximum was 256% (Table 6.3). Temperature optimum was 28 °C, and maximum was 34 °C. The fungus survived for 4 hours in dry wood of 95 °C (Table 6.4). There is no spread by strands and thus no growth through brickwork (Table 6.2). Thus, it may be assumed that refurbishment only needs drying and exchange of destroyed timber.

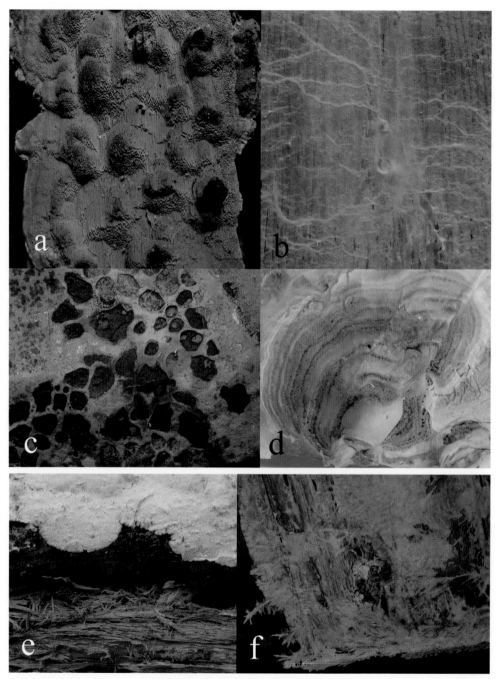

*Figure 6.7.* Donkioporia expansa *and* Diplomitoporus lindbladii. *(a)* D. expansa *fruit body, (b)* D. expansa *mycelium in laboratory culture, (c)* D. expansa *dry guttation drops, (d)* D. expansa *colored mycelium, (e)* D.s lindbladii *fruit body, (f)* D. lindbladii *mycelium.*

The trimitic, perennial, resupinate, first white, then ochre to reddish-tobacco-brown (to grey when old), firmly attached fruit bodies (Figure 6.7a) are to 10 cm thick and become widely effused to a few square meters. They are wavy to stairs-like on walls, often multi-layered (to 9 layers), tough-elastic with silvery surface when fresh, hard and brittle when dry, easily separable when old, mainly made up of long tubes (4-5 circular to angular pores per mm), often with amber guttation drops, which leave behind small black pits when dry, and produce ellipsoid spores (4.5-7 × 3.2-3.7 μm).

The mycelium (Figures 6.7b and 6.7d) is first white to cream, then light-yellowish, grey to brown, when old often luxuriant, firm and tough, frequently with paper-like brown crusts, grows frequently in wood shakes and cavities, usually with amber guttation drops or with brown to black spots (remainders of dried guttation; Figure 6.7c). The white mycelium contains hyaline to brown, not very thick-walled fibers, which are hardly separable from clamped vegetative hyphae (0.7-1.5 μm in diameter). The colored mycelium contains light-brown to brown fibers (1.5-3 μm thick), hyaline to brown, thick-walled, sometimes rarely clamped and branched vegetative hyphae (2-6 μm thick) and vessels (to 11 μm thick). At high air humidity, mycelium also grows on free wood surfaces with thin, skin-like mycelial flaps with bizarre seeds, later as thick, brownish surface mycelium and partly with distinct margin and poroid fruit bodies. Lemon-shaped, hyaline, thick-walled arthrospores (5-9 × 7-12 μm) are found in surface and substrate mycelium. Black demarcation lines occur between mycelium and wood.

Strands have not yet been observed in buildings (Table 6.2), but short, root-like, cream, yellowish to grey-brown strand-like structures have been found hidden under mycelium on wood samples in laboratory culture.

## Diplomitoporus lindbladii

*Diplomitoporus lindbladii* (Coriolaceae, Aphyllophorales) grows outdoors circumpolar in the conifers zone, in Europe rarely in the Mediterranean region, on dead trees (*Abies*) and also on hardwoods. Indoor decay as a white-rot occurred several times in the subfloor, roof area and on half-timbering (Table 6.1).

The trimitic, annual to biannual, resupinate, easily separable fruit bodies (to 6 mm thick, biannual basidiomes thicker; Figure 6.7e) become widely effused (a few decimeters) and have a frayed margin. The white-cream, grey when old, upper surface with 2-4 circular-angular pores (to 3 mm deep) per mm produces allantoid to cylindrical, hyaline spores (5-7 × 1.5-2 μm). White, yellowing when dry, root-like, iceflower-like strands, similar to those of *Antrodia vaillantii*, containing fibers similar to *A. vaillantii*, but soluble in 5% KOH, have been found in laboratory culture. The fungus is bipolar.

## *Asterostroma* species

### Asterostroma cervicolor

*Asterostroma cervicolor* (Hymenochaetaceae, Aphyllophorales) has been often found in buildings (Table 6.1), producing a white-rot of softwoods, often limited in extent, in cellar, subfloor, roof and walls, on windows, on skirting, ceiling, fiber and gypsum boards, and in half-timbering.

The monomitic, resupinate, sheet-like, thin, whitish to ochre or cinnamon fruit bodies (Figure 6.8a) without pores are hardly distinguishable from mycelium, produce subglobose, tuberculate, warty spores and may be found also on brickwork.

The crème-brown, up to 1 mm wide, root and ice-flower-like (similar to *A. vaillantii*) strands with a rough appearance contain clampless, vegetative hyphae (1.5-3 μm thick) and remain flexible when dry. Figure 6.8b shows strands of this white-rot fungus growing on brown-rotted wood. Strands are often present next to a fruit body, embedded in white mycelium, and are sometimes also found across and inside brickwork over a long distance. The surface mycelium (Figure 6.8c) is first white, then brown and grows sometimes only as small mycelial plugs. The rarely branched and to 190 μm sized stellar setae occur within fruit bodies, mycelium and strands.

### Asterostroma laxum

*Asterostroma laxum* (Figure 6.8d) is similar to *A. cervicolor* and known from Europe and Asia on dead softwoods and was found on staircases (Table 6.1). The dichotomously branched stellar setae (Figure 6.8e) are of up to 90 μm diameter. The subglobose spores are without warts. The cream-colored to red-brown strands with fibrous surface are partly embedded in white mycelium and fruit bodies, and occur also on and in brickwork. The clampless vegetative hyphae are 2-4 μm thick.

## *Tapinella panuoides (Paxillus panuoides)*, stalkless paxillus, oyster rollrim

*Tapinella panuoides* (Paxillaceae, Boletales) (Bavendamm 1953) grows mostly on conifers producing a brown-rot, on the basis of living pines, on stumps, stored wood, structural timber outdoors (sleepers, bridges, balconies), garden furniture, mine timber, and in buildings, sometimes associated with *Coniophora* species, on moist

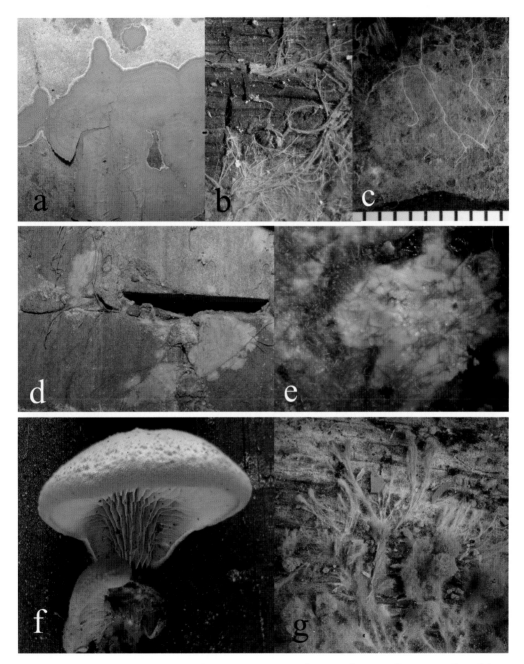

*Figure 6.8.* Asterostroma cervicolor, Asterostroma laxa *and* Tapinella panuoides. *(a)* A. cervicolor *fruit body, (b)* A. cervicolor *strands overgrowing brown-rotten wood, (c)* A. cervicolor *mycelium, (d)* A. laxa *fruit body, (e)* A. laxa *stellar setae, (f)* T. panuoides *fruit body, (g)* T. panuoides *mycelium.*

places in cellars, below bath-rooms, on windows and half-timbering, in cow-sheds and greenhouses.

The monomitic, annual, thin, small (2-12 cm), shell- or bell-shaped fruit bodies (Figure 6.8f) with small eccentric stipe or attached are solitary or in groups and also tile-like. The upper surface is pale-yellow to olive brown, the lower surface has saffron-orange gills. Normal fructification occurs in the dark (Kreisel 1961).

The surface mycelium (Figure 6.8g) is first dirty-white to yellowish, then loam-yellow, brownish to ochre, and near fruit body partly violet. Strands are first crème to loam-yellow, then brownish to ochre, up to 3 mm wide, root-like branched (similar to those of *Coniophora puteana*), however not becoming black, and grow on timber and brickwork. They contain refractive, partly thickened ("blown-up") vegetative hyphae (2.5-6 μm in diameter), indistinct fibers (1.5-5 μm in diameter), often only in darker strands, and hyaline septate, clamped vessels without bars, mostly with "blown-up" hyphal segments, up to 25 μm in diameter.

## *Trechispora* species

*Trechispora* species (Hydnodontaceae, Aphyllophorales) occur in the forest on dead soft- and hardwoods remainders and on stumps in Europe, North America and Asia. *Trechispora farinacea* and *T. mollusca* have been found in buildings, producing a white-rot on damp timber in cellars, windows, behind roof insulation, below bathrooms (defect sanitary facilities), and in half-timbering (Table 6.1). The two fungi are not distinguishable on the basis of mycelia and strands. For identification see Jülich (1984) and Hansen *et al.* (1997).

The resupinate, thin, poroid, grandinioid or smooth, fragile, crème-white fruit bodies (Figure 6.9a) produce warty, translucent and small spores (4-5.5 × 3-4.5 μm). The snow-white to cream-colored, fragile, often only short strands (0.2-1 mm in diameter; Figure 6.9b) are often found near a fruit body. The contain clamped vegetative hyphae, partly with bubble-like swellings (1-4 μm), no fibers, sometimes vessels (4-9 μm in diameter) with small clamps.

## *Dacrymyces stillatus*, orange jelly

*Dacrymyces stillatus* (Dacrymycetaceae, Dacrymycetales, Heterobasidiomycetes) produces a brown-rot in soft- and hardwoods (Reid 1974). Decay is commonly patchy with small pockets of rot and often restricted to interior of timber. The species is common outdoors on windows (appearing through the paint), facades and along

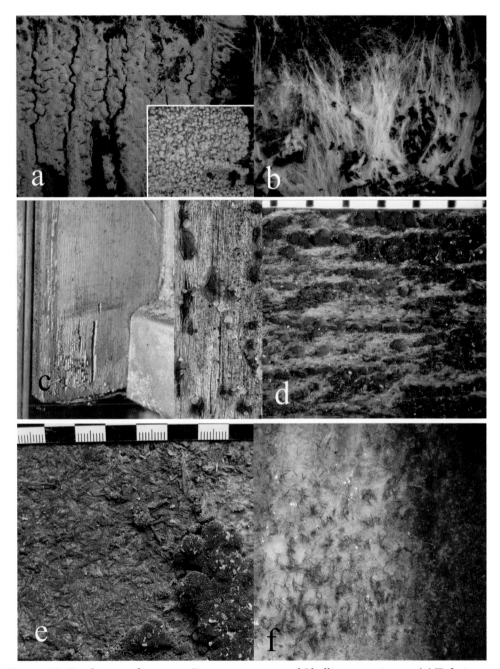

*Figure 6.9.* Trechispora farinacea, Dacrymyces *sp. and* Phellinus contiguus. *(a)* T. farinacea *fruit body, (b)* T. farinacea *strands, (c)* Dacrymyces *fruit bodies on window timber, (d)* Dacrymyces *older fruit bodies, (e)* P. contiguus *fruit body, (f)* P. contiguus *mycelium and setae.*

the gable board of the roof (Alfredsen *et al.* 2005), on half-timbering, indoors on windows and doorframes (Table 6.1).

The fruit bodies (Figures 6.9c and 6.9d) are whitish, yellow-orange-red, dark orange when dry, button-shaped, lenticular to mug- and plate-like, 1-15 mm wide, gelatinous-elastic, slimy melting when old, solitary and in groups, with often two different forms on the same place, a brighter form with basidiospores and a darker form with arthrospores.

## Phellinus contiguus

*Phellinus contiguus* (Hymenochaetaceae, Aphyllophorales) occurs circumglobal in warmer and tropical zones on dead hard- and softwoods, producing a white-rot, in buildings (windows, half-timbering, shingle roofs) usually on hardwoods like oak, ash, false acacia, elm and beech, rarely on softwoods (fir and spruce).

The light- to red-brown, corky-tough fruit bodies (few centimeters to several decimeters; Figure 6.9e) are resupinate on substrate undersides with stairs-like pore formation on vertical substrates, when growing with yellowish, matted margin (to 1.2 cm thick). Round-angular to slitted pores (2-3 per mm, to 8 mm deep) produce elliptical, smooth, hyaline spores (3-3.5×6-7 µm). Simple, dark-brown, to 180 µm long setae (Figure 6.9f) are found within mycelium, strand and fruit body. Mycelium is downy, loam-yellow to brown, also white when young. Strand-like structures are up to 4 mm wide and 0.5 mm thick, firmly attached, often finger-shaped branched and contain pale yellow, thin-walled, rarely branched fibers (2-3 µm in diameter), and hyaline, clampless vegetative hyphae (2 µm in diameter).

## Identification and characterization

### Traditional methods

For identification of indoor wood-decay fungi, fruit bodies are preferentially used (Grosser 1985, Breitenbach and Kränzlin 1986, Jahn 1990, Ryvarden and Gilbertson 1993/1994, Bravery *et al.* 2003, Schmidt 2006, Huckfeldt and Schmidt 2006a). Some species, however, can only rarely be found in buildings as fruit bodies. Fruit body formation after isolation in laboratory culture is rare, e.g. in *Antrodia vaillantii*. However, several indoor species form mycelial strands (cords). The classical strand diagnosis by Falck (1912) includes only a few species. A diagnostic key with color photographs based on strand samples from infected buildings and on samples from laboratory culture comprises 20 strand-forming species (Huckfeldt and Schmidt 2006b).

Classical methods might be rapid if there are fruit bodies or strands. If neither fruit bodies nor strands are found, but only vegetative mycelia in buildings or in culture, keys and books for identification are available (Nobles 1965, Stalpers 1978, Lombard and Chamuris 1990). However, some genera among house-rot fungi can hardly or not at all be distinguished at species level in culture, as is true of *Antrodia*, *Coniophora*, and *Leucogyrophana*. Thus, molecular methods (Schmidt 2000, 2009) are currently used for identification.

## Molecular methods

Molecular methods applied for the characterization, identification, and classification of organisms are independent of subjective judgment and based on objective information (molecules) revealed by the target organism. Consequently, molecular methods have been used since the 1980s for indoor wood-decay fungi (Jellison and Goodell 1988, Palfreyman *et al.* 1988, Schmidt and Kebernik 1989). The following overview describes methods and some results that are related to dentification, characterization and phylogeny.

### Protein-based techniques

#### SDS-PAGE

In SDS-PAGE (sodium dodecyl sulfate polyacrylamide gel electrophoresis), the whole cell protein is extracted from fungal tissue, denatured and negatively charged with mercaptoethanol and sodium dodecyl sulphate (SDS). The proteins are separated according to size on acrylamide gels and visualized by Coomassie blue or other stainings. The banding pattern obtained discriminates organisms at the species level and below. This technique has been used to distinguish a number of wood-decay fungi (Vigrow *et al.* 1991a, Palfreyman *et al.* 1991, Schmidt and Moreth 1995). For example, the closely related *Serpula lacrymans* and *S. himantioides* could be distinguished by their respective specific protein banding pattern (Schmidt and Kebernik 1989; Figure 6.10).

SDS-PAGE is a rapid technique when the sample originates from a pure culture. The technique is advantageous to basic research. However, due to the many bands on the gel, SDS-PAGE has no practical significance for identification.

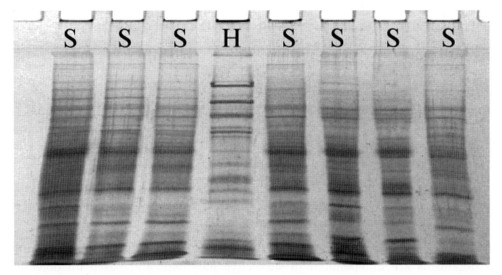

*Figure 6.10. Protein bands of isolates of* Serpula lacrymans *(S) after SDS-PAGE. Culture H was identified by ITS sequencing to be* S. himantioides *(Schmidt and Kebernik 1989).*

## Immunological methods

Immunological methods use polyclonal antisera or monoclonal antibodies. Antisera produced by animals like mice and rabbits as answer to the injection of mycelial fragments, extracts ore culture filtrates are investigated by Western blotting, enzyme-linked immunosorbent assay (ELISA) or immunofluorescence (Clausen 2003). Assays have been performed on several house-rot fungi, including *Coniophora puteana, Gloeophyllum trabeum, Lentinus lepideus, Oligoporus placenta* and *S. lacrymans* (e.g. Jellison and Goodell 1988, Palfreyman *et al.* 1988, Glancy *et al.* 1990, Clausen *et al.* 1991, Kim *et al.* 1991a,b, Vigrow *et al.* 1991b, Toft 1992, 1993, Clausen 1997). The diagnostic potential lies in the identification of species without the need of a prior isolation and pure culturing and in the detection of fungi at early stages of decay (Clausen and Kartal 2003). However, the experiments may exhibit cross-reactions with non-target organisms.

### DNA-based techniques

Most molecular investigations on indoor wood-decay fungi use the polymerase chain reaction (PCR) to amplify the DNA of the sample, either in the traditional form or as multiplex PCR and real-time PCR (e.g. Vainio and Hantula 2000, Hietala *et al.* 2003,

Eikenes *et al.* 2005). Subsequent gel electrophoresis is applied to visualize the PCR products and to assess their quality.

## RAPD analysis

The technique of RAPD (randomly amplified polymorphic DNA) analysis uses only one, short (often 10mer) and randomly chosen PCR primer which anneals several times to the genome. Thus, RAPD analysis discriminates fungi by a polymorphic banding pattern at low taxonomical level, namely isolates, intersterility groups and species. Theodore *et al.* (1995) showed polymorphism among eight isolates of *S. lacrymans*. Similarity between other *S. lacrymans* isolates (Figure 6.11) was found by Schmidt and Moreth (1998). *Serpula himantioides* (Figure 6.11) and *C. puteana* showed isolate polymorphism. Special isolates of *C. puteana* used for wood preservative tests were identified by Göller and Rudolph (2003).

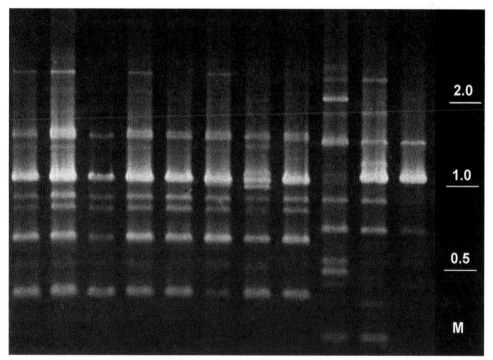

*Figure 6.11. RAPD-pattern of isolates of* Serpula lacrymans *(8 left-most lanes) and* S. himantioides *(3 right-most lanes) obtained with primer GGACTCCACG. M=Marker (kb) (Schmidt and Moreth 1998).*

RAPD analysis does not require prior information of the target DNA and is a rapid technique when starting from pure cultures. However, short primers imply an increased sensitivity to contamination. The technique is unsuited for species identification from unknown samples by comparison, as other, as not yet investigated fungi may by chance share a similar banding pattern.

## Use of rDNA

The investigation of rDNA (ribosomal DNA) has become popular approach (Figure 6.12). The conserved rDNA-regions 18S and 28S are preferentially selected for phylogenetic analyses of genera, families, and higher taxonomic groups. The rapidly evolving internal transcribed spacers (ITS 1 and 2) are suitable for closely related species. The intergenic spacers (IGS 1 and 2) show the highest intraspecific diversity among all rDNA regions due to length polymorphism. Depending on the intention, the RNA genes or the spacers are used for analyses. These domains are either restricted by endonucleases for RFLP analyses or they are sequenced.

## RFLP analyses of rDNA

RFLP analysis after restriction of the ITS region with restriction endonucleases differentiated single isolates of *C. puteana*, *G. trabeum* and *O. placenta* (Zaremski *et al.* 1998). Various isolates of the closely related *S. lacrymans* and *S. himantioides* exhibited distinct fragment profiles after double digestion with the endonucleases *Hae*III/*Taq*I (Schmidt and Moreth 1999). *Taq*I differentiated *S. lacrymans*, *S. himantioides*, *Donkioporia expansa*, *C. puteana*, *Antrodia vaillantii*, *O. placenta*, and *Gloeophyllum sepiarium* (Figure 6.13). ITS-RFLP analysis is currently a favored database for the identification of wood decay and associated fungi (Zaremski *et al.* 1998, Adair *et al.* 2002, Diehl *et al.* 2004, Råberg *et al.* 2004).

The rapid and inexpensive technique does not require prior information on the target DNA. However, the limited ITS size of only 600-700 bases prevents a separation of all

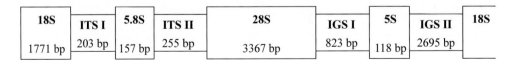

| 18S | ITS I | 5.8S | ITS II | 28S | IGS I | 5S | IGS II | 18S |
|---|---|---|---|---|---|---|---|---|
| 1771 bp | 203 bp | 157 bp | 255 bp | 3367 bp | 823 bp | 118 bp | 2695 bp | |

*Figure 6.12. Schematic diagram of one rDNA unit. The number in the boxes is the size in base pairs for* Antrodia vaillantii*.*

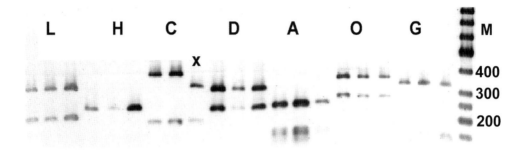

*Figure 6.13. ITS-RFLP-pattern of* Serpula lacrymans *(L),* S. himantioides *(H),* Coniophora puteana *(C),* Donkioporia expansa *(D),* Antrodia vaillantii *(A),* Oligoporus placenta *(O) and* Gloeophyllum sepiarium *(G) after restriction with restriction endonuclease* TaqI. *Culture (x) subsequently identified by sequencing to be* Coniophora olivacea. *M=Marker (bp) (Schmidt 2006).*

approximately 100 basidiomycetes that have been found in buildings (Table 6.1). An as yet unanalyzed species may feign another fungus by exhibiting similar fragments. Isolate variation in the ITS, removing a recognition site of a restriction enzyme or providing a new restriction site, results in an unspecific fragment pattern.

In T-RFLP (terminal restriction fragment length polymorphism) analysis, one or both primers are connected with a fluorescent dye. When the end fragments with the dye undergo sequence analysis, only these fragments are determined. Råberg *et al.* (2005) applied the technique to *C. puteana* and *O. placenta*.

**SSPP**

Based on the sequence divergence among species, oligonucleotide sequences may be the basis for species-specific PCR primers, which in turn may be used for SSPP (species-specific priming PCR). Specific oligonucleotide sequences located in the ITS II region of seven indoor basidiomycetes were designed for subsequent identification of unknown samples (Moreth and Schmidt 2000, Schmidt and Moreth 2000). Using the ITS 1 primer of White *et al.* (1990) as a forward primer the amplified ITS fragments exhibit a distinct and predictable size for the fungi (Figure 6.14).

SSPP is a precise and rapid technique. At first sight, the technique seems to be a powerful molecular identification tool for fungi. Subsequent restriction of the PCR amplicon as well as the use of fungal pure cultures, axenically obtained samples, and precautions to exclude DNA from the laboratory or from contaminated field

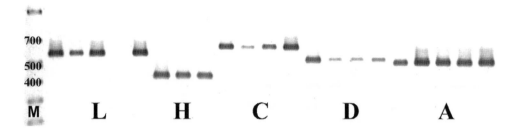

*Figure 6.14. Positive PCR reaction within the ITS region by specific reverse primers.* L=Serpula lacrymans: *ATGTTTCTTGCGACAACGAC*, H=S. himantioides: *TCCCACAACCGAAAC-AAATC*, C=Coniophora puteana: *AGTAGCAAGTAAGGCATAGA*, D=Donkioporia expansa: *TCGCCAAAACGCTTCACGGT*, A=Antrodia vaillantii: *CACCGATAAGCCGACTCATT. Three to five mycelial samples of these species are shown. M=Marker (bp) (Moreth and Schmidt 2000).*

material are not required. The technique is utilized in Germany for commercial fungal diagnosis. Limitations are: The ITS size of only 600-700 nucleotides prevents the design of specific primers for all approximately 100 indoor basidiomycetes. An as yet unanalyzed species may react with the primer of another fungus which would indicate a wrong name. The ITS sequences of closely related species from the genera *Antrodia* (Schmidt and Moreth 2003), *Coniophora* (Schmidt *et al.* 2002/2003) and *Gloeophyllum* (Schmidt *et al.* 2002) are too similar for designing species-specific primers.

### Sequencing of rDNA

Sequencing of ribosomal DNA domains avoids the main limitations of RFLP and SSPP analyses because the complete information of the sequence of the target DNA is obtained. Usually, sequences are deposited in one of the international databases, which, however, exchange all data: EMBL-EBI (European Molecular Biology Laboratory, European Bioinformatics Institute, www.ebi.ac.uk), NCBI-GenBank (National Center for Biotechnology Information, USA, www.ncbi.nlm.nih.gov), DDBJ (DNA data bank of Japan, www.ddbj.nig.ac.jp).

### Use of rDNA sequences for identification

The ITS sequences of a great number of wood-decay fungi are known. The sequence of isolate S7 was in 1999 the first ITS deposition from *S. lacrymans* in the databases (AJ 245948, Schmidt and Moreth 2000). Table 6.5 shows the accession numbers of rDNA-

Table 6.5. Sequenced and deposited rDNA regions of indoor wood-decay basidiomycetes (after Schmidt 2006 and Schmidt and Moreth 2008).

| | 18S | ITS I | 5.8S | ITS II | 28S | IGS I | 5S | IGS II |
|---|---|---|---|---|---|---|---|---|
| Serpula lacrymans | AJ440945 AJ440946 | AJ245948 AJ249268 AJ419907 AJ419908 AJ419909 AJ419910 | | | AJ440939 AJ440940 AJ440941 | AM946622 | | |
| | AM946629 | | | | | | | |
| Serpula himantioides | AJ440947 AJ440948 | AJ245949 AJ419911 | | | AJ440942 AJ440943 AJ440944 | AM649623 | | |
| | AM946630 | | | | | | | |
| Meruliporia incrassata | | AJ419912 AJ419913 | | | | AM946624 | | |
| Leucogyrophana mollusca | | AJ419914 AJ419915 | | | | | | |
| Leucogyrophana pinastri | | AJ419916 AJ419917 | | | | AM946625 | | |
| Coniophora puteana | AJ488581 | AJ249502 AJ249503 AJ344109 AJ344110 | | | AJ583426 | AM946627 | | |
| | AM946631 | | | | | | | |
| Coniophora marmorata | AJ540306 | AJ518879 AJ518880 | | | AJ583427 | AM946628 | | |
| | AM946632 | | | | | | | |
| Coniophora arida | AJ488582 | AJ345007 AJ344113 | | | | | | |
| Coniophora olivacea | AJ488905 | AJ344112 AJ345009 | | | | | | |
| Antrodia vaillantii | AJ488583 | AJ249266 AJ344140 AJ421007 AJ421008 | | | AJ583429 | | | |
| | AM286436 | | | | | | | |
| Antrodia sinuosa | AJ488906 | AJ345011 AJ416068 | | | | | | |
| Antrodia serialis | | AJ344139 AJ345010 | | | | | | |
| Antrodia xantha | AJ488584 | AJ345012 AJ415569 | | | AJ583430 | | | |

*Table 6.5. Continued.*

| | 18S | ITS I | 5.8S | ITS II | 28S | IGS I | 5S | IGS II |
|---|---|---|---|---|---|---|---|---|
| *Oligoporus placenta* | | AJ249267 AJ416069 | | | | | | |
| *Gloeophyllum abietinum* | AJ560802 | AJ420947 AJ420948 | | | AJ583431 | | | |
| *Gloeophyllum sepiarium* | AJ540308 | AJ344141 AJ420946 | | | AJ583432 | | | |
| *Gloeophyllum trabeum* | | AJ420949 AJ420950 | | | | | | |
| *Donkioporia expansa* | AJ540307 | AJ249500 AJ249501 | | | AJ583428 | | | |

Six-digit number: EMBL accession number.

ITS sequences of 18 indoor wood-decay species (Schmidt and Moreth 2002/2003). The list comprises many of the common house-rot basidiomycetes (cf. Table 6.1). The sizes, including the ITS 1 and ITS 4 primers of White *et al.* (1990), ranged from 625 bp in *Gloeophyllum sepiarium* to 734 bp in *Leucogyrophana pinastri*. Intraspecific variation is neglectable. There was only one variable nucleotide position and a gap among six isolates of *S. lacrymans*. A maximum of five variations were found among 17 isolates of *C. puteana*. Thus, ITS sequences can be used to identify unknown fungal samples through sequence comparison using the Basic Local Alignment Search Tool (BLAST) (e.g. www.ncbi.nlm.nih.gov/blast). BLAST has revealed the ITS-sequence identity of a "wild" *S. lacrymans* isolate from the Himalayas by comparing it with indoor isolates (White *et al.* 2001, also Palfreyman *et al.* 2003), detected misnamed isolates of *S. lacrymans* (Horisawa *et al.* 2004), identified *Antrodia* spp. and *Serpula* spp. isolations (Högberg and Land 2004), confirmed *C. puteana* isolates (Råberg *et al.* 2004), and re-identified fungi from inoculated wood samples (Zaremski *et al.* 2005).

Limitations of ITS-sequencing are that identification of unknown samples is time-consuming because each sample must be sequenced. Samples can be only identified by BLAST, if the correct control sequence has been deposited in the databases. Therefore, more than one deposited sequence (if possible from different laboratories) should be considered for identification by BLAST. As advantage, sequencing of the ITS is currently the best molecular tool for species identification because a sample is identified by its unique series of about 650 nucleotides.

**Use of rDNA sequences for phylogenetics**

ITS sequences and those from the 18S, 28S and IGS are also suitable for phylogenetic analyses. A phylogenetic tree based on the ITS sequences of the indoor Coniophoraceae is shown in Figure 6.15 (Moreth and Schmidt 2005). A phylogenetic ITS tree of *Serpula lacrymans* isolates showed that isolates collected in nature in Czech Republic, India, Pakistan and Russia group in the branch of indoor isolates but differ from wild Californian isolates (Kauserud *et al.* 2004a).

Partial 28S rDNA sequences were used by Bresinsky *et al.* (1999) and Jarosch and Besl (2001) for *S. lacrymans, S. himantioides, Meruliporia incrassata* and for *Coniophora* and *Leucogyrophana* species. Complete 18S and 28S rDNA sequences are known for some indoor wood-decay fungi (Moreth and Schmidt 2005, Table 6.5). The size of the 18S rDNA for the different fungi amounts to 1,800 bp, and the 28S rDNA is around 3,300 bp. Albeit limited to maximum three investigated isolates per species, intraspecific variation was low.

The complete sequence of the intergenic spacer (IGS 1 and IGS 2) was available for few basidiomycetes, namely the pathogen *Filobasidiella neoformans* (Fan *et al.* 1995) and the ectomycorrhizal fungus *Laccaria bicolor* (Martin *et al.* 1999). Complete IGS sequences of some common house-rot basidiomycetes are meanwhile known (Table 6.5). In *Antrodia vaillantii*, the IGS comprises 3,636 bp and consists of the shorter IGS 1 (823 bp) and the longer IGS 2 (2,695 bp) sequences. The intergenic spacers

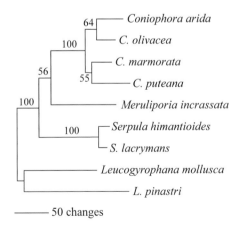

*Figure 6.15. Most parsimonious phylogenetic tree of indoor wood-decay Coniophoraceae based on rDNA-ITS sequences. Bootstrap support values over 50% are shown. (Moreth and Schmidt 2005).*

of *A. vaillantii* are separated by the short 5S rDNA (118 bp), whose transcription occurs in the same direction as the 18S-5.8S-28S genes, which is the case in most basidiomycetes. In *Serpula lacrymans, S. himantioides, Meruliporia incrassata, Leucogyrophana pinastri, Coniophora puteana* and *C. marmorata*, which all belong to the Coniophoraceae, the 5S rDNA is transcribed in the reverse direction (Schmidt and Moreth 2008).

The complete sequence of one rDNA unit was available for the eukaryotic parasite *Encephalitozoon cuniculi* (Peyretaillade *et al.* 1998). If ongoing whole genome sequencing analyses are not taken into account, the whole rDNA sequence of *A. vaillantii* was the first basidiomycetous deposition (AM286436; cf. Figure 6.12). The complete rDNA sequence of the above Coniophoraceae is meanwhile deposited (cf. Table 6.5).

## AFLP analysis

AFLP (amplified fragment length polymorphism) analysis is based on (1) total genomic restriction, (2) ligation of primer adapters, and (3) unselective followed by selective PCR of anonymous DNA fragments from the entire genome.

AFLP analysis of 19 European isolates of *S. lacrymans* indicated that the fungus in Europe is genetically extremely homogeneous by observing that only five out of 308 scored AFLP fragments were polymorphic (Kauserud *et al.* 2004b). A worldwide sample of 91 isolates was divided into two distantly related main groups: isolates from natural environments in California (*S. lacrymans* var. *shastensis*) and isolates from buildings in Japan, Europe, northeast America and Oceania (*S. lacrymans* var. *lacrymans*) (Kauserud *et al.* 2007a).

AFLP markers are more reproducible compared to RAPD and microsatellites (see paragraph on "SSR analysis") analysis and give a higher resolution.

## SSR analysis

Microsatellites or simple sequence repeats (SSR) are variable genomic regions of tandem repeats of up to seven nucleotides.

Microsatellite markers were developed for *S. lacrymans* by Högberg. *et al.* (2006). Kauserud *et al.* (2007a) showed with these markers that a worldwide collection of 84 isolates of *S. lacrymans* var. *lacrymans* was divided into three main groups: mainland

Asia as presumable origin of the variety; mainly Japan; and a cosmopolitan group – predominantly Europe, North America and Oceania.

The variability of the number of repeats at a particular locus and the conservation of the flanking sequences make microsatellites valuable genetic markers, providing information for the identification and on genetic diversity and relationship among genotypes.

### DNA-arrays (DNA-chips, microarrays)

DNA-arrays are chips carrying up to 10,000 different DNA probes, e.g. oligonucleotides, which are raster-like bound on their surface. DNA molecules of unknown samples hybridize specifically with the corresponding DNA probe, and the hybridized chip area is detected colorimetrically. A chip, which can identify 25 wood-decay basidiomycetes (mainly indoor species), is available (cf. Müller *et al.* 2009).

### Additional molecular methods

In the technique of MALDI-TOF MS (matrix-assisted laser desorption/ionization time-of-flight mass spectrometry), biomolecules or whole cells are embedded in a crystal of matrix molecules, which absorb the energy of a laser. The sample is ionized and transferred to the gas phase. The ions are accelerated in an electric field, and their time of flight is determined in a detector.

Figure 6.16 shows the first MALDI-TOF MS fingerprints of basidiomycetes, namely, those of the closely related sister taxa *S. lacrymans*, *S. himantioides* and *C. puteana*, *C. marmorata* (Schmidt and Kallow 2005). The obtained spectra may be used for subsequent diagnosis of unknown fungal samples by comparison. The technique is rapid, but it needs expensive equipment.

MVOCs (microbial volatile organic compounds) such as pinenes, acrolein, and ketones were found in *S. lacrymans*, *C. puteana* and *O. placenta* (Korpi *et al.* 1999, also Bjurman 1992). Blei *et al.* (2005) distinguished pure cultures of *A. sinuosa*, *C. puteana*, *Donkioporia expansa*, *G. sepiarium*, *S. lacrymans* and *S. himantioides*. Field experiments, however, were influenced by the distance of sampling from the infested and/or destroyed wood, and also by the rate of air change in buildings.

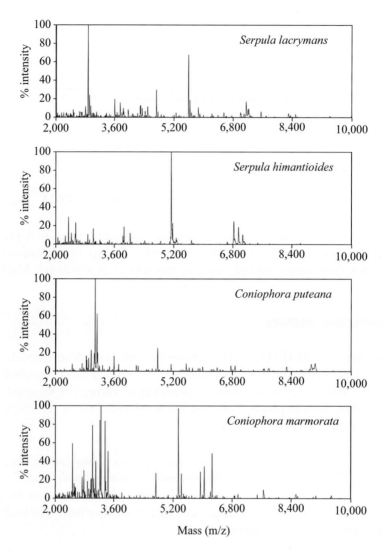

*Figure 6.16. MALDI-TOF mass spectra of mycelia of each two closely related* Serpula *and* Coniophora *species (Schmidt and Kallow 2005).*

## Current work on Serpula lacrymans

The USA Department of Energy Joint Genome Institute has sequenced the whole genome of *S. lacrymans* (Joint Genome Institute 2010). Two monokaryons derived from breeding experiments (Schmidt and Moreth-Kebernik 1991) were used.

# References

Abou Heilah AN and Hutchinson SA (1977) Range of wood-decaying ability of different isolates of *Serpula lacrymans*. Trans Br Mycol Soc 68: 251-257.

Adair S, Kim SH and Breuil C (2002) A molecular approach for early monitoring of decay basidiomycetes in wood chips. FEMS Microbiol Lett 211: 117-122.

Alfredsen G, Solheim H and Jenssen KM (2005) Evaluation of decay fungi in Norwegian buildings. IRG/WP/10562: 1-12.

Bagchee K (1954) *Merulius lacrymans* (Wulf.) Fr. in India. Sydovia 8: 80-85.

Bavendamm W (1951a) *Merulius lacrimans* (Wulf.) Schum. ex Fries. Holz Roh-Werkstoff 9: 251-252.

Bavendamm W (1951b) *Coniophora cerebella* (Pers.) Duby. Holz Roh-Werkstoff 9: 447-448.

Bavendamm W (1952a) *Poria vaporaria* (Pers.) Fr. Holz Roh-Werkstoff 10: 39-40.

Bavendamm W (1952b) *Lenzites abietina* (Bull.) Fr. Holz Roh-Werkstoff 10: 261-262.

Bavendamm W (1953) *Paxillus panuoides* Fr. Holz Roh-Werkstoff 11: 331-332.

Bavendamm W (1969) Der Hausschwamm und andere Bauholzpilze. Fischer, Stuttgart, Germany.

Bech-Andersen J (1985) Alkaline building materials and controlled moisture conditions as causes for dry rot *Serpula lacrymans* growing only in houses. IRG/WP/1272.

Bech-Andersen J (1987a) The influence of the dry rot fungus (*Serpula lacrymans*) *in vivo* on insulation materials. Mater Org 22: 191-202.

Bech-Andersen J (1987b) Production, function and neutralization of oxalic acid produced by the dry rot fungus and other brown-rot fungi. IRG/WP/1330.

Bech-Andersen J (1995) The dry rot fungus and other fungi in houses. Hussvamp Laboratoriet Forlag, Gl. Holte, Denmark.

Bech-Andersen J and Andersen C (1992) Theoretical and practical experiments with eradication of the dry rot fungus by means of microwaves. IRG/WP/1577: 1-4.

Bech-Andersen J, Elborne SA, Goldie F, Singh J, Singh S and Walker B (1993) The true dry rot fungus (*Serpula lacrymans*) found in the wild in the forests of the Himalayas. IRG/WP/10002.

Bernicchia A (2005) Polyporaceae s. l. Fungi Europaei Vol. 10. Ed. Candusso, Alassio, Italy, 808 pp.

Bjurman J (1992) Analysis of volatile emissions as an aid in the diagnosis of dry rot. IRG/WP/2393.

Bjurman J (1994) Determination of microbial activity in moulded wood by the use of fluorescein diacetate. Mater Org 28: 1-16.

Blei M, Fiedler K, Rüden H and Schleibinger HW (2005) Differenzierung von Holz zerstörenden Pilzen mittels ihrer mikrobiellen flüchtigen organischen Verbindungen (MVOC). In: Keller R, Senkpiel K, Samson RA and Hoekstra ES (eds.) Mikrobielle allergische und toxische Verbindungen. Schriftenr Inst Medizin Mikrobiol Hygiene Univ Lübeck 9, pp. 163-178.

Bravery AF and Grant C (1985) Studies on the growth of *Serpula lacrymans* (Schumacher ex Fr.) Gray. Mater Org 20: 171-192.

Bravery AF, Berry RW, Carey JK and Cooper DE (2003) Recognising wood rot and insect damage in buildings. 2nd edn. BRE, Watford, UK.

Breitenbach J and Kränzlin F (1986) Pilze der Schweiz. Vol. 2. Nichtblätterpilze. Mykologia, Luzern, Switzerland.

Bresinsky A, Jarosch M, Fischer M, Schönberger I and Wittmann-Bresinsky B (1999) Phylogenetic relationships within *Paxillus* s.l. (Basidiomycetes, Boletales): separation of a southern hemisphere genus. Plant Biol 1: 327-333.

Burdsall HH (1991) *Meruliporia (Poria) incrassata*: occurrence and significance in the United States as a dry rot fungus. In: Jennings DH and Bravery AF (eds.) *Serpula lacrymans*. Wiley, Chichester, UK, pp 189-191.

Cartwright KStG and Findlay WPH (1958) Decay of timber and its prevention. 2$^{nd}$ edn. HMSO, London, UK.

Clausen C (1997) Immunological detection of wood decay fungi – an overview of techniques developed from 1986 to present. Int Biodeter Biodegrad 39: 133-143.

Clausen CA (2003) Detecting decay fungi with antibody-based tests and immunoassays. In: Goodell B, Nicholas DB and Schultz TP (eds.) Wood deterioration and preservation. ACS Symp Ser 845, Am Chem Soc, Washington, DC, USA, pp. 325-336.

Clausen CA and Green F, Highley TL (1991) Early detection of brown-rot decay in southern yellow pine using immunodiagnostic procedures. Wood Sci Technol 26: 1-8.

Clausen CA and Kartal SN (2003) Accelerated detection of brown-rot decay: comparison of soil block test, chemical analysis, mechanical properties, and immunodetection. Forest Prod J 53: 90-94.

Clausen CA, Green F and Highley TL (1993) Characterization of monoclonal antibodies to wood-derived $\beta$-1,4-xylanase of *Postia placenta* and their application to detection of incipient decay. Wood Sci Technol 27: 219-228.

Cockcroft R (ed.) (1981) Some wood-destroying basidiomycetes, vol. 1 of a collection of monographs. IRG/WP, Boroko, Papua New Guinea, pp. 1-186.

Coggins CR (1980) Decay of timber in buildings. Dry rot, wet rot and other fungi. Rentokil, East Grinstead, UK.

Coggins CR (1991) Growth characteristics in a building. In: Jennings DH and Bravery AF (eds.) *Serpula lacrymans*. Wiley, Chichester, UK, pp. 81-93.

Collett O (1992a) Comparative tolerance of the brown-rot fungus *Antrodia vaillantii* (DC.: Fr.) Ryv. isolates to copper. Holzforsch 46: 293-298.

Collett O (1992b) Variation in copper tolerance among isolates of the brown-rot fungi *Postia placenta* (Fr.) M. Lars. & Lomb. and *Antrodia xantha* (Fr.) Ryv. Mater Org 27: 263-271.

Cooke WB (1955) Fungi of Mount Shasta (1936-1951). Sydowia 9: 94-215.

Cymorek S and Hegarty B (1986a) A technique for fructification and basidiospore production by *Serpula lacrymans*. (Schum. ex Fr.) SF Gray in artificial culture. IRG/WP/2255.

Cymorek S and Hegarty B (1986b) Differences among growth and decay capacities of 25 old and new strains of the dry rot fungus *Serpula lacrymans* using a special test arrangement. Mater Org 21: 237-249.

Da Costa EWB (1959) Abnormal resistance of *Poria vaillantii* (D.C. ex Fr.) Cke. strains to copper-chrome-arsenate wood preservatives. Nature 183: 910-911.

Da Costa EWB and Kerruish RM (1964) Tolerance of *Poria* species to copper-based wood preservatives. Forest Prod J 14: 106-112.

Da Costa EWB and Kerruish RM (1965) The comparative wood-destroying ability and preservative tolerance of monokaryotic and dikaryotic mycelia of *Lenzites trabea* (Pers.) Fr. and *Poria vaillantii* (DC ex Fr.) Cke. Ann Bot 29: 241-252.

Diehl SV, McElroy TC and Prewitt ML (2004) Development and implementation of a DNA-RFLP database for wood decay and wood associated fungi. IRG/WP/10527: 1-8.

Doi S (1989) Evaluation of preservative-treated wooden sills using a fungus cellar with *Serpula lacrymans* (Fr.) Gray. Mater Org 24: 217-225.

Doi S (1991) *Serpula lacrymans* in Japan. In: Jennings DH and Bravery AF (eds.) *Serpula lacrymans*. Wiley, Chichester, UK, pp 173-187.

Doi S and Jamada A (1991) Antagonistic effect of *Trichoderma* spp. against *Serpula lacrymans* in the soil treatment test. J Hokkaido Forest Prod Res Inst 6: 1-5.

Doi S and Togashi I (1989) Utilization of nitrogenous substance by *Serpula lacrymans*. IRG/WP/1397.

Domański S (1972) Fungi. Polyporaceae I. Natl Center Sci Econ Inform, Warszawa, Poland.

Donk MA (1974) Check list of European polypores. North-Holland Publ, Amsterdam, the Netherlands.

Eaton RA and Hale MDC (1993) Wood: decay, pests and protection. Chapman & Hall, London, UK.

Eikenes M, Hietala A, Alfredsen G, Fossdal CG and Solheim H (2005) Comparison of quantitative real-time PCR, chitin and ergosterol assays for monitoring colonization of *Trametes versicolor* in birch wood. Holzforsch 59: 568-573.

Eslyn E and Lombard FF (1983) Decay in mine timbers. II. Basidiomycetes associated with decay of coal mine timbers. Forest Prod J 33: 19-23.

Falck R (1909) Die Lenzites-Fäule des Coniferenholzes. Hausschwammforsch 3: 1-234.

Falck R (1912) Die Meruliusfäule des Bauholzes. Hausschwammforsch 6: 1-405.

Falck R (1927) Gutachten über Schwammfragen. Hausschwammforsch 9: 12-64

Fan M, Chen L-C, Ragan MA, Gutell RR, Warner JR, Currie BP and Casadevall A (1995) The 5S rRNA and the rRNA intergenic spacer of the two varieties of *Cryptococcus neoformans*. J Med Vet Mcol 33: 215-221.

Findlay WPK (1967) Timber pests and diseases. Pergamon, Oxford, UK.

Gersonde M (1958) Untersuchungen über die Giftempfindlichkeit verschiedener Stämme von Pilzarten der Gattungen *Coniophora*, *Poria*, *Merulius* und *Lentinus*. I. *Coniophora cerebella* (Pers.) Duby. Holzforsch 12: 73-83.

Gilbertson RL and Ryvarden L (1987) North American Polypores, Vol 2. Fungiflora, Oslo, Norway.

Ginns J (1978) *Leucogyrophana* (Aphyllophorales): identification of species. Can J Bot 56: 1953-1973.

Ginns J (1982) A monograph of the genus *Coniophora* (Aphyllophorales, Basidiomycetes). Opera Botanica 61: 1-61.

Glancy H, Palfreyman JW, Button D, Bruce A and King B (1990) Use of an immunological method for the detection of *Lentinus lepideus* in distribution poles. J Inst Wood Sci 12: 59-64.

Göller K and Rudolph D (2003) The need for unequivocally defined reference fungi – genomic variation in two strains named as *Coniophora puteana* BAM Ebw. 15. Holzforsch 57: 456-458.

Goodell B, Nicholas DD and Schulz TP (eds.) (2003) Wood deterioration and preservation. Advances in our changing world. ACS Symp Series 845. American Chemical Society, Washington DC, USA.

Göttsche R and Borck HV (1990) Wirksamkeit Kupfer-haltiger Holzschutzmittel gegenüber *Agrocybe aegerita* (Südlicher Schüppling). Mater Org 25: 29-46.

Green F and Clausen CA (2003) Copper tolerance of brown-rot fungi: time course of oxalic acid production. Int Biodet Biodegrad 51: 145-149.

Grinda M and Kerner-Gang W (1982) Prüfung der Widerstandsfähigkeit von Dämmstoffen gegenüber Schimmelpilzen und holzzerstörenden Basidiomyceten. Mater Org 17: 135-156.

Grosser D (1985) Pflanzliche und tierische Bau- und Werkholzschädlinge. DRW Weinbrenner, Leinfelden-Echterdingen, Germany.

Guillitte O (1992) Epidémiologie des attaques. In: La mérule et autres champignons nuisable dans les bâtiments. Jardin Bot Nat Belg, Domaine Bouchout, Belgium, pp 34-42.

Hansen L, Knudsen H, Dissing H, Ahti T, Ulvinen T, Gulden G, Ryvarden L, Persson O and Strid A (1997) Nordic macromycetes. Heterobasidioid, aphyllophoroid and gastromycetoid basidiomycetes. Vol 3. Nordsvamp, Copenhagen, Denmark.

Harmsen L (1953) *Merulius tignicola* Harmsen, eine neue Hausschwamm-Art in Dänemark. Holz Roh-Werkstoff 11: 68-69.

Harmsen L (1960) Taxonomic and cultural studies on brown spored species of the genus *Merulius*. Friesia 6: 233-277.

Harmsen L (1978) Draft of a monographic card for *Serpula himantioides* (Fr.) Karst. IRG/WP/174: 1-8.

Harmsen L, Bakshi BK and Choudhury TG (1958) Relationship between *Merulius lacrymans* and *M. himantioides*. Nature 181: 1011.

Hartig R (1885) Die Zerstörung des Bauholzes durch Pilze. Springer, Berlin, Germany.

Harz CO (1888) Bergwerkpilze. Botanisches Centralblatt 36: 375-380.

Hastrup ACS, Green F, Clausen CA and Jensen B (2005) Tolerance of *Serpula lacrymans* to copper-based wood preservatives. Int Biodeter Biodegrad 56: 173-177.

Hegarty B (1991) Factors affecting the fruiting of the dry rot fungus *Serpula lacrymans*. In: Jennings DH and Bravery AF (eds.) *Serpula lacrymans*. Wiley, Chichester, UK, pp. 39-53.

Hegarty B and Schmitt U (1988) Basidiospore structure and germination of *Serpula lacrymans* and *Coniophora puteana*. IRG/WP/1340.

Hegarty B and Seehann G (1987) Influence of natural temperature variation on fruitbody formation by *Serpula lacrymans* (Wulfen: Fr.) Schroet. Mater Org 22: 81-86.

Hegarty B, Buchwald G, Cymorek S and Willeitner H (1986) Der echte Hausschwamm – immer noch ein Problem? Mater Org 21: 87-99.

Hietela A, Eikenes M, Kvaalen H, Solheim H and Fossdal C (2003) Multiplex real-time PCR for monitoring *Heterobasidion annosum* colonization in Norway spruce clones that differ in disease resistance. Appl Environ Microbiol 69: 4413-4420.

Highley TL, Dashek WV (1998) Biotechnology in the study of brown- and white-rot decay. In: Bruce A, Palfreyman JW (eds.) Forest products biotechnology. Taylor & Francis, London, pp. 15-36.

Hof T (1981a) *Gloeophyllum abietinum* (Bull. ex Fr.) Karst. In: Cockcroft R (ed.) Some wood-destroying basidiomycetes. IRG/WP, Boroko, Papua New Guinea, pp. 55-66.

Hof T (1981b) *Gloeophyllum sepiarium* (Wulf. ex Fr.) Karst. In: Cockcroft R (ed.) Some wood-destroying basidiomycetes. IRG/WP, Boroko, Papua New Guinea, pp. 67-79.

Hof T (1981c) *Gloeophyllum trabeum* (Pers. ex Fr) Murrill. In: Cockcroft R (ed.) Some wood-destroying basidiomycetes. IRG/WP, Boroko, Papua New Guinea, pp. 81-94.

Högberg N and Land CJ (2004) Identification of *Serpula lacrymans* and other decay fungi in construction timber by sequencing of ribosomal DNA – a practical approach. Holzforsch 58: 199-204.

Högberg N, Svegården IB and Kauserud H (2006) Isolation and characterization of fifteen polymorphic microsatellite markers for the devasting dry rot fungus *Serpula lacrymans*. Molec Ecol Notes 6: 1022-1024.

Horisawa S, Sakuma Y, Takata K and Doi S (2004) Detection of intra- and interspecific variation of the dry rot fungus *Serpula lacrymans* by PCR-RFLP and RAPD analysis. J Wood Sci 50: 427-432.

Huckfeldt T and Schmidt O (2006a) Hausfäule- und Bauholzpilze. Rudolf Müller, Köln, Germany.

Huckfeldt T and Schmidt O (2006b) Identification key for European strand-forming house-rot fungi. Mycologist 20: 42-56.

Huckfeldt T, Kleist G and Quader H (2000) Vitalitätsansprache des Hausschwammes (*Serpula lacrymans*) und anderer holzzerstörender Pilze. Z. Mykol 66: 35-44.

Huckfeldt T, Schmidt O and Quader H (2005) Ökologische Untersuchungen am Echten Hausschwamm und weiteren Hausfäulepilzen. Holz Roh-Werkstoff 63: 209-219.

Humar M, Petrič M and Pohleven F (2001) Changes of the pH value of impregnated wood during exposure to wood-rotting fungi. Holz Roh-Werkstoff 59: 288-293.

Jahn H (1990) Pilze an Bäumen. Patzer, Berlin, Germany.

Jarosch M and Besl H (2001) *Leucogyrophana*, a polyphyletic genus of the order Boletales (Basidiomycetes). Plant Biol 3: 443-448.

Jellison J and Goodell B (1988) Immunological detection of decay in wood. Wood Sci Technol 22: 293-297.

Jellison J, Howell C, Goodell B and Quarles SL (2004) Investigations into the biology of *Meruliporia incrassata*. IRG/WP/10508: 1-9.

Jennings DH (1987) Translocation of solutes in fungi. Biol Rev 62: 215-243.

Jennings DH (1991) The physiology and biochemistry of the gegetative mycelium. In: Jennings DH, Bravery AF (eds.) *Serpula lacrymans*. Wiley, Chichester, pp 55-79.

Jennings DH and Bravery AF (eds.) (1991) *Serpula lacrymans*: fundamental biology and control strategies. Wiley, Chichester, UK.

Jennings DH and Lysek G (1999) Fungal biology. Understanding the fungal lifestyle. 2nd ed. BIOS Sci Pub, Oxford, UK.

Joint Genome Institute (2010) Available at: http://genome.jgi-psf.org.

Jordan CR, Dashek WW and Highley TL (1996) Detection and quantification of oxalic acid from the brown-rot decay fungus *Postia placenta*. Holzforschung 50: 312-318.

Jülich W (1984) Basidiomyceten 1. Teil. Die Nichtblätterpilze, Gallertpilze und Bauchpilze (Aphyllophorales, Heterobasidiomycetes, Gastromycetes). In: Gams H (ed.) Kleine Kryptogamenflora, vol. 2b/1. Fischer, Stuttgart, Germany.

Jülich W and Stalpers J (1980) The resupinate non-poroid Aphyllophorales of the temperate northern hemisphere. KNAW, North-Holland Publ, Amsterdam, the Netherlands.

Käärik A (1981) *Coniophora puteana* (Schum. ex Fr.) Karst. In: Cockcroft R (ed.) Some wood-destroying basidiomycetes. IRG/WP, Boroko, Papua New Guinea, pp. 11-21.

Kartal SN, Munir E, Kakitani T and Imamura Y (2004) Bioremediation of CCA-treated wood by brown-rot fungi *Fomitopsis palustris*, *Coniophora puteana* and *Laetiporus sulphureus*. J Wood Sci 50: 182-188.

Kauserud H, Högberg N, Knudsen H, Elbornes SA and Schumacher T (2004a) Molecular phylogenetics suggest a North American link between the anthropogenic dry rot fungus *Serpula lacrymans* and its wild relative *S. himantioides*. Molec Ecol 13: 3137-3146.

Kauserud H, Sætre G-P, Schmidt O, Decock C, Schumacher T (2006a) Genetics of self/nonself recognition in *Serpula lacrymans*. Fungal Genet Biol 43: 503-510.

Kauserud H, Schmidt O, Elfstrand M and Högberg N (2004b) Extremely low AFLP variation in the European dry rot fungus (*Serpula lacrymans*): implications for self/nonself-recognition. Mycol Res 108: 1264-1270.

Kauserud H, Stensrud Ø, Decock C, Shalchian-Tabrizi K and Schumacher T (2006b) Multiple gene genealogies and AFLP sugest cryptic speciation and long-distance dispersal in the basidiomycete *Serpula himantioides* (Boletales). Molec Ecol 15: 421-431.

Kauserud H, Svegården IB, Decock C and Hallenberg N (2007b) Hybridization among cryptic species of the cellar fungus *Coniophora puteana* (Basidiomycota). Molec Ecol 16: 389-399.

Kauserud H, Svegården IB, Sætre G-P, Knudsen H, Stensrud Ø, Schmidt O, Doi S, Sugiyama T and Högberg N (2007a) Asian origin and rapid global spread of the destructive dry rot fungus *Serpula lacrymans*. Molec Ecol 16: 3350-3360.

Kim YS, Goodell B and Jellison J (1991a) Immuno-electron microscopic localization of extracellular metabolites in spruce wood decayed by brown-rot fungus *Postia placenta*. Holzforsch 45: 389-393.

Kim YS, Jellison J, Goodell B, Tracy V and Chandhoke V (1991b) The use of ELISA for the detection of white- and brown-rot fungi. Holzforsch 45: 403-406.

Kleist G and Seehann G (1999) Der Eichenporling, *Donkioporia expansa* – ein wenig bekannter Holzzerstörer in Gebäuden. Z Mykol 65: 23-32.

Koch A-P (1991) The current status of dry rot in Denmark and control strategies. In: Jennings DH and Bravery AF (eds.) *Serpula lacrymans*. Wiley, Chichester, UK, pp 147-154.

Koch AP, Kjerulf-Jensen C and Madsen B (1989) New experiments with the dry rot fungus in Danish buildings, heat treatment and viability tests. IRG/WP/1423.

Koch P (1985) Wood decay in Danish buildings. IRG/WP/1261.

Korpi A, Pasanen AL and Viitanen H (1999) Volatile metabolites of *Serpula lacrymans*, *Coniophora puteana*, *Poria placenta*, *Stachybotrys chartarum* and *Chaetomium globosum*. Building Environ 34: 205-211.

Kotlaba F (1992) Nalezy drevomorky domaci – *Serpula lacrymans* v prirode. Ceska Mykologie 46: 143-147.

Kreisel H (1961) Die phytopathogenen Großpilze Deutschlands. Fischer, Jena.

Krieglsteiner GJ (1991) Verbreitungsatlas der Großpilze in Deutschland (West), vol. 1A. Ulmer, Stuttgart .

Krieglsteiner GJ (2000) Die Großpilze Baden-Württembergs. Vol. 1. Ulmer, Stuttgart.

Leithoff H, Stephan I, Lenz MT and Peek R-D (1995) Growth of the copper tolerant brown rot fungus Antrodia vaillantii on different substrates. IRG/WP/10121: 1-10.

Liese J (1950) Zerstörung des Holzes durch Pilze und Bakterien: In. Mahlke F, Troschel R and Liese J (eds.) Handbuch der Holzkonservierung. 3$^{rd}$ ed. Springer, Berlin, Germany.

Lohwag K (1954) Der Hausschwamm *Gyrophana lacrymans* (Wulf.) Pat. und seine Begleiter. Sydowia II, 6: 268-283.

Lombard FF (1990) A cultural study of several species of *Antrodia* (Polyporaceae, Aphyllophorales). Mycologia 82: 185-191.

Lombard FF and Chamuris GP (1990) Basidiomycetes. In: Wang CJK and Zabel RA (eds.) Identification manual for fungi from utility poles in the eastern United States. Am Type Culture Collection, Rockville, USA, pp 21-104.

Lombard FF and Gilbertson GP (1965) Studies on some western *Porias* with negative or weak oxidase reaction. Mycologia 57: 43-76.

Martin F, Selosse M-A and Le Tacon F (1999) The nuclear rDNA intergenic spacer of the ectomycorrhizal basidiomycete *Laccaria bicolor*: structural analysis and allelic polymorphism. Microbiol 145: 1605-1611.

Mez C (1908) Der Hausschwamm und die übrigen holzzerstörenden Pilze der menschlichen Wohnungen. Lincke, Berlin, Germany.

Miric M and Willeitner H (1984) Lethal temperatures for some wood-destroying fungi with respect to eradication by heat treatment. IRG/WP/1229.

Moreth U and Schmidt O (2000) Identification of indoor rot fungi by taxon-specific priming polymerase chain reaction. Holzforsch 54: 1-8.

Moreth U and Schmidt O (2005) Investigations on ribosomal DNA of indoor wood decay fungi for their characterization and identification. Holzforsch 59: 90-93.

Müller D, Rangno N, Jacobs J, Hiller C, Tusche D, Scheiding W and Brabetz W (2009) Schnelle und zuverlässige Differentialdiagnose aller relevanten Hausfäulepilze mittels DNA-Chiptechnologie. 3$^{rd}$ Mykol Kolloq, Inst Holztechnol Dresden: 10 pp.

Nobles MK (1965) Identification of cultures of wood-inhabiting Hymenomycetes. Can J Bot 43: 1097-1139.

Nuss I, Jennings DH and Veltkamp CJ (1991) Morphology of *Serpula lacrymans*. In: Jennings DH and Bravery AF (eds.) *Serpula lacrymans*. Wiley, Chichester, UK, pp. 9-38.

Paajanen L and Viitanen H (1989) Decay fungi in Finnish houses on the basis of inspected samples from 1978 to 1988. IRG/WP/1401: 1-4.

Paajanen LM (1993) Iron promotes decay capacity of *Serpula lacrymans*. IRG/WP/10008: 1-3.

Palfreyman JW and Low G (2002) Studies of the domestic dry rot fungus *Serpula lacrymans* with relevance to the management of decay in buildings. Res Rep, Historical Scotland, Edinburgh, UK.

Palfreyman JW, Gartland JS, Sturrock CJ, Lester D, White NA, Low GA, Bech-Andersen J and Cooke DEL (2003) The relationship between „wild" and „building" isolates of the dry rot fungus *Serpula lacrymans*. FEMS Microbiol Lett 228: 281-286.

Palfreyman JW, Glancy H, Button D, Bruce A, Vigrow A, Score A and King B (1988) Use of immunoblotting for the analysis of wood decay basidiomycetes. IRG/WP/2307: 1-8.

Palfreyman JW, Phillips EM and Staines HJ (1996) The effect of calcium ion concentration on the growth and decay capacity of *Serpula lacrymans* (Schumacher ex Fr.) Gray and *Coniophora puteana* (Schumacher ex Fr.) Karst. Holzforsch 50: 3-8.

Palfreyman JW, Vigrow A, Button D, Hegarty B and King B (1991) The use of molecular methods to identify wood decay organisms. 1. The electrophoretic analysis of *Serpula lacrymans*. Wood Protect 1: 15-22.

Palmer JG, Eslyn WE (1980) Monographic information on *Serpula (Poria) incrassata*. IRG/WP/160: 1-61.

Pegler DN (1991) Taxonomy, identification and recognition of Serpula lacrymans. In: Jennings DH, Bravery AF (eds.) *Serpula lacrymans*. Wiley, Chichester, pp. 1-7.

Peyretaillade E, Biderre C, Peyret P, Duffieux F, Méténier G, Gouy M, Michot B, Vivarès CP (1998) Microsporidian *Encephalitozoon cuniculi*, a unicellular eukaryote with an unusual chromosomal dispersion of ribosomal genes and a LSU rDNA reduced to the universal core. Nucleic Acids Res 26: 3513-3520.

Rabanus A (1939) Über die Säure-Produktion von Pilzen und deren Einfluß auf die Wirkung von Holzschutzmitteln. Mittlg dtsch Forstver 23: 77-89.

Råberg U, Högberg N and Land CJ (2004) Identification of brown-rot fungi on wood in above ground conditions by PCR, T-RFLP and sequencing. IRG/WP/10512: 1-6.

Råberg U, Högberg N and Land CJ (2005) Detection and species discrimination using rDNA T-RFLP for identification of wood decay fungi. Holzforsch 59: 696-702.

Rayner ADM and Boddy L (1988) Fungal decomposition of wood. Its biology and ecology. Wiley, Chichester, UK.

Reid DA (1974) A monograph of the British Dacrymycetales. Trans Br mycol Soc 62: 433-494.

Ridout B (2000) Timber decay in buildings. The conservation approach to treatment. E & FN Spon, London, UK.

Ridout BV (2007) Understanding decay in building timbers. In: Noldt U, Michels H (eds.) Wood-destroying organisms in focus. Merkur, Detmold, Germany, pp. 225-230.

Rypáček V (1966) Biologie holzzerstörender Pilze. Fischer, Jena.

Ryvarden L and Gilbertson RL (1993/1994) European polypores. 2 Parts. Synopsis Fungorum 6, Fungiflora Oslo, Norway.

Savory JG (1964) Dry rot – a re-appraisal. Rec Br Wood Preserv Assoc 1964: 69-76.

Schmidt O (2000) Molecular methods for the characterization and identification of the dry rot fungus *Serpula lacrymans*. Holzforsch 54: 221-228.

Schmidt O (2003) Molekulare und physiologische Charakterisierung von Hausschwamm-Arten. Z Mykol 69: 287-298.

Schmidt O (2006) Wood and tree fungi. Biology, damage, protection, and use. Springer, Berlin, Germany.

Schmidt O (2007) Indoor wood-decay basidiomycetes: damage, causal fungi, physiology, identification and characterization, prevention and control. Mycol Progress 6: 261-279.

Schmidt O (2009) Molecular identification and characterization of indoor wood decay basidiomycetes. In: Gherbawy Y, Mach RL and Rai M (eds.) Current advances in molecular mycology. Nova Sci Publ, New York, USA, pp. 333-348.

Schmidt O and Huckfeldt T (2005) Gebäudepilze. In: Müller J (ed.) Holzschutz im Hochbau. Fraunhofer IRB, Stuttgart, Germany, pp. 44-72.

Schmidt O and Kallow W (2005) Differentiation of indoor wood decay fungi with MALDI-TOF mass spectrometry. Holzforsch 59: 374-377.

Schmidt O and Kebernik U (1989) Characterization and identification of the dry rot fungus *Serpula lacrymans* by polyacrylamide gel electrophoresis. Holzforsch 43: 195-198.

Schmidt O and Moreth U (1995) Detection and differentiation of *Poria* indoor brown-rot fungi by polyacrylamide gel electrophoresis. Holzforsch 49: 11-14.

Schmidt O and Moreth U (1996) Biological characterization of *Poria* indoor brown-rot fungi. Holzforsch 50: 105-110.

Schmidt O and Moreth U (1998) Characterization of indoor rot fungi by RAPD analysis. Holzforsch 52: 229-233.

Schmidt O and Moreth U (1999) Identification of the dry rot fungus, *Serpula lacrymans*, and the wild merulius, *S. himantioides*, by amplified ribosomal DNA restriction analysis (ARDRA). Holzforsch 53: 123-128.

Schmidt O and Moreth U (2003) Molecular identity of species and isolates of internal pore fungi *Antrodia* spp. and *Oligoporus placenta*. Holzforsch 57: 120-126.

Schmidt O and Moreth U (2008) Ribosomal DNA intergenic spacer of indoor wood-decay fungi. Holzforsch 62: 1759-764.

Schmidt O and Moreth-Kebernik U (1990) Biological and toxicant studies with the dry rot fungus *Serpula lacrymans* and new strains obtained by breeding. Holzforsch 44: 1-6.

Schmidt O and Moreth-Kebernik U (1991a) A simple method for producing basidiomes of *Serpula lacrymans* in culture. Mycol Res 95: 375-376.

Schmidt O and Moreth-Kebernik U (1991b) Monokaryon pairings of the dry rot fungus *Serpula lacrymans*. Mycol Res 95: 1382-1386.

Schmidt O, Grimm K and Moreth U (2002) Molekulare und biologische Charakterisierung von *Gloeophyllum*-Arten in Gebäuden. Z Mykol 68: 141-152.

Schmidt O, Grimm K and Moreth U (2002/2003) Molecular identity of species and isolates of the *Coniophora* cellar fungi. Holzforsch 56: 563-571 and Addendum in Holzforsch 57: 228.

Schmidt O, Moreth U (2000) Species-specific PCR primers in the rDNA-ITS region as a diagnostic tool for *Serpula lacrymans*. Mycol Res 104: 69-72.

Schmidt O, Moreth U (2002/2003) Data bank of rDNA-ITS sequences from building-rot fungi for their identification. Wood Sci Technol 36: 429-433 and revision in Wood Sci Technol 37: 161-163.

Schultze-Dewitz G (1985) Holzschädigende Organismen in der Altbausubstanz. Bauztg 39: 565-566.

Schultze-Dewitz G (1990) Die Holzschädigung in der Altbausubstanz einiger brandenburgischer Kreise. Holz-Zbl 116: 1131.

Schulze B and Theden G (1948) Zur Kenntnis des Gelbrandigen Hausschwammes *Merulius pinastri* (Fries) Burt 1917. Nachrichtenbl Dtsch Pflanzenschutzdienst 2: 1-5.

Seehann G (1984) Monographic card on *Antrodia serialis*. IRG/WP/1145: 1-11.

Seehann G (1986) Butt rot in conifers caused by Serpula himantioides (Fr.) Karst. Eur J Forest Pathol 16: 207-217.

Seehann G and Hegarty BM (1988) A bibliography of the dry rot fungus, *Serpula lacrymans*. IRG/WP/1337.

Seehann G and Von Riebesell M (1988) Zur Variation physiologischer und struktureller Merkmale von Hausfäulepilzen. Mater Org 23: 241-257.

Segmüller J and Wälchli O (1981) *Serpula lacrymans* (Schum. ex Fr.) S.F. Gray. In: Cockcroft R (ed.) Some wood-destroying basidiomycetes. IRG/WP, Boroko, Papua New Guinea, pp. 141-159.

Siepmann R (1970) Artdiagnose einiger holzzerstörender Hymenomyceten an Hand von Reinkulturen. Nova Hedwigia 20: 833-863.

Silverborg SB (1953) Fungi associated with the decay of wooden buildings in New York State. Phytopath 43: 20-22.

Stalpers JA (1978) Identification of wood-inhabiting Aphyllophorales in pure culture. Stud Mycol 16. Centraalbureau Schimmelcultures, Baarn, the Netherlands.

Steinfurth A (2007) Possibilities and limitations of thermal control of dry rot (*Serpula lacrymans*). In: Noldt U and Michels H (eds.) Wood-destroying organisms in focus. Merkur, Detmold, Germany, pp. 207-210.

Stephan I and Peek R-D (1992) Biological detoxification of wood treated with salt preservatives. IRG/WP/3717.

Stephan I, Leithoff H and Peek R-D (1996) Microbial conversion of wood treated with salt preservatives. Mater Org 30: 179-199.

Sutter H-P (2003) Holzschädlinge an Kulturgütern erkennen und bekämpfen. 4th edn. Haupt, Bern, Switzerland.

Sutter H-P, Jones EBG and Wälchli O (1983) The mechanism of copper tolerance in *Poria placenta* (Fr.) Cke. and *Poria vaillantii* (Pers.) Fr. Mater Org 18: 241-262.

Sutter H-P, Jones EBG and Wälchli O (1984) Occurrence of crystalline hyphal sheats in *Poria placenta* (FR.) CKE. J Inst Wood Sci 10: 19-25.

Theden G (1941) Untersuchungen über die Feuchtigkeitsansprüche der wichtigsten in Gebäuden auftretenden holzzerstörenden Pilze. Angew Bot 23: 189-253.

Theden G (1972) Das Absterben holzzerstörender Pilze in trockenem Holz. Mater Org 7: 1-10.

Theodore ML, Stevenson TW, Johnson GC, Thornton JD and Lawrie AC (1995) Comparison of *Serpula lacrymans* isolates using RAPD PCR. Mycol Res 99: 447-450.

Thörnqvist T, Kärenlampi P, Lundström H, Milberg P and Tamminen Z (1987) Vedegenskaper och mikrobiella angrepp i och på byggnadsvirke. Swed Univ Agric Sci Uppsala 10.

Thornton JD (1989) The restricted distribution of *Serpula lacrymans* in Australian buildings. IRG/WP/1382: 1-12.

Thornton JD (1991) Australian scientific research on *Serpula lacrymans*. In: Jennings DH and Bravery AF (eds.) *Serpula lacrymans*. Wiley, Chichester, UK, pp. 155-171.

Toft L (1992) Immuno-fluorescence detection of basidiomycetes in wood. Mater Org 27: 11-17.

Toft L (1993) Immunological identification *in vitro* of the dry rot fungus *Serpula lacrymans*. Mycol Res 97: 290-292.

Vainio EJ and Hantula J (2000) Direct analysis of wood-inhabiting fungi using denaturing gradient gel electrophoresis of amplified ribosomal DNA. Mycol Res 104: 927-936.

Van Acker J and Stevens M (1996) Laboratory culturing and decay testing with *Physosporinus vitreus* and *Donkioporia expansa* originating from identical cooling tower environments show major differences. IRG/WP/10184.

Van Acker J, Stevens M and Rijchaert V (1995) Highly virulent wood-rotting basidiomycetes in cooling towers. IRG/WP/10125.

Verrall AF (1968) *Poria incrassata* rot: prevention and control in buildings. USDA Forest Serv Tech Bull 1385.

Vigrow A, King B and Palfreyman JW (1991b) Studies of *Serpula lacrymans* mycelial antigens by Western blotting techniques. Mycol Res 95: 1423-1428.

Vigrow A, Palfreyman JW and King B (1991a) On the identity of certain isolates of *Serpula lacrymans*. Holzforsch 45: 153-154.

Viitanen H and Ritschkoff A-C (1991) Brown rot decay in wooden constructions. Effect of temperature, humidity and moisture. Swed Univ Agric Sci Dept Forest Prod 222.

Wälchli O (1973) Die Widerstandsfähigkeit verschiedener Holzarten gegen Angriffe durch den echten Hausschwamm (*Merulius lacrymans* (Wulf.) Fr.). Holz Roh-Werkstoff 31: 96-102.

Wälchli O (1976) Die Widerstandsfähigkeit verschiedener Holzarten gegen Angriffe durch *Coniophora puteana* (Schum. ex Fr.) Karst. (Kellerschwamm) und *Gloeophyllum trabeum* (Pers. ex. Fr.) Murrill (Balkenblättling). Holz Roh-Werkstoff 34: 335-338.

Wälchli O (1980) Der echte Hausschwamm – Erfahrungen über Ursachen und Wirkungen seines Auftretens. Holz Roh-Werkstoff 38: 169-174.

Watkinson SC, Davison EM and Bramah J (1981) The effect of nitrogen availability on growth and cellulolysis by *Serpula lacrimans*. New Phytol 89: 295-305.

Ważny H and Czajnik M (1963) On the occurrence of indoor wood-decay fungi in Poland (Polish). Fol Forest Polonica 5: 5-17.

Ważny J and Thornton JD (1989) Comparative laboratory testing of strains of the dry rot fungus *Serpula lacrymans* (Schum. ex Fr.) S.F. Gray. IV. The action of CCA and NaPCP in an agar-block test. Holzforsch 43: 231-233.

Ważny J, Krajewski KJ and Thornton JD (1992) Comparative laboratory testing of strains of the dry rot fungus *Serpula lacrymans* (Schum. ex Fr.) S.F.Gray. VI. Toxic value of CCA and NaPCP preservatives by statistical estimation. Holzforsch 46: 171-174.

Weigl J and Ziegler H (1960) Wassergehalt und Stoffleitung bei *Merulius lacrimans* (Wulf.) Fr. Arch Mikrobiol 37: 124-133.

White NA, Dehal PK, Duncan JM, Williams NA, Gartland JS, Palfreyman JW and Cooke DEL (2001) Molecular analysis of intraspecific variation between building and "wild" isolates of the dry rot fungus *Serpula lacrymans* and their relatedness to *S. himantioides*. Mycol Res 105: 447-452 .

White TJ, Bruns T, Lee S and Taylor J (1990) Amplification and direct sequencing of fungal ribosomal genes for phylogenetics. In: Innis MA, Gelfand DH, Sninisky JJ and White TJ (eds.) PCR protocols. Academic Press, San Diego, CA, USA, pp. 315-322.

Wilcox WW and Dietz M (1997) Fungi causing above-ground wood decay in structures in California. Wood Fiber Sci 29: 291-298.

Zabel RA and Morrell JJ (1992) Wood microbiology. Decay and its prevention. Academic Press, San Diego, CA, USA.

Zaremski A, Ducousso M, Domergue O, Fardoux J, Rangin C, Fouquet D, Joly H, Sales C, Dreyfus B and Prin Y (2005) *In situ* molecular detection of some white-rot and brown-rot basidiomycetes infecting temperate and tropical woods. Can J For Res 35: 1256-1260.

Zaremski A, Prin Y, Ducousso M and Fouquet D (1998) Caractérisation moléculaire des champignons lignivores. Utilisation d'une nouvelle technique [Molecular characterization of wood-decaying fungi]. Bois Forêts Tropiques 257: 63-69.

Zoberst W (1952) Die physiologischen Bedingungen der Pigmentbildung von *Merulius lacrymans domesticus* Falck. Arch Mikrobiol 18: 1-31.

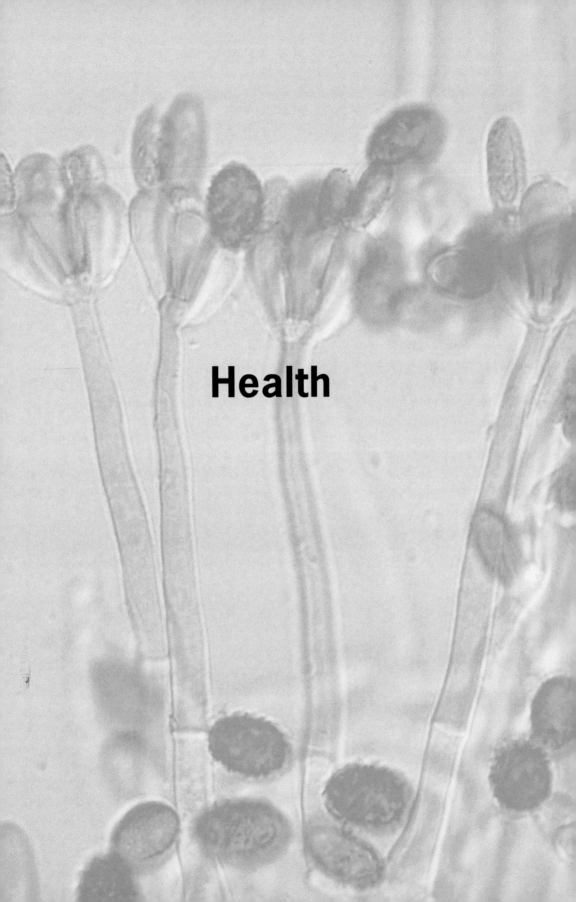

Health

# 7 Health effects from mold and dampness in housing in western societies: early epidemiology studies and barriers to further progress

*J. David Miller*
*Ottawa-Carleton Institute of Chemistry, Carleton University, Ottawa ON K1S 5B6, Canada*

## Introduction

Asthma is a serious respiratory condition that can be triggered by exposure to a variety of allergens, such as pollen, animal dander, cockroaches, dust mites and molds as well as various air pollutants and irritants, such as environmental tobacco smoke and other factors such as cold air and exercise. There is a large difference in asthma rates in different countries depending and the reasons for this are not clear (Weinmayr *et al.* 2007).

The rise in asthma in the latter part of the 20th century (NAS 2000) has engendered much scientific discussion (although there is some evidence that the rates have reached a plateau (Pearce *et al.* 2007). Since it is improbable that there have been genetic changes in human populations, a number of competing hypotheses have been put forward to account for this rise. The principal ideas fall into three areas: (1) increased exposure to perennial allergens (indoor allergens) resulting from housing changes, (2) decreased exposure to bacterial and viral infections due to improved hygiene and immunization, as well as alteration in the gut flora because of antibiotics and dietary changes, and (3) loss of a lung specific protective effect in the last 50 years. The last hypothesis is largely only a theoretical possibility, but the so-called hygiene hypothesis has attracted a lot of attention. However, it is important to note that by 1900 seasonal hay fever became a problem in Germany and England and by 1935 in New York. This all preceded the changes associated with the hygiene hypothesis. By the late 1950s, hay fever was common and a major problem in New York City. From studies of diverse populations (Platts-Mills 2005, Ring 2005), the best evidence is that the rise of asthma started around the late 1960s in Europe and the USA. The earliest record of the relationship of fungi to asthma and hay fever in *indoor air* comes from the end of the 17th century, when in his 1698 text book "A treatise on the asthma" Sir John Floyer noted asthmatic symptoms in individuals who had visited a wine cellar (cited in Miller and Day 1997). Other changes in population health such in obesity that affect asthma management may also be pertinent for asthma prevalence, although this is also not reliably known (Selgrade *et al.* 2006).

There is no ambiguity that since World War II there have been major changes in housing conditions, leading to increased exposure to indoor allergens such as dust mites, cockroaches and fungi, depending on the climate. In Western European cities, indoor temperatures were low (7-12 °C) and the night-time carbon dioxide concentration ranged up to twice that of outdoor air, and homes were damp (Carnelley *et al.* 1887). [By the mid 1980s, average winter temperatures in Edinburgh homes were 17-18 °C, but many homes were damp (Strachan and Sanders 1989).]

Although traditional plaster and wood are very resistant to fungal growth, the use of wallpaper in middle class homes was not uncommon in Western Europe, Colonial America (Lynn 1980) and early Canada. Preservation of these materials from fungal growth remains an issue when buildings are cool (Miller and Holland 1981). The iconic building-associated fungus, *Stachybotrys chartarum*, was isolated from wallpaper from a house in Prague (Corda 1837). There is a large literature associated with the use of arsenicals as green dyes, and later as insecticides and for their antifungal properties in the built environment (Dudley 1938) and even in clothes (Anonymous 1904). Although there was some fungal growth in the built environment, the widespread use of traditional material and mixed arsenical preparations with undoubted antifungal activity in building surface treatments (DaCosta 1972) certainly limited fungal damage. Such was their use that the release of these arsenicals from fragments and dusts has been associated with a number of illnesses and deaths in older houses from the 18th to the 20th century (Richardson 2000, Cullen and Bentley 2005). Some fungi can liberate arsine gas when grown on arsenicals of the type common in early materials (Huss 1913) as well as on copper-chrome-arsenic (CCA, Cullen *et al.* 1984). There has been speculation that mold-produced arsine gas was involved, but compared to particulate sources of arsenic exposure this seems unlikely (Cullen and Bentley 2005).

Because (1) the allergen is known and both exposure and subsequent IgE antibodies can be measured in human sera and (2) an exposure-response can be demonstrated in many populations in diverse environments (NAS 2000), house dust mites are a well known cause of asthma. Increased exposure to house dust mites in the built environment has been documented in many countries (NAS 2000; Platts-Mills 2005). This is firmly associated with tighter homes, better heated but increasingly under-ventilated. Haysom and Reardon (1998) have stated that, while homes in eastern and western Canada had very high rates of natural ventilation because of construction techniques and the need to heat the homes in winter by combustion (wood stoves and coal and then oil furnaces). This means that ventilation needs to be improved in more modern, tight homes by mechanical means.

The first record of house dust mites in Canada was in 1970 when approx. 20% of dust samples collected from patients from patients "allergic to house dust" contained *Dermatophagoides farinae* and *D. pteronyssinus*. Approximately 5% of homes (some samples from Ottawa, Ontario; 45.40N, 75.72W; mean 1970 winter outdoor temperature -10 °C, Environment Canada) had what were described as high numbers of mites (Sinha *et al.* 1970). The first reports from US were in some homes in Columbus, Ohio (40.0N 83.1W; mean 1970 winter temperature *c.* 0° C; McCloskey *et al.* 1986) contained rather more house dust mites (Warton 1970) than the Canadian samples. In comparison, at the same time the prevalence and numbers of house dust mite infestations in London, England, was high (along with the Netherlands and other countries; Maunsell *et al.* 1968, Platts-Mills 2005). In 1984, house dust mites remained uncommon in the first formal studies of homes, mainly in Ontario and Quebec (unpublished data, described in Miller *et al.* 1988). By 1995, mites were common in samples from homes in Southern Ontario (Dales and Miller 1999) and Vancouver (Chan-Yeung *et al.* 1995), but compared with the previous decade the percentage of samples above the estimated sensitizing concentration had increased, although remaining low, more so in Winnipeg (Chan-Yeung *et al.* 1995). All of this was concurrent with strong efforts to reduce the air change rates in Canadian homes to save energy (Miller 2007). House dust samples collected in Prince Edward Island Canada (Dales *et al.* 2006) and Ottawa (Miller *et al.* 2007), among other Canadian centers, showed the same high prevalence of dust mites, but much higher percentages of samples above the estimated sensitizing levels. As noted by Platts-Mills (2005), there is no doubt that changes in Canadian homes have produced this important difference in exposure over the past several decades. With the cold climate and single family home lifestyle of Canada the rise of dust mites lagged behind Europe.

The same changes in housing conditions, i.e. the elimination of pervasive use of chemicals with antifungal activity in susceptible building materials seen in Europe during the latter part of the 20th century has been seen in the US and Canada. In addition, the ventilation rate has been reduced over the last 25 years has occurred particularly in the northern US and in most of Canada (Miller 2007). In Western Europe, there is a problem with under-heating or lack of insulation increasing dampness (Bonnefoy 2007; Howden-Chapman *et al.* 2007; Thompson and Petticrew 2007). North America, these changes have also included the shift from plaster to paper-faced gypsum board that had begun in earnest by the mid-1960s. This material is very susceptible to fungal growth (Flannigan and Miller 2011). By the late 1980s, a large percentage of the housing stock included considerable quantities of this building material, along with many other changes in constructional details that reduced resistance to water intrusion and led to a large increase in air-conditioning in summer (Treschel 1994, Odom and DuBose 1996, Lawton *et al.* 1998, Lawton 1999, Lawrence and Martin 2001, NAS 2004, Foto *et al.* 2005, Kercsmar *et al.* 2006, Loftness *et al.* 2006, Miller *et al.* 2007). As with house dust mites, starting from the mid- to late 1960s, there has been an increase in the prevalence of homes that are damp and moldy except with a lag of around 20 years.

Although it is widely said that mold and associated respiratory disease is a new issue, this is incorrect. As noted above, the first textbook in English on asthma published in 1698 comments on asthma from mold in a wine cellar. This possibly relates to the well known occupational asthma associated with moldy cork (Lacey 1994, Winck *et al.* 2004) or from exposure to dry rot spores which had become serious by the later 17th century (Frankland and Hay 1951, Singh 1999). The text book that for many years was the standard text for the teaching of allergy in the US from 1946, Feinberg's "Allergy in Practice" noted fungi as a cause of asthma. Papers giving the early exposure data on house dust mites in Europe noted that indoor fungi were associated with asthma, and some of the fungi reported are familiar today in Canadian homes (Sinha *et al.* 1970). This is because interest in fungi indoors as allergens preceded the study of house dust mites by nearly two decades. Important papers on molds and rot fungi in indoor air were published from 1948 in Scandinavia and the early 1950s in England and Holland. These included some relating visible mold to exposures (reviewed by Hyde 1973). The importance of indoor sinks in porous materials and settled dust was reported in 1952 in US and England (Maunsell 1952, Swaebly and Christensen 1952), and the influence of repairs on indoor mold exposures in 1954 (Maunsell *et al.* 1954). The Dutch researcher Van Der Werff published a book "Mould fungi and bronchial asthma" in 1958. By 1970, more formal studies on sampling and epidemiology were reported and this continued into the early 1970s, in US by Kozak *et al.* (1980) and Solomon (1975) and in the Netherlands (discussed in Verhoeff 1994), and by the late 1970s clear recommendations to reduce mold exposures in residential housing made by allergy researchers (e.g. Gravesen 1978) and in the medical community (Kozak *et al.* 1985).

The modern era of mold in dampness and disease began with the publication of three studies of similar design that examined respiratory disease, damp and mold involving increasingly large numbers of children. The first study from David Strachan involved *c.* 900 children was conducted in Scotland (Strachan 1988). The second involved *c.* 4,600 children from the Harvard Six Cities Studies from Douglas Dockery and John Spengler (Brunekreef *et al.* 1989). A similar investigation based on the Health Canada Long Range Transport of Air Pollution Studies that involved nearly 15,000 children and 18,000 adults was begun by Harry Zwanenburg with a working group and completed by Robert Dales (Dales *et al.* 1991a,b). These three studies began for similar reasons and against the backdrop of increasing home dampness resulting from different causes but with the same outcome.

## Early large-scale studies linking dampness and mold to health

From the written record, Strachan and his colleagues began by undertaking a review of 1983 records relating to 7 year old children in a health practice serving a "socially deprived area" of Edinburgh comprising *c.* 15,000 patients at the time. Most of the housing would have been poorly insulated pre- and post-World War II structures lacking central heating along with some Victorian flats (apartments), with modest floor areas by North American standards. This led to the issuance of a questionnaire about school absences and nocturnal coughing of the 7-year-old child then followed by a questionnaire from which the information was obtained (Table 7.1; Strachan and Elton 1986).

Of parameters, reported dampness and mold were associated with increased wheeze, school absences and night time cough. The prevalence of mold and dampness was relatively comment as was parental smoking (Table 7.1). This survey was followed up by a questionnaire/objective medical test administered to circa 900 7-year-old

*Table 7.1. Risk factors for respiratory symptoms derived from a questionnaire of c. 200 children in Edinburgh in 1984 (adapted from Strachan and Elton 1986).*

| Risk factor | Prevalence (%) |
|---|---|
| family history of wheeze | 48 |
| family size >4 | 82 |
| siblings | 84 |
| > one person per room | 67 |
| others sleeping in bedroom | 58 |
| bedroom unheated | 60 |
| bedroom window open at night | 28 |
| gas appliance | 69 |
| coal appliance | 14 |
| parental smoking | 75 |
| damp | 30 |
| mold | 21 |

children within Edinburgh through their schools. This study found that objective bronchial reactivity was reported more often when the parents reported visible mold in the home. There was a possibility of reporting bias, i.e. that parents might associate mold and dampness with the symptoms of their child. Notably, whether the respondents were tenants or owners their reporting of dampness and mold were similar (Strachan 1988). The issue of reporting bias has been studied in greater detail in the Canadian studies (see below) and, although misclassification is high, homeowner reports are generally reliable. Strachan later noted that dampness and mold related to wheeze in children but not in adults, and suggested that "atopy was probably not an important mechanism" (Strachan 1989). Subsequent studies were carried out by Strachan and colleagues on indoor RH in relation to observed mold and dampness, but these were generally inclusive (Strachan and Sanders 1989). In retrospect, this finding was not unexpected since surface moisture, a function of wall temperature and/or building defects is much more important. Mycological studies were also pursued on a small number of homes some considerable time after the administration of the questionnaire, and again these were inconclusive (Strachan *et al.* 1990).

The second, larger study in the critical early series came from the group working on the Harvard Six Cities studies. Spengler had commented on the possible importance of microbials indoors (Sexton and Spengler 1983, Spengler 1985) and had conducted investigations in housing which had been influential in raising for him the question of microbial involvement (Spengler, personal communication).

As was the case with the above studies, this 1988 project was school-based, and involved children aged 10. The questionnaire was adapted from the Harvard Six Cities questionnaire, with additional questions mold and dampness indicators (visible mold, water damage, basement water; a composite measure "dampness" was derived from the individual variables). The height and forced expiratory capacity were measured for each child. In this study, wheeze and cough were significantly associated with molds and dampness indicators. Measurements of indoor RH in a random sample of 1,800 homes were conducted, but the authors noted that this is "less important than the dampness of specific surfaces" (Brunekeef *et al.* 1989). Mycological studies were subsequently performed on approximately 40% of the homes in one of the communities. Like the earlier air sampling done on the Edinburgh homes, air samples were taken by impaction on agar medium semi-selective for culturable fungi with high water activity optima. The data tended to support the visual inspection results (Su *et al.* 1992) but are inconclusive for many reasons, not least because of the use of selective media. This very carefully done study indicated that the presence of mold and dampness increased respiratory disease.

The third of the early larger studies, i.e. by Health Canada, was also school-based, involving 30 communities from the interior of British Columbia to Nova Scotia. This was conducted using a questionnaire similar to that used for the previously noted Harvard study (Table 7.2). Approx. 18,000 questionnaires were distributed, of which nearly 15,000 were returned (Dales 1991a,b). The key questions asked are described in Table 7.2. The largest proportion of dwellings (81%) was of one-family detached homes (single family dwellings). Of the reminder, 13% were small apartment buildings and 6% were one family attached homes (duplexes). Molds were reported in nearly 35% of homes, flooding in 24%, and moisture in 14%.

For children, all respiratory symptoms and separately for bronchitis and for cough, were found to be significantly higher in homes with reported mold or dampness (Dales *et al.* 1991a). For adults, the presence of home dampness and mold increased the prevalence of lower respiratory symptoms (any cough, phlegm, wheeze, or wheeze with dyspnea) among those reporting dampness or mold compared with those not current smokers, [i.e. both] ex-smokers and non-smokers (all *P*-values <0.001). This association persisted after adjusting for several socio-demographic variables, including age, sex, and region, and several other exposure variables, including active and passive cigarette smoke, natural gas heating, and wood stoves. The odds ratio between symptoms and dampness was 1.62 (95% confidence interval, 1.48-1.78) in the final model chosen. This association persisted despite stratification by the presence of allergies or asthma. This latter observation echoed the comment from Strachan (1989) of the possible existence of a non-allergenic (i.e. toxic) population health response associated with mold and dampness.

The reliability of this questionnaire has been the subject of considerable further investigation by the group in Ottawa. Their first effort was to test whether the results were repeatable. One of the largest communities in the original study (Wallaceburg, 30 km from Windsor, Ontario, adjacent to Detroit, MI) was chosen in order to study the questionnaire. When this was sent to the parents of 1,600 children on two occasions, one month apart, just under half completed the questionnaire both times. For mold and dampness indicators (Table 7.2), agreement between the two administrations ranged from 87% for visible mold growth to 95% for basement flooding. For adult symptoms, agreement ranged from 80% for upper respiratory symptoms to 99% for physician-diagnosed asthma; for children's symptoms, agreement ranged from 81% for upper respiratory symptoms to 97% for current asthma. For all symptoms, statistical analysis indicated moderate to high reproducibility, meaning that the questionnaires administered in Canada for collecting information on respiratory symptoms was reliable (Dales *et al.* 1994).

Table 7.2. Key questions on Health Canada survey asked over the previous 12 months except where otherwise noted.

Housing conditions
1. Moisture: have you ever had wet or damp spots on surfaces inside your present home other than in the basement (for example, on walls, wallpaper, ceilings or carpets)?
2. Visible mold: have you ever had mold or mildew growing on any surface inside your present home? (For example – on walls, wallpaper, ceilings, carpets, shower curtain, etc. In the basement? In the shower area(s)? In other areas of your house?)
3. Flooding: have you had a leak, flooding, or water damage in your basement?
4. Dampness/mold: the presence of moisture, mold, or flooding as defined earlier with the exception that mold in the shower area was not included.

Health variables for adults
1. Persistent cough: do you usually cough for 3 consecutive months or more during the year?
2. Persistent phlegm: do you usually bring up phlegm for 3 consecutive months or more during the year?
3. Wheeze: do you occasionally have wheezing or whistling in your chest apart from colds or on most days or nights?
4. Wheeze with dyspnea: have you ever had an attack of wheezing that has made you feel short of breath?
5. Lower respiratory symptoms: the presence of persistent cough, persistent phlegm, wheeze, wheeze with dyspnea, as defined earlier.
6. Asthma: has a doctor ever said that you had asthma?
7. Chronic respiratory disease: has a doctor ever said that you had chronic bronchitis or emphysema?
8. Upper respiratory tract symptoms: have you experienced (any one of the following) on three or more separate occasions during the past 3 months: nose irritation, runny or stuffy nose, sneezing, throat irritation?
9. Eye irritation: have you experienced itchy eyes on three or more separate occasions during the past 3 months?

Health variables for children
1. Persistent cough: does he or she usually cough for as much as 3 months of the year?
2. Persistent wheeze: does a wheezing or whistling in the chest occur occasionally apart from colds or on most days or nights?
3. Current asthma: has a doctor ever said that this child had asthma and does he or she still have asthma?
4. Bronchitis: within the past year has this child had bronchitis?
5. Upper respiratory symptoms: indicate if this child has experienced (any one of the following) on three or more separate occasions during the past 3 months: nose irritation, runny or stuffy nose, sneezing, throat irritation?
6. Eye irritation: indicate if this child has experienced itchy eyes on three or more separate occasions during the past 3 months?

A further series of studies was conducted on the same community to examine the reliability of the environmental data. There were two components of this work: (1) a study of 400 of the homes from which questionnaire data had been obtained in duplicate, and (2) a more detailed study of a subset of these homes, more versus less moldy, based on the environmental data collected.

Environmental measurements were taken in the largest living area and the child's bedroom. In both, 18 hour air particulate samples were analyzed for endotoxin and ergosterol. Swab samples were taken of visible mold. In the living area, a dust sample was collected and analyzed for xerophilic and hydrophilic fungi, ergosterol, endotoxin, dust mite allergens and species, cat allergen and cell-line cytotoxicity. In the child's bedroom, dust mite allergens were determined in the mattress dust (Miller 1995).

Approx. 270 species of molds were recovered from the dust, although only 20 were common (Miller and Day 1997). Excepting phylloplane species, this group included several moderately xerophilic *Penicillium*, *Aspergillus* and *Eurotium* species (determined at University of Toronto) based on the then prevailing taxonomy (Scott *et al.* 2000). The xerophile *Wallemia sebi* was among the top 15 species in the homes when expressed on a colony forming units per gram (cfu/g) settled dust basis (data summarized in Miller and Day 1997; see also Amend *et al.* 2010). The ratio of the sum of the cfu/g dust for species of *Penicillium*, *Aspergillus* and *Eurotium* divided by the sum of phylloplane species indicated that about 20% of the houses have unusually large exposure to the latter group. A similar proportion of houses had visible non-phylloplane mold growth from swab sample results (Miller 1995).

Notional concentrations of *Aspergillus*, *Eurotium* and *Penicillium* were twice as high when mold or mildew was reported than when not mentioned ($P=0.01$). Reported mold, water damage and moldy odors were associated with elevated levels of indoor fungi as defined above. However, there was some evidence of reporting bias. In the presence of low concentrations of viable fungi in dust, respondents reporting allergies were more almost twice as likely to report visible mold growth. In homes with elevated concentrations of dust fungi, visible mold growth was half as likely to be reported among smoking occupants as among non-smoking. The 12-50% relative increase in symptom prevalence associated with reported mold growth remained significant both before and after adjusting for subject characteristics, dust mite antigens and endotoxins (Dales and Miller 1999).

From among the 400 homes, 39 with high objective measures of mold in the dust and ergosterol in the 18 hour air samples were chosen and 20 with correspondingly low values. Inspection revealed 0.42 m$^2$ and 1.2 m$^2$ visible mold in the low and high-mold homes, respectively, the method of documenting visible mold being similar to that in Foto *et al.* (2005). Associated with this there was a 25-fold increase in cfu/g dust on high water activity media and a 10-fold increase in the proportion of *Aspergillus* and *Penicillium* spp. to phylloplane species in the dust and detectable airborne ergosterol. Moisture source strength measurements were highly correlated to mold cfu/g in dust. House age, heating system, general construction, occupant density, temperature and relative humidity in the two groups were similar. Air change rate was higher in the high-mold homes. There were no significant differences in Der p + Der f1, cat antigen, dust and air endotoxin values, rates of combustion spillage, smoking prevalence and TVOC concentrations (Anonymous 1995, Lawton *et al.* 1998). The visible mold data were incomplete because destructive testing was not done. However, subsequent studies have revealed that inspections involving the methods used provide information that is usually proportional to the extent of hidden mold. Further, tests of the distance that mold growth is present on building materials beyond the line visible to the naked eye under field conditions show that this is typically another 0.5 m (Miller *et al.* 2000).

Objective health measures were made with the children in the subset of 39 homes. Peak flows, respiratory symptoms, nasal lavage and extensive blood immunological measures were examined. Gross symptoms followed the pattern of the previously described epidemiological studies. Living in more, compared to less contaminated homes was associated with a larger number of lymphocyte CD3$^+$ T cells expressing CD45RO and CD20$^+$ B cells expressing CD5$^+$ and a reduced CD4/CD8 ratio. These relate to the children being challenged by allergens and small, but clinically significant changes (on a population level) in relation to time to recovery from infectious disease, respectively. The levels of statistical significance were $P=0.02$, $P=0.07$ and $P=0.06$ respectively after controlling for child's age, dust mite antigens and the presence of furry or feathered pets or a humidifier. This represented two studies of the population taken within one year (Dales *et al.* 1998).

The third Canadian study was a prospective infant respiratory health study (Dales *et al.* 2006). This provided evidence of a bottom threshold for the extent of damage and influence on a population level for visible area of mold and dampness similar to that seen in the Wallaceburg studies (Miller *et al.* 1999, Dales *et al.* 2010).

Three cognizant authority reviews addressing mold and asthma (NAS 2000) and mold and dampness (Health Canada 2004, NAS 2004) reviewed the available literature on mold and health to the time of publication. After many studies carried out in different populations, in different countries and by different researchers there remain two important population health effects of mold and dampness (1) an effect associated with allergic mechanisms, and (2) an effect based on a non-allergic (toxic) mechanism. The possible existence of a toxic mechanism when dampness is associated with mold has not been reported in the studies of house dust mite exposure.

Since the publication of these reviews, a number of large studies have appeared. A very large asthma centre study was reported comprising *c.* 17,000 people in cities throughout the EU. In a subset of *c.* 1,100 of these, the frequency of sensitisation to molds increased significantly with increasing asthma severity for either for severe versus mild asthma) across all study areas. Mold sensitization was greater in areas where there was a higher prevalence of damp housing (Zureik *et al.* 2002). Two large Finnish studies have been published that have substantially increased the evidence base that indoor mold might cause asthma. The first study was of case-control study of adult asthmatics and mold in the workplace involving *c.* 1,400 people. The risk of asthma was related to the presence of visible mold and/or mold odor in the workplace and the fraction of asthma attributable to mold exposure to be 35% among the exposed (Jaakkola *et al.* 2002). Secondly, the same group conducted a longitudinal study of the independent and combined effects of parental atopy and exposure to molds in homes on the development of asthma in childhood in a 6-year prospective cohort study of 1,984 children 1-7 years of age without asthma at recruitment. Approx. 7% of the children developed asthma, which was significantly related to the extent of visible mold and the presence of mold odor in the home reported at baseline. This effect was not confounded by, and did not interact with, parental atopy (Jaakkola *et al.* 2005).

The exposure information was gathered by careful assessment of the extent of visible mold (Haverinen-Shaughnessy *et al.* 2006), so far the only exposure measure that has proved reliable for health assessment (Health Canada 2004). From the exposure perspective, visible mold reflects both superficial and more serious damage. Air samples from mold-damaged homes had greater numbers of *Aspergillus* and *Penicillium* spores as well as spore and mycelial fragments. Long-duration air samples of these metabolites (5 days) were highly correlated to the area of visible mold in an analysis of data from 110 homes so far in the PEI infant health study (Foto *et al.* 2005). Both measures were strongly correlated to the area of visible mold and this held for the entire dataset of some 300 homes (unpublished data). However, the correlation for ergosterol was somewhat higher than for fungal glucan. This proved to be because there is some fungal glucan from yeasts in settled dust not related to mold damage. In this study, the molecular weights of the glucans were measured by molecular sieve chromatography and the majority of the glucan was fungal (i.e. representing both filamentous fungi and yeasts) in origin (Foto *et al.* 2005). The homes where the indoor fungal burden was mainly derived from outdoors and those where it was mainly derived from moldy building materials contained less glucan and ergosterol in the air, but nonetheless both were present in similar amounts. However, as the degree of mold and water damage gets more serious, there is greater potential for misclassification. That is when the visible mold does not proportionally reflect the extent of actual damage (Miller *et al.* 2000, Berghout *et al.* 2005).

Cho *et al.* (2006) conducted a prospective study of 640 infants in a high risk population (one atopic parent) in Cincinnati. In this study, mold and water damaged were assessed by trained inspectors by methods similar to that from the previously mentioned studies in Finland and Canada (Foto *et al.* 2005). More than half of the homes had little visible mold, but 5% had 0.2 m$^2$ visible mold or more. This was associated with increased risk of recurrent wheezing – nearly 2´ in infants and 6´ in aeroallergen-sensitized infants. Infants in the high-income group had a significantly lower risk of recurrent wheezing than those in the low-income group. Iossifova *et al.* (2009) used similar data in a group of circa 500 infants followed for three years to demonstrate that exposure to more versus less mold increased the probability causing asthma.

Pekkanen *et al.* (2007) studied matched new cases of asthma in children aged 12-84 months (n=121) in a prospective study with randomly-selected population controls (n=241). Home inspections were conducted as described in Haverinen-Shaughnessy *et al.* (2006). The risk of asthma increased with the severity of moisture damage and presence of visible mold in the main living quarters, but not in other areas of the house. Associations were comparable for atopic and non-atopic asthma and for children aged >30 months or ≥30 months.

These more recent studies have generally strengthened the argument for both allergic and non-allergic based population health outcomes identified by the National Academy of Science and Health Canada panels. This has also resulted in clear guidance from the US Centers for Disease control in terms of the need to eliminate moisture and remediate mold damage (Krieger *et al.* 2010). In addition, mold damage is included in recommendations based on intervention data for multi-allergens reduction and control as beneficial for population health have been supported by the US CDC Community Guide process (CDC 2009) and in a Cochrane review (Maas *et al.* 2009) As identified by the National Academy of Science panel on asthma (NAS 2000), the critical problem in resolving the percentage of attributable risk and causality for the allergic and toxic population health effects remains the lack of useful exposure assessment tools. The term "exposure assessment" in this context means internal dose.

# Biomarkers

## *Antigens/allergens*

To resolve the causality questions for allergy, the National Academy of Sciences panel stated that "standardized methods for assessing exposure to fungal allergens are essential, preferably based on measurement of allergens rather than culturable or countable fungi" and that the identification of more fungal allergens and patterns of cross-reactivity was a priority (NAS 2004). These allergens persist regardless of spore viability and are found on spores, in spore and mycelial fragments and in smaller particles now known to comprise the large majority of exposure to fungi indoors (see Green *et al.*, Chapter 8). Importantly, after discovery they can be used to measure external (antigens/allergens) and internal exposure (reactive human antibodies) in populations. These are the types of data needed to gather information on causality for allergic-based disease such as asthma.

Although human antigens and allergens are known from many fungi, the majority of well-characterized human allergens come from fungi that occur outdoors (*Alternaria, Cladosporium*) and from a few disease-causing fungi (e.g. *Aspergillus fumigatus*; Horner *et al.* 1994, Simon-Nobbe *et al.* 2008).

For the existing human antigens from mold fungi associated with building materials, most information exists concerning *Stachybotrys chartarum sensu lato* and *Penicillium chrysogenum*. Both of these fungi share a common biological feature that has not been widely appreciated in the allergy community, namely that they have historically represented a collection of genotypes. The broad species *Stachybotrys chartarum* was split into two taxa, *S. chartarum* and *S. chlorohalonata* (Andersen *et al.* 2003). It is closely related to *Stachybotrys echinata* (previously named *Memnoniella echinata*). This occupies a similar environmental niche albeit in warmer climates (Miller *et al.* 2003). *S. chartarum* has two chemotypes, one of which overlaps with the metabolite spectrum of *S. cholorohalonata*. In the case of *Fusarium graminearum* chemotypes because there are important genetic differences (Miller *et al.* 1991, O'Donnell *et al.* 2000). Xu *et al.* (2007) examined chemotypes of strains of *S. chartarum* and of strains of *S. chlorohalonata* collected from all over North America and were able to identify a 34 kDa human antigen identified using sera from many patients from many locations. This antigen is produced by both *S. chartarum* and *S. chlorohalonata*. The protein is produced in quantity in on the natural substrate of the fungus (straw) and on paper-faced gypsum wallboard and suitable monoclonals have been produced to detect the antigen in building samples (Xu *et al.* 2008). The concentrations of the antigen on spores are high (Rand and Miller 2008) and when the *S. chartarum* and *S.*

*chlorohalonata* grow on gypsum wallboard, it comprises as much as 1% of excreted protein (Xu *et al.* 2007). The complete sequence of the protein has been determined and it has been expressed in *Escherichia coli* (Shi *et al.* 2011).

There are other reports of spore-borne protein(s) from *S. chartarum sensu lato* being immunomodulatory in mice, but neither their occurrence in nature nor their human antigenicity has been reported (Schmechel *et al.* 2006). Polyclonal antibodies have also been produced to an uncharacterized hemolysin-like protein, stachylysin (Vesper *et al.* 2001). The polyclonal antibodies reported for the assay recognise proteins from other unrelated species such as *P. chrysogenum* and may cross react more broadly (Van Emon *et al.* 2005, Rand and Miller 2008).

Allergens from *P. chrysogenum* have been more extensively studied, and five have been identified and accepted by the International Union of Immunological Societies Allergen Nomenclature Sub-Committee. Various reports indicate there are many more antigenic proteins (Shen *et al.* 2003, and references cited therein). As with *S. chartarum sensu lato*, *P. chrysogenum* is known to be a species complex. In 2004, molecular studies identified the existence of 4 groups within this species (Scott *et al.* 2004), one of which (group 4) dominated indoors in the Wallaceburg (ON) homes. A study involving strains from across Canada demonstrated that chemotypes occurred in these strains. More importantly, all but a few of the strains isolated from building materials across Canada were in group 4 (De la Campa *et al.* 2007). This molecular species is different genetically and based on polyphasic taxonomy can probably be defined as a new species (Scott, personal communication).

Most of the known allergens from *P. chrysogenum* have been isolated principally from Asian strains of *P. chrysogenum*. Based on strains of unknown clade and sera collected in Taiwan, *P. chrysogenum* produces 32, 34 and 60 kDa antigens (Shen *et al.* 2003, and references reported therein). More than 90% of these strains were observed by immunoblotting to excrete a 52 and a 40 kDa protein antigenic to humans. Canadian clade 4 strains produced different combinations of fungal proteins, recognized on western blot using human sera from previously screened patients (similar to Xu *et al.* 2007). From this work, a strongly and consistently antigenic 52 kDa protein was isolated from group 4 strains, but not reliably detected in the strains tested from the three other non building-associated clades. This protein is also produced in nature including on paper-faced wallboard. As with the *S. chartarum* 34 kDa antigen, the 52 kDa antigen is also present in high concentrations on spores and is excreted to a similar extent. The 52 kDa protein had previously been suspected as being an allergen based on human and animal data but was ignored subsequent, possibly because it is not present in non clade 4 strains (Wilson *et al.* 2009, Luo *et al.* 2010). As noted previously, to allow the allergen exposure outcome questions to be addressed much more work similar to this with authentic and characterized strains isolated from building materials is needed.

Using a similar process, related 43 and 41 kDa human antigens isolated from *Aspergillus versicolor* antigen not present in other damp building fungi has been reported by Liang *et al.* (2011). *A. versicolor* is one of the most common fungi in damp buildings in the UK, various European and Scandinavian countries as well as the United States and Canada. Based on studies from Finland, Norway and Germany, it is among the common species resulting in an IgE reaction. Benndorf *et al.* (2008) studied allergens found in *A. versicolor* spores which were all common fungal proteins. Using washed spore fragments, Schmechel *et al.* (2005) produced antibodies that did not show species specificity, possibly because, as with Bennedorf *et al.* (2008), the extracts contained mainly common fungal proteins.

**Glucan**

As noted, there are two accepted effects on population health, one associated with allergic reactions and one associated with toxicity. Miller (1990, 1992) noted that since absolute amounts of low molecular weight were very low in spores and fragments, the putative effect must be a direct effect from inhalation exposure to fragments directly on lung biology. There are at least two possible candidates: $\beta$-1$\rightarrow$3-D-glucan and low molecular weight toxins. Additionally, since damp buildings have a number of communities of fungi associated with different types of moisture failures, the response to low molecular weight compounds of greatly diverse chemical structures has necessarily to be a general response, which was perplexing.

Some basidiomycetes produce $\beta$-1$\rightarrow$3-D-glucan and $\beta$-1$\rightarrow$6-D-glucans (Panchero-Sanchez *et al.* 2006). Although not especially well studied, different fungi produce (different) amounts of $\beta$-1$\rightarrow$3-D-glucan relative to other glucans (Odabasi *et al.* 2006). Some yeasts make a $\beta$-1$\rightarrow$3-D-glucan with a short $\beta$-1$\rightarrow$6-D-glucan chain at a ratio of 1: 1 (zymosan; Ohno *et al.* 2001). However, the anamorphic Trichocomaceae, i.e. *Penicillium, Aspergillus* and related hyphomycetes associated with damp building materials, contain predominantly $\beta$-1$\rightarrow$3-D-glucan in the triple helical form (Foto *et al.* 2005, Odabasi *et al.* 2006).

Pulmonary inflammation in rodents resulting from the glucan form dominating in cleistothecial ascomycetes indicate that the partially opened helical conformation of $\beta$-1$\rightarrow$3-D-glucan has greater immunoactivating potency than alternative conformations (Young *et al.* 2006). In Sprague Dawley rats, intratracheally instilled zymosan has been shown to have a variety of immunotoxic and immunomodulatory properties (Young and Castranova 2005). This includes increased key regulatory molecules such as TNF-α as well as IL-6, IL-10, and IL-12p70 (Young *et al.* 2006).

The C-type lectin-like receptor, Dectin-1, was identified as the major receptor for fungal β-glucans on murine macrophages. It plays a role in the cellular response to these carbohydrates (Willment *et al.* 2001). The glucan receptor (bGR) of human Dectin-1 is expressed in all monocyte populations as well as macrophages, DC, neutrophils and eosinophils. This receptor is also expressed on B cells and some T cells. The two bGR isoforms – bGR-A and bGR-B are expressed by these cell populations in peripheral blood, but expression of the latter is in a cell-specific fashion. Dectin-1 expression is not significantly modulated on macrophages during inflammation, but is decreased on recruited granulocytes. This means that exposure to mold glucan has important immunomodulatory effects (Willment *et al.* 2001, 2005, Willment and Brown 2008).

When blood leucocytes from healthy volunteers and patients allergic to house dust mite were incubated with a number of forms of fungal glucans, histamine release was not observed on exposure to only the glucans. In the presence of anti-immunoglobulin E (IgE) antibody or specific antigens, all the glucans investigated led to an enhancement of the IgE-mediated histamine release. Two glucans, β-1→3-D and β-1→6-D, induced a significant potentiation of the mediator release at *c.* 0.2 μM. These two glucans increase IgE-mediated histamine release (Holck *et al.* 2007).

From a biochemistry perspective, the response of a glucan-sensitive factor in the blood of the horseshoe crab (*Limulus polyphemus*) is best understood. This ancient marine animal has an effective immune system based on very sensitive receptors for endotoxin and fungal glucan (factor G). These receptors stimulate a series of factors that result in blood clotting around the invading bacterium or fungus. The system that recognizes glucan in *Limulus* Amoebocyte Lysate (LAL) is most sensitive to the triple helical form of glucan that predominates in anamorphic ascomycetes as well as the linear β-1→3-D-glucan in triple helical form, curdlan (Tanaka *et al.* 1991, Aketagawa *et al.* 1993).

Inhalation exposure β-1→3-D-glucan causes a number of effects, including changes in neutrophils, macrophages, complement and eosinophils as well as changes in blood levels of inflammatory markers in both humans and animals. Early studies of small populations in residential housing suggested an exposure-response relationship between concentrations of airborne fungal glucan and eye, nose and throat irritation, dry cough, itching skin, hoarseness and fatigue (Rylander and Lin 2000, Thorn *et al.* 2001, Beijer *et al.* 2002, Douwes 2005, Williams *et al.* 2005).

There are two human exposure studies, both of which have limitations in terms of size and available description. A group of 16 non-atopic, asymptomatic subjects were exposed to aerosolized curdlan ($\beta$-1$\rightarrow$3-D-glucan; 161 kDa) for 4 h at approx. 200 ng/m$^2$. These subjects had a small but statistically-significant reduction in $FEV_1$ that remained significant for three days after exposure. In the entire population in this study (26), there was a small but statistically significant increase in severity of nasal and throat irritation (Rylander 1993, 1996).

There are also two studies involving nasal instillation of curdlan. In a double-blinded crossover study, five garbage workers with occupational airway symptoms and five healthy garbage workers were intranasally exposed to 1 mg/l curdlan in saline (and the saline diluent as a control) for 15 min (along with similar treatments involving endotoxin, compost waste dust and *Aspergillus fumigatus* spores). Nasal cavity volume and nasal lavage tests were performed at zero time and 3, 6, and 11 h post exposure. Curdlan induced an increase in albumin and a slight increase in IL-1$\beta$ 6-11 h post exposure. The most pronounced effect was on total nasal volume was seen after challenge with curdlan with a significant increase 6 h after the challenge (Sigsgaard *et al.* 2000). More recently, 36 volunteers were randomly exposed to clean air, 332 µg/m$^3$ office dust and 379 µg/m$^3$ dust spiked with around 4 mg glucan (dust spiked with aldehyde was a positive control). Acoustic rhinometry, rhinostereometry, nasal lavage, and lung function tests were carried out. After the exposures to dust spiked with the glucan, nasal volume had decreased relative to clean air or office dust exposure ($P=0.036$). After 3 h, glucan-spiked dust produced a measurable swelling ($P=0.039$). There was no significant change in cytokines and interleukin-8. Nasal eosinophil cell concentration increased on exposure to dust spiked with glucan ($P=0.045$; Bønløkke *et al.* 2006).

There are also two challenge studies that exposed volunteers to the Basidiomycete form, grifolan (1$\rightarrow$6,1$\rightarrow$3-$\beta$-D-glucan, 50,000 MW; Beijer *et al.* 2002; Thorn *et al.* 2001). In both studies, small immunomodulatory effects were observed.

From two studies in Canada, exposure to fungal glucan in air from long-duration samples was greater inside than out (Foto *et al.* 2005, Miller *et al.* 2007). Glucan is present in similar concentrations in spores of species that are common in outdoor air and migrate indoors as well as those that grow on building materials. It is also present in yeasts which accumulate indoors (Foto *et al.* 2004, 2005). An important variable, however, is that when houses have visible mold, spores and hyphal fragments are a larger percentage of the total intact spores (Foto *et al.* 2005). This means that the glucan (and allergens and low molecular weight toxins) are able to penetrate deeper into the lungs in greater relative amounts.

As noted, the triple helical form of glucan induces a variety of inflammatory responses. Rand *et al.* (2010) showed that installation of as little as 4 ng glucan/animal results in dectin-1 mRNA transcription and expression in bronchiolar epithelium, alveolar macrophages, and alveolar type II cells. Compared to controls, 54 of 83 genes assayed were significantly modulated. Nine gene mRNA transcripts (Ccl3, Ccl11, Ccl17, Ifng, Il1α, Il-20, TNF-α, Tnfrsf1b, and CD40lg ) were expressed indicating the possibility of a central role. Immunohistochemistry revealed Ccl3, Il1-alpha, and TNF-alpha expression in bronchiolar epithelium, alveolar macrophages and alveolar type two cells. This illustrated the role these cells have in the recognition of, and response to glucan. Further studies of this type should be useful in explaining the irritant and inflammatory responses. In addition, potentially other responses may be discovered from more study of the human and rodent exposure.

**Low molecular weight compounds**

Most of what was known about the effects of spore borne toxins still comes from one fungus, *S. chartarum sensu latto* (Miller *et al.* 2003). Most experiments reported to date involved intratracheal instillation of spores, which results in massive injury. Rodent studies showed that the low dose response to spores of an atranone-producing *S. chartarum* strain was different from spores of macrocyclic trichothecene-producing strains (Flemming *et al.* 2004, Hudson *et al.* 2005). The latter strains resulted in a rapid response, sustained for some 24 h post-instillation and followed by decline to control values. The atranone-producing strain stimulated inflammatory and cytotoxic responses that increased from control levels 3-24 h post-instillation to significantly-different values 24-96 h post-instillation.

Rand *et al.* (2005, 2006) have examined dose-response relationships associated with pure brevianamide A, mycophenolic acid, roquefortine C (the former from *P. brevicompactum* and the latter from *P. chrysogenum*) as well as atranones A and C. Albeit with different time courses, potencies, and somewhat different patterns of response, instilled doses at the nM/g body weight level induced significant manifest inflammatory responses. These included differentially elevated macrophage and neutrophil numbers and differential changes in MIP-2, TNF-α and IL-6 concentrations in the bronchioalveolar lavage fluid of the mice. Pure atranone essentially replicated the response of the instillation of low dose of the spores of the atranone chemotype of *S. chartarum* (Rand *et al.* 2005, 2006). These experiments demonstrated that, regardless of the diverse array of structures of these toxins, all activated the C-X-C pathway in the pertinent lung cells.

In contrast, *in vitro* studies employing RAW 264.7 macrophages (Nielsen *et al.* 2002, Huttunen *et al.* 2004) indicated that the production of inflammatory mediators in immortalized macrophages was not triggered by pure atranones. These findings are not supported by subsequent *in vivo* studies and in primary cell culture studies. Working with very pure toxins from fungi common on damp building materials and which have been reported on moldy building materials, Rand *et al.* (2010) studied their effects in a murine model by installation. The toxins included atranone C, brevianamide, cladosporin, mycophenolic acid, neoechinulin A and B, sterigmatocystin and TMC-120A in a murine model. The dose used was used was within the estimated range of possible human exposure. Histology and histochemistry revealed that toxin exposed lungs resulted in inflammation. Data from both gene expression and translation of certain genes in the treatment groups suggested a central roles of toxin-induced pro-inflammatory lung responses. Hierarchical cluster analysis revealed significant patterns of gene transcription linked in three toxin groupings: (1) brevianamide, mycophenolic acid and neoechinulin B, (2) neoechinulin A and sterigmatocystin, and (3) cladosporin, atranone C and TMC-120 (Miller *et al.* 2010). The results further confirm the inflammatory nature of metabolites/toxins. The idea that a common molecular pathway is induced by low exposures of various fungal toxins *in vivo* provides a plausible hypothesis to explain a major, recognized population health effect. A door has been opened to studies on potential biomarkers of this effect.

Primary cell cultures and *in vivo* models rather than immortalized cell lines are the tools of choice as further efforts are made to develop *in silico* models and develop markers of human disease (Collins *et al.* 2008, *Rand et al.* 2011, and references cited therein).

## Conclusions

In the past 30 years, there has been an apparent change in the percentage of asthmatics sensitized to mold. Based on data acquired in the late 1980s, the maximum attributable risk appeared to be 20% of total asthmatics (Dekker *et al.* 1991). European studies indicate that sensitization to mold is a significant risk factor for asthma and for admission to hospital for asthma (Zureik *et al.* 2002, O'Driscoll *et al.* 2005). A recent analysis of fungal allergy suggests these percentages are rising further (Simon-Nobbe *et al.* 2008). Since all buildings have some mold, and because a large percentage of the existing housing stock has dampness and mold problems, more effort is needed to resolve the thresholds of effect for mold. This cannot be done until true markers of exposure (internal dosemetry) are identified.

## Acknowledgements

I thank the Natural Research & Engineering Research Council, Canada Mortgage and Housing Corporation and Health Canada for support. Dr. Robert Dales, Mr. Jim White and Mr. Ken Ruest for their insights over the past 20 years and Dr. Brian Flannigan for advice on the Edinburgh studies discussed and many useful comments on the manuscript.

## References

Aketegawa J, Tanaka S, Tamura H and Shibatav YH (1993) Activation of limulus coagulation factor G by several 1,3 β D-glucans: comparison of the potency of glucans with identical degree of polymerization but different conformations. J Biochem 113: 683-686.

Amend AS, Seifert KA, Samson R and Bruns TD (2010) Indoor fungal composition is geographically patterned and more diverse in temperate zones than in the tropics. Proc Natl Acad Sci USA 107: 13748-13753.

Andersen B, Nielsen KF, Thrane U, Cruse M, Taylor J and Jarvis BB (2003) Molecular and phenotypic descriptions of *Stachybotrys chlorohalonata sp. nov.* and two chemotypes of *Stachybotrys chartarum* found in water-damaged buildings. Mycologia 95: 1227-1238.

Anonymous (1904) The danger from arsenic in clothing. JAMA 43: 1235.

Anonymous (1995) Moldy houses: why they are & why we care. Canada Mortgage & Housing Corporation, Ottawa, Canada, 66 pp.

Beijer L, Thorn J and Rylander R (2002) Effects after inhalation of 1,3 β D glucans and relation to mould exposure in the home. Med Inflam 11: 149-153.

Benndorf D, Muller A, Bock K, Manuwald O, Herbarth O and Von Bergen M (2008) Identification of spore allergens from the indoor mould *Aspergillus versicolor*. Allergy 63: 454-460.

Berghout J, Miller JD, Mazerolle R, O'Neill L, Wakelin C, Mackinnon B, Maybee K, Augustine D, Levi CA, Levi C, Levi T and Milliea B (2005) Indoor environmental quality in homes of asthmatic children on the Elsipogtog Reserve (NB), Canada. Int J Circumpol Health 64: 77-85.

Bønløkke JH, Stridh G, Sigsgaard T, Kjærgaard SK, Löfstedt H, Andersson K, Bonefeld-Jørgensen EC, Jayatissa MN, Bodin L, Juto J-E and Mølhave L (2006) Upper-airway inflammation in relation to dust spiked with aldehydes or glucan. Scand J Work Environ Health 32: 374-382.

Bonnefoy X (2007) Inadequate housing and health: an overview. Int J Environ Pollution 30: 411-429.

Brunekreef B, DW Dockery, FE Speizer, JH Ware , JD Spengler and BG Ferris (1989) Home dampness and respiratory morbidity in children. Am Rev Resp Dis 140: 1363-1367.

Carnelly D, Haldane JS and Anderson AM (1887) The carbonic acid, organic matter and micro-organisms in air, more especiallyin dwellings and schools. Phil Trans Royal Society, series B, 178: 61-111.

CDC (2009) Asthma control. United States Centers for Disease Control, Atlanta, GA, USA. Available at: http://www.thecommunityguide.org/asthma/index.html.

Chan-Yeung M, Becker A, Lam H, Dimich-Ward H, Ferguson A, Warren P, Simons E. Broder I and Manfred J (1995) House dust mite allergen levels in two cities in Canada: effects of season, humidity, city and home characteristics. Clin Experl Allergy 25: 240-246.

Cho SH, Reponen T, LeMasters G, Levin L, Huang J, Meklin T, Ryan P, Villareal M and Bernstein D (2006) Mold damage in homes and wheezing in infants. Ann Allergy Asthma Immunol 97: 539-545.

Collins FS, Gray GM and Bucher JR (2008) Toxicology. transforming environmental health protection. Science 319: 906-907.

Corda AC (1837) *Icones fungorum hucusque cognitorum* I. Prague, Czechoslovakia, p. 21.

Cullen WR and Bentley R J (2005) The toxicity of trimethylarsine: an urban myth. Environ Monitor 7: 11-15.

Cullen WR, McBride BC, Pickett WA and Reglinski, J (1984) The wood preservative chromated copper arsenate is a substrate for trimethylarsine biosynthesis. Appl Environ Microbiol 47: 443-444.

DaCosta EWB (1972) Variation in the toxicity of arsenic compounds to microorganisms and the suppression of the inhibitory effects by phosphate. Appl Microbiol 23: 46-53.

Dales RE, H Zwanenburg, R Burnett and CA Franklin (1991a) Respiratory health effects of home dampness and molds among children. Am J Epidemiol 134: 196-293.

Dales RE, I Schweitzer, S Bartlett, M Raizenne and R Burnett (1994) Indoor air quality and health: reproducibility of respiratory symptoms and reported home dampness and molds using a self-administered questionnaire. Indoor Air 4: 2-7.

Dales R, JD Miller, JM White, C Dulberg and AI Lazarovitis (1998). The influence of residential fungal contamination on peripheral blood lymphocyte populations in children. Arch Environ Health 53: 190-195.

Dales RE and Miller JD (1999) Residential fungal contamination and health: microbial cohabitants as covariates. Environ Health Perspect 107 S3: 481-483.

Dales RE, Miller JD, Ruest K, Guay M and Judek S (2006). Airborne endotoxin is associated with respiratory illness in the first two years of life. Environ Health Perspect 114: 610-614.

Dales RE, R Burnett and H Zwanenburg (1991b) Adverse effects in adults exposed to home dampness and molds. Am Rev Resp Dis 143: 505-509.

Dales RE, Ruest K, Guay M, Marro K and Miller JD (2010) Residential fungal growth and incidence of respiratory illness during the first two years of life. Environ Res 110: 692-698.

De La Campa R, Seifert K and Miller JD (2007) Toxins from strains of *Penicillium chrysogenum* isolated from buildings and other sources. Mycopathologia 163: 161-168.

Dekker C, Dales R, Bartlett S, Brunekreef B and Zwanenburg H (1991) Childhood asthma and the indoor environment. Chest 100: 922-926.

Douwes J (2005) Health effects of 1,3 β D glucans: The epidemiological evidence. In: Young, S-H and Castranova V (eds.) Toxicology of 1→3 beta-glucans. CRC Press, Boca Raton, FL, USA, pp. 35-52.

Dudley HC (1938) Dangers in the use of arsenic compounds in the preservation of wood and paper used in the construction and management of dwellings. Bull Office Intern Hyg Pub 30: 2612-2613.

Feinberg SM (1946) Allergy in practice. Year Book Publishers, Chicago, IL, USA.

Flannigan B and Miller JD (2011) Microbial growth in indoor environments. In: Flannigan B, Samson RA and Miller JD (eds.) Microorganisms in home and indoor work environments: diversity, health impacts, investigation and control, second edition. Taylor & Francis, New York, NY, USA, pp. 57-107.

Flemming J, Hudson B and Rand TG (2004) Comparison of inflammatory and cytotoxic lung responses in mice after intratracheal exposure to spores of two different *Stachybotrys chartarum* isolates. Toxicol Sci 78: 267-275.

Foto M, Plett J, Berghout J and Miller JD (2004) Modification of the Limulus Amebocyte Lysate assay for the analysis of glucan in indoor environments. Anal Bioanal Chem 379: 156-162.

Foto M, Vrijmoed LLP, Miller JD, Ruest K, Lawton M and Dales RE (2005) A comparison of airborne ergosterol, glucan and Air-O-Cell data in relation to physical assessments of mold damage and some other parameters. Indoor Air 15: 257-266.

Frankland AW and Hay MJ (1951) Dry rot as a cause of allergic complaints. Acta Allergologica 4: 186-200.

Gravesen S (1978) Identification and prevalence of culturable mesophilic microfungi in house dust from 100 Danish homes. Allergy 33: 268-272.

Haverinen-Shaughnessy U, Pekkanen J, Hyvärinen A, Nevalainen A, Putus T, Korppi M and Moschandreas D (2006) Children's homes – determinants of moisture damage and asthma in Finnish residences. Indoor Air 16: 248-255.

Haysom JC and Reardon ST (1998) Why houses need mechanical ventilation. Construction Update 14. IRC, National Research Council of Canada, Ottawa, Canada.

Health Canada (2004) Fungal contamination in public buildings: health effects and investigation methods. Health Canada, Ottawa, Canada.

Holck P, Sletmoen M, Stokke BT, Permin H and Norn S (2006) Potentiation of histamine release by microfungal (1→3)- and (1→6)-β-D-glucans. Basic & Clin Pharm & Tox 101: 455-458.

Horner WE, Helbling A, Salvaggio JE and Lehrer SB (1995) Fungal allergens. Clin Microbiol Rev 8: 161-179.

Howden-Chapman P, Matheson A, Crane J, Viggers H, Cunningham M, Blakely T, Cunningham C, Woodward A, Saville-Smith K, O'Dea D, Kennedy M, Baker M, Waipara N, Chapman R and Davie G (2007) Effect of insulating existing houses on health inequality: cluster randomized study in the community. BMJ 334: 460-469.

Hudson B, Flemming J, Sun G and Rand TG (2005) Comparison of immunomodulator mRNA expression and concentration in lungs of *Stachybotrys chartarum* spore exposed mice. J Toxicol Environ Health A 68: 1321-1335.

Huss H (1913) Microorganisms forming arsine. Svensk Farmaceutisk Tidskrift 17: 289-294.

Huttunen K, Pelkonen J, Nielsen KF, Nuutinen U, Jussila J and Hirvonen MR (2004) Synergistic interaction in simultaneous exposure to *Streptomyces californicus* and *Stachybotrys chartarum*. Environ Health Perspect 112: 659-665.

Hyde HA (1973) Atmospheric pollen grains and spores in relation to allergy. II. Clinical Allergy 3: 109-126.

Iossifova YY, Reponen T, Ryan PH, Levin L, Bernstein DI, Lockey JE, Hershey GK, Villareal M and LeMasters G (2009) Mold exposure during infancy as a predictor of potential asthma development. Ann Allergy Asthma Immunol 102: 131-137.

Jaakkola JJ, Hwang BF and Jaakkola N (2005) Home dampness and molds, parental atopy, and asthma in childhood: a six-year population-based cohort study. Environ Health Perspect 113: 357-361.

Jaakkola MS, Nordman H, Piipari R, Uitti J, Laitinen J, Karjalainen A, Hahtola P and Jaakkola JJ (2002) Indoor dampness and molds and development of adult-onset asthma: a population-based incident case-control study. Environ Health Perspect 110: 543-547.

Kercsmar CM, Dearborn DG, Schluchter M, Xue L, Kirchner HL, Sobolewski J, Greenberg SJ, Vesper SJ and Allan T (2006) Reduction in asthma morbidity in children as a result of home remediation aimed at moisture sources. Environ Health Perspect 114: 1574-1580.

Kozak PP, Gallup J, Cummins LB and Gillman SA (1985). Endogenous mold exposure: environmental risk to atopic and non-atopic patients. In: Gammage RB and Kaye SV (eds.). Indoor air and human heath. Lewis Publishers, Chelsea, MI, USA, pp. 149-170.

Kozak PP, Gallup J, Cummins LH and Gillman SA (1980) Currently available methods for home mold surveys. II. Examples of problem homes surveyed. Annals Allergy 45: 167-176.

Krieger J, Jacobs DE, Ashley PE, Baeder A, Chew CL, Dearborn H, Hynes HP, Miller JD, Morley R and Rabito F (2010) Housing interventions and control of asthma-related indoor biologic agents: a review of the evidence. J Public Health Man & Practice 16: s11-s20.

Lacey J (1994) Microorganisms in organic dusts. In: Rylander R and Jacobs RR (eds.) Handbook of organic dusts. Lewis Publishers, Boca Raton, FL, USA, pp. 1-17.

Lawrence R and Martin JD (2001) Moulds, moisture and microbial contamination of First Nations housing in British Columbia, Canada. In J Circumpolar Health 60: 150-156.

Lawton MD (1999) Reacting to durability problems with Vancouver buildings. In: Lacasse MA and Vanier DJ (eds.). Durability of building materials and components 8 (part 2). NRC Research Press, Ottawa, Canada, pp. 990-998.

Lawton MD, Dales RE and White J (1998) The influence of house characteristics in a Canadian community on microbiological contamination. Indoor Air 8: 2-11.

Liang L, Zhao W, Xu J and Miller JD (2011) Characterization of two related exoantigens from the biodeteriogenic fungus *Aspergillus versicolor*. Int Biodegrad Biodet (in press).

Loftness V, Hakkinen B, Adan O and Nevalainen (2006) Elements that contribute to healthy building design. Environ Health Perspect 115: 965-970.

Luo W, Wilson AM and Miller JD (2010) Characterization of a 52 kDa exoantigen of *Penicillium chrysogenum* and monoclonal antibodies suitable for its detection. Mycopathologia 169: 15-26.

Lynn C (1980) Wallpaper in America from the seventeenth century to World War I. Norton & Company, New York, NY, USA.

Maas T, Kaper J, Sheikh A, Knottnerus JA, Wesseling G, Dompeling E, Muris JW and Van Schayck CP (2009) Mono and multifaceted inhalant and/or food allergen reduction interventions for preventing asthma in children at high risk of developing asthma. Cochrane Database Syst Rev 2009 Jul 8; CD006480.

Maunsell K (1952) Air-borne fungal spores before and after raising dust; sampling by sedimentation. Int Arch Allergy Appl Immunol 3: 93-102.

Maunsell K (1954) Concentration of airborne spores in dwellings under normal conditions and under repair. Int Arch Allergy Appl Immunol 5: 373-376.

Maunsell K, Wraith DG and Cunnington AM (1968) Mites and house-dust allergy in bronchial asthma. Lancet 291: 1267-1270.

McCloskey JW (1986) Seasonal temperature patterns of selected cities in and around Ohio. Ohio J Sci 86: 8-10.

Miller JD (1990) Fungi as contaminants of indoor air. In: Proceedings 5th international conference on indoor air quality and climate, Toronto, Canada, 29 July – 3 August 1990, pp. 51-64.

Miller JD (1992) Fungi as contaminants of indoor air. Atmosph Environ 26A: 2163-2172.

Miller JD (1995) Quantification of health effects of combined exposures: a new beginning. In: Morawska L, Bofinger ND and Maroni M (eds.) Indoor air – an integrated approach. Elsevier, Amsterdam, the Netherlands, pp. 159-168.

Miller JD (2007) Indoor air quality and occupant health in the residential built environment: future directions. In: Yoshino H (ed.) Proceedings IAQVEC 2007, Sendai, Japan, pp. 15-22.

Miller JD and Day JH (1997) Indoor mold exposures: epidemiology, consequences and immune therapy. Can J Allergy Clinical Immunol 2: 25-32.

Miller JD and Holland H (1981) Biodeteriogenic fungi in two Canadian historic houses subject to different environmental controls. Int Biodetn Bull 17: 39-45.

Miller JD, Dales RE and White J (1999) Exposure measures for studies of mold and dampness and respiratory health. In: Johanning E (ed.) Bioaerosols, fungi and mycotoxins: health effects, assessment, prevention and control. Eastern New York Occupational and Environmental Health Center, Albany, NY, USA, pp. 298-305.

Miller JD, Dugandzic R, Frescura A-M and Salares, V (2007) Indoor and outdoor-derived contaminants in urban and rural homes in Ottawa, Canada. J Air Waste Management Association 57: 297-302.

Miller JD, Greenhalgh R, Wang YZ and Lu M (1991) Mycotoxin chemotypes of three *Fusarium* species. Mycologia 83: 121-130.

Miller JD, Haisley PD and Reinhardt JH (2000) Air sampling results in relation to extent of fungal colonization of building materials in some water damaged buildings. Indoor Air 10: 146-151.

Miller JD, Laflamme AM, Sobol Y, Lafontaine P and Greenhalgh R (1988) Fungi and fungal products in some Canadian houses. Int Biodetn 24: 103-120.

Miller JD, Rand TG and Jarvis BB (2003) *Stachybotrys chartarum*: cause of human disease or media darling? Medical Mycology 41: 271-291.

Miller JD, Sun M, Gilyan M, Roy J and Rand TG (2010) Inflammation-associated gene transcription and expression in mouse lungs induced by low molecular weight compounds from fungi from the built environment. Chemico-Biological Int 183: 113-124.

National Academy of Science (NAS) (2000) Clearing the air: asthma and indoor air exposures. National Academies Press, Washington, DC, USA.

National Academy of Science (NAS) (2004) Damp indoor spaces and health. National Academies Press, Washington, DC, USA.

Nielsen KF, Huttunen K, Hyvarinen A, Andersen B, Jarvis BB and Hirvonen M-R (2002) Metabolite profiles of *Stachybotrys* isolates from water-damaged buildings and their induction of inflammatory mediators and cytotoxicity in macrophages. Mycopathologia 154: 201-205.

O'Donnell K, Kistler HK, Tacke HK and Casper HK (2000) Gene genealogies reveal global phylogeographic structure and reproductive isolation among lineages of *Fusarium graminearum*, the fungus causing wheat scab. Proc Natl Acad Sci USA 97: 7905-7910.

Odabasi Z, Paetznick VL, Rodriguez JR, Chen E, McGinnis MR and Ostrosky-Zeichner L (2006) Differences in beta-glucan levels in culture supernatants of a variety of fungi. Med Mycol 44: 267-272.

Odom JD and DuBose G (1996) Preventing indoor air quality problems in hot, humid climates. C2MHill, Orlando, FL, USA.

O'Driscoll BR, Hopkinson LC and Denning DW (2005) Mold sensitization is common amongst patients with severe asthma requiring multiple hospital admissions. BMC Pulm Med 5: 4.

Ohno N, Miura T, Miura NN, Adachi Y and Yadomae T (2001) Structure and biological activities of hypochlorite oxidized zymosan. Carbohydr Polymer 44: 339-349.

Pacheco-Sanchez M, Boutin Y, Angers P, Gosselin A and Tweddell RJ (2006) A bioactive $\beta$ 1$\rightarrow$3, $\beta$, 1$\rightarrow$4 $\beta$ D glucan from *Collybia dryophila* and other mushrooms. Mycologia 98: 180-185.

Pearce N, Aït-Khaled N, Beasley R, Mallol J, Keil U, Mitchell E, Robertson C and the ISAAC Phase Three Study Group (2007) Worldwide trends in the prevalence of asthma symptoms: phase III of the International Study of Asthma and Allergies in Childhood (ISAAC). Thorax 62: 758-766.

Pekkanen J, Hyvärinen A, Haverinen-Shaughnessy U, Korppi M, Putus T and Nevalainen A (2007) Moisture damage and childhood asthma: a population-based incident case-control study. Eur Respir J 29: 509-515.

Platts-Mills TA (2005) Asthma severity and prevalence: an ongoing interaction between exposure, hygiene, and lifestyle. PLoS Med 2: e34.

Rand TG and Miller JD (2008) Immunohistochemical and immunocytochemical detection of SchS34 antigen in *Stachybotrys chartarum* spores and spore impacted mouse lungs. Mycopathologia 165: 73-80.

Rand TG and Miller JD (2011) Toxins and inflammatory compounds. In: Flannigan B, Samson RA and Miller JD (eds) Microorganisms in home and indoor work environments: diversity, health impacts, investigation and control. second edition. Taylor & Francis, New York, NY, USA, pp. 209-306.

Rand TG, Flemming J, Giles S, Miller JD and Puniani E (2005) Inflammatory and cytotoxic responses in mouse lungs exposed to purified toxins from building isolated *Penicillium brevicompactum* Dierckx and *P. chrysogenum* Thom. Tox Sci 87: 213-222.

Rand TG, Flemming J, Miller JD and Womiloju TO (2006) Comparison of inflammatory and cytotoxic responses in mouse lungs exposed to atranone A and C from *Stachybotrys chartarum*. Tox Environl Health 68: 1321-1235.

Rand TG, Sun M, Gilyan A, Downey J and Miller JD (2010) Dectin-1 and inflammation-associated gene transcription and expression in mouse lungs by a toxic (1,3)-β-D glucan. Arch Toxicol 84: 205-220.

Richardson B (2000) Defects and deterioration in buildings, 2$^{nd}$ edition. Routledge, Abingdon, UK.

Ring J (2005). Clinical manifestation and classification of allergic diseases. Allergy in Practice. Springer, Berlin, Germany, pp. 1-7.

Rylander R (1993) Experimental exposures to 1,3 beta D glucan. ASHRAE Transactions 1993: 338-340.

Rylander R (1996) Airway responsiveness and chest symptoms after inhalation of endotoxin or (1→3)-β D glucan. Indoor Built Environ 5: 106-111.

Rylander R and Lin RH (2000) (1,3)-beta-D-glucan – relationship to indoor air-related symptoms, allergy and asthma. Toxicology 152: 47-52.

Schmechel D, Simpson JP and Lewis DM (2005) The production and characterization of monoclonal antibodies to the fungus *Aspergillus versicolor*. Indoor Air 15 S9: 11-19.

Schmechel D, Simpson JP, Beezhold D and Lewis DM (2006) The development of species-specific immunodiagnostics for *Stachybotrys chartarum*: the role of cross-reactivity. J Immunol Meth 309: 150-159.

Scott J, Malloch D, Wong B, Sawa T and Straus N (2000) DNA heteroduplex fingerprinting in *Penicillium*. In: Samson RA and Pitt JI (eds.): Integration of modern taxonomic methods for *Penicillium* and *Aspergillus* classification. Harwood Academic Publishers, Amsterdam, the Netherlands, pp. 225-238.

Scott J, Untereiner WA, Wong B, Straus NA and Malloch D (2004). Genotypic variation in *Penicillium chrysogenum* from indoor environments. Mycologia 96: 1095-1105.

Selgrade MK, Lemanske RF Jr, Gilmour MI, Neas LM, Ward MD, Henneberger PK, Weissman DN, Hoppin JA, Dietert RR, Sly PD, Geller AM, Enright PL, Backus GS, Bromberg PA, Germolec DR and Yeatts KB (2006) Induction of asthma and the environment: what we know and need to know. Environ Health Perspect 114: 615-619.

Sexton K, Letz R and Spengler JD (1983) Indoor air pollution: a public health perspective. Science 221: 9-17.

Shen H-D, Chou H, Tam MF, Chang C-Y, Lai H-Y and Wang S-R (2003) Molecular and immunological characterization of Pen ch 18, the vacuolar serine protease major allergen of *Penicillium chrysogenum*. Allergy 58: 993-1002.

Shi C, Smith ML and Miller JD (2011) Characterization of human antigenic proteins SchS21 and SchS34 from *Stachybotrys chartarum*. Int Arch Allergy Immunol 155: 74-85.

Sigsgaard T, Bonefeld-Jùrgensen EC, Kjñrgaard SK, Mamas S and Pedersen OF (2000) Cytokine release from the nasal mucosa and whole blood after experimental exposures to organic dusts. Eur Respir J 16: 140-145.

Simon-Nobbe B, Denk U, Pöll V, Rid R and Breitenbach M (2008) The spectrum of fungal allergy. Int Arch Allergy Immunol 145: 58-86.

Singh, J (1999) Review: dry rot and other wood-destroying fungi: their occurrence, biology, pathology and control. Indoor Built Environ 18: 3-20.

Sinha RN, Van Bronswijk JEH and Wallace HAH (1970) House dust allergy, mites andtheir fungal associations. CMAJ 103: 300-301.

Solomon WR (1975) Assessing fungus prevalence in domestic interiors. J Allergy Clin Immunol 56: 235-242.

Spengler JD (1985) Indoor air pollution. N Engl Reg Allergy Proc 6: 126-134.

Strachan D (1989) Damp housing and ill health. BMJ 299: 325.

Strachan DP (1988) Damp housing and childhood asthma: validation of reporting of symptoms. BMJ 297: 1223-1226. Erratum in BMJ 297: 1500.

Strachan DP and Elton RA (1986) Relationship between respiratory morbidity in children and the home environment. Family Practice 3: 137-142.

Strachan DP and Sanders CH (1989) Damp housing and childhood asthma; respiratory effects of indoor air temperature and relative humidity. J Epidemiol Community Health 43: 7-14.

Strachan DP, Flannigan B, McCabe EM and McGarry F (1990) Quantification of airborne moulds in the homes of children with and without wheeze. Thorax 45: 382-387.

Su HJ, Rotnitzky A, Burge HA and Spengler JD (1992) Examination of fungi in domestic interiors by using factor analysis: correlations and associations with home factors. Appl Environ Microbiol 58: 181-186.

Swaebly M and Christensen CM (1952) Moulds in house dust, furniture stuffing and the air within houses. J Allergy 23: 370-374.

Tanaka S, Aketagawa J, Takahashi S and Shibata Y (1991) Activation of *Limulus* coagulation factor G by (1,3)-b-d-glucans. Carb Res 218: 167-174.

Thomson H and Petticrew M (2007) Housing and health. BMJ 334: 434-435.

Thorn J, Beiker L and Rylander R (2001) Effects after inhalation of (1→ 3) – β D glucan in healthy humans. Mediat Inflamm 10: 1730178.

Trechsel HR (1994) Moisture control in buildings. ASTM manual series; MNL 18. ASTM, Philadelphia, PA, USA.

Van der Werff PJ (1958) Mould fungi and bronchial asthma I. A mycological and clinical study. Stenfert Kroes, Leiden, the Netherlands.

Van Emon JM, Reed AW, Yike I, Vesper SJ (2005). Measurement of Stachylysin™ in serum to quantify human exposures to the indoor mold *Stachybotrys chartarum*. J Occup Environ Med 45: 582-591.

Verhoeff AP (1994) Home dampness, fungi and house dust mites and respiratory symptoms in children. PhD Thesis, Erasmus University, Rotterdam, the Netherlands.

Vesper SJ, Magnuson S, Dearborn DG, Yike, I and Haugland RA. (2001) Initial characterization of the hemolysin stachylysin from *Stachybotrys chartarum*. Infect Immun 69: 912-916.

Weinmayr G, Weiland SK, Björkstén B, Brunekreef B, Büchele G, Cookson WO, Garcia-Marcos L, Gotua M, Gratziou C, Van Hage M, Von Mutius E, Riikjärv MA, Rzehak P, Stein RT, Strachan DP, Tsanakas J, Wickens K, Wong GW and ISAAC Phase Two Study Group (2007) Atopic sensitization and the international variation of asthma symptom prevalence in children. Am J Respir Crit Care Med 15: 565-574.

Wharton GW (1970) Mites and commercial extracts of house dust. Science 167: 1382-1383.

Williams DL, Lowman DW and Ensley HE (2005) Introduction to the chemistry and immunobiology of β glucans. In: Young, S-H and Castranova V (eds.) Toxicology of 1→3 beta-glucans. CRC Press, Boca Raton, FL, USA, pp. 1-34.

Willment JA and Brown GD (2008) C-type lectin receptors in antifungal immunity.Trends Microbiol 16: 27-32.

Willment JA, Gordon S and Brown GD (2001) Characterization of the human beta -glucan receptor and its alternatively spliced isoforms. J Biol Chem 276: 43818-43123.

Willment JA, Marshall AS, Reid DM, Williams DL, Wong SY, Gordon S and Brown GD (2005) The human beta-glucan receptor is widely expressed and functionally equivalent to murine Dectin-1 on primary cells. Eur J Immunol 35: 1539-1547.

Wilson AM, Luo W and Miller JD (2009) Using human sera to identify a 52-kDa exoantigen of *Penicillium chrysogenum* and implications of polyphasic taxonomy of anamorphic ascomycetes in the study of antigenic proteins. Mycopathologia 168: 213-226.

Winck JC, Delgado L, Murta R, Lopez M and Marques JA (2004) Antigen characterization of major cork moulds in Suberosis (cork worker's pneumonitis) by immunoblotting. Allergy 59: 739-745.

Xu J, Jensen JT, Liang Y, Belisle D and Miller JD (2007) The biology and immogenicity of a 34 kDa antigen of *Stachybotrys chartarum sensu latto*. Int Biodeg Biodet 60: 308-318.

Xu J, Liang Y, Belisle DP and Miller JD (2008) Characterization of monoclonal antibodies to an antigenic protein from *Stachybotrys chartarum* and its measurement in house dust. J Immun Meth 332: 121-128.

Young SH and Castranova V (2005). Animal model of (1→3)-β glucan-induced pulmonary inflammation in rats. In: Young, S-H and Castranova V (eds.) Toxicology of 1→3 beta-glucans. CRC Press, Boca Raton, FL, USA, pp. 65-93.

Young SH, Roberts JR and Antonini JM (2006) Pulmonary exposure to 1→3-beta-glucan alters adaptive immune responses in rats. Inhal Toxicol 18: 865-874.

Zureik M, Neukirch C, Leynaert B, Liard R, Bousquet J and Neukirch F (2002) Sensitisation to airborne moulds and severity of asthma: cross sectional study from European Community respiratory health survey. BMJ 325: 411-414.

# 8   Aerosolized fungal fragments

Brett J. Green[1], Detlef Schmechel[1] and Richard C. Summerbell[2]
[1]Centers for Disease Control and Prevention, Morgantown, WV, USA; [2]Sporometrics
Inc., Toronto ON M6K 1Y9, Canada

## Introduction

Airborne fungal conidia derived from environmentally abundant and morphologically discernible fungal genera, including *Alternaria* Nees, *Aspergillus* P. Micheli ex Link, *Cladosporium* Link and *Penicillium* Link have been traditionally acknowledged as the etiological agents responsible for personal fungal exposure. The contribution of these fungal conidia to indoor, outdoor and occupational environments is well documented. Exposure assessment studies have shown that airborne particulate concentrations vary widely and are strongly influenced by climatic conditions, geographical location and the type of disturbance (Luoma and Batterman 2001, Mitakakis *et al.* 2001b, Ferro *et al.* 2004, Tovey and Green 2004, Chen and Hildemann 2009). Epidemiological studies have identified associations between personal exposure to fungal conidia and exacerbations of respiratory disease in persons suffering from seasonal rhinitis (Li and Kendrick 1995a) and asthma (Fung *et al.* 2000, Downs *et al.* 2001, Zureik *et al.* 2002). Even mortality (Licorish *et al.* 1985, Targonski *et al.* 1995) has been attributed to conidial exposure in subjects previously sensitized to fungi. As a result of these studies and the conclusions of a recent report on damp indoor spaces and health by the Institute of Medicine (IOM 2004), the indoor biodeterioration and health effects associated with fungal contamination in indoor environments has become a research priority.

Scientific understanding of the significance of fungal exposure in these environments has been confounded by methodological difficulties in enumerating and identifying various fungal components in environmental samples (Green *et al.* 2006c). In particular, the extent to which common but often overlooked fungal bioaerosols contribute to the aeroallergen load has remained unclear. Recent surveillance studies utilizing modern detection methodologies have provided new insight into the complex diversity of fungal bioaerosols (Pitkäranta *et al.* 2008, Fröhlich-Nowoisky *et al.* 2009, Green *et al.* 2009). Basidiomycota and Ascomycota spores and fragments have been shown to be prevalent in much greater concentrations in indoor and outdoor environments than previously considered (Pitkäranta *et al.* 2008, Fröhlich-Nowoisky *et al.* 2009, Green *et al.* 2009). Many of these fungi have been overlooked because macroscopic and microscopic identification is difficult and common nutrient media does not facilitate colony growth. On a more basic biological level, the contribution of airborne hyphal fragments (including airborne chlamydospores as well as small fragments of plant host material containing fungal hyphae or chlamydospores) to fungal dissemination is also very poorly known.

Fungal particulates can be divided into those explicitly differentiated as separable, dispersive reproductive structures (spores, conidia, etc.) and those that may have become mechanically severed from the parent mycelium but were not programmatically differentiated as separable. There is no convenient terminology distinguishing these two categories of objects, both of which may be functionally reproductive. To facilitate discussion, we propose here the terms gonomorphic and non-gonomorphic to designate these reproductively differentiated and non-differentiated forms.

The contribution of non-gonomorphic particles, such as fragmented hyphae, to the lower atmosphere was first described by Meier (1935) following Charles A. Lindberg's bioaerosol collection expeditions over the Arctic and Atlantic Ocean in 1933. Ensuing studies identified hyphal fragments as common atmospheric bioaerosols in numerous locations including the Tasman Sea (Newman 1948), the Atlantic Ocean (Pady and Kapica 1955), the Mediterranean Sea (Sreeramula 1958), England (Pady and Gregory 1963, Harvey 1970), Wales (Harvey 1970), the United States (Pady 1957, 1959, Kramer *et al.* 1959a,b, Pady and Kramer 1960; Pady *et al.* 1962, Sinha and Kramer 1971), Canada (Pady and Kapica 1956), and the Canadian Arctic (Pady and Kapica 1953). Recent studies located in Australia (Green *et al.* 2005c, 2006b), India (Atluiri *et al.* 1995, Verma *et al.* 2006), Norway (Halstensen *et al.* 2007), and the United States (Green *et al.* 2009, Levetin *et al.* 2009) have also documented the presence of hyphal fragments in air samples. Hyphal fragments were demonstrated to be within the size range 7-100 µm (Pady and Kramer 1960, Pady and Gregory 1963) and were characterized in terms of their wall thickness, melanization, and septation, as well as of any conidiophore features recognized in the examined materials (Figure 8.1A) (Pady 1957, Pady and Kramer 1960, Green *et al.* 2005c). In particular, these non-gonomorphic particles were shown to be of dematiaceous origin and to belong to species in the orders Capnodiales, Pleosporales and Eurotiales. This finding was of great clinical and agricultural significance at the time (Sinha and Kramer 1971). Between 1933 and 1971, airborne hyphal fragments were considered important fungal dispersal agents: viable hyphae were shown to germinate and produce new fungal colonies as readily as conidia did (Pathak and Pady 1965, Harvey 1970). In addition, airborne hyphal fragments were shown to be present in high concentrations throughout the year, including winter, and to generally show diurnal periodicity (Pady *et al.* 1962, Harvey 1970). In spite of these findings, the interest in hyphal fragments decreased with time and these non-gonomorphic

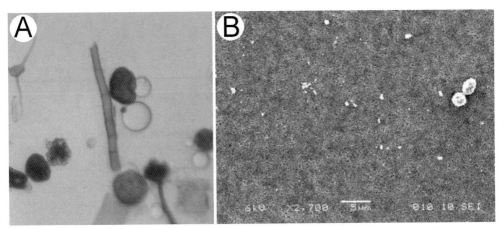

*Figure 8.1. Photomicrographs of (A) an environmental hyphal fragment with visible septation and melanization and (B) aerosolized Aspergillus versicolor fungal fragments. Scale bar 20 μm. Figure 8.1B reproduced from Green et al. (2006c) with permission from Informa Medical and Pharmaceutical Science – Journals; Figure 8.1A reproduced from Gòrny (2004) with permission from the Institute of Agricultural Medicine, Lublin, Poland.*

particles were not included in later aerometric surveys due to difficulties associated with enumeration and identification. Other types of non-gonomorphic particles such as amorphous fungal wall fragments were not considered at all.

Allergic sensitization to particulate matter of biological origin including, fungi, bacteria, plant pollen and non-gonomorphic particles derived from fungi was first proposed by Salvaggio and Aukrust (1981). Fungal derived particulate matter was first discussed during the fifth Indoor Air Quality and Climate Conference held in Toronto in 1990, however, it was not until a Danish-Finnish workshop on fungi in buildings in 1999 that a significant release of non-gonomorphic particles (<2.5 μm) from fungal cultures was experimentally disclosed (Kildeso *et al.* 1999, Madsen *et al.* 2005). In *in vitro* studies, submicron non-gonomorphic particles derived from cultures of *Aspergillus ustus* (Bainier) Thom & Church, *A. versicolor* (Vuill.) Tirab., *Chaetomium globosum* Kunze, *Cladosporium sphaerospermum* Penz., *Penicillium chrysogenum* Thom, *Trichoderma harzianum* Rifai, *Ulocladium* sp. Preuss and a fungus identified under the ambiguous name *Verticillium lecanii* (Zimm.) Viégas (now split into multiple species in the genus *Lecanicillium*) were shown to be aerosolized following exposure to various airflows (Kildeso *et al.* 1999, Madsen *et al.* 2005). These particulates were initially termed fungal micro-particles, however, Gòrny *et al.* (2002, 2004) renamed them fungal fragments. This terminology has been retained in several recent *in vitro* chamber experiment studies that have further explored fungal fragmentation (Kanaani *et al.* 2008, 2009). As the term implies, fungal fragments are morphologically unrecognizable non-gonomorphic particles of biological origin that are derived from fragmented (<0.1-2.5 µm) intracellular and extracellular structures of spores, conidia, hyphae, chlamydospores, yeasts, fruiting bodies, etc., following biotic or abiotic stress (Figure 8.1B).

Personal exposure to non-gonomorphic particles is of great clinical interest. These particulates contain antigens (Gòrny *et al.* 2002), allergens (Green *et al.* 2005c, 2009), mycotoxins (Sorenson *et al.* 1987, Brasel *et al.* 2005) and (1→3)-β-D-glucans (Seo *et al.* 2007, Reponen *et al.* 2008). They remain suspended in the air significantly longer than spores do (Gòrny *et al.* 2002, Madsen *et al.* 2005, Kanaani *et al.* 2008, Reponen *et al.* 2008) and their presence in indoor air samples is an indication of fungal contamination (Yang and Heinsohn 2007). Moreover, they are small enough to be inhaled deep into the nasal, pharyngeal, laryngeal, tracheobronchial and alveolar regions of the respiratory tract (Miller *et al.* 2003, Cho *et al.* 2005). Much of the interest in this material has stemmed from personal exposure studies that have established pollen and fungal fragments to be aeroallergen sources, and, in particular, to be the principal etiological agents of thunderstorm asthma (Schappi *et al.* 1999a,b, Staff *et al.* 1999, Taylor *et al.* 2002, 2004; Taylor and Jonsson 2004; Nasser and Pulimood 2009). However, the difficulties associated with collecting and enumerating within this size fraction has limited this analysis to a handful of fungal species investigated under experimental conditions.

The process of distinguishing fungal and non-fungal particulate matter in environmental air samples is complex. Until recently, it has been subjective, primarily restricted to direct microscopic identification of relatively large, morphologically characterizable non-gonomorphic particles (Figure 8.1A; >2.5-100 µm) that have been collected by impaction (Pady and Kramer 1960, Pady and Gregory 1963, Harvey 1970, Sinha and Kramer 1971, Li and Kendrick 1995b, Green *et al.* 2005c, 2006c, 2009, Levetin *et al.* 2009). Recent studies utilizing size-selective impaction sampling have enabled the quantification of several non-gonomorphic particle size ranges using a variety of molecular, immunological and biochemical techniques (Menetrez *et al.* 2001, Womiloju *et al.* 2003, Foto *et al.* 2005, Lau *et al.* 2006, Womiloju *et al.* 2006, Blachere *et al.* 2007, Reponen *et al.* 2008, Madsen *et al.* 2009, Seo *et al.* 2009). An increased interest in non-gonomorphic particles has led to the development of new sampling and quantification methods (Graham *et al.* 2000, Green *et al.* 2005a,b,c, 2006a,b,d, 2009, Lindsley *et al.* 2006, Sercombe *et al.* 2006a, Rydjord *et al.* 2007, Seo *et al.* 2007) enabling the detection and quantification of these particles in experimental chamber studies and field studies (Chen *et al.* 2004, Lindsley *et al.* 2006, Reponen *et al.* 2008, Kanaani *et al.* 2009, Seo *et al.* 2009). The future use of these methodologies in fungal exposure studies will allow detailed insight to be gained about the health effects associated with non-gonomorphic particle exposure.

In the present review, the state of knowledge of non-gonomorphic fungal fragments will be summarized and the known or suspected implications for human health will be addressed. New detection and quantification technologies designed to deal with these particles will also be discussed.

## The process of fungal fragmentation

In the health related literature, there has been a recognition that the initiation of fungal fragmentation is brought about by fungal autolysis in association with several *in situ* and environmental factors. Fungal autolysis is a process of self digestion of aged hyphae; it involves the release of intracellular material following the permeabilization of the cell wall (White *et al.* 2002, Madsen *et al.* 2005). This process initiates vacuolation. It may be triggered by intrinsic factors such as fungal ageing, cell death and hydrolase activity, as well as by extrinsic factors including nutrient limitation, dehydration and abiotic stress (Harvey 1970, Papagianni *et al.* 1999, Green *et al.* 2006c). Specifically, autolysis disrupts organelles and weakens the tensile strength and rigidity of the hyphae (Paul *et al.* 1994, Li *et al.* 2002a,b). As a result, abiotic stresses, such as those arising from wind, substrate vibration and disturbance lead to the separation of hyphae at septal junctions, which directly increases hyphal fragmentation (White *et al.* 2002, Green *et al.* 2006c). In contrast, the processes through which conidia become reduced to fragments have not been investigated; some authors have suggested that mechanisms may include the rupturing of didymo-, dictyo- or phragmoconidia along cross walls through mechanical stress or by osmotic pressure differences arising from differences in moisture levels (Green *et al.* 2006c, Nasser and Pulimood 2009).

To date, the study of fungal fragmentation has primarily been restricted to *in vitro* exposure chamber studies that have focused on the release of fine fungal particulates. The earliest studies of this nature were conducted by Kildeso *et al.* (1999, 2003) and Madsen *et al.* (2005), who demonstrated that fungal autolysis accounted directly for the finding that 42-day old *T. harzianum* cultures yielded two- to tenfold higher fragment levels than younger cultures (Kildeso *et al.* 1999, 2003, Madsen *et al.* 2005). These findings were put in context in experiments indicating that similar concentrations of fragments were released from colonized gypsum boards treated with an autolysis-inducing buffer (Madsen *et al.* 2005). Recent *in vitro* exposure chamber studies have also confirmed these earlier findings by showing that greater quantities of submicrometer-sized non-gonomorphic particles were released from older cultures (Seo *et al.* 2009). These studies demonstrated that although fungal autolysis of older growth hyphae may be important in the release of fungal fragments, various other abiotic factors including mechanical and physical stresses are also required for fragmentation and dissemination.

The influence of mechanical stress on the release of non-gonomorphic particles in exposure chamber situations was first described by Gòrny *et al.* (2002). In this study, the concentration of fungal fragments released from *A. versicolor, Cladosporium cladosporioides* (Fresen.) G.A. de Vries and *Penicillium melinii* Thom colonies grown on ceiling tile surfaces was shown to be significantly higher than the concentrations released from agar surfaces. The latter increased with each tested air velocity; for example, $24 \times 10^3$ particles were released at 0.3 m/s and $5.7 \times 10^5$ particles at 29 m/s. Similarly, Kildeso *et al.* (1999, 2003) demonstrated that particulates 1-2 µm in size were aerosolized from *T. harzianum* cultures grown on gypsum boards and these were released in significantly increased concentrations as the airflow increased to 3 m/s. Recently, fungal fragmentation of several common indoor fungal contaminants has also been demonstrated in exposure chamber studies (Kanaani *et al.* 2008, 2009). In these studies, the initiation of fragmentation was observed at 1.8-3.3 m/s, and fragmentation increased by up to 400 fold with increasing air velocity. Compared to previous studies, no fragmentation was observed at air speeds that represented typical indoor ventilation environments (0.1-0.4 m/s, Kanaani *et al.* 2009).

Vibration stress was also shown to increase the concentration of *C. cladosporioides* and *P. melinii* fragments significantly beyond the level of *A. versicolor* fragments, even when relatively low airflow (0.3 m/s) was used (Gòrny *et al.* 2002). Much higher concentrations of fragments were released from ceiling tiles than from agar surfaces, which was directly related to air velocity. Compared to agar treatments, where a significant proportion of the hyphal mass grows into the ideal conditions of the nutrient medium, fungal colonies growing on or in building materials are exposed to many types and sources of microenvironmental stress. These extrinsic factors include desiccation stress, airflow, turbulence, temperature fluctuations and zones of convection (Menetrez *et al.* 2001, Green *et al.* 2006c, McGinnis 2007). They collectively accelerate autolysis and the release of fungal fragments. In addition, increased air turbulence and vibration effects of the ceiling tile surface and cavities were adduced by Gòrny *et al.* (2002, 2004) as factors that could explain much of the variation seen in their *in vitro* studies, above and beyond that attributed to differences in phenotype of the fungi studied. Variations in the morphology of fungal species contaminating building surfaces may be an overlooked ecological aspect influencing the aerosolization of non-gonomorphic particles in indoor environments. For instance, *A. versicolor* and *P. melinii* colonies are characterized by extended conidiophores and phialides that are exposed to more extrinsic stresses than are conidiogenous structures of *C. cladosporioides* (Gòrny *et al.* 2002, Gorny 2004). Spore morphology has also been proposed to influence fragmentation, however, this hypothesis has not been confirmed by microscopic analysis (Kanaani *et al.* 2009). To date, variations in the morphology of conidiophores, the arrangement of conidia, and the age of the mycelium may account for the differences in non-gonomorphic particle concentrations observed among species in these studies; however this remains understudied and requires further investigation.

Various mechanisms of fungal fragmentation other than those detailed previously have been identified. Recently, personal exposure studies have shown that particulate matter concentrations were directly related to the type and intensity of personal substrate disturbance (Mitakakis *et al.* 2000, 2001b; Buttner *et al.* 2002, Green *et al.* 2006b, Sercombe *et al.* 2006b, Chen and Hildemann 2009). Using nasal air samplers, Mitakakis *et al.* (2000) and Green *et al.* (2006b) specifically demonstrated that when non-gonomorphic particles, including hyphae, were quantified, inhaled concentrations varied among individuals and among disturbance regimes. Other recent studies have shown that the schizolytic and rhexolytic separation of conidia from fertile cells produces subcellular fragments (Cole and Samson 1979, Sigler 1989). For example, the outer wall surrounding the base of *Sporodesmium ehrenbergii* M.B. Ellis conidia ruptures, and fragments are then liberated (Cole and Samson 1979). This process may lead to the dissemination of fine and ultrafine non-gonomorphic particles. A further source of non-gonomorphic particles that is rarely considered is the grazing or other comminution of conidia, hyphae, yeast cells, etc., by prokaryotes, protozoans and microarthropods. Patrick and colleagues (Clough and Patrick 1972, Old and Patrick 1976, Anderson and Patrick 1978) have shown that soil-inhabiting microorganisms such as bacteria and mycophagous amoebae predate upon various fungal species, starting by perforating the outer conidial and hyphal wall. Similarly, microarthropods such as *Archegozetes longisetosus, Acarus siro, Folsomia candida, Oniscus asellus, Blattella germanica, Ctenolepisma longicaudata, Tyrophagus putrescentiae, Dermatophagoides pteronyssinus, D. farinae, Liposcelis* sp, and *Glycyphagus* sp. have been shown to graze and fragment hyphae of vesicular arbuscular mycorrhizal or saprotrophic fungi, including, in the latter category, *Alternaria alternata* (Fr.) Keissl., *Aspergillus fumigatus* Fresen., *A. penicilloides* Speq., *Eurotium repens* de Bary and *P. chrysogenum* (Van de Lustgraaf 1978, Van Bronswijk 1981, Moore *et al.* 1985, Klironomos and Kendrick 1996, Klironomos and Ursic 1998, Van Asselt 1999, Gange 2000, Hanlon and Anderson 2004, Smrz and Norton 2004, Schneider *et al.* 2005). These studies demonstrate that ecological considerations are often overlooked as modes of fungal fragmentation. The liberation of non-gonomorphic particles is a complex process that can only be studied well through interdisciplinary collaboration. It involves the combination of numerous microenvironmental and ecological variables, many of which remain misunderstood or ignored. Future studies investigating fungal fragmentation should specifically focus on fungi that frequently colonize damp or water-damaged building materials, including various phylloplane fungi such as *Alternaria*, *Epicoccum* Link and *Curvularia* Boedijn, xerophiles and xerotolerant fungi such as members of *Wallemia* Johan-Olsen, *Eurotium* Link, *Aspergillus* and *Penicillium* and hydrophiles such as *Chaetomium*, *Ulocladium*, *Stachybotrys* Corda and *Acremonium* Link (Horner 2003).

## Contributions of hyphal fragments and particulates to the environment

The extent to which airborne non-gonomorphic particles have been found to contribute to the overall bioaerosol load is summarized in Table 8.1. In the studies, impacted structures were visually recognized as hyphal fragments in direct microscopy mainly based on their possession of (1) hypha-like morphology within the size range 2.5-100 µm, (2) visible septation and (3) melanization (Figure 8.1A) (Pady and Kramer 1960, Pady and Gregory 1963, Harvey 1970, Sinha and Kramer 1971, Green *et al.* 2005c, 2006c). The first documented identification of airborne hyphal fragments based on use of these criteria was reported by Meier (1935) from air samples collected several thousand feet above the Arctic. Similar observations were made from air samples collected during later flights over the Tasman Sea (Newman 1948), the Atlantic Ocean (Pady and Kapica 1955) and Canada (Pady and Kapica 1953, 1956). Significant concentrations of hyphae were detected in all cases. Following these preliminary studies, the interest of researchers in the contribution of airborne hyphae to the bioaerosol load intensified, and complementary quantitative aerometric surveys were undertaken (Table 8.1).

The most comprehensive airborne surveys that focused on the contribution of hyphal fragments to the bioaerosol load were initiated by Pady *et al.* in Manhattan, Kansas (Pady 1957), and Rothamsted, England (Pady and Gregory 1963) as well as by Harvey in Cardiff, Wales (Harvey 1970). In these quantitative volumetric studies, it was determined that hyphal fragments accounted for 0.2-16% of the total fungal count. Hyphal fragments were present throughout the entire year and their numbers showed diurnal periodicity (Pady 1959) with the highest counts reported between 2 and 5 pm (Pady *et al.* 1962, Harvey 1970) during spring, summer and fall (Pady 1957). Highest concentrations in these studies were recorded during summer months and ranged from 172-1,700 hyphae/m$^3$ (Pady and Kramer 1960, Pady and Gregory 1963). In winter, hyphal concentrations were recorded within the range of 1-206 hyphae/m$^3$ (Pady and Kramer 1960, Pathak and Pady 1965) and in Montréal, Québec, hyphae

*Table 8.1. Concentrations of non-gonomorphic particles (>2.5 mm) in indoor, outdoor and occupational environments.*

| Study | Collection method | Location[1] | Indoor/outdoor | Hyphal concentration[2] | Frequency |
|---|---|---|---|---|---|
| Meier 1935 | Passive sedimentation | Artic circle 2,500-12,500 ft ASL | O | hyphae identified | N/A |
| Newman 1948 | Passive sedimentation | Tasman Sea 2,100-4,500 ft ASL | O | 29 hyphae identified | N/A |
| Pady et al. 1953 | McGill GE Sampler | Quebec to Yukon, Canada, 9,000 ft ASL | O | 1.4-17.3 ft³ | 2-19% |
| Pady et al. 1955 | Bourdillon Slit Sampler | Atlantic Ocean 8,000-9,000 ft ASL | O | 0.03-1.36 ft³ | N/A |
| Pady et al. 1956 | Bourdillon Slit Sampler | Montreal, Canada 150 ft AGL | O | 0.2-16 ft³ | 0.01-32% |
| Sreeramulu 1958 | Burkard Trap (BT) | Mediterranean Sea, Malta 20 m ASL | O | 3.7 m³ (0-16.2m³) | 6.6% |
| Kramer et al. 1959b, Pady 1957, Kramer 1959a, Pady 1960, Pathak 1965 | Pady-Rittis Slit Sampler | Manhattan, KS, USA 150 ft AGL | O | S 172-1,700 m³ W 1-206 m³ | 2.7% |
| Pady et al. 1963 | Cascade Impactor | Rothamsted, England 20 m AGL | O | 126 m³ (10-599 m³) | N/A |
| Harvey 1970 | BT | Cardiff, Wales 14 m AGL | O | 30 m³ (<10-176 m³) | N/A |
| Atluri et al. 1995 | Verticle Cylinder Trap | Rajahmundry, India | O | hyphae identified | N/A |
| Li and Kendrick 1995b,c | Samplair – MK1 | Kitchener-Waterloo, Canada 50-80 cm AGL | I/O | I=146 m³ O=112 m³ | I=6.3% O=3.2% |
| Delfino et al. 1997 | BT | San Diego County, CA, USA 4 meters AGL | O | 203 m³ (21-416 m³) | 6.9% |
| Foto et al. 2005 | Air-O-Cell | Prince Edward Island, Canada 50-80 cm AGL | I | 4-33% | 8.6% |
| Green et al. 2005c | IOM Personal Air Sampler (PAS) | Sydney, Australia 50-80 cm AGL | I | 30 hyphae AU (0-190 hyphae) | N/A |

Table 8.1. Continued.

| Study | Collection method | Location[1] | Indoor/ outdoor | Hyphal concentration[2] | Frequency |
|---|---|---|---|---|---|
| Verma et al. 2006 | Rotorod Sampler | Jabalpur, India | O | N/A | 7.3% |
| Sercombe et al. 2006b | Nasal Lavage | Sydney, Australia 1-2 m AGL | I/O | O=0-51 hyphae AU | N/A |
| Green et al. 2006b | 1. BT<br>2. IOM PAS<br>3. Intra Nasal Air Sampler | Casino, Australia<br>BT – 4 m AGL<br>IOM – 1-2 m AGL<br>INAS – 1-2 m AGL | O | BT 19 hyphae AU<br>IOM 800 hyphae AU<br>INAS 241 hyphae AU | BT=0.5%<br>IOM=55.2%<br>INAS=56.8% |
| Halstensen et al. 2007 | PAS-6 Personal Air Sampler Cassette | River Glomma, Mjosa, Trondheim Fjord, Norway 1-2 m AGL | O | $3\times10^6$ m$^3$ | 4.6% |
| Green et al. 2009 | Button Air Sampler Cassette | New York City, NY, USA 1-2 m AGL | I | I=3-215 hyphae AU | N/A |
| Levetin et al. 2009 | BT | Tulsa, OK, USA<br>New Orleans, LO, USA | O | Tulsa = 257 m$^3$<br>New Orleans = 81 m$^3$ | Tulsa = 3.4%<br>New Orleans = 1.2% |

[1] ASL and AGL denotes above sea level and above ground level, respectively.
[2] S denotes summer and W denotes winter; AU denotes arbitrary units.

Fundamentals of mold growth in indoor environments

accounted for up to 11.9 and 16% of the total fungal count in January and February, respectively (Pady and Kapica 1956). Hyphal fragments were also shown by Pady *et al.* (1960, 1962) and Harvey (1970) to retain viability. Given the appropriate substrate and environmental conditions, 16-90% of fungal fragments germinated and formed new colonies (Pady and Kramer 1960, Pathak and Pady 1965, Harvey 1970).

Recent aerometric surveys employing a variety of air sampling strategies have attempted to quantify hyphal fragments in indoor and outdoor environments, including occupational environments (Table 8.1). These studies showed that hyphal fragments represented a significant proportion of the aerospora and made up 0.01-56% of the total fungal count (Table 8.1). The prevalence of hyphal fragments in indoor air was shown to be considerably higher than in outdoor air (Table 8.1). This pattern was demonstrated in Li and Kendrick's (1995b) assessment of the aeromycota in Kitchener-Waterloo, Ontario, where the average prevalence of hyphal fragments in indoor environments was 6.3% (146 hyphae/m$^3$), compared to only 3.2% (112 hyphae/m$^3$) outdoors. In an indoor assessment, Foto *et al.* (2005) showed that fragments of hyphae, spores and conidia comprised 16% of the total fungal count; these counts were shown to not correlate with total counts for spores and conidia. Li and Kendrick (1996) suggested that indoor hyphal fragments were principally derived from outdoor sources in summer, but that during winter months, hyphal fragment counts were mainly derived from common fungi on indoor surfaces, such as *Alternaria* species, as well as from diverse contaminated substrates and from house plants (Li and Kendrick 1996). As mentioned earlier, Yang *et al.* (2007) proposed that the presence of hyphal fragments in indoor environments was an indicator of indoor fungal contamination. Data supporting this contention were reported by Foto *et al.* (2005) that demonstrated that air samples in buildings with more versus less visible fungal damage had increased levels of spore and hyphal fragments. The studies cited in Table 8.1 highlight the significance of the hyphal fragment component of the indoor aerospora. Future microbial exposure assessments should include the quantification of airborne spore and hyphal fragments.

Occupational exposure to airborne hyphal fragments has been recently reported by Halstensen and colleagues (2007). In this exposure assessment of Norwegian grain farmers, aerosolized hyphae were identified in 76% of personal air samples. Airborne concentrations (3×10$^6$/m$^3$) were significantly greater than values previously reported for other occupational environments (Verma *et al.* 2006). Grain threshing was shown to directly elevate the airborne concentration of hyphae; this finding paralleled the results of a previous study showing that *Alternaria* conidia were released during local crop processing activities (Mitakakis *et al.* 2001b). Other extrinsic variables that have been reported to influence the daily concentration of hyphal fragments include temperature, rainfall and wind speed (Harvey 1970, Li and Kendrick 1994).

In contrast to airborne hyphae, fungal subcellular fragments are relatively amorphous particulates derived from intracellular and extracellular structures that have become aerosolized following biotic and abiotic stress. Such fragments have few if any discernible morphological features and are less than 2.5 µm in size (Figure 8.1B), which makes their enumeration and quantification in light microscopy subjective and difficult. As a consequence, the majority of studies analyzing this non-gonomorphic size fraction have been *in vitro* aerosolization chamber studies.

Aerosolization of non-gonomorphic particles from nutrient agar and from building material surfaces was reported in all fungi tested by Gòrny *et al.* (2002), Cho *et al.* (2005), Kildeso *et al.* (1999, 2003), Madsen *et al.* (2005), Brasel *et al.* (2005), Seo *et al.* (2009) and Kanaani *et al.* 2009). These aerosolization chamber studies demonstrated that fungal fragments were released at levels up to 514 times higher than conidia (Kildeso *et al.* 1999, 2003, Gòrny *et al.* 2002, Cho *et al.* 2005, Madsen *et al.* 2005, Kanaani *et al.* 2009). Antigens (Gòrny *et al.* 2002) and mycotoxins (Brasel *et al.* 2005) could also be detected in this size fraction. The extent that parallel phenomena occurred in the environment remained unknown until the nonviable bioaerosols in outdoor air were studied by Menetrez *et al.* (2001), Lau *et al.* (2006), Womiloju *et al.* (2003) and most recently Reponen *et al.* (2008) and Madsen *et al.* (2009). These studies are summarized in Table 8.2. In them, indoor and outdoor bioaerosol sources were collected and fractionated according to size using a variety of methods including cascade impaction (Menetrez *et al.* 2001), the Graseby high volume sampler (Lau *et al.* 2006), the semi-hivol/MiniVol sampler (Womiloju *et al.* 2003) and a fragment sampling system (Menetrez *et al.* 2001, Cho *et al.* 2005, Reponen *et al.* 2008, Madsen *et al.* 2009). Each size fraction of impacted particles was analyzed for various fungal antigens and macromolecules including molecules characteristic of *A. alternata* (Menetrez *et al.* 2001) and *Aspergillus niger* Tiegh (Menetrez *et al.* 2001), as well as glycerophospholipids (Womiloju *et al.* 2003), $(1\rightarrow3)$-β-D-glucan (Reponen *et al.* 2008, Madsen *et al.* 2009), and N-acetyl-β-D-glucosaminidase (Madsen *et al.* 2009) (Table 8.2). Although particle bounce could be a potential confounder in each of these collection methodologies, and the extent to which various fragment types may bounce is unclear, the results of the studies showed that the quantity of antigen detected in the fragment fraction was equal to or greater than that in the spore fraction in both indoor and outdoor environments (Table 8.2). These data demonstrate that fragments of fungi are found in significant airborne concentrations and that they add an antigen burden to the environment beyond that previously recognized in studies based on gonomorphic propagules. These field studies confirm the applicability of the previous

*Table 8.2. Analyte concentrations of non-gonomorphic particles in indoor and outdoor environments.*

| Study | Collection method | Location | Indoor/ outdoor | Analyte detected | Concentration |
|---|---|---|---|---|---|
| Menetrez *et al.* 2001 | 8-stage Cascade Impaction | Research Triangle Park, North Carolina | I/O | *Alternaria alternata* antigen | $I<2.5_{\mu m}$=9.95 ng $I>2.5_{\mu m}$=10.4 ng $O<2.5_{\mu m}$=26.1 ng $O>2.5_{\mu m}$=6.2 ng |
| Menetrez *et al.* 2001 | 8-stage Cascade Impaction | Research Triangle Park, North Carolina | I/O | *Aspergillus niger* antigen | $I<2.5_{\mu m}$=0.07 ng $I>2.5_{\mu m}$=2.91 ng $O<2.5_{\mu m}$=0.95 ng $O>2.5_{\mu m}$=4.1 ng |
| Womiloju *et al.* 2003 | MiniVol | Toronto, Ontario Canada | O | Glycerophospho- lipids | L1 $PM_{2.5w}$=53.65 mg[a] L1 $PM_{2.5g}$=700 ng L2 $PM_{2.5w}$= 11.16 mg L2 $PM_{2.5g}$=210 ng L2 $PM_{2.5w}$= 41.05 mg L3 $PM_{2.5g}$=600 ng |
| Lau *et al.* 2006 | Graseby High Volume Sampler | Hong Kong | O | Ergosterol | L1 $PM_{2.5}$=55.2 pg/m$^3$ [b] L1$PM_{>2.5}$=50.2 pg/m$^3$ L2$PM_{2.5}$=56.8 pg/m$^3$ L2$PM_{>2.5}$=33.6 pg/m$^3$ |
| Reponen *et al.* 2008 | Fragment Sampling System | New Orleans, Louisiana and Southern Ohio | I/O | $(1\rightarrow3)$-β-D- glucan | $SF_{NO}$=4,610.9 pg/m$^3$ [c] $FF_{NO}$=247 9 pg/m$^3$ $SF_{SO}$=199.8 pg/m$^3$ $FF_{SO}$=106.3 pg/m$^3$ |
| Madsen *et al.* 2009 | Triplex Cyclone Sampler | Danish Biofuel Plants (n=14) | I | $(1\rightarrow3)$-β-D- glucan | $PM_1$=5.6 ng/m$^3$ (0.68-27 ng/m$^3$) |

[a] $LPM_{2.5w/g}$ denotes the location, weight ($_w$) and glycerophospholipid content ($_g$) of $PM_{2.5}$ particulates collected in the Greater Toronto Area, specifically Egbert (L1), Downsview (L2) and Yonge (L3).
[b] LPM denotes the location of $PM_{2.5}$ particulates collected in Hong Kong, specifically Hong Kong University of Science and Technology (L1) and Hong Kong Science Museum (L2).
[c] SF denotes spore fraction and FF denotes fragment fraction in New Orleans (NO) and Southern Ohio (SO).

culture-based studies. However, the origin of most of the fungal fragments found outdoors remains unknown and requires further investigation.

## Collection, enumeration and quantification of hyphal and fungal fragments

The analysis of non-gonomorphic particles has been confounded by the obvious difficulties associated with collection, enumeration and quantification of morphologically diverse, relatively amorphous particles. In the first reported observations of airborne hyphae over the Arctic (Meier 1935) and Tasman Sea (Newman 1948), fungi were collected by impaction onto agar- or oil-coated glass slides. Impacted hyphae showing visible septation, melanization and sizes in the range of 5-100 µm were subjectively identified and quantified by a trained mycologist (Meier 1935, Newman 1948). In modern assessments of indoor, outdoor and occupational environments there is an extensive array of commercially available gravimetric impaction methods that are based on variations used in these initial studies (Dillon *et al.* 1999). Although collection efficiencies have improved and new sampling advances allow the monitoring of personal exposure, one of the key limitations associated with the analysis of these samples is the subjective identification of fungi based on the morphology of the collected particles (Green *et al.* 2006c). Furthermore, non-gonomorphic particles <2.5 µm with no morphologically distinctive features can not be identified when these methods of collection are used.

The recent development of immunodiagnostic techniques and monoclonal (mAb)/ polyclonal (pAb) antibodies has enabled the detection and quantification of non-gonomorphic particles and other particulate matter of biological origin. The indirect immunodetection of airborne fungi collected by impaction was first demonstrated by Popp *et al.* (1988). In this proof-of-concept study, collected fungal conidia from indoor environments were incubated with serum from fungal sensitized subjects. Specific immunoglobulin E (IgE) fractions were found to bind to the outer wall conidial antigens and, following incubation with an anti-human IgE-FITC conjugate, to form fluorescent immune complexes, which indicated patient sensitization. This technique enabled the demonstration of specific IgE to various fungi including *Cladosporium, Penicillium, Aspergillus* and *Alternaria* species. However, there were several limitations to the technique, including the loss of up to 10% of fungal conidia from the collection tape during staining and rinsing steps, as well as the continued difficulties associated with identifying morphologically nondescript particulates.

To mitigate the methodological limitations of indirect immunodetection of fungi, a technique termed the Halogen Immunoassay (HIA), was developed (Tovey *et al.* 2000). It was derived from an immunodiagnostic press blotting technique that had been used to quantify grass and Japanese cedar pollen allergens (Takahashi *et al.* 1993, Takahashi and Nilsson 1995). The HIA enabled the simultaneous visualization of individual particles collected by volumetric air sampling and the antigens mobilized on these particles. Antigen visualization was brought about by doing an enzymatic immunostaining with human serum IgG/IgE (Green *et al.* 2006d), or pAbs or mAbs (O'Meara *et al.* 1998, De Lucca *et al.* 1999b, 2000, Poulos *et al.* 1999, 2002, Mitakakis *et al.* 2001a). The HIA was originally developed to detect a range of common perennial and seasonal aeroallergen sources including house dust mite (De Lucca *et al.* 1999a, 2000, Poulos *et al.* 1999), cat danders (O'Meara *et al.* 1998), cockroach (De Lucca *et al.* 1999b), latex (Poulos *et al.* 2002) and pollen (Razmovski *et al.* 2000). Later, it was successfully adapted to the detection of conidial and hyphal antigens of common indoor and outdoor fungal species (Mitakakis *et al.* 2001a, Green *et al.* 2003, 2005c, 2006d).

The utilization of HIA in fungal exposure assessment studies has provided new insights into the nature of personal exposure to airborne fungi. Such advancements include elucidation of the role of conidial germination and the allergen expression it involves on respiratory mucosal surfaces (Mitakakis *et al.* 2001a, Green *et al.* 2003). Also, light has been shed on the contribution of previously ignored fungal genera and subcellular fungal fragments to the aeroallergen load (Green *et al.* 2005c). These preliminary studies demonstrated the utility of this technique, but difficulties with the enumeration and differentiation of non-gonomorphic particles in standard light microscopy has limited the application of this technique to fungal bioaerosols over 5 μm in size (Green *et al.* 2005c, 2006d). To overcome this limitation, the HIA was further refined to allow for the dual immunostaining of samples. This improvement enabled the identification of relevant morphologically nondescript fungal particulates to proceed simultaneously with the detection of patient sensitization. Particle identification was conducted using immunostaining with a specific mAb or pAb, while sensitization was investigated using serum IgE/IgG immunostaining (Green *et al.* 2005a,b). Recent modifications based on visualizing fluorescent-conjugated secondary antibodies in confocal microscopy have greatly improved the sensitivity of detecting immune complexes formed during the HIA (Green *et al.* 2006a, 2009). A schematic outlining the fluorescent HIA procedure yielding products visible in confocal microscopy is presented in Figure 8.2.

The enumeration of impacted fine and ultrafine non-gonomorphic particles obtained in volumetric air sampling was long considered a technically formidable challenge. In a series of developmental experiments conducted in 2006, the utility of the fluorescent HIA as a technique to detect aerosolized fungal particulates was

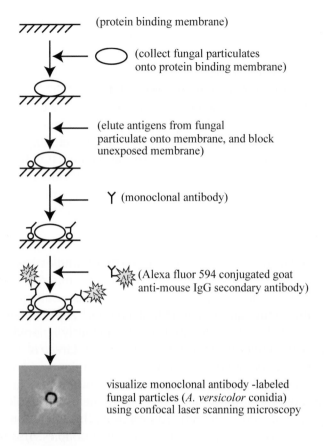

///// (protein binding membrane)

(collect fungal particulates onto protein binding membrane)

(elute antigens from fungal particulate onto membrane, and block unexposed membrane)

Y (monoclonal antibody)

(Alexa fluor 594 conjugated goat anti-mouse IgG secondary antibody)

visualize monoclonal antibody -labeled fungal particles (*A. versicolor* conidia) using confocal laser scanning microscopy

*Figure 8.2. Experimental outline of the fluorescent HIA for the detection of airborne fungi using confocal microscopy. Briefly, the HIA provides an innovative approach to visualize proteins released from biological derived particulates, such as airborne fungi. Airborne fungal particulates, such as this example of an amerospore are impacted onto mixed cellulose ester (MCE) protein binding membranes using volumetric air sampling. Impacted fungal propagules are then permanently laminated using an adhesive-coated glass coverslip. The MCE-coverslip sandwich is then immersed in buffer for several hours to allow the secretion and immobilization of antigens in close proximity to the fungal amerospore on the MCE. Following blocking of vacant binding sites on the MCE, adsorbed antigens are detected with primary mAb/pAb for particle identification or human IgE for characterizing patient sensitization. The resulting immune complexes such as this example of an mAb immunostained* Aspergillus versicolor *spore, are then labeled with Alexafluor conjugated secondary antibodies and visualized using confocal laser scanning microscopy.*

assessed. These experiments have not been previously described and hence detailed methods are given here.

Non-gonomorphic particles derived from *A. versicolor* cultures were aerosolized and collected onto mixed cellulose ester (MCE) protein binding membranes using the fragment sampling system (Seo *et al.* 2007, 2009, Reponen *et al.* 2008). The air sampling method separated aerosolized *A. versicolor* particles into three size fractions, >2.25 μm, 1.05-2.25 μm and <1.0 μm. The MCE membranes with impacted *A. versicolor* conidia and fragments were permanently laminated by overlaying the sample with an optically-clear-adhesive-coated glass cover slip and immunostained using the fluorescent HIA as previously described (Green *et al.* 2006a). Adsorbed *A. versicolor* antigens were detected with primary monoclonal antibody 18G2 (IgG1) diluted 1:50 in 1% bovine serum albumin (BSA) in phosphate buffered saline (PBS) and 0.5% Tween 20. Negative control treatments were processed in parallel by substituting the primary mAb with either hybridoma tissue culture medium diluted 1:50 or control mAb 9B4 (IgG1) diluted 1:50. The control mAb was shown previously to react with conidia of *Stachybotrys chartarum* (Ehrenb.) Wallr. but not to conidia or mycelia of *A. versicolor* (Schmechel *et al.* 2005). For the fluorescent detection of mAb 18G2 and negative control treatments, the MCE was incubated for 1 hour with Alexa Fluor 488 goat anti-mouse IgG (green fluorescence; Molecular Probes, Inc, Eugene, OR, USA) diluted 1:500 in 5% BSA/PBS/0.05% Tween 20. Each sample was then mounted on a microscope slide in Prolong Gold antifade reagent (Molecular Probes, Inc.). Confocal laser scanning images were captured using a Zeiss LSM 510 laser scanning confocal system (Carl Zeiss, Thornwood, NY, USA) with an Axioplan 2 microscope, 40X X-Apochromat water immersion objective lens and argon and HeNe lasers as previously described (Green *et al.* 2006a).

Figure 8.3 shows the immunodetection of *A. versicolor* antigens using confocal microscopy. Compared to previous HIA studies using enzymatic immunostaining, the fluorescent HIA resulted in greatly increased resolution for particles smaller than 2.5 μm. It also increased immunostaining sensitivity and improved the brightness of antibody staining. Immunostaining using mAb 18G2 resulted in distinct haloes around intact *A. versicolor* conidia (Figure 8.3A and B) and fragments as small as 0.5 μm (Figure 8.3C and D). The results of this study demonstrated the utility of the HIA to detect conidia in addition to fine and ultrafine non-gonomorphic particles of *A. versicolor*. Given the future availability of species-specific mAbs for fungi important in indoor air hygiene and medicine, the utilization of the HIA in exposure assessment studies is expected to help in clarifying the nature of the agents causing adverse health effects in situations where persons are exposed to aerosolized fungal bioaerosols, in particular non-gonomorphic particles in contaminated indoor environments.

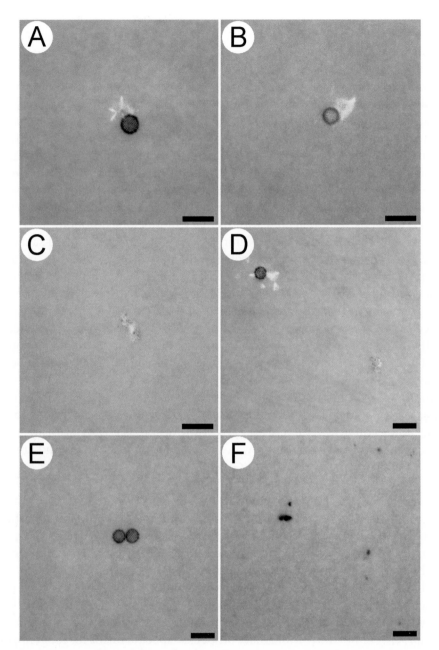

Figure 8.3. Fluorescent HIA immunostaining of aerosolized Aspergillus versicolor (A and B) conidia, (C) fragments, (D) conidia and fungal fragments using mAb 18G2 (IgG1). Negative controls (E and F) using mAb 9B4 (IgG1) showed no specific localized immunostaining. No immunostaining was observed in negative controls using hybridoma tissue culture medium (data not shown). Scale bar 5 μm.

An alternative methodology for the detection of impacted non-gonomorphic particles is scanning electron microscopy (SEM). This method has been used in fungal exposure studies (Skogstad *et al.* 1999, Eduard *et al.* 2001) to characterize airborne fungi including hyphal fragments in occupational environments (Halstensen *et al.* 2007). Compared to other microscopic techniques, SEM provides enhanced resolution and magnification of impacted fungal conidia and particulates (Wittmaack *et al.* 2005). As an analytical tool, it is limited by the need for subjective interpretation of fungal spore morphology as well as by the requirement for a trained mycologist and an SEM technician to perform the analysis. Recent developments integrating field-emission SEM with the HIA has facilitated the immunodetection of ultrafine particulates from environmental sources (Sercombe *et al.* 2006a). However, unlike the fluorescent HIA, field emission SEM of immunostained samples requires extensive sample preparation, which may result in the loss of particles and their antigens from the surface of the membrane (Sercombe *et al.* 2006a).

Other recent immunodiagnostic developments include the detection of fungi using flow cytometry and SEM (Rydjord *et al.* 2007). Rydjord *et al.* (2007) used flow cytometry to quantify fungus-specific IgG bound to surface antigens of *A. versicolor*, *P. chrysogenum* and *C. herbarum* (Pers.) Link. The localization of IgG antibody binding sites was additionally confirmed by SEM (Rydjord *et al.* 2007). Compared to the previous SEM immunolabelling method produced by Sercombe *et al.* (2006a), IgG antibody binding and particle retention was significantly increased due to the pre-fixing of membranes in 0.1% glutaraldehyde (Rydjord *et al.* 2007). The utilization of flow cytometry for the quantification of fungal materials marked with labeled antibodies is an exciting immunodiagnostic development. This will potentially enable the quantification of various size fractions including fungal conidia as well as coarse, fine and ultrafine non-gonomorphic particles. This high resolution will be of great utility to future fungal exposure studies. In addition, this technique may potentially be used in combination with recently developed monoclonal antibodies (Schmechel *et al.* 2005, Xu *et al.* 2008) as a tool to identify sources of fungal contamination in indoor environments.

In contrast to the limited range of techniques that can be combined with single stage volumetric sampling devices, the recent development of size-selective or cyclone-based cascade samplers has enabled the quantification of fungal conidia and non-gonomorphic particles using a combination of biochemical, immunological and molecular techniques. mAbs have been used in an enzyme-linked immunoadsorbent assay format to quantify the levels of non-gonomorphic particles collected in *in vitro* chamber aerosolization experiments (Gòrny *et al.* 2002). Similarly, Menetrez *et al.* (2001) quantified the levels of *A. alternata* and *A. niger* antigens in various size fractions that were collected with a cascade sampler in indoor and outdoor environments. ELISA detection of macrocyclic trichothecene mycotoxins derived from size-fractionated *S. chartarum* particulates smaller than conidia has also been demonstrated (Brasel *et al.* 2005). Furthermore, other quantitative methods such as gas/liquid chromatography mass spectrometry and the limulus amebocyte lysate chromogenic assay are commercially available and have been used in conjunction with fine particle samples to quantify various fungal biomarkers including ergosterol (Miller and Young 1997, Foto *et al.* 2005, Lau *et al.* 2006), fungus-specific phospholipids (Womiloju *et al.* 2003, 2006) and (1-3)-β-D-glucans (Foto *et al.* 2004, 2005, Reponen *et al.* 2008, Seo *et al.* 2009).

Recent advances in cyclone based air sampling methodologies have enabled the size fractionation and subsequent detection and quantification of non-gonomorphic particles. Seo *et al.* (2007) developed a field-compatible collection system for fungal fragments. The system was primarily designed to overcome the particle bounce phenomenon observed in cascade impaction. The method fractionates airborne fungal material into a conidial fraction, a fraction containing a mixture of conidia and non-gonomorphic particles, and a submicron non-gonomorphic particle fraction (Seo *et al.* 2007). Preliminary aerosolization experiments have demonstrated that non-gonomorphic particles could be separated from intact conidia. This was recently confirmed in field studies of homes affected by hurricane Katrina (Reponen *et al.* 2008). Also recently, Chen *et al.* (2004) and Lindsley *et al.* (2006) at the National Institute for Occupational Safety and Health fabricated both a one-stage and a two-stage cyclone-based personal bioaerosol sampler. In contrast to the field-compatible collection system developed by Seo *et al.* (2007), the two-stage sampler draws airborne particulates into two microcentrifuge tubes and a backup filter. The destination of each particle is dependent on its aerodynamic diameter. The first tube (T1) of the sampler collects fungal conidia, pollen and other particles greater than 1.8 μm in diameter. The second (T2) collects particles from 1.0-1.8 μm in diameter. The backup filter (F) collects submicron particulates. This personal air sampling device is expected to be of great utility in environmental and occupational exposure assessment studies. The fractionation of particles according to size and their deposition directly into microcentrifuge tubes facilitates the rapid and direct processing of samples for downstream applications such as ELISA, HIA and quantitative polymerase chain reaction (qPCR) analysis. In microscopic examination of the materials collected, the sampler was found to be extremely efficient at separating fungal conidia from non-gonomorphic particles. Very few spores penetrated to the backup filter (Chen *et al.* 2004, Lindsley *et al.* 2006). In preliminary studies, conidia and fragments of *A. versicolor* as well as aerosolized influenza viral particles were detected and quantified in all stages using qPCR (Blachere *et al.* 2007). Non-gonomorphic particles of *A.*

*parasiticus* Speare were also detected in the second microcentrifuge tube and on the backup filter when an aflatoxin specific ELISA was used for detection (data not shown). These data demonstrated the utility of this type of sampling to collect and size fractionate airborne particulates in general indoor as well as occupational environments. Another advantage of the technique is that direct processing of samples in each collection microcentrifuge tube limits sample loss and contamination. Ultimately, this enables the quantification of fungal particulates using a variety of biochemical, immunological and molecular detection techniques.

## Implications for human health

Fungal contamination of moisture-damaged building materials in homes, schools, office buildings and hospitals has been a recognized public health concern for centuries (Feinberg 1946, Swaebly and Christensen 1952, Maunsell 1954, Townsend 1967, Tobin *et al.* 1987, Miller *et al.* 1988, Brunekreef *et al.* 1989, Dales *et al.* 1991, Miller *et al.* 2000, IOM 2004, Portnoy *et al.* 2005). Personal exposure to aerosolized fungi in these environments has been demonstrated to exacerbate various adverse health effects. Excluding invasive disease including opportunistic infections caused by fungi such as *A. fumigatus*, these effects range from subjective symptoms such as fatigue, cognitive difficulties and memory loss (Baldo *et al.* 2002, Etzel 2002, Gordon *et al.* 2004, Fox *et al.* 2005) to more definable respiratory diseases including airway inflammation (Douwes *et al.* 2000, Douwes 2005), hypersensitivity pneumonitis (Fink *et al.* 1973, Yoshida *et al.* 1996, Wright *et al.* 1999, Moreno-Ancillo *et al.* 2004, Yoshimoto *et al.* 2004), seasonal rhinitis (Li and Kendrick 1995a, Andersson *et al.* 2003), allergic sensitization (Bush and Portnoy 2001, Nolles *et al.* 2001, Green *et al.* 2003, 2005c, 2009, Helbling 2003), asthma (Fung *et al.* 2000, Downs *et al.* 2001, Zureik *et al.* 2002) and even, in cases of severe asthmatic attacks, death (Licorish *et al.* 1985; Targonski *et al.* 1995). Mycotoxins, microbial volatile organic compounds, $(1-3)$-$\beta$-D-glucan and allergens derived from several environmentally abundant, well characterized and morphologically identifiable fungal genera have been shown to exacerbate a number of these conditions. However, the influence of non-gonomorphic particles on personal exposure and adverse health effects has remained unclear, as has the influence of other gonomorphic propagules that are not readily morphologically recognized in air samples.

Clinical interest in the role of aerosolized non-gonomorphic particles in non-infectious respiratory disease increased following the previously described aerometric surveys. In these surveys, non-gonomorphic particles were primarily derived from known aeroallergens including *Alternaria*, *Cladosporium* Link, *Helminthosporium* Link, *Curvularia* and *Penicillium* species (Pady and Kapica 1953, Sinha and Kramer

1971). As mentioned earlier, hyphal fragments are non-gonomorphic particles that tend to be primarily derived from outdoor sources during the summer and from indoor sources during the winter (Li and Kendrick 1996). These findings are of clinical significance given the relatively high proportion of individuals sensitized to these fungal species and the findings that major allergens such as Alt a 1 and Asp f 1 are present in greater quantities in hyphae as compared to conidia (Sporik *et al.* 1993, Mitakakis *et al.* 2001a, Green *et al.* 2003). In a recent study, allergen release from non-gonomorphic particles collected from an indoor environment was confirmed using the HIA (Green *et al.* 2005c). The particles were dematiaceous and were higher in concentration than conidia of any individual fungal genus. Only around 25% of the non-gonomorphic particles demonstrated detectable immunostaining. This low value was attributed either to the specificity of the human serum IgE used for detection or to the degradation of antigens following exposure to ultraviolet light, temperature and moisture (Green *et al.* 2005c). Interestingly, immunostaining was localized around the terminal ends of hyphal fragments and septal junctions (Figure 8.4A), as well as at the sites of conidial fragmentation (Figure 8.4B) (Green *et al.* 2005c). These studies also demonstrated, for the first time, that previously unrecognized fungal aeroallergen sources accounted for 8% of the total fungal count and that these sources were immunostained in the HIA. The intense immunostaining of aerosolized hyphae facilitates studies demonstrating that these particulates are important aeroallergen sources in indoor environments. The importance of these materials was further attested in longitudinal epidemiological studies when the addition of non-gonomorphic particle counts to other fungal counts improved the statistical association of fungal aeroallergen counts with the severity of asthma attacks (Delfino *et al.* 1997). Such results demonstrate the need to incorporate counts of hyphae and of a broadened taxonomic spectrum of conidia into fungal exposure assessment studies.

In contrast to airborne hyphae, the contribution of fine and ultrafine non-gonomorphic particles to adverse health effects has remained unclear. This may be due to the aforementioned difficulties associated with collection and quantification. Although epidemiological studies have demonstrated clinical associations between adverse health effects and personal exposure to particulate matter (Simoni *et al.* 2002), the influence of subcellular non-gonomorphic particles on health has only recently been proposed. In a computer-generated respiratory deposition model, Cho and colleagues (2005) demonstrated that non-gonomorphic particles of *A. versicolor, P. melinii* and *S. chartarum* penetrate into the respiratory tract much more deeply and in higher concentrations than conidia. The significance of these findings is reinforced by the results of Menetrez *et al.* (2001), Brasel *et al.* (2005), Sorenson *et al.* (1987), Lau *et al.* (2006) Reponen *et al.* (2008), Madsen *et al.* (2009), and Seo *et al.* (2009).

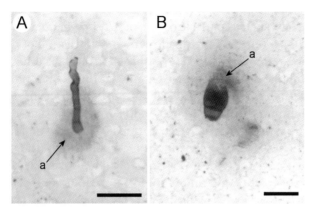

*Figure 8.4. Halogen immunostaining of airborne hyphal and conidial fragments using a human serum IgE pool from fungal sensitive subjects. Resultant immunostaining of (A) airborne hyphal fragment with immunostaining restricted around the terminal ends of hyphal tips and (B) immunostaining restricted to the site of conidial fragmentation. Scale bar 20 μm. Figures reprinted from Green* et al. *(2005c) with permission from Elsevier.*

In these studies, *A. alternata* antigen, *A. niger* antigen, macrocyclic trichothecenes, satratoxin G, satratoxin H, ergosterol and (1→3)-β-D-glucan were found to subsist in equal or highly elevated concentrations in the non-gonomorphic particle fraction as compared to the conidial fraction (Menetrez *et al.* 2001, Reponen *et al.* 2008). These findings emphasize that fine and ultrafine non-gonomorphic particles make significant contributions to the bioaerosol load.

Non-gonomorphic particles in air are largely derived from species belonging to the orders Capnodiales, Eurotiales and, in particular, Pleosporales (Sinha and Kramer 1971, Fröhlich-Nowoisky *et al.* 2009, Levetin *et al.* 2009). Given the extent of cross reactivity amongst the Pleosporalean fungi, including *Alternaria* (Schmechel *et al.* 2008), these findings may account for the unexpectedly high prevalence of sensitization to *Alternaria* species in atopic populations.

In contrast to morbidity based on exposures to irritants, toxins and antigens, morbidity based on infectious disease requires viable particles. In relation to aspergillosis, chronic rhinosinusitis, zygomycosis, and other opportunistic diseases caused by common environmental fung bioaerosols (Summerbell 2001), the subcellular fraction of non-gonomorphic particles is of low potential consequence, except perhaps as a carrier of immunodepressants such as gliotoxin (Pahl *et al.* 1996) in cases where infectious particles and non-viable particles are inhaled together. On the other hand, larger non-gonomorphic particles and other comminuted material

*Brett J. Green, Detlef Schmechel and Richard C. Summerbell*

containing living cells, such as partial conidia of *Alternaria*, may be just as infectious as conidia, or perhaps more so in hyphal fragments consisting of multiple cells (due to the elevated inoculum potential of several cells together). Whether or not fungal materials are in the respirable size range for inhalation deep into the lungs depends largely on their diameter in their smallest axis. Short sections of slender hypha may also have greater pulmonary penetration potential than conidia in some fungal groups such as Pleosporales. This matter has been extensively reviewed for *S. chartarum* by Miller *et al.* (2003), however, the contribution of viable non-gonomorphic particles to infectious disease remains to be investigated. Further research on this topic would be of interest to ensure that the modeling of infectious disease risk due to non-gonomorphic particles is correctly estimated, and that sampling is optimized for the types of particles actually involved in disease causation.

## Conclusions and future perspectives

Further fungal exposure assessment studies are required to elucidate the adverse health effects associated with non-gonomorphic fungal particle exposure. The recent developments in air sampling and analytical detection techniques will further enable the quantification of these particulates in indoor, outdoor and occupational environments. Future investigations should focus on the quantification of these materials using the discussed methodological advances or later improvements that are made on them. The correlation of reliably established antigen quantities with adverse health effects will ultimately improve patient management as well as scientific understanding.

## Acknowledgements

The findings and conclusions in this report are those of the authors and do not necessarily represent the views of the National Institute for Occupational Safety and Health, Centers for Disease Control and Prevention. The data presented in Figure 8.3 was supported in part by the Inter-Agency Agreement NIEHS Y1-ES0001-06. The monoclonal antibody 9B4 is currently being patented (USA Patent Application No. 10/483, 921) by the Centers for Disease Control and Prevention. In addition, the authors would like to thank Tiina Reponen from the University of Cincinnati for collecting fragmented *Aspergillus versicolor* that was utilized for Halogen Immunoassay analysis as presented in Figure 8.3.

# References

Anderson TR and Patrick ZA (1978) Mycophagous amoeboid organisms from soil that perforate spores of *Thielaviopsis basicola* and *Cochliobolus sativus*. Phytopathology 68: 1618-1626.

Andersson M, Downs S, Mitakakis T, Leuppi J and Marks G (2003) Natural exposure to *Alternaria* spores induces allergic rhinitis symptoms in sensitized children. Pediatr Allergy Immunol 14: 100-105.

Atluiri JB, Devi DL and Rao GN (1995) Atmospheric fungal spores at Rajahmundry in the context of respiratory allergy. J Environ Biol 16: 37-43.

Baldo JV, Ahmad L and Ruff R (2002) Neuropsychological performance of patients following mold exposure. Appl Neuropsychol 9: 193-202.

Blachere FM, Lindsley WG, Slavin JE, Green BJ, Anderson SA, Chen BT and Beezhold DH (2007) Bioaerosol sampling for the detection of aerosolized influenza virus Influenza. Other Respir Viruses 1: 113-120.

Brasel TL, Douglas DR, Wilson SC and Straus DC (2005a) Detection of airborne *Stachybotrys chartarum* macrocyclic trichothecene mycotoxins on particulates smaller than conidia. Appl Environ Microbiol 71: 114-122.

Brunekreef B, Dockery DW, Speizer FE, Ware JH, Spengler JD and Ferris BG (1989) Home dampness and respiratory morbidity in children. Am Rev Respir Dis 140: 1363-1367.

Bush RK and Portnoy JM (2001) The role and abatement of fungal allergens in allergic diseases. J Allergy Clin Immunol 107: 430-440.

Buttner MP, Cruz-Perez P, Stetzenbach LD, Garrett PJ and Luedtke AE (2002) Measurement of airborne fungal spore dispersal from three types of flooring materials. Aerobiologia 18: 1-11.

Chen BT, Feather GA, Maynard A and Rao CY (2004) Development of a personal sampler for collecting fungal spores. Aerosol Sci Technol 38: 926-937.

Chen Q and Hildemann LM (2009) The effects of human activities on exposure to particulate matter and bioaerosols in residential homes. Environ Sci Technol 43: 4641-4646.

Cho SH, Seo SC, Schmechel D, Grinshpun SA and Reponen T (2005) Aerodynamic characteristic and respiratory deposition of fungal particles. Atmos Environ 39: 5454-5465.

Clough KS and Patrick ZA (1972) Naturally occurring perforations in chlamydospores of *Thielaviopsis basicola* in soil. Can J Bot 50: 2251-2253.

Cole GT and Samson RA (1979) Patterns of development in *conidial fungi*. Pitman Publishing Limited, London, UK.

Dales RE, Zwanenburg H, Burnett R and Franklin CA (1991) Respiratory health effects of home dampness and molds among children. Am J Epidemiol 134: 196-203.

De Lucca S, Sporik R, O'Meara TJ and Tovey ER (1999a) Mite allergen (Der p 1) is not only carried on mite feces. J Allergy Clin Immunol 103: 174-175.

De Lucca SD, O'Meara TJ and Tovey ER (2000) Exposure to mite and cat allergens on a range of clothing items at home and the transfer of cat allergen in the workplace. J Allergy Clin Immunol 106: 874-879.

De Lucca SD, Taylor DJ, O'Meara TJ, Jones AS and Tovey ER (1999b) Measurement and characterization of cockroach allergens detected during normal domestic activity. J Allergy Clin Immunol 104: 672-80.

Delfino RJ, Zeiger RS, Seltzer JM, Street DH, Matteucci RM, Anderson PR and Koutrakis P (1997) The effect of outdoor fungal spore concentrations on daily asthma severity. Environ Health Perspect 105: 622-635.

Dillon HK, Miller JD, Sorenson WG, Douwes J and Jacobs RR (1999) Review of methods applicable to the assessment of mold exposure to children. Environ Health Perspect 107: 473-480.

Douwes J (2005) (1-3)-b-D-glucans and respiratory health: a review of the scientific evidence. Indoor Air 15: 160-169.

Douwes J, Zuidhof H, Doekes G, Van der Zee SC, Wouters I, Boezen MH and Brunekreef B (2000) 1-3)-b-D-Glucan and endotoxin in house dust and peak flow variability in children. Am J Respir Crit Care Med 162: 1348-1354.

Downs SH, Mitakakis TZ, Marks GB, Car NG, Belousova EG, Leuppi JD, Xuan W, Downie SR, Tobias A and Peat JK (2001) Clinical importance of *Alternaria* exposure in children. Am J Respir Crit Care Med 164: 455-9.

Eduard W, Douwes J, Mehl R, Heederik D and Melbostad E (2001) Short term exposure to airborne microbial agents during farm work: exposure-response relations with eye and respiratory symptoms. Occup Environ Med 58: 113-118.

Etzel RA (2002) Mycotoxins. JAMA 287: 425-427.

Feinberg SM (1946) Allergy in practice. The Year Book Publishers, Chicago, IL, USA.

Ferro AR, Kopperud RJ and Hildemann L (2004) Elevated personal exposure to particulate matter from human activities in a residence. J Expos Anal Environ Epidemiol 14: S34-S40.

Fink JN, Schlueter DP and JJ Barboriak (1973) Hypersensitivity pneumonitis due to exposure to *Alternaria.* Chest 6, suppl 4: S49.

Foto M, Plett J, Berghout J and Miller JD (2004) Modification of the Limulus amebocyte lysate assay for the analysis of glucan in indoor environments. Anal Bioanal Chem 379: 156-162.

Foto M, Vrijmoed LLP, Miller JD, Ruest K, Lawton M and Dales RE (2005) A comparison of airborne ergosterol glucan and Air-O-Cell data in relation to physical assessments of mold damage and some other parameters. Indoor Air 15: 257-266.

Fox DD, Greiffenstein MF and Lees-Haley PR (2005) Commentary on cognitive impairment with toxigenic fungal exposure. Appl Neuropsychol 12: 129-133.

Fröhlich-Nowoisky J, Pickersgill DA, Després VR and Pöschl U (2009) High diversity of fungi in air particulate matter. Proc Natl Acad Sci USA 106: 12814-12819.

Fung F, Tappen D and Wood G (2000) *Alternaria*-associated asthma. Appl Occup Environ Hyg 15: 924-927.

Gange A (2000) Arbuscular mycorrhizal fungi Collembola and plant growth. Trends Ecol Evol 15: 369-372.

Gordon WA, Cantor JB, Johanning E, Charatz HJ, Ashman TA, Breeze JL, Haddad L and Abramowitz S (2004) Cognitive impairment associated with toxigenic fungal exposure: a replication and extension of previous findings. Appl Neuropsychol 11: 65-74.

Gòrny RL (2004) Filamentous microorganisms and their fragments in indoor air – a review. Ann Agric Environ Med 11: 185-197.

Gòrny RL, Reponen T, Willeke K, Schmechel D, Robine E, Boissier M and Grinshpun SA (2002) Fungal fragments as indoor air biocontaminants. Appl Environ Microbiol 68: 3522-3531.

Graham JA, Pavlicek PK, Sercombe JK, Xavier ML and Tovey ER (2000) The nasal air sampler: a device for sampling inhaled aeroallergens. Ann Allergy Asthma Immunol 84: 599-604.

Green BJ, Millechia L, Blachere FM, Tovey ER, Beezhold DH and Schmechel D (2006a) Dual fluorescent Halogen Immunoassay for bioaerosols using confocal microscopy. Anal Biochem 354: 151-153.

Green BJ, Mitakakis TZ and Tovey ER (2003) Allergen detection from 11 fungal species before and after germination. J Allergy Clin Immunol 111: 285-289.

Green BJ, O'Meara TJ, Sercombe JK and Tovey ER (2006b) Measurement of personal exposure to outdoor aeromycota in northern New South Wales Australia. Ann Agric Environ Med 13: 225-234.

Green BJ, Schmechel D, Sercombe JK and Tovey ER (2005a) Enumeration and detection of aerosolized *Aspergillus fumigatus* and *Penicillium chrysogenum* conidia and hyphae using a novel double immunostaining technique. J Immunol Methods 307: 127-134.

Green BJ, Schmechel D and Tovey ER (2005b) Detection of *Alternaria alternata* conidia and hyphae using a novel double immunostaining technique. Clin Diagn Lab Immunol 12: 1114-1116.

Green BJ, Sercombe JK and Tovey ER (2005c) Fungal fragments and undocumented conidia function as new aeroallergen sources. J Allergy Clin Immunol 115: 1043-1048.

Green BJ, Tovey ER, Beezhold DH, Perzanowski MS, Acosta LM, Divjan AI and Chew GL (2009) Surveillance of fungal allergic sensitization using the fluorescent Halogen Immunoassay. J Mycol Med 19: 253-261.

Green BJ, Tovey ER, Sercombe JK, Blachere FM, Beezhold DH and Schmechel D (2006c) Airborne fungal fragments and allergenicity. Med Mycol 44: S245-S255.

Green BJ, Yli-Panula E and Tovey ER (2006d) Halogen immunoassay a new method for the detection of sensitization to fungal allergens; comparisons with conventional techniques. Allergol Int 55: 131-139.

Halstensen AS, Nordby KC, Wouters IM and Eduard W (2007) Determinants of microbial exposure in grain farming. Ann Occup Hyg 51: 581-592.

Hanlon RDG and Anderson JM (2004) The effects of collembola grazing on microbial activity in decomposing leaf litter. Oecologia 38: 93-99.

Harvey R (1970) Air-spora studies at Cardiff III Hyphal fragments. Trans Brit Mycol Soc 54: 251-254.

Helbling A (2003) Fungi as allergens. Allergologie 26: 482-489.

Horner WE (2003) Assessment of the indoor environment: evaluation of mold growth indoors. Immunol Allergy Clin North Am 23: 519-531.

Institute of Medicine (2004) Damp indoor spaces and health. The National Academies Press, Washington, DC, USA.

Kanaani H, Hargreaves M, Ristovski Z and Morawska L (2008) Deposition rates of fungal spores in indoor environments factors effecting them and comparison with non-biological aerosols. Atmos Env 42: 7141-7154.

Kanaani H, Hargreaves M, Ristovski Z and Morawska L (2009) Fungal spore fragmentation as a function of airflow ratesand fungal generation methods. Atmos Env 43: 3725-3735.

Kildeso J, Wurtz H and Nielsen KF (1999) Quantification of the release of fungal spores from water damaged plasterboards. In: DAMBIB-group (ed.) Molds in buildings, Danish-Finnish Workshop on Molds in Buildings 7-8 October 1995.

Kildeso J, Wurtz H, Nielsen KF, Kruse P, Wilkins K, Thrane U, Gravesen S, Nielsen PA and Schneider T (2003) Determination of fungal spore release from wet building materials. Indoor Air 13: 148-155.

Klironomos JN and Kendrick WB (1996) Palatability of microfungi to soil arthropods in relation to the functioning of arbuscular mycorrhizae. Biol Fertil Soils 21: 43-52.

Klironomos JN and Ursic M (1998) Density dependent grazing on the extraradical hyphal network of the arbuscular mycorrhizal fungus *Glomus intraradices* by the collembolan *Folsomia candida*. Biol Fertil Soils 26: 250-253.

Kramer CL, Pady SM and Rogerson CT (1959a) Kansas aeromycology IV: *Alternaria*. Trans Kans Acad Sci 62: 252-256.

Kramer CL, Pady SM and Rogerson CT (1959b) Kansas aeromycology II Materials methods and general results. Trans Kans Acad Sci 62: 184.

Lau APS, Lee AKY, Chan CK and Fang M (2006) Ergosterol as a biomarker for the quantification of the fungal biomass in atmospheric aerosols. Atmos Environ 40: 249-259.

Levetin E, Owens C, Weaver H and Davis W (2009) Airborne fungal fragments: are we overlooking an important source of aeroallergens? J Allergy Clin Immunol 123: S231.

Li DW and Kendrick B (1994) Functional relationships between airborne fungal spores and environmental factors in Kitchener Waterloo Ontario as detected by canonical correspondence analysis. Grana 33: 166-176.

Li DW and Kendrick B (1995a) Indoor aeromycota in relation to residential characteristics and allergic symptoms. Mycopathologia 131: 149-57.

Li DW and Kendrick B (1995b) A year-round comparison of fungal spores in indoor and outdoor air. Mycologia 87: 190-195.

Li DW and Kendrick B (1995c) A year-round outdoor aeromycological study in Waterloo, Ontario, Canada. Grana 34: 199-207.

Li DW and Kendrick B (1996) Functional and causal relationships between indoor and outdoor airborne fungi. Can J Bot 74: 194-209.

Li ZJ, Bhargava S and Marten MR (2002a) Measurements of the fragmentation rate constant imply that the tensile strength of fungal hyphae can change significantly during growth. Biotechnol Lett 24: 1-7.

Li ZJ, Shukla V, Wenger K, Fordyce A, Pedersen AG and Marten M (2002b) Estimation of hyphal tensile strength in production-scale *Aspergillus oryzae* fungal fermentations. Biotech Bioeng 87: 190-195.

Licorish K, Novey HS, Kozak P, Fairshter RD and Wilson AF (1985) Role of *Alternaria* and *Penicillium* spores in the pathogenesis of asthma. J Allergy Clin Immunol 76: 819-825.

Lindsley WG, Schmechel D and Chen BT (2006) A two-stage cyclone using microcentrifuge tubes for personal bioaerosol sampling. J Environ Monit 8: 1136-1142.

Luoma M and Batterman SA (2001) Characterization of particulate emissions from occupant acitivities in offices. Indoor Air 11: 35-48.

Madsen AM, Wilkins K and Poulsen OM (2005) Micro-particles from fungi. In: Johanning E (ed.) Bioaerosols, fungi, bacteria, mycotoxins and human health: patho-physiology clinical effects exposure assessment prevention and control in indoor environments and work. Fungal Research Group Inc, Albany, NY, USA, pp. 276-291.

Madsen AM, Schlünssen V, Olsen T, Sigsgaard T and Avci H (2009) Airborne fungal and bacterial components in PM1 dust from biofuel plants. Ann Occup Hyg 53: 749-757.

Maunsell K (1954) Concentration of airborne spores in dwellings under normal conditions and under repair. Int Arch Allergy Appl Immunol 5: 373-376.

McGinnis MR (2007) Indoor mould development and dispersal. Med Mycol 45: 1-9.

Meier FC (1935) Collecting microorganisms in the arctic atmosphere: with field notes and material by CA Lindbergh. Sci Mon 40: 5-20.

Menetrez MY, Foarde KK and Ensor DS (2001) An analytical method for the measurement of nonviable bioaerosols. J Air Waste Manage Assoc 51: 1436-1442.

Miller JD and Young JC (1997) The use of ergosterol to measure exposure to fungal propagules in indoor air. AIHA J 58: 39-43.

Miller JD, Haisley PD and Reinhardt JH (2000) Air sampling results in relation to extent of fungal colonization of building materials in some water-damaged buildings. Indoor Air 10: 146-151.

Miller JD, Laflamme AM, Sobol Y, Lafontaine P and Greenhalgh R (1988) Fungi and fungal products in some Canadian houses. Int Biodeterior 24: 103-120.

Miller JD, Rand TG and Jarvis BB (2003) *Stachybotrys chartarum*: Cause of human disease or media darling? Med Mycol 41: 271-292.

Mitakakis TZ, Barnes C and Tovey ER (2001a) Spore germination increases allergen release from *Alternaria*. J Allergy Clin Immunol 107: 388-390.

Mitakakis TZ, Clift A and McGee PA (2001b) The effect of local cropping activities and weather on the airborne concentration of allergenic *Alternaria* spores in rural Australia. Grana 40: 230-239.

Mitakakis TZ, Tovey ER, Xuan W and Marks GB (2000) Personal exposure to allergenic pollen and mould spores in inland New South Wales Australia. Clin Exp Allergy 30: 1733-1739.

Moore JC, St John TV and Coleman DC (1985) Ingestion of vesicular-arbuscular mycorrhizal hyphae and spores by soil microarthropods. Ecology 66: 1979-1986.

Moreno-Ancillo A, Dominguez-Noche C, Gil-Adrados AC and Cosmes PM (2004) Hypersensitivity pneumonitis due to occupational inhalation of fungi-contaminated corn dust. J Investig Allergol Clin Immunol 14: 165-167.

Nasser SM and Pulimood TB (2009) Allergens and thunderstorm asthma. Curr Allergy Asthma Rep 9: 384-390.

Newman IV (1948) Aerobiology of commercial air routes. Nature 161: 275-276.

Nolles G, Hoekstra MO, Schouten JP, Gerritsen J and Kauffman HE (2001) Prevalence of immunoglobulin E for fungi in atopic children. Clin Exp Allergy 31: 1564-1570.

O'Meara T, DeLucca S, Sporik R, Graham A and Tovey E (1998) Detection of inhaled cat allergen. Lancet 351: 1488-1489.

Old KM and Patrick ZA (1976) Perforation of lysis of spores of *Cochliobolus sativus* and *Thielaviopsis basicola* in natural soils. Can J Bot 54: 2798-2809.

Pady SM (1957) Quantitative studies of fungus spores in the air. Mycologia 49: 339-353.

Pady SM (1959) A continuous spore sampler. Phytopathology 46: 757-760.

Pady SM and Gregory PH (1963) Numbers and viability of airborne hyphal fragments in England. Trans Brit Mycol Soc 46: 609-613.

Pady SM and Kapica L (1953) Air-borne fungi in the artic and other parts of Canada. Can J Bot 31: 309-323.

Pady SM and Kapica L (1955) Fungi in air over the Atlantic Ocean. Mycologia 47: 34-50.

Pady SM and Kapica L (1956) Fungi in air masses over Montreal during 1950 and 1951. Can J Bot 34: 1-15.

Pady SM and Kramer CL (1960) Kansas aeromycology VI: hyphal fragments. Mycologia 52: 681-687.

Pady SM, Kramer CL and Wiley BJ (1962) Kansas aeromycology XII: materials methods and general results of diurnal studies 1959-1960. Mycologia 54: 168-180.

Pahl HL, Krauss B, Schulze-Osthoff K, Decker T, Traenckner EB, Vogt M, Myers C, Parks T, Warring P, Mühlbacher A, Czernilofsky AP and Baeuerle PA (1996) The immunosuppressive fungal metabolite gliotoxin specifically inhibits transcription factor NF-kappa. BJ Exp Med 183: 1829-1840.

Papagianni M, Mattey M and Kristiansen B (1999) Hyphal vacuolation and fragmentation in batch and fed-batch culture of *Aspergillus niger* and its relation to citric acid production. Process Biochem 35: 359-366.

Pathak VK and Pady SM (1965) Numbers and viability of certain airborne fungus spores. Mycologia 57: 301-310.

Paul GC, Kent CA and Thomas CR (1994) Hyphal vacuolation and fragmentation in *Penicillium chrysogenum*. Biotechnol Bioeng 44: 655-660.

Pitkäranta M, Meklin T, Hyvarinen A, Paulin L, Auvinen P, Nevalainen A and Rintala H (2008) Analysis of fungal flora in indoor dust by ribosomal DNA sequence analysis quantitative PCR and culture. Appl Environ Microbiol 74: 233-244.

Popp W, Zwick H and Rauscher H (1988) Indirect immunofluorecent test on spore sampling preparations: a technique for diagnosis of individual mold allergies. Stain Technol 63: 249-253.

Portnoy JM, Kwak K, Dowling P, VanOsdol T and Barnes C (2005) Health effect of indoor fungi. Ann Allergy Asthma Immunol 94: 313-319.

Poulos LM, O'Meara TJ, Hamilton RG and Tovey ER (2002) Inhaled latex allergen (Hev b 1). J Allergy Clin Immunol 109: 701-706.

Poulos LM, O'Meara TJ, Sporik R and Tovey ER (1999) Detection of inhaled Der p 1. Clin Exp Allergy 29: 1232-1238.

Razmovski V, O'Meara TJ, Taylor DJM and Tovey ER (2000) A new method for simultaneous immunodetection and morphologic identification of individual sources of pollen allergens. J Allergy Clin Immunol 105: 725-731.

Reponen T, Seo SC, Grimsley F, Lee T, Crawford C and Grinshpun SA (2008) Fungal fragments in moldy houses: a field study in homes in New Orleans and Southern Ohio. Atmos Environ 41: 8140-8149.

Rydjord B, Namork E, Nygaard UC, Wiker HG and Hetland G (2007) Quantification and characterisation of IgG binding to mould spores by flow cytometry and scanning electron microscopy. J Immunol Methods 323: 123-131.

Salvaggio J and Aukrust L (1981) Mold induced asthma. J Allergy Clinical Immunol 68: 327-346.

Schappi GF, Taylor PE and Pain MCF (1999a) Concentrations of major grass group 5 allergens in pollen grains and atmospheric particles: implications for hay fever and allergic asthma sufferers sensitized to grass pollen allergens. Clin Exp Allergy 29: 633-641.

Schappi GF, Taylor PE, Staff IA, Rolland JM and Suphioglu C (1999b) Immunologic significance of respirable atmospheric starch granules containing major birch allergen Bet v 1. Allergy 54: 478-483.

Schmechel D, Green BJ, Blachere FM, Janotka E and Beezhold DH (2008) Analytical bias of cross-reactive polyclonal antibodies for environmental immunoassays of *Alternaria alternata*. J Allergy Clin Immunol 121: 763-768.

Schmechel D, Simpson JP, Beezhold D and Lewis DM (2005) The development of species-specific immunodiagnostics for *Stachybotrys chartarum*: the role of cross-reactivity. J Immunol Methods 309: 150-159.

Schneider K, Renker C and Maraun M (2005) Oribatid mite (Acari Oribatida) feeding on ectomycorrhizal fungi. Mycorrhiza 16: 67-72.

Seo SC, Grinshpun SA, Iossifova Y, Schmechel D, Rao C and Reponen T (2007) A new field-compatible methodology for the collection and analysis of fungal fragments. Aerosol Sci Technol 41: 794-803.

Seo SC, Reponen T, Levin L and Grinshpun SA (2009) Size-fractionated (1-3)-$\beta$-D-glucan concentrations aerosolized from different moldy building materials. Sci Total Environ 407: 806-814.

Sercombe JK, Eduard W, Romeo T, Green BJ and Tovey ER (2006a) Detection of allergens from *Alternaria alternata* by gold-conjugated anti-human IgE and field emission scanning electron microscopy. J Immunol Methods 316: 167-170.

Sercombe JK, Green BJ and Tovey ER (2006b) Recovery of germinating fungal conidia from the nasal cavity following environmental exposure. Aerobiologia 22: 295-304.

Sigler L (1989) Problems in application of the terms "blastic" and "thalic" to modes of conidiogenesis in some onygenalean fungi. Mycopathologia 106: 155-161.

Simoni M, Carrozzi L, Baldacci S, Scognamiglio A, Di Pede F, Sapigni T and Viegi G (2002) The Po River Delta (north Italy) indoor epidemiological study: effects of pollutant exposure on acute respiratory symptoms and respiratory function in adults. Arch Environ Health 57: 130-136.

Sinha RJ and Kramer CL (1971) Identifying hyphal fragments in the atmosphere. Trans Kans Acad Sci 74: 48-51.

Skogstad A, Madso L and Eduard W (1999) Classification of particles from the farm environment by automated sizing counting and chemical characterisation with scanning electron microscopy-energy dispersive spectroscopy. J Environ Monit 1: 379-382.

Smrz J and Norton RA (2004) Food selection and internal processing in *Archegozetes longisetosus* (Acari: Oribatida). Pedobiologia 48: 111-120.

Sorenson WG, Frazer DG, Jarvis BB, Simpson J and Robinson VA (1987) Trichothecene mycotoxins in aerosolized conidia of *Stachybotrys atra*. Appl Environ Microbiol 53: 1370-1375.

Sporik RB, Arruda LK, Woodfolk J, Chapman MD and Platts-Mills TAE (1993) Environmental exposure to *Aspergillus fumigatus* allergen (*Asp f* I). Clin Exp Allergy 23: 326-331.

Sreeramula T (1958) Spore content of air over the Mediterranean sea. J Indian Bot Soc 27: 220-228.

Staff IA, Schappi G and Taylor PE (1999) Localisation of allergens in ryegrass pollen and in airborne micronic particles. Protoplasma 208: 47-57.

Summerbell RC (2001) Respiratory tract infections caused by indoor fungi. In: Flannigan B Samson RA and Miller JD (eds.). Microorganisms in home and indoor work environments: diversity, health impacts, investigation and control. Taylor & Francis New York, NY, USA, pp. 195-217.

Swaebly M and Christensen CM (1952) Moulds in house dust furniture stuffing and the air within houses. J Allergy 23: 370-374.

Takahashi Y, Nagoya T, Watanabe M, Inouye S, Sakaguchi M and Katagiri SA (1993) A new method of counting airborne Japanese cedar (*Cryptomeria japonica*) pollen allergens by immunoblotting. Allergy 48: 94-98.

Takahashi Y and Nilsson S (1995) Aeroallergen immunoblotting with human IgE antibody. Grana 34: 357-360.

Targonski PV, Persky VW and Ramekrishnan V (1995) Effect of environmental molds on risk of death from asthma during pollen season. J Allergy Clin Immunol 95: 955-961.

Taylor PE and Jonsson H (2004) Thunderstorm asthma. Curr Allergy Asthma Rep 4: 409-413.

Taylor PE, Flagan RC, Miguel AG, Valenta R and Glovsky MM (2004) Birch pollen rupture and the release of aerosols of respirable allergens. Clin Exp Allergy 34: 1591-1596.

Taylor PE, Flagan RC, Valenta R and Glovsky MM (2002) Release of allergens as respirable aerosols: a link between grass pollen and asthma. J Allergy Clin Immunol 109: 51-56.

Tobin RS, Baranowski E, Gilman AP, Kuiper-Goodman T, Miller JD and Giddings M (1987) Significance of fungi in indoor air: report of a working group. Can J Public Health 78: S1-S32.

Tovey E, Taylor D, Graham A, O'Meara T, Lovborg U, Jones A and Sporik R (2000) New immunodiagnostic system. Aerobiologia 16: 113-118.

Tovey ER and Green BJ (2004) Measuring environmental fungal exposure. Med Mycol 43: S67-S70.

Townsend GL (1967) Sir John Floyer (1649-1734) and his study of pulse and respiration. J Hist Med Allied Sci 22: 286-316.

Van Asselt L (1999) Interactions between domestic mites and fungi. Indoor Built Environ 8: 216-220.

Van Bronswijk JEMH (1981) House dust biology for allergists acarologists and mycologists. NIB Publishers, Zoelmond, the Netherlands.

Van de Lustgraaf B (1978) Ecological relationships between xerophilic fungi and house-dust mites (Acari: Pyroglyphidae). Oecologia 33: 351-359.

Verma KS, Haninder M, Payal S and Pratibha V (2006) Evaluating poultry workers susceptibility against fungal aeroallergens using Anderson and Rotorod sampler. J Phytol Res 19: 15-18.

White S, McIntyre M, Berry DR and McNeil B (2002) The autolysis of industrial filamentous fungi. Crit Rev Biotechnol 22: 1-14.

Wittmaack K, Wehnes H, Heinzmann U and Agerer R (2005) An overview on bioaerosols viewed by scanning electron microscopy. Sci Total Environ 346: 244-255.

Womiloju TO, Miller JD and Mayer PM (2006) Phospholipids from some common fungi associated with damp building materials. Anal Bioanal Chem 384: 972-979.

Womiloju TO, Miller JD, Mayer PM and Brook JR (2003) Methods to determine the biological composition of particulate matter collected from outdoor air. Atmos Environ 37: 4335-4344.

Wright RS, Dyer Z, Liebhaber MI, Kell DL and Harber P (1999) Hypersensitivity pneumonitis from *Pezizia domiciliana*: a case of El Nino lung. Am J Respir Crit Care Med 160: 1758-1761.

Xu J, Liang Y, Belisle D and Miller JD (2008) Characterization of monoclonal antibodies to an antigenic protein from *Stachybotrys chartarum* and its measurement in house dust. J Immunol Methods 332: 121-128.

Yang CS and Heinsohn PA (2007) Sampling and analysis of indoor microorganisms. John Wiley and Sons Inc, Hoboken, NJ, USA.

Yoshida K, Suga M, Yamasaki H, Nakamura K, Sato T, Kakishima M, Dosman JA and Ando M (1996) Hypersensitivity pneumonitis induced by a smut fungus *Ustilago esculenta*. Thorax 51: 650-651.

Yoshimoto A, Ichikawa Y, Waseda Y, Yasui M, Fujimura M, Hebisawa A and Nakao S (2004) Chronic hypersensitivity pneumonitis caused by *Aspergillus* complicated with pulmonary Aspergilloma. Internal Med 43: 982-985.

Zureik M, Neukirch C, Leynaert B, Liard R, Bousquet J, Neukirch F and European Community Respiratory Health S (2002) Sensitisation to airborne moulds and severity of asthma: cross sectional study from European Community respiratory health survey. BMJ 325: 411-414.

# 9 Mycotoxins on building materials

*Kristian Fog Nielsen and Jens C. Frisvad*
*Technical University of Denmark, Lyngby, Denmark*

## Introduction

Even though indoor fungal growth has been considered a problem for at least 5,000 years (the bible, Leviticus Chapter 14, 33-48), we are facing increasing levels of buildings with mold problems. This is highly frustrating since the problem is easily avoided. However due to poor planning in the building process as a whole, more complex building constructions, an accelerated building process, and decreasing levels of education among construction workers, mold problems are on the rise.

Mycotoxins have been suggested as one of the major possible causes of the health problems observed in moldy buildings, thus it is necessary to understand the metabolic potential of indoor fungi to understand potential exposure. Microfungi have the ability to produce a high number of secondary metabolites, which they probably need for survival in their natural habitat (Williams *et al.* 1989, Frisvad *et al.* 1998). Most of these are produced as a response to other organisms, especially bacteria and fungi (Gloer 1995). The metabolites which can cause an adverse health response in vertebrates when introduced by a natural route, are referred to as mycotoxins (Smith and Solomons 1994). Lots of data on trichothecenes from *Fusarium* (deoxynivalenol, T-2 toxins, etc.) being produced in indoor materials can be found on the internet; unfortunately, all these data are incorrect, based as they are on the use of poorly performed analytical chemistry.

Since the production of secondary metabolites and mycotoxins is highly species-specific (Larsen *et al.* 2005), identification of the indoor air associated fungi to species level is vital. If possible, identifications should be based on a polyphasic approach including chemical profiling methods, as the metabolic potential of the organism can also be assessed in the same go. DNA-based methods must be used with care, since many of the deposited sequences in databases can come from incorrectly identified cultures. Furthermore, ITS sequences generally do not provide enough resolution between species and should be avoided (Seifert *et al.* 2007). It is also dangerous to use data from papers describing novel metabolites, as the identification of the producing culture(s) is often wrong in these papers. Finally, one should be aware that in many studies of indoor fungi, species identifications are incorrect.

Just because a fungus can produce a metabolite in the laboratory does not mean that it will be produced *in situ* in wet buildings. When assessing which mycotoxins may be produced on a material, one needs to consider both humidity conditions as well as possible species succession in addition to the material and nutrients available from the material.

It is convenient to divide the indoor fungi into three categories based on (1) their water activity $a_w$, (2) requirements regarding *laboratory substrates*, and (3) responses to non-stationary water activity ($a_w$) (Nielsen 2002):

- *Primary colonizers* or *storage molds.* These molds are capable of growing at $a_w$ <0.8, with *Penicillium chrysogenum* and *Aspergillus versicolor* as the absolutely most common ones followed by species such as: *P. brevicompactum, P. corylophilum, P. commune, A. niger, A. fumigatus, A. insuetus, A. sydowii, A. calidoustus, Paecilomyces variotii,* several *Eurotium* species and *Wallemia sebi,* although many of them have optimal growth rates at "water damage conditions" ($a_w$ close to 1).
- *Secondary colonizers (phylloplane fungi).* These fungi grow at a minimal $a_w$ between 0.8 and 0.9, comprising the genera *Cladosporium, Ulocladium, Alternaria* and *Phoma.* These are all phylloplane fungi and seem to be able to grow at non-stationary humidity.
- *Tertiary colonizers* or *water damaged molds.* These fungi need an $a_w$>0.9, and include many of the most toxic species such as *Stachybotrys chartarum, Chaetomium globosum, Stachybotrys (Memnoniella) echinata* and *Trichoderma* spp.

Numerous reviews on mycotoxigenic fungi in the indoor environment exist, but very few deal with the actual mycotoxins and metabolites that are produced in the buildings. To predict the mycotoxins that may be present in buildings a number of challenges have to be faced:

1. In larger genera such as *Penicillium, Aspergillus, Alternaria* and *Fusarium* only a few specialists are able to identify cultures to species level (Frisvad *et al.* 1998), and the many incorrect sequences from incorrectly identified cultures, in e.g. Gene bank, has made this even more impossible for non-experts.
2. Incorrect findings of numerous mycotoxins from most species can be found (Frisvad 1989, Frisvad *et al.* 2006).
3. Even in extracts of fungal species, which have been extensively studied for their metabolites, unknown active metabolites, synergistic effects and masked mycotoxins can still be observed (Sorenson *et al.* 1984, Gareis *et al.* 1990, Jarvis 1992, Miller 1992, Stoev *et al.* 2002).

4. Very few secondary metabolites described have been tested in more than a very few assays, fewer in full animal studies and only a handful in inhalation studies (Rand *et al.* 2005).
5. Metabolite production varies considerably between laboratory media, temperature and water activity, so molds may produce very different metabolites when growing on building materials (Frisvad and Gravesen 1994, Ren *et al.* 1999, Nielsen 2002) as compared to laboratory media. In most cases the nutrient-poor building materials will allow much poorer metabolite production than seen in the lab.
6. On naturally infested materials molds grow in mixed cultures with other molds and bacteria (Räty *et al.* 1994, Anderson *et al.* 1997, Peltola *et al.* 2001), which may induce production of other metabolites than in pure cultures.

## Analytical methods

Specific detection of most mycotoxins is a difficult task requiring dedicated analytical chemists and highly specialized equipment. From an analytical point of view, building materials is a very difficult "matrix" with an unlimited number of combinations of materials, wallpapers, paints, and dust which can interfere with the analytical methods. This combined with the numerous metabolites produced by fungi makes indoor mycotoxin analysis very complicated.

Techniques like thin layer chromatography (TLC) or HPLC with single UV detection are not specific enough, for example the finding of trichothecenes in dust from a ventilation system (Smoragiewicz *et al.* 1993) is based on false positive results (Nielsen 2002).

With the developments in analytical techniques and instrumentation, it is now recommended to use a chromatographic step (HPLC, CE or GC) with tandem mass spectrometry (3 qualifier ions) or high resolution mass spectrometry (<5 ppm mass accuracy) with at least two ions detected.

However, even with HPLC-MS/MS mistakes occurs, such as for Tuomi *et al.* (2000), who clearly report false positive results for most of the mycotoxins claimed to be detected in the paper. Presumably only the detection of satratoxins and sterigmatocystin was correct. Besides the fact that there were no apparent producers of most of the toxins detected, there were also "odd" results with respect to biosynthetic families. For example, it is not usual to find 3-acetyl-DON and not deoxynivalenol (DON), nor T-2 tetraol without T-2 toxin (Nielsen 2002). Recently, Polizzi *et al.* (2009) published the tentative identification of 42 fungal metabolites based on LC-QTOF-MS, however only accurate mass was used and not retention time or adduct formation. Nor was the fact that many fungal metabolites have the same elementary composition taken into account, thus must all these results be considered false positives.

For some reason many authors have used the PEI columns (polyethylene imine) suggested by Hinkley (Hinkley and Jarvis 2000, Hinkley *et al.* 2003), and indeed these columns work well for the sterigmatocystins, atranones, and macrocyclic trichothecenes, but not for the spirocyclic drimanes, which the PEI was actually originally intended to remove. Moreover, this weak cation exchanger absorbs many ionic and highly polar substances including many mycotoxins, and is therefore not suitable as a routine purification column.

Consequently the best way to get clean samples is actually to avoid getting them dirty. Thus for material samples it is highly recommended to avoid extracting the infected building material and only sample the biomass for extraction. This can be done, for example, by collecting the biomass onto a 0.2 μm filter using vacuum. Teflon filters work fine, but are usually highly cytotoxic and cannot be used in combination with cell-based assays. From an exposure point of view, this sampling technique makes more sense as it is only biomass and the micro-particles liberated from it over time which can become airborne and could be considered dangerous.

Analyses for trichothecenes in the urine of those presumably exposed to *Stachybotrys* have also been published by Croft (2002) and even patented. However, this test is most certainly invalid, as subjects with amounts of trichothecenes detectable by TLC or GC would have levels of toxins sufficient to cause massive internal bleeding and death. Moreover, the large amounts of interfering material in urine would make it impossible to use TLC and/or GC for analysis of trichothecenes, as these techniques will not be able to differentiate trichothecene spots/peaks from the urine matrix components. It would seem that this method is used to extract money from desperate homeowners and people with inexplicable illnesses.

A method for detecting mycotoxin antibodies in serum was published by Vojdani *et al.* (Vojdani *et al.* 2003a,b, Brasel *et al.* 2004). An examination of these papers indicates that: (1) their purification methodology will not have yielded pure satratoxin H for antibody experiments; (2) the method was not validated; (3) the technique used to prepare the toxin- conjugates would not yield same serum albumin adducts as are common in humans; (4) humans in developing countries chronically exposed to high dietary deoxynivalenol and aflatoxins amounts are reliably known not to produce antibodies to mycotoxins. Their license to perform these clinical tests was removed by the US Department of Health and Human Services (Vinson *et al.* 2005).

Cytotoxicity testing using various cell lines has been used as a very successful research tool for cytotoxic compounds such as trichothecenes and gliotoxin (Hanelt *et al.* 1994, Ruotsalainen *et al.* 1998, Murtoniemi *et al.* 2001, Johanning *et al.* 2002). However, this is a specific research tool and mycotoxins can have many other effects than cytotoxicity. So this methodology has high risks for both false negatives and false positives and is therefore not recommended as a routine investigation tool for possible problem buildings.

For instance, *Stachybotrys chlorohalonata* and *S. chartarum* chemo-type A are still considered as dangerous as the chemo-type S, which produces the highly cytotoxic macrocyclic trichothecenes, however the chemo-type S will be 100-10,000 times more potent in a cytoxicity-based cell assay.

## The fungi

### Stachybotrys

*S. chartarum* is the most well-known and feared fungus in indoor environments. Since this genus require high water activities ($a_w \gg 0.95$) and it is rare in nature, it is a very powerful indicator organism for indoor mold problems. In addition to *S. chartarum*, *S. chlorohalonata* is also found in buildings; presumably in a ratio of 4 to 1 respectively. *Stachybotrys* growth is very common on soaking wet gypsum boards, which it can cover within 2-3 weeks (Nielsen *et al.* 1998b, Gravesen *et al.* 1999).

The link to indoor air appeared in 1986, when Croft *et al.* (1986) described a household in Chicago, where the occupants suffered from symptoms similar to stachybotryotoxicosis. Filters after air sampling in this home were black due to the many *Stachybotrys* spores (BB Jarvis, personal communication).

Its reputation was boosted dramatically when it was associated with the idiopathic pulmonary hemosiderosis cases in infants in Cleveland (Montana *et al.* 1995, Jarvis *et al.* 1996, Dearborn *et al.* 1997). From a chemical point of view, it makes good sense to consider *Stachybotrys* as a big problem, as the biomass released from materials infested with this genus contains significantly higher quantities of secondary metabolites than other indoor molds (Nielsen 2002).

*S. chartarum* is mainly known for the production of the highly cytotoxic macrocyclic trichothecenes, satratoxins H, G, F, iso-F, roridin L-2, roridin E epimers, hydroxy-roridin E, verrucarins J and B (selected structures can be seen in Figure 9.1). However, when analyzing building-derived strains only 30-40% (chemo-type S) actually produces the macro-cyclic trichothecenes, mainly as satratoxin H, roridins E and L-2 (Croft *et al.* 1986, Jarvis *et al.* 1998, Nielsen *et al.* 1998a, Vesper *et al.* 2000, Andersen *et al.* 2001, 2002b). Reports of roridin A are false positive findings due to poor chromatography and interference from other macrocyclic trichothecenes.

Instead the remaining isolates produce the atranones and their dolabellane precursors (Hinkley *et al.* 2000). These isolates can be further split into two types, *S. chartarum* chemo-type A and *S. chlorohalonata* (Andersen *et al.* 2002b, 2003, Cruse *et al.* 2002, Peltola *et al.* 2002). These two types can also produce low quantities of simple trichothecenes (trichodermin and trichodermol).

Strangely, very little attention has been paid to the huge quantities of 10-40 different spirocyclic drimanes (Figure 9.1) and their precursors, which on a quantitative basis are the most important metabolites of this fungus (Hinkley and Jarvis 2000, Andersen *et al.* 2002b, Nielsen 2002). This class of metabolites covers activities such as complement inhibition (*Stachybotrys* lactones, lactams, and di-aldehydes), inhibitors of various enzymatic systems, enhancers of fibrinolytical, plasminogen and anti-thrombotic activity (Kohyama *et al.* 1997, Hasumi *et al.* 1998), neurotoxic effects (Nozawa *et al.* 1997) and inhibitors of TNFa liberation and cytotoxicity (Kaneto *et*

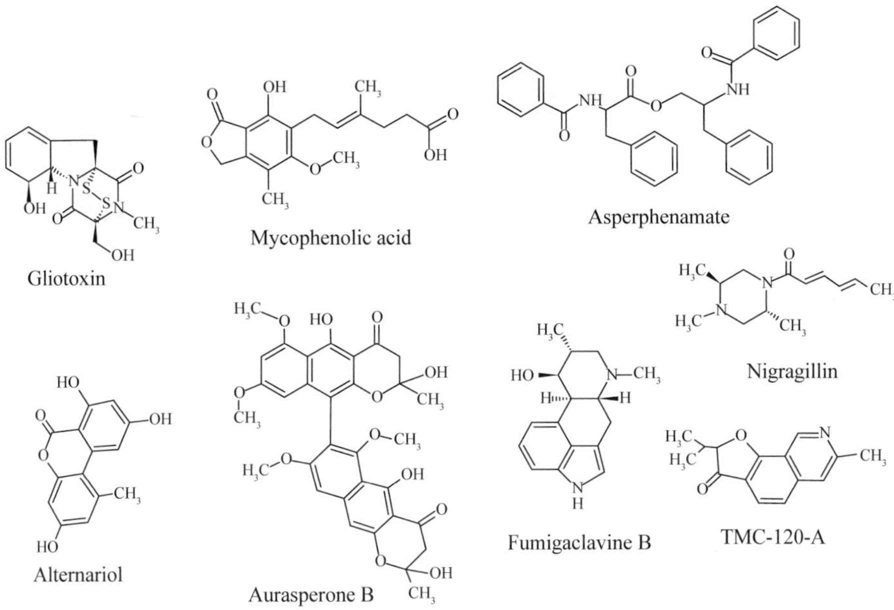

*Figure 9.1. Selected structures for secondary metabolites and mycotoxins found in moldy buildings.*

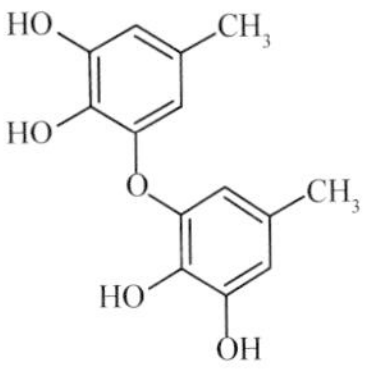

Violaceol I

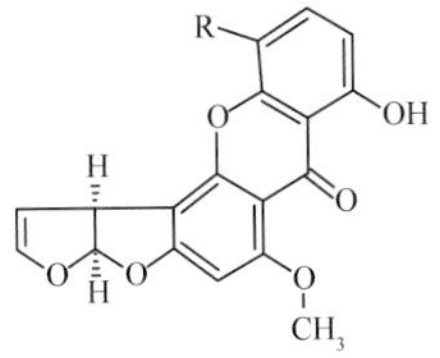

R= H: sterigmatocystin
R = CH₃-O: 5-Methoxysterigmatocystin

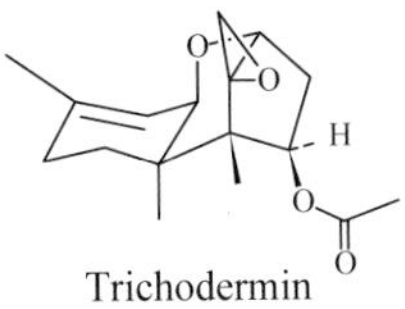

Trichodermin

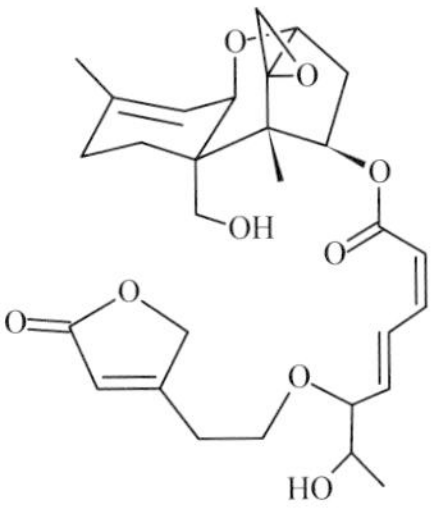

Roridin L-2

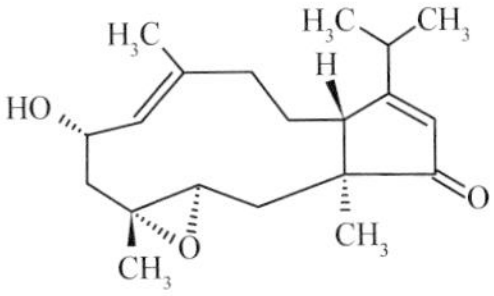

3,4-Epoxy-6-hydroxy-dolabella-7E,12-dien-14-one

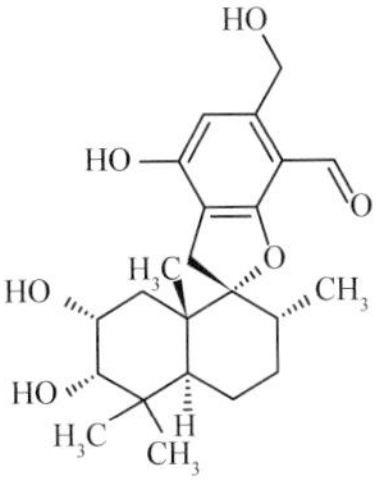

Mer-NF-5003-B

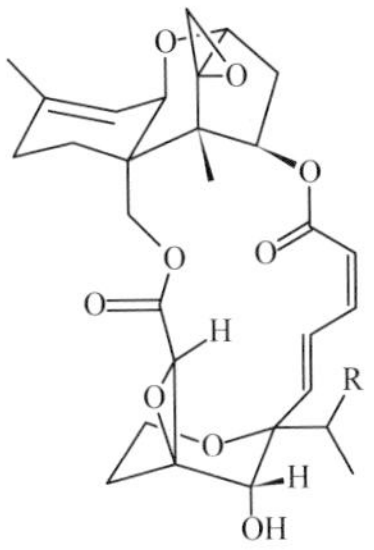

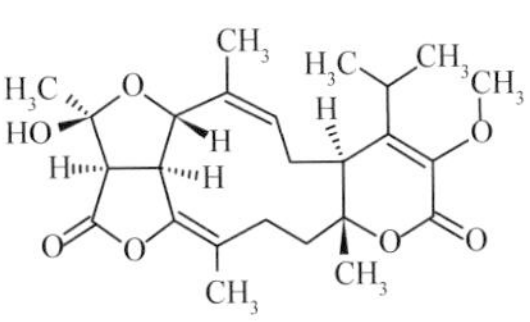

Atranone-B

Stachybotrydial

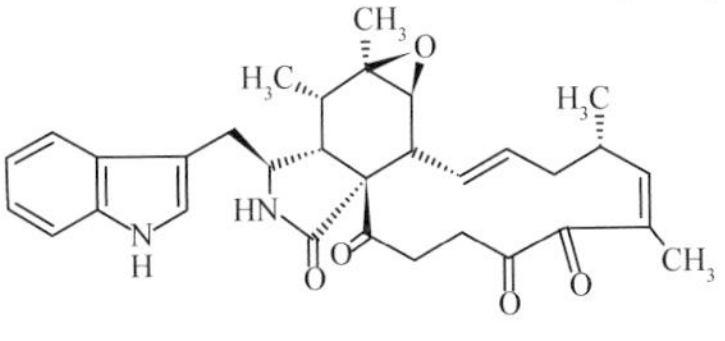

Satratoxin G(R=OH) & F(R=O)

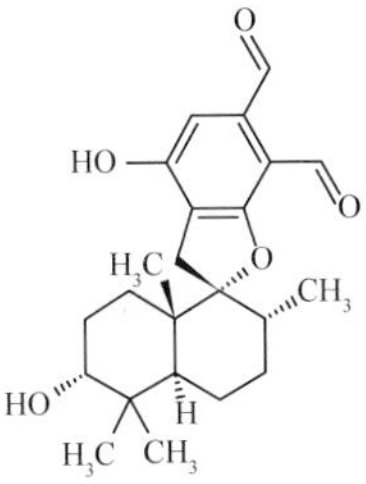

Chaetoglobosin C

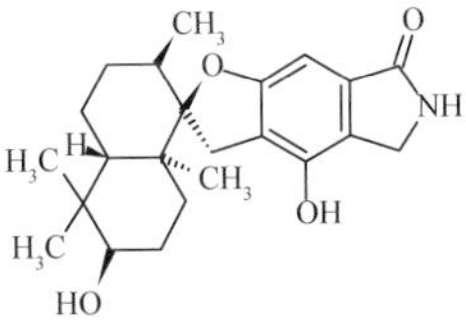

Spirodihydrobenzofuranlactam I

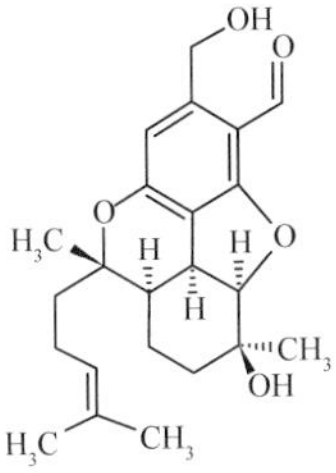

Bisabosqual B

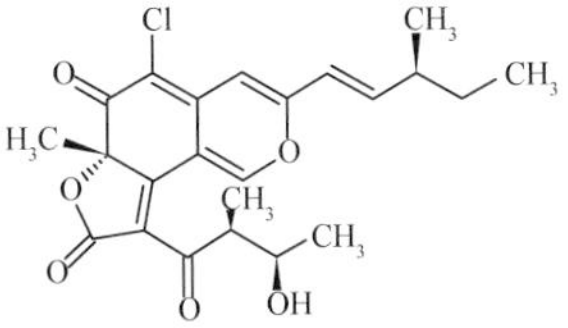

Chaetoviridin A

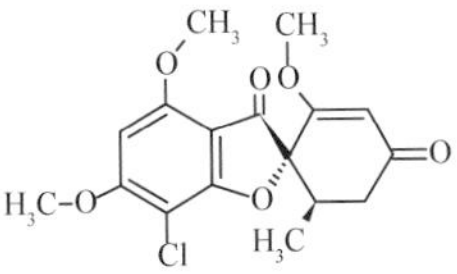

Griseofulvin

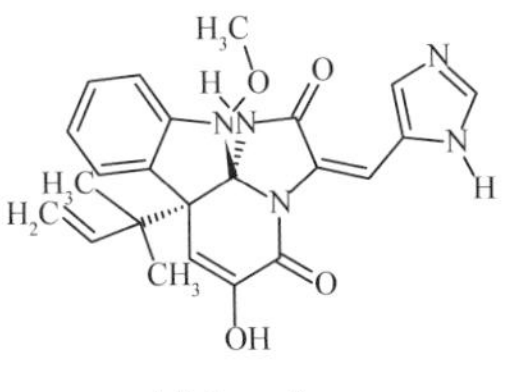

Meleagrine

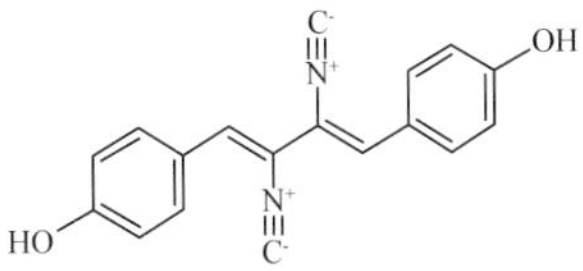

Xanthocillin-X

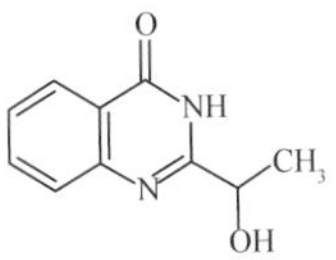

Chrysogine

*al.* 1994). A number of precursors, such as SMTP-1, where the drimane part is not cyclized, has also been described (Hasumi *et al.* 1998). Much information on these metabolites can be found in numerous patents and a more comprehensive list can be found in Nielsen (2002). The spirocyclic drimanes are called mero-terpenoids, as they are produced partly by the terpenoid pathway (e.g. lower rings under the spirio bond in stachybotrydial in Figure 9.1), and partly by the polyketide pathway (part above the spirio bond in stachybotrydial in Figure 9.1).

Other interesting metabolites include stachylysin, which is claimed to be associated with the idiopathic pulmonary hemosiderosis case homes (Vesper *et al.* 1999, Vesper and Vesper 2002, Gregory *et al.* 2002, Van Emon *et al.* 2003). However, the claims of stachylysin concentrations of 371 ng/ml in the blood of exposed persons (Van Emon *et al.* 2003), is unrealistically high, as this corresponds to approx. 2 mg of stachylysin in the (approx. 6 l) blood of an exposed person, which again corresponds to 200 mg inhaled *Stachybotrys* biomass, if it is assumed that 1% of the biomass is stachylysin. Spores of *P. chrysogenum* may also contain stachylysin (Yike *et al.* 2006).

On gypsum boards and paper materials quite a stable chemical profile of 10-40 different spirocyclic drimanes is produced (Nielsen 2002) followed by traces of trichodermin, and depending on the chemo-type, either atranones (as the two dolabellanes and atranones B and C) (Vesper *et al.* 2000) or macrocyclic trichothecenes are observed (Croft *et al.* 1986, Hodgson *et al.* 1998, Jarvis *et al.* 1998, Knapp *et al.* 1999). The primary spirocyclic drimanes have UV spectra identical to Mer-NF5003B (Kaneto *et al.* 1994, Roggo *et al.* 1996, see Figure 1 in Andersen *et al.* 2002b), and bisabosquals components described by Minagawa *et al.* (2001), but LC-MS analyses have revealed that the molecular masses are 50-200 Da higher (KF Nielsen, unpublished). Other important drimanes detected are stachybotryamide, stachybotrylactams and di-aldehydes. The quantities of metabolites based on LC-UV, LC-MS and bioassays are significantly higher than detected in other molds (Johanning *et al.* 1996, Anderson *et al.* 1997, Nielsen *et al.* 1998a,b, Nielsen 1999, 2002).

Studies on the *in vivo* and *in vitro* activity of *Stachybotrys* spores and biomass have revealed significant differences in the two chemo-types (Korpi *et al.* 1997, 2002, Nikulin *et al.* 1997b, Ruotsalainen *et al.* 1998, Nielsen *et al.* 2001), where chemo-type A induces inflammation and chemo-type S, high cytotoxicity and in low concentrations also inflammation. Atranones have proved to be very inflammagenic (Rand *et al.* 2006). Rao *et al.* (2000) have verified that the inflammation was caused by methanol extractable metabolites, however this includes all non-protein metabolites described until now, as all metabolites shown in Figure 9.1 should be highly soluble in methanol.

In naturally infested materials macrocyclic trichothecenes have been detected by Croft *et al.* (1986) who found satratoxin H, verrucarin B, verrucarin J and the trichoverrins. Several later studies have verified these findings (Anderson *et al.* 1997, Flappan *et al.* 1999, Tuomi *et al.* 2000, Gottschalk *et al.* 2008). From an exposure point of view, Sorenson *et al.* (1987) have shown that the spores from macrocyclic trichothecenes producing isolates contained approx. 10-40 ppm of the macrocyclic trichothecenes ≈40 fg/spore, whereas Nikulin *et al.* (1997a) found 140 fg/spore.

Recently micro particles from *S. chartarum* strains have been shown to contain satratoxin (Brasel *et al.* 2005a), which was expected since 50% of the satratoxin content can be extracted from mycelium using water, indicating that satratoxins are found on the outside of the spores and hyphae are presumably in the slime/ carbohydrate secreted by the fungus.

Satratoxins, stachybotryamide, stachybotrylactams, and the metabolites resembling the Mer-NF5003B and/or bisabosquals, were detected by combined LC-DAD and GC-MS/MS in air samples from a house with severe indoor air problems. The samples were also highly cytotoxic to cell cultures (MTT test) due to the satratoxins (Johanning *et al.* 2002). Strong indications of macrocyclic trichothecenes have also been demonstrated using ELISA (Brasel *et al.* 2005b). Recently satratoxin and roridin E have been detected in air samples in several countries (Bloom *et al.* 2007, Gottschalk *et al.* 2008, Polizzi *et al.* 2009).

The most important finding, however, is the reporting of satratoxin-G-albumin adducts (Yike *et al.* 2006), which is by far the most promising exposure marker for revealing the effects of mycotoxins in the indoor climate.

In buildings *Stachybotrys (Memnoniella) echinata* is found in the same places as the other species of *Stachybotrys,* and on laboratory media *S. echinata* produces high quantities of griseofulvin, dechlorogriseofulvins, spirocyclic drimanes (different from the ones from *Stachybotrys*), components related to mycophenolic acid and the two simple trichothecenes: trichodermin and trichodermol (Jarvis *et al.* 1996, Hinkley *et al.* 1998, 1999, Nielsen 2002). A naturally infested sample of plasterboard contained griseofulvin, dechlorogriseofulvin, and epi-dechlorogriseofulvin (Nielsen 2002).

Chemical analysis on 11 isolates, grown on six substrates showed that trichodermin, trichodermol and the three griseofulvins where consistently produced on all media and in significant quantities compared to other metabolites (Nielsen 2002). Combined with data from rice cultures (Jarvis *et al.* 1998) and a few naturally infected material samples (Jarvis *et al.* 1998, Nielsen 2002), it must be concluded that these components seem to be produced on all solid substrates. Spirocyclic drimanes, related to the ones produced by *S. chartarum*, were also consistently produced; however in significantly (ca. 10×) lower quantities than the griseofulvins.

To conclude on this species, it seems to be isolated in the same places as *S. chartarum* although less frequently, which may be due to difficulties of distinguishing it from *S. chartarum*. However, studies on its *in vitro* and *in vitro* pulmonary toxicity and inflammatory potential is needed (when grown on relevant materials), as well as knowledge on what is actually liberated from infested materials.

## Chaetomium globosum

This genus has a very characteristic appearance due to its distinct black or dark green to brown, hairy perithecia visible to the naked eye. It is common on very wet wallpaper, gypsum board and wood products. On the artificially infested water-damaged material growth was very fast, covering the surface with substantial amounts of biomass after two weeks (Gravesen *et al.* 1999, Nielsen *et al.* 1999).

In buildings *C. globosum* is the most common *Chaetomium* species (Andersen and Nissen 2000). It is known to produce the highly cytotoxic chaetomins and especially chaetoglobosins, which inhibit cytoplasmatic division and glucose transport in tissue (Ueno 1985); all 30 indoor-derived isolates produced chaetoglobosins on agar media (Fogle *et al.* 2007). Other *Chaetomium* species also produce sterigmatocystins (Sekita *et al.* 1981) but these species have not been found in buildings (Nielsen 2002).

In one study of wallpapered plasterboard ($a_w$ close to 1) all six building-derived *C. globosum* isolates produced high quantities of chaetoglobosin A and C (up to 50 and 7 µg/cm$^2$) and numerous unknown metabolites (Nielsen *et al.* 1999), of which the major ones have now been identified as chaetoviridins A, B and C. Later analyses of naturally infested plasterboards, wood and textile have shown the same profile as well as numerous hydrolysis products of the chaetoglobosins (Nielsen 2002). Polizzi *et al.* (2009) also found these compounds in moldy buildings. The chaetoglobosins belong to the cytohalasins, which are also produced by several *Penicillium* and *Aspergillus* species. However, the toxicity of these compounds during inhalation is not known.

In some tests we observed that *C. globosum* grew very well together with gram-negative bacteria on various building materials (Nielsen 2002) and showed the highest inflammatory response seen. However, when determining the inflammatory potential of this mixture in mice macrophages, huge quantities of TNFα and NO were liberated due to endotoxin from the bacteria. This observation may point to more studies on the interaction between bacteria and molds, as the bacteria may need the powerful enzyme-systems of molds. However, spores from the pure cultures did not induce any inflammatory response in macrophages, but was cytotoxic to the cells (Nielsen 2002).

As the spores of this fungus are packed in the perithecia, the exposure to airborne spores may be rather limited and smaller micro-particles are thus more likely to be the main problem, as has been shown for other fungi (Gorny *et al.* 2002, Kildesø *et al.* 2003).

## Aspergillus versicolor

*A. versicolor* is, together with *P. chrysogenum*, the most common indoor species, and is able to grow on very nutrient-poor materials such as concrete and plaster. Together with several Streptomycetes (Sunesson *et al.* 1997) and *Chaetomium globosum* (Kikuchi *et al.* 1981) *A. versicolor* may be one of the major producers of the moldy smelling component, geosmin, in buildings (Bjurman and Kristensson 1992).

*A. versicolor* has a highly variable macro-morphology, but a consistent chemical profile on laboratory substrates, usually producing high quantities of the carcinogenic mycotoxin, sterigmatocystin and other members of its biosynthetic pathway (versicolorins, etc.) (Frisvad and Gravesen 1994, Frisvad and Thrane 2002). On yeast extract sucrose agar it will produce low quantities of sterigmatocystins (Nielsen *et al.* 1998b), and violaceol I and II as well as violaceic acid (called versicolins in (Nielsen *et al.* 1998b)). Reports of production of cyclopiazonic acid, ochratoxin A, griseofulvin, dechlorogriseofulvin, and 3,8-dihydroxy-6-methoxy-1-methylxanthone by *A. versicolor* are due to contaminated cultures and/or misidentification (Frisvad 1989, Frisvad and Thrane 2002).

This species is capable of growing on almost any kind of water-damaged material ($a_w$~1), and chipboard and wallpaper supported production of the most visible biomass (Ezeonu *et al.* 1995, Nielsen *et al.* 1998b, Nielsen 2002). It has also been shown that *A. versicolor* grows well on various paints under constant humidity but very slowly under transient humidity conditions (Nielsen *et al.* 1998b, Nielsen 2002). During long-term low humidity trials on materials with a broad array of indoor fungi (Nielsen *et al.* 2004) we notice that *A. versicolor* was one of the most frequently observed fungi capable of growing at 5 °C. This is contradictory to Smith and Hill (1982), who observed that *A. versicolor* requires 9 °C for growth on a high-nutrient substrate such as malt extract agar. On wallpaper paste 50% of the isolates contained sterigmatocystin in their conidia (Larsen and Frisvad 1994).

On various building materials, and at $a_w$ close to 1, *A. versicolor* produces sterigma-tocystin and 5-methoxy-sterigmatocystin in quantities up to 20 and 7 µg/cm² respectively (up to 1% of biomass) (Nielsen *et al.* 1998b, 1999). However, below $a_w$ 0.9 sterigmatocystins were not produced in detectable levels as measured by HPLC-UV (Nielsen 2002).

Interestingly, red-colored areas infested with non- or poorly-sporulating *A. versicolor* biomass contained the largest quantities of sterigmatocystins, whereas areas with many conidia contained very small quantities (Nielsen *et al.* 1999). Other metabolites detected from *A. versicolor* on wet materials include violaceol I and II (at quite high levels) as well as violaceic acid and occasionally also penigequinolone (Nielsen 2002).

During the low humidity experiments (Nielsen *et al.* 2004) and (Nielsen 2002), where the sterigmatocystins and their analogues were almost never detected, a number of compounds previously only detected in extracts of *A. ochraceus* were found.

Sterigmatocystin has been detected in levels up to 4 ng/g in carpet dust from moldy buildings in about 20% of the samples analyzed (Engelhart *et al.* 2002), and recently also in a handful of settled dust and airborne dust samples showing actual exposure (Bloom *et al.* 2007). Fungal fragments which are significantly smaller than the spores are liberated from *A. versicolor* infested building materials (Kildesø *et al.* 2000) and may account for the health problems in buildings where it is the predominant species on the materials without being so in cultivation-based air samples.

No data has been published on the related *A. sydowii*, which produces very different metabolites.

### Aspergillus ustus *section* Usti

*Aspergillus ustus* sensu stricto is very rare and only one strain of this cereal-borne fungus has been described from indoor sources (Houbraken *et al.* 2007, Slack *et al.* 2009). In contrast, two other members of the complex *A. insuetus* and *A. calidoustus* are quite common in damp- and water-damaged buildings (Nielsen *et al.* 1999, Houbraken *et al.* 2007). Differentiation between *A. insuetus* and *A. calidoustus* versus *A. ustus* is important since the latter produces several aflatoxin precursors, the austocystins (Steyn and Vleggaar 1974) and versicolorin C, as well as austalides.

*A. insuetus* and *A. calidoustus* produce totally different metabolites like TMC-120A, B and C, pergillin and ophiobolins (Slack *et al.* 2009). These were also detected along with kotanin when 5 indoor-derived *A. insuetus* isolates were grow on various gypsum and wood-based materials with and without wallpaper (Nielsen *et al.* 1999). In these tests austamide, austdiol, or austocystins were not detected, although reference standards were available (Nielsen *et al.* 1999).

When *A. insuetus/A. calidoustus* (Nielsen *et al.* 1999) were incubated on a number of materials, they grew very slowly on pure gypsum boards and chipboard, whereas growth was much faster on the wallpapered boards covering the surface after 2 weeks. On the gypsum boards solid structures, 1-2 mm in diameter, containing up to hundred Hülle cells were observed.

## Aspergillus niger *section* Nigri

*Aspergillus niger* is difficult to differentiate from the other members *Aspergillus* section *Nigri*, however the building-derived isolates we have examined have all been *A. niger sensu stricto* (Nielsen *et al.* 1999). Until recently the one real mycotoxin which was considered from *A. niger*, was ochratoxin A that is produced by about 3-10% of the isolates (Schuster *et al.* 2002). Other metabolites include naphtha-γ-pyrones, nigragillin, kotanin, and the malformins A, B and C (Nielsen *et al.* 2009). However, very recently it was shown that *A. niger sensu stricto* also produces fumonisins (Frisvad *et al.* 2007).

On building materials ($a_w$ close to 1) only orlandin, nigragillin, and >20 unidentified metabolites, including naphtho-γ-pyrones (especially aurasperones), were detected from the two indoor-derived isolates examined (Nielsen *et al.* 1999). Two indoor-derived isolates could not produce ochratoxin A (LC with fluorescence detection) when grown on materials or on CYA or YES agar (Nielsen 2002). The analytical methods used would not have detected malformins or fumonisins, which should be analyzed by LC-MS.

When *A. niger* is found in buildings, there are usually significant quantities of biomass but it is unusual for a full overview of its preferred materials to have been established, although it seems to grow very poorly on gypsum boards (Nielsen 2002). Extracts of *A. niger* always contain significant quantities of the slightly toxic (Kobbe *et al.* 1977) naphtha-γ-pyrones (Samson *et al.* 2002). The distribution of indoor strains producing the malformins (Kim *et al.* 1993), ochratoxin A or fumonisins is not yet known.

## Aspergillus flavus

*Aspergillus flavus* is occasionally isolated from infested building materials, and since it is a producer of the most potent natural carcinogen aflatoxin $B_1$ (Davis *et al.* 1966,

Frisvad and Thrane 2002), several groups have examined its ability to do this on building materials. *A. flavus* can also produce kojic acid, aspergillic acid, and the two other mycotoxins, 3-nitropropionic acid and cyclopiazonic acid (Frisvad and Thrane 2002). It should be mentioned that far from all strains produce aflatoxins; strains from cold and temperate climates have a particularly low prevalence of aflatoxins producers.

When aflatoxinogenic strains were inoculated on various building materials they did not produce aflatoxins (Rao *et al.* 1997, 1999). Later, when examining materials contaminated with *A. flavus* neither kojic acid, aspergillic acid, cyclopiazonic acid nor aflatoxin $B_1$ could be detected (Nielsen *et al.* 2004), even when using the hemiacetal derivatization for HPLC-FLD detection (Scott 1999). However, the *in vitro* and *in vivo* toxic and inflammatory potential of material cultures should still be investigated as *A. flavus* produces many other metabolites when growing on materials, and hence may still be a problem.

Recent results based on LC-MS/MS show that very small amounts of aflatoxins may be found in dust, presumably as a result of *A. flavus* growth in dust (Polizzi *et al.* 2009). However, these amounts seem to be insignificant compared to exposure to aflatoxins and related compounds from food.

## Aspergillus fumigatus

*Aspergillus fumigatus* is frequently isolated in moldy buildings, especially from dust. It has an amazing arsenal of biological active metabolites, such as gliotoxins, verruculogen, fumitremorgins, fumitoxins, and fumigaclavines, etc. (Frisvad *et al.* 2009).

When cultivated on wood pieces ($a_w$ close to 1) it produces compounds tremorgenic to rats (presumable fumitremorgins) (Land *et al.* 1987), whereas isolates capable of producing verruculogen and helvolic acid, failed to produce these on ceiling tiles and plasterboards (Ren *et al.* 1999). Gliotoxin (up to 40 ng/cm$^2$) and several gliotoxin analogues are produced by *A. fumigatus* inoculated on wood, plasterboard and chipboard at $a_w$ close to 1 (Nieminen *et al.* 2002). When grown on paint, several fumigaclavines were detected in the spores (Panaccione and Coyle 2005).

## Aspergillus ochraceus/A. westerdijkiae

*Aspergillus ochraceus* is occasional isolated from infested materials, and has quite an arsenal of mycotoxins, such as ochratoxin A, penicillic acid, xanthomegnin, viomellein and vioxanthin (Frisvad and Thrane 2002). No studies on growth experiments with this species on building materials have been found in the literature. However, high levels of ochratoxin A (up to 1,500 ppb) have been detected in dust from a house (Richard *et al.* 1999). The findings were confirmed by several highly selective methods, but it was not shown if the ochratoxin A originated from growth on building materials or food scraps.

### Other aspergilli

A number of other aspergilli are also found in buildings especially *Eurotium repens* (*A. glaucus* group in Raper and Fennell (1965)), often isolated from wood containing products and from roof constructions which get very hot during the day. *Eurotium* is not recognized as a mycotoxin-producing genus (Samson *et al.* 2002), but has nonetheless never been tested for its inhalation toxicity. Many of them can grow at 37 °C and may cause infections in immunosuppressed people (Samson *et al.* 2010).

### Penicillium chrysogenum

This is the most common species found in moldy buildings, where its ability to grow in dry habitats on quite a broad array of substrates helps it survive. When isolating moldy material and dust, 3-4 types of *P. chrysogenum* will often be isolated, with one type with a usually yellow reverse, being a xanthocillin producer (De la Campa *et al.* 2007). Because of the difficult taxonomy and identification of *Penicillium* species, it is also possible that these types may represent different species with different extrolite profiles.

The most toxic metabolite isolated is probably roquefortine C (Rand *et al.* 2005) but it may also produce secalonic acid D, which is produced by the related *P. oxalicum* (Frisvad *et al.* 2004). PR toxin has been reported from this species (Frisvad and Filtenborg 1983, Dai *et al.* 1993).

Other metabolites include ω-hydroxyemodin, pyrovoylaminobenzamides, chrysogine, meleagrin, and xanthocillin X (Frisvad and Filtenborg 1989). A more comprehensive list can be found in Nielsen (2002) and Frisvad *et al.* (2004).

*P. chrysogenum* grew very fast on the wallpapered gypsum boards and chipboard, seen as gray to light-green discolored patches. The materials were overgrown in two to three weeks (Gravesen *et al.* 1999, Nielsen *et al.* 1999). On painted samples it grew faster than *A. versicolor*, and was visible after 2 weeks; here it appeared light blue, but changed color to gray over time (Nielsen 2002). On naturally and artificially infested building materials ($a_w$ close to 1), chrysogine, 2-pyrovoylaminobenzamide (Nielsen *et al.* 1999, Nielsen 2002) and meleagrin (Vishwanath *et al.* 2009) have been detected but in very low quantities compared with the extracted biomass. We and others have also detected roquefortine C in low levels when using LC-MS/MS (Polizzi *et al.* 2009, Vishwanath *et al.* 2009). Overall, *P. chrysogenum* produced very low quantities of secondary metabolites compared with *Aspergillus* species, e.g. even when excessive areas of infected material was sampled and analyzed (Nielsen 2002).

## Penicillium brevicompactum *and* P. bialowiezense

*Penicillium brevicompactum* is also very common (Frisvad and Gravesen 1994) and is often isolated from wood (Seifert and Frisvad 2000). Mycophenolic acid and its analogues are the most well known and most consistently produced (Stolk *et al.* 1990); it was the first fungal metabolite to be purified and crystallized, as early as in 1893 (Bentley 2000). The most toxic component is the mutagenic botryodiploidin (Fujimoto *et al.* 1980, Frisvad and Filtenborg 1989). Other metabolites include asperphenamate, brevianamides and Raistrick phenols (Andersen 1991). A more comprehensive list can be found in Nielsen (2002) and Frisvad *et al.* (2004).

All isolates grew well on water-damaged wallpapered chipboard and chipboard (Nielsen *et al.* 1999), although growth was slower on the wallpapered gypsum-boards, and no growth was observed on gypsum boards. This observation suggests that growth is stimulated by components present in wood, which corresponds well to field observations where it is frequently isolated from wooden materials (Seifert and Frisvad 2000).

On chipboard (Nielsen *et al.* 1999) 1/3 of the isolates produced mycophenolic acid, and 50% of the strains also produce asperphenamate, quinolactacin (only produced by *P. bialowiezense* (Frisvad *et al.* 2004)), asperphenamate, and several unidentified metabolites (Nielsen *et al.* 1999, Nielsen 2002).

## Paecilomyces

Two species of *Paecilomyces* are often found in water-damaged buildings, *P. variotii* Bain. and *P. lilacinus* (Thom) Samson, with the first being the most abundant. The last species does not belong to *Paecilomyces* and will be accommodated in a new genus (Samson *et al.* 2010). *P. variotii* is capable of growing on very low nutrient substrates such as optical lenses, whereas *P. lilacinus* seems to prefer protein-rich media (Domsch *et al.* 1980). *P. variotii* can produce the strongly cytotoxic viriditoxin

(Samson *et al.* 2002) as well the metabolites ferrirubin, variotin, fusigen and indole-3-acetic acid (Gravesen *et al.* 1994).

## Alternaria

The *Alternaria tenuissima* species group is the predominant *Alternaria* found in buildings (Nielsen *et al.* 1999, Andersen *et al.* 2002a) and not the rare *A. alternata*, which unfortunately has been given the designation of any "black *Alternaria*" (Andersen *et al.* 2001). On laboratory media the *Alternaria tenuissima* species group produces alternariols, tentoxin, tenuazonic acids, altertoxin I, and numerous unknown metabolites (Andersen *et al.* 2002a).

*Alternaria* metabolites have been reviewed thoroughly by Woody and Chu (1992). The acute toxicity ($LD_{50}$) of alternariols in rodents seems to be in the range of 400 mg/kg, which is quite low even though they are reported to be teratogenic at 10 times lower levels. Alternariols are mutagenic in the Ames test, but there are only a few data on higher animals, and none on inhalation studies.

Ren *et al.* (1998) showed that only alternariols were produced on various building materials, whereas the altertoxins were not produced in detectable quantities. All 5 isolates, except of the *A. infectoria* group isolate, inoculated on different materials (Nielsen *et al.* 1999) grew well and produced alternariol, its mono methyl ether and three to five unknown analogues with identical UV-spectra. No altertoxins, tentoxin or tenuazonic acid were detected in any of the extracts, confirming other results (Ren *et al.* 1998). None of the methods had a good sensitivity for tenuazonic acid however.

The most potent mycotoxins produced by this genus are the AAL toxins, which are related to the fumonisins (Shier *et al.* 1991, Winter *et al.* 1996) *Alternaria arborescens* infecting tomato plants, but such strains have not been detected in buildings. It has not been shown if the AAL toxins are produced by *Alternaria* in buildings. These metabolites are currently the only really relevant mycotoxins known from this genus, but require LC-MS detection.

Interestingly, we were unable to obtain naturally infected materials with substantial amounts of *Alternaria* biomass, as the isolates recovered from the materials always ended up being *Ulocladium* which does not produce any known mycotoxins (Nielsen *et al.* 1999). Thus, the presence of this fungus in buildings only seems to pose a very limited risk from a mycotoxin point of view, especially as the main source of exposure comes from outdoors, where the air spore concentrations are magnitudes higher than indoors.

## Ulocladium

*Ulocladium* Preuss is often isolated from water-damaged materials together with *Stachybotrys*, and generally it is *Ulocladium chartarum* and *U. atrum* which are two commonly occurring species of this genus in buildings (Frisvad and Gravesen 1994, Gravesen *et al.* 1997). Although this genus is morphologically similar to *Alternaria* and shares the same major allergen, almost no secondary metabolites have been isolated from it.

All four isolates inoculated on materials (Nielsen *et al.* 1999) grew well, but did not produce any detectable quantities of secondary metabolites, which is in accordance with the literature where few metabolites from this genus have been described. Analyses of natural samples with excessive growth have not revealed any metabolites in the polarity range we have looked at to date.

To conclude on this genus, it seems to be isolated in the same places as *S. chartarum* although more frequently, as it is presumably able to grow under the same humidities as *Alternaria*. The low production of metabolites indicates, however, that *Ulocladium* is not a major toxicological problem. Nevertheless, this should be confirmed by testing its *in vitro* and *in vitro* pulmonary toxicity and inflammatory potential.

*Ulocladium* is a good indicator for wet environments and hence water damage, but significant microscopic training is required for it to be differentiated from *Alternaria*.

## Cladosporium

*Cladosporium* Link ex Fr. is a phylloplane genus and during the summer there can be as many as 15,000 spores per $m^3$ air of *C. cladosporioides* and *C. herbarum* (Gravesen *et al.* 1994), both containing many of the same allergenic proteins (Achatz *et al.* 1995). *C. cladosporioides* is very common in bathrooms, window frames, and window regions, environments which often have very different humidities during the day, and where it can easily outgrow other non-phyloplanes. Presumable this is due to its ability to restart growth from the hyphal tip much faster than other fungi (Park 1982). Nielsen and Adan (unpublished results) observed that *C. sphaerospermum* on various gypsum-based materials (with and without paint and wallpaper) in a mixed culture was able to outgrow *P. chrysogenum* under non-stationary humidity conditions even in a ratio of 1 to 100 (Nielsen 2002).

There are no described mycotoxins from this genus, and the overall number of non-volatile secondary metabolites is approx. 100 (Anonymous 2007), which is low compared to genera such as *Aspergillus* and *Penicillium*, from which approx. 1,250 have been described from each genus. The main activities of metabolites described from *Cladosporium* have been antifungal and/or plant growth inhibitors.

## Trichoderma

*Trichoderma* Pers. ex Fr. is common on water-damaged materials containing wood, and six different species have been isolated from Danish buildings: *T. longibrachiatum, T. harzianum, T. citrinoviride, T. atroviride, T. viride* and *T. harmatum* (Lübeck *et al.* 2000). *Trichoderma* is known for producing the mycotoxins trichodermin and trichodermol, originally isolated from a strain identified as *T. viride* (Godtfredsen and Vangedal 1964, 1965), and recently re-identified as *T. brevicompactum* (Nielsen *et al.* 2005). This genus is taxonomically distinct from the species found in buildings, although its micro-morphology is close to *T. harzianum*. Some extremely closely related species to *T. brevicompactum*: *T. arundinaceum, T. turrialbense,* and *T. protrudens* (Degenkolb *et al.* 2008a) are responsible for production of the trichothecene harzianum A, a compound originally described from *T. harzianum* (Corley *et al.* 1994) and *Hypocrea* sp. (Lee *et al.* 2005).

The non-production of trichothecenes by indoor *Trichoderma* was further backed up by GC-MS analysis of 8 indoor-derived strains grown on various building materials (Nielsen *et al.* 1998b), as well as screening of numerous other strains (Nielsen and Thrane 2001), including 36 indoor strains (Lübeck *et al.* 2000). There are other studies claiming trichothecene production by *Trichoderma*, even one on indoor-derived strains (Cvetnic and Pepelnjak 1997), however all studies have been conducted using unsound analytical chemical techniques.

Species in the genus are known for production of peptiabiotics (Degenkolb *et al.* 2008b), which are generally considered non-toxic to animals. However, none of them have been tested by inhalation and, due to their ionophoric properties, could therefore be a serious problem in lung cells. There is only one study dealing with peptiabiotics in relation to indoor strains and in this study a sperm cell assay was used for toxicity assessment (Peltola *et al.* 2001).

A large number of metabolites from *Trichoderma* are small α-pyrones and lactones (<300 g/mol), which are often volatile. Many of these are anti-fungal. Production of gliotoxin and viridin have also been reported from several species, but these have always been shown to be *T. virens* (Dennis and Webster 1971) which has not been found in indoor environments (Samson *et al.* 2002).

Recently it was shown that double-blinded exposure of 8 "sensitive" humans to $6 \times 10^5$ spores/$m^3$ of *T. harzianum,* grown on wallpapered gypsum boards, for 6 minutes, did not induce any significant changes in self-reported symptoms, clinical measurements or blood analyses (Meyer *et al.* 2005). The set-up was (for economic reasons) partly built using plastic material, and during characterization of the set-up it was shown that particles smaller than the spores were absorbed into the walls of the set-up. This is a problem, since *T. harzianum* produces micro-particles when drying out (Kildesø *et al.* 2003).

## Fusarium

Trichothecene-producing Fusaria have not been found in buildings, which is why simple type B trichothecenes are not an indoor air problem. However, recently Vishwanath *et al.* (2009) found enniatins in buildings; these are probably produced by *Fusarium oxysporum.*

## Exposure

For many years it was assumed that the spores and perhaps large mycelial fragments were the source of exposure (Sorenson 1999), and that spore counting could be used for exposure assessment. However, recently this was shown not to be the case, as fragments significantly smaller than the spores (down to 0.3 µm) are released from the mycelia of infested materials (Kildesø *et al.* 2000, Gorny *et al.* 2002). The fragments can be liberated in amounts hundreds of times higher than the number of spores, with no correlation between the number of released fragments and spores (Gorny *et al.* 2002).

Toxins have also been detected in fine particle matter from artificially prepared micro-particles (Brasel *et al.* 2005a) as well as fine dust from air-sampling (Johanning *et al.* 2002, Bloom *et al.* 2007).

The most efficient sampling method must be based on toxins or even better their adducts in blood samples, as has been done for satratoxin-G-albumin. This strategy was also the one which really solved the true aflatoxin exposure problem and provided a valid link to epidemiology (Autrup *et al.* 1991). It might also be interesting to examine the sterigmatocystin-guanine or lysine adducts in order to determine the exposure from *A. versicolor.*

## Conclusion

Mycotoxin production on materials is occurring at high $a_w$ (>0.9, in the material surface), but significant mycotoxin production will only occur above $a_w$ 0.95. However, exposure is highest from dry materials and decaying biomass. Thus the worst-case scenario is consecutive water damage, where large quantities of biomass and mycotoxins will be formed followed by desiccation of the biomass, causing many spores and micro-particular fragments to be aerosolized and deposited all over the building *and* the building envelope.

Compared with other indoor fungi, *S. chartarum* produces very high amounts of secondary metabolites when it grows in buildings. Likewise *Chaetomium globosum* should also be considered with caution when it is dry, as micro-fragments will contain considerable quantities of chaetoglobosin mycotoxins. The penicillia investigated until now produce very few and low quantities of secondary metabolites and mycotoxins when growing on building materials. The same goes for aspergilli, except for *A. versicolor*, which may produce up to 1% of its biomass as sterigmatocystins, but then exposure to sterigmatocystins should occur via micro-particles as almost no spores are formed on indoor material.

Working with mycotoxins in buildings is a multidisciplinary task, demanding knowledge of chemotaxonomy, fungal metabolism and biosynthetic pathways, and fungal identification. For the latter an internationally acknowledged specialist of the fungal genera investigated should consulted. Furthermore, knowledge on fungal physiology and growth on building materials, advanced analytical chemistry, particle physics, cell biology and toxicology is necessary. Therefore a taskforce of specialists from these disciplines is needed for future research.

The most promising research tool for mycotoxins exposure would seem to be LC-MS/MS to look for mycotoxins and metabolites of these in human blood or more likely as conjugates in blood.

## Acknowledgements

KFN would like to thank the Danish Technical Research Council (26-04-0050) for its financial support and J.D. Miller and Bruce B. Jarvis for their fruitful discussions.

# References

Achatz G, Oberkofler H and Lechenauer E (1995) Molecular cloning of major and minor allergens of *Alternaria alternata* and *Cladosporium herbarum*. Mol Immunol 32: 213-227.

Andersen B (1991) Consistent production of phenolic compounds by *Penicillium brevicompactum* from chemotaxonomic characterization. Ant van Leeuwenhoek 60: 115-123.

Andersen B, Krøger E and Roberts RG (2001) Chemical and morphological segregation of *Alternaria alternata, A. gaisen* and *A. longipes*. Mycol Res 105: 291-299.

Andersen B, Krøger E and Roberts RG (2002a) Chemical and morphological segregation of *Alternaria aborescens, A. infectoria* and *A. tenuissima* species groups. Mycol Res 106: 170-182.

Andersen B, Nielsen KF and Jarvis BB (2002b) Characterization of *Stachybotrys* from water-damaged buildings based on morphology, growth and metabolite production. Mycologia 94: 392-403.

Andersen B, Nielsen KF, Thrane U, Szaro T, Taylor JW and Jarvis BB (2003) Molecular and phenotypic description of *Stachybotrys chlorohalonata* sp. nov. and two chemotypes of *Stachybotrys chartarum* found in water-damaged buildings. Mycologia 95: 1227-1238.

Andersen B and Nissen A (2000) Evaluation of media for detection of *Stachybotrys* and *Chaetomium* species associated with water-damaged buildings. Int Biodet Biodegr 46: 111-116.

Anderson MA, Nikulin M, Köljalg U, Anderson MC, Rainey F, Reijula K, Hintikka E-L and Salkinoja-Salonen MS (1997) Bacteria, molds, and toxins in water-damaged building materials. Appl Environ Microbiol 63: 387-393.

Anonymous (2007) Antibase 2007, the Natural Compound Identifier. Database. Wiley-VCH Verlag, Weinheim, Germany.

Autrup JL, Schmidt J, Seremet T and Autrup H (1991) Determination of exposure to aflatoxins among Danish worker in animal-feed production through the analysis of aflatoxin $B_1$ adducts to serum albumin. Scand J Work Env Health 17: 436-440.

Bentley R (2000) Mycophenolic acid: a one hundred year odyssey from antibiotic to immunosuppressant. Chem Rev 100: 3801-3825.

Bjurman J and Kristensson J (1992) Volatile production by *Aspergillus versicolor* as a possible cause of odor in houses affected by fungi. Mycopathologia 118: 173-178.

Bloom E, Bal K, Nyman E, Must A and Larsson L (2007) Mass Spectrometry-based strategy for direct detection and quantification of some mycotoxins produced by *Stachybotrys* and *Aspergillus* spp. in indoor environments. Appl Environ Microbiol 73: 4211-4217.

Brasel TL, Campbell AW, Demers RE, Ferguson BS, Fink J, Vojdani A, Wilson SC and Straus DC (2004) Detection of trichothecene mycotoxins in sera from individuals exposed to *Stachybotrys chartarum* in indoor environments. Arch Env Health 59: 317-323.

Brasel TL, Douglas DR, Wilson SC and Straus DC (2005a) Detection of airborne *Stachybotrys chartarum* macrocyclic trichothecene mycotoxins on particulates smaller than conidia. Appl Environ Microbiol 71: 114-122.

Brasel TL, Martin JM, Carriker CG, Wilson SC and Straus DC (2005b) Detection of airborne *Stachybotrys chartarum* macrocyclic trichothecene mycotoxins in the indoor environment. Appl Environ Microbiol 71: 7376-7388.

Corley DG, Miller-Wideman M and Durley RC (1994) Isolation and structure of harzianum A: a new trichothecene from *Trichoderma harzianum*. J Nat Prod 57: 442-425.

Croft WA, Jarvis BB and Yatawara CS (1986) Airborne outbreak of trichothecene mycotoxicosis. Atmos Environ 20: 549-552.

Croft WA, Jastromski BM, Croft AL and Peters HA (2002) Clinical confirmation of trichothecene mycotoxicosis in patient urine. J Environ Biol 23: 301-320.

Cruse M, Teletant R, Gallagher T, Lee T and Taylor JW (2002) Cryptic species in *Stachybotrys chartarum*. Mycologia 94: 814-822.

Cvetnic Z and Pepelnjak S (1997) Distribution and mycotoxin-producing ability of some fungal isolates from the air. Atmos Environ 31: 491-495.

Dai MC, Tabacchi R and Saturnin C (1993) Nitrogen-containing aromatic compounds from the culture medium of *Penicillium chrysogenum* Thom. Chimia 47: 226-229.

Davis ND, Diener UL and Eldridge DW (1966) Production of aflatoxins $B_1$ and $G_1$ by *Aspergillus flavus* on a semisynthetic medium. Appl Microbiol 14: 378.

De la Campa R, Seifert K and Miller JD (2007) Toxins from strains of *Penicillium chrysogenum* isolated from buildings and other sources. Mycopathologia 163: 161-168.

Dearborn DG, Infeld MD and Smidth PG (1997) Pulmonary hemorrhage/hemosiderosis among infants 1993-1996. MMWR 46: 33-35.

Degenkolb T, Dieckmann R, Nielsen KF, Graefenhan T, Theis C, Zafari D, Chaverri P, Ismaiel A, Bruckner H, von Dohren H, Thrane U, Petrini O and Samuels GJ (2008a) The *Trichoderma brevicompactum* clade: a lineage with new species, new peptaibiotics and mycotoxins. Mycol Progr 7: 177-219.

Degenkolb T, Van Dohren H, Nielsen KF, Samuels GJ and Brückner H (2008b) Recent advances and future prospects in peptaibiotics, hydrophobin and mycotoxin research and their importance for chemotaxonomy of *Trichoderma* and *Hypocrea*. Chemistry & Biodiversity 5: 671-680.

Dennis C and Webster J (1971) Antagonistic properties of species-groups of *Trichoderma*. I. Production of non-volatile antibiotics. Trans Brit Mycol Soc 57: 25-39.

Domsch KH, Gams W and Anderson T-H (1980) Compendium of soil fungi, 1st ed. IHW-Verlag, Regensburg, Germany.

Engelhart S, Loock A, Skutlarek D, Sagunski H, Lommel A, Farber H and Exner M (2002) Occurrence of toxigenic *Aspergillus versicolor* isolates and sterigmatocystin in carpet dust from damp indoor environments. Appl Environ Microbiol 68: 3886-3890.

Ezeonu IM, Prince DL, Crow SA and Ahearn DG (1995) Effects of extracts of fiberglass insulation on the growth of *Aspergillus fumigatus* and *A. versicolor*. Mycopathologia 132: 65-69.

Flappan SM, Portnoy J, Jones P and Barnes C (1999) Infant pulmonary hemorrhage in a suburban home with water damage and mold (*Stachybotrys atra*). Env Health Persp Supp 107: 927-930.

Fogle MR, Douglas DR, Jumper CA and Straus DC (2007) Growth and mycotoxin production by *Chaetomium globosum*. Mycopathologia 164: 49-56.

Frisvad JC (1989) The connection between the penicillia and aspergilli and mycotoxins with special emphasis on misidentified isolates. Arch Env Contam Tox 18: 452-467.

Frisvad JC and Filtenborg O (1983) Classification of terverticillate *Penicillia* based on profile of mycotoxins and other secondary metabolites. Appl Environ Microbiol 46: 1301-1310.

Frisvad JC and Filtenborg O (1989) Terverticillate penicillia: chemotaxonomy and mycotoxin production. Mycologia 81: 837-861.

Frisvad JC and Gravesen S (1994) *Penicillium* and *Aspergillus* from Danish homes and working places with indoor air problems: identification and mycotoxin determination. In: Samson RA, Flannigan B, Flannigan ME and Verhoeff AP (eds.) Health implications of fungi in indoor air environment. Elsevier, Amsterdam, the Netherlands, pp. 281-290.

Frisvad JC, Nielsen KF and Samson RA (2006) Recommendations concerning the chronic problem of misidentification of mycotoxinogenic fungi associated with foods and feeds. Adv Exp Med Biol 571: 33-46.

Frisvad JC, Rank C, Nielsen KF and Larsen TO (2009) Metabolomics of *Aspergillus fumigatus*. Med Mycol 47: S71.

Frisvad JC, Smedsgaard J, Larsen TO and Samson RA (2004) Mycotoxins, drugs and other extrolites produced by species in *Penicillium* subgenus *Penicillium*. Stud Mycol 49: 201-241.

Frisvad JC and Thrane U (2002) Mycotoxin production by common filamentous fungi. In: Samson RA, Hoekstra ES, Frisvad JC and Filtenborg O (eds.) Introduction to food- and airborne fungi. 6th ed. Centraalbureau voor Schimmelcultures, Utrecht, the Netherlands, pp. 321-330.

Frisvad JC, Thrane U and Filtenborg O (1998) Role and use of secondary metabolites in fungal taxonomy. In: Frisvad JC, Bridge PD and Arora DK (eds.) Chemical fungal taxonomy. Marcel Dekker, New York, NY, USA, pp. 289-319.

Frisvad JC, Smedsgaard J, Samson RA, Larsen TO and Thrane U (2007) Fumonisin $B_2$ production by *Aspergillus niger*. J Agric Food Chem 55: 9727-9732.

Fujimoto Y, Kamiya M, Tsunoda H, Ohtsubo K and Tatsuno T (1980) Reserche toxicologique des substances metaboliques de *Penicillium carneo-lutescens*. Chem Pharm Bull 248: 1062-1067.

Gareis M, Bauer J, Thiem J, Pland G, Brabley S and Gedek B (1990) Cleavage of zearalenone glycoside, a "masked" mycotoxin, during digestion on swine. J Vet Med B 37: 236-240.

Gloer JB (1995) The chemistry of fungal antagonism and defense. Can J Bot 73: S1265-S1274.

Godtfredsen WO and Vangedal S (1964) Trichodermin, a new antibiotic, related to trichothecin. Proc Chem Soc 6: 188-189.

Godtfredsen WO and Vangedal S (1965) Trichodermin, a new sesquiterpene antibiotic. Acta Chem Scand 19: 1088-1102.

Gorny RL, Reponen T, Willeke K, Schmechel D, Robine E, Boissier M and Grinshpun SA (2002) Fungal fragments as indoor air biocontaminants. Appl Environ Microbiol 68: 3522-3531.

Gottschalk C, Bauer J and Meyer K (2008) Detection of satratoxin G and H in indoor air from a water-damaged building. Mycopathologia 166: 103-107.

Gravesen S, Frisvad JC and Samson RA (1994) Microfungi, 1st ed. Munksgaard, Copenhagen, Denmark.

Gravesen S, Nielsen PA, Iversen R and Nielsen KF (1999) Microfungal contamination of damp buildings – examples of risk constructions and risk materials. Env Health Persp 107 (Suppl. 3): 505-508.

Gravesen S, Nielsen PA and Nielsen KF (1997) Skimmelsvampe i vandskadede bygninger. SBI rapport 282, Statens Byggeforskningsinstitut, Hørsholm, Denmark [In Danish].

Gregory L, Rand TG, Dearborn D, Yike I and Vesper S (2002) Immunocytochemical localization of stachylysin in *Stachybotrys chartarum* spores and spore-impacted mouse and rat lung tissue. Mycopathologia 156: 109-117.

Hanelt MM, Gareis M and Kollarczik B (1994) Cytotoxicity of mycotoxin evaluated by the MTT cell-culture assay. Mycopathologia 128: 167-174.

Hasumi K, Ohyama S, Kohyama T, Ohsaki Y, Takayasu R and Endo A (1998) Isolation of SMTP-3, 4, 5 and -6, novel analogs of staplabin, and their effects on plasminogen activation and fibrinolysis. J Antibiot1059-1068.

Hinkley SF, Fettinger JC, Dudley K and Jarvis BB (1999) Memnobotrins and memnoconols: novel metabolites from *Memnoniella echinata*. J Antibiot 52: 988-997.

Hinkley SF and Jarvis BB (2000) Chromatographic method for *Stachybotrys* toxins. In: Pohland A and Trucksess MW (eds.) Methods in molecular biology 157. Mycotoxin protocols. Humana Press, Totowa, NJ, USA, pp. 173-194.

Hinkley SF, Mazzola EP, Fettinger JC, Lam YKT and Jarvis BB (2000) Atranones A-G, from the toxigenic mold *Stachybotrys chartarum*. Phytochemistry 55: 663-673.

Hinkley SF, Moore JA, Squillari J, Tak H, Oleszewski R, Mazzola EP and Jarvis BB (2003) New atranones from the fungus *Stachybotrys chartarum*. Magn Reson Chem 41: 337-343.

Hodgson MJ, Morey PR, Leung WY, Morrow L, Miller JD, Jarvis BB, Robbins H, Halsey JF and Storey E (1998) Building-associated pulmonary disease from exposure to *Stachybotrys chartarum* and *Aspergillus versicolor*. J Occup Env Med 40: 241-249.

Houbraken J, Due M, Varga J, Meijer M, Frisvad JC and Samson RA (2007) Polyphasic taxonomy of *Aspergillus* section *Usti*. Stud Mycol 59: 107-128.

Jarvis BB (1992) Macrocyclic trichothecenes from Brazilian *Baccharis* species: from microanalysis to large-scale isolation. Phytochem Anal 3: 241-249.

Jarvis BB, Sorenson WG, Hintikka E-L, Nikulin M, Zhou Y, Jiang J, Wang S, Hinkley SF, Etzel RA and Dearborn DG (1998) Study of toxin production by isolates of *Stachybotrys chartarum* and *Memnoniella echinata* isolated during a study of pulmonary hemosiderosis in infants. Appl Environ Microbiol 64: 3620-3625.

Jarvis BB, Zhou Y, Jiang J, Wang S, Sorenson WG, Hintikka E-L, Nikulin M, Parikka P, Etzel RA and Dearborn DG (1996) Toxigenic molds in water-damaged buildings: dechlorogriseofulvins from *Memnoniella echinata*. J Nat Prod 59: 553-554.

Johanning E, Biagini RE, Hull D, Morey PR, Jarvis BB and Landsbergis P (1996) Health and immunology study following exposure to toxigenic fungi (*Stachybotrys chartarum*) in a water-damaged office environment. Int Arch Occup Health 68: 207-218.

Johanning E, Gareis M, Chin YS, Hintikka E-L, Jarvis BB and Dietrich R (1998) Toxicity screening of materials from buildings with fungal indoor air quality problems (*Stachybotrys chartarum*). Mycotoxin Res 14: 60-73.

Johanning E, Gareis M, Nielsen KF, Dietrich R and Märtlbauer E (2002) Airborne mycotoxin sampling and screening analysis. In: Levin H, Bendy G and Cordell J (eds.) Indoor air 2002, The 9th International Conference on Indoor air quality and Climate, June 30-July 5, 2002, Monterey, California. Vol. 5. The International Academy of Indoor Air Sciences, Santa Cruz, CA, USA, pp. 1-6.

Kaneto R, Dobashi K, Kojima I, Sakai K, Shibamoto N, Yoshioka T, Nishid AH, Okamoto R, Akagawa H and Mizuno S (1994) Mer-nf5003b, mer-nf5003e and mer-nf5003f, novel sesquiterpenoids as avian-myeloblastosis virus protease inhibitors produced by *Stachybotrys* sp. J Antibiot 47: 727-730.

Kikuchi T, Shigetoshi K, Suehara H, Nishi A and Tsubaki K (1981) Odorous metabolite of a fungus, *Chaetomium globosum*, Kinze ex Fr. Identification of geosmin, a musty-smelling compound. Chem Pharm Bull 29: 1782-1784.

Kildesø J, Würtz H, Nielsen KF, Kruse P, Wilkins CK, Thrane U, Gravesen S, Nielsen PA and Schneider T (2003) Determination of fungal spore release from wet building materials. Indoor Air 13: 48-55.

Kildesø J, Würtz H, Nielsen KF, Wilkins CK, Gravesen S, Nielsen PA, Thrane U and Schneider T (2000) The release of fungal spores from water damaged building materials. In: Seppänen O and Säteri J (eds) Proceedings of Healthy Buildings 2000, August 6-10, 2000, Espoo, Finland. SYI Indoor Air Information Oy, Helsinki, pp. 313-318.

Kim K-W, Sugawara F, Yoshida S, Murofushi N, Takahashi N and Curtis RW (1993) Structure of malformin B, a phytotoxic metabolite produced by *Aspergillus niger*. Biosci Biotechnol Biochem 57: 787-791.

Knapp JF, Michael JG, Hegenbartk MA, Jones PE and Black PG (1999) Case records of the Children's Mercy Hospital, case 02-1999: a 1-month-old infant with respiratory distress and chock. Ped Emerg Care 15: 288-293.

Kobbe B, Cushman M, Wogan GN and Demain AL (1977) Production and antibacterial activity of malformin C, a toxic metabolite of *Aspergillus niger*. Appl Environ Microbiol 33: 996-997.

Kohyama T, Hasumi K, Hamanaka A and Endo A (1997) SMTP-1 and -2, novel analogs of staplabin produced by *Stachybotrys microspora* IFO30018. J Antibiot 50: 172-174.

Korpi A, Pasanen A-L, Pasanen P and Kalliokoski P (1997) Microbial growth and metabolism in house dust. Int Biodet Biodegr 40: 19-27.

Korpi A, Kasanen JP, Raunio P, Kosma VM, Virtanen T and Pasanen A-L (2002) Effects of aerosols from nontoxic *Stachybotrys chartarum* on murine airways. Inhal Toxicol 14: 521-540.

Land CJ, Hult K, Fuchs R, Hagelberg S and Lundström H (1987) Tremorgenic mycotoxins from *Aspergillus fumigatus* as a possible occupational health problems in sawmills. Appl Environ Microbiol 53: 787-790.

Larsen TO and Frisvad JC (1994) Production of volatiles and presence of mycotoxins in conidia of common indoor penicilia and aspergillii. In: Samson RA, Flannigan B, Flannigan ME and Verhoeff AP (eds.) Health implications of fungi in indoor air environment. Elsevier, Amsterdam, the Netherlands, pp. 251-279.

Larsen TO, Smedsgaard J, Nielsen KF, Hansen ME and Frisvad JC (2005) Phenotypic taxonomy and metabolite profiling in microbial drug discovery. Nat Prod Rep 22: 672-695.

Lee HB, Kim Y, Jin HZ, Lee JJ, Kim CJ, Park JY and Jung HS (2005) A new *Hypocrea* strain producing harzianum A cytotoxic to tumour cell lines. Lett Appl Microbiol 40: 497-503.

Lübeck M, Poulsen SK, Lübeck PS, Jensen DF and Thrane U (2000) Identification of *Trichoderma* strains from building materials by ITS1 ribotypning, UP-PCR fingerprinting and UP-PCR cross hybridization. FEMS Microbiol Lett 185: 129-134.

Meyer HW, Jensen KA, Nielsen KF, Kildesø J, Norn S, Permin H, Poulsen L, Malling HJ, Gravesen S and Gyntelberg F (2005) Double blind placebo controlled exposure to molds: exposure system and clinical results. Indoor Air 15: 73-80.

Miller JD (1992) Fungi as contaminant of indoor air. Atmos Environ 26A: 2163-2172.

Minagawa K, Kouzuki S, Nomura K, Kawamura Y, Tani H, Terui Y, Nakai H and Kamigauchi T (2001) Bisabosquals, novel squalene synthase inhibitors – II. Physico-chemical properties and structure elucidation. J Antibiot 54: 896-903.

Montana E, Etzel RA, Dearborn DG, Sorenson WG and Hill R (1995) Acute pulmonary hemorrhage in infancy associated with *Stachybotrys atra* – Cleveland Ohio, 1993-1995. Am J Epidem 141: S83.

Murtoniemi T, Nevalainen A, Suutari M, Toivola M, Komulainen H and Hirvonen M-R (2001) Induction of cytotoxicity and production of inflammatory mediators in RAW264.7 macrophages by spores grown on six different plasterboards. Inhalation Toxicology 13: 233-247.

Nielsen KF (2002) Mold growth on building materials. Secondary metabolites, mycotoxins and biomarkers. PhD Thesis. BioCentrum-DTU, Technical University of Denmark.

Nielsen KF, Graefenhan T, Zafari D and Thrane U (2005) Trichothecene production by *Trichoderma brevicompactum*. J Agric Food Chem 53: 8190-8196.

Nielsen KF, Gravesen S, Nielsen PA, Andersen B, Thrane U and Frisvad JC (1999) Production of mycotoxins on artificially and naturally infested building materials. Mycopathologia 145: 43-56.

Nielsen KF, Hansen MØ, Larsen TO and Thrane U (1998a) Production of trichothecene mycotoxins on water damaged gypsum boards in Danish buildings. Int Biodet Biodegr 42: 1-7.

Nielsen KF, Huttunen K, Hyvärinen A, Andersen B, Jarvis BB and Hirvonen M-R (2001) Metabolite profiles of *Stachybotrys* spp. isolates from water damaged buildings, and their capability to induce cytotoxicity and production of inflammatory mediators in RAW 264.7 macrophages. Mycopathologia 154: 201-205.

Nielsen KF, Mogensen JM, Johansen M, Larsen TO and Frisvad JC (2009) Review of secondary metabolites and mycotoxins from the *Aspergillus niger* group. Anal Bioanal Chem 395: 1225-1242.

Nielsen KF, Nielsen PA, Holm G and Uttrup LP (2004) Mold growth on building materials under low water activities. Influence of humidity and temperature on fungal growth and secondary metabolism. Int Biodet Biodegr 54: 325-336.

Nielsen KF and Thrane U (2001) A fast method for detection of trichothecenes in fungal cultures using gas chromatography-tandem mass spectrometry. J Chromatogr A 929: 75-87.

Nielsen KF, Thrane U, Larsen TO, Nielsen PA and Gravesen S (1998b) Production of mycotoxins on artificially inoculated building materials. Int Biodet Biodegr 42: 8-17.

Nieminen SM, Karki R, Auriola S, Toivola M, Laatsch H, Laatikainen R, Hyvarinen A and Von Wright A (2002) Isolation and identification of *Aspergillus fumigatus* mycotoxins on growth medium and some building materials. Appl Environ Microbiol 68: 4871-4875.

Nikulin M, Reijula K, Jarvis BB and Hintikka E-L (1997a) Experimental lung mycotoxicosis induced by *Stachybotrys atra*. Int J Experim Pathol 77: 213-218.

Nikulin M, Reijula K, Jarvis BB, Veijalaninen P and Hintikka E-L (1997b) Effects of intranasal exposure to spores of *Stachybotrys atra* in mice. Fund Appl Tox 35: 182-188.

Nozawa Y, Yamamoto K, Ito M, Sakai N, Mizoue K, Mizobe F and Hanada K (1997) Stachybotrin C and parvisporin, novel neuritogenic compounds 1. Taxonomy, isolation, physico-chemical and biological properties. J Antibiot 50: 635-640.

Panaccione DG and Coyle CM (2005) Aboundant respirable ergot alkaoids from the common airborne fungus *Aspergillus fumigatus*. Appl Environ Microbiol 71: 3106-3111.

Park D (1982) Phylloplane fungi: tolerance of hyphal tips to drying. Trans Brit Mycol Soc 79: 174-179.

Peltola J, Anderson MA, Haahtela T, Mussalo-Rauhamaa H, Rainey FA, Koppenstedt RM, Samson RA and Salkinoja-Salonen M (2001) Toxic metabolite producing bacteria and fungus in an indoor environment. Appl Environ Microbiol 67: 3269-3274.

Peltola J, Niessen L, Nielsen KF, Jarvis BB, Andersen B, Salkinoja-Salonen M and Möller EM (2002) Toxigenic diversity of two different RAPD groups of *Stachybotrys chartarum* isolates analyzed by potential for trichothecene production and for boar sperm cell motility inhibition. Can J Microbiol 48: 1017-1029.

Polizzi V, Delmulle B, Adams A, Moretti A, Susca A, Picco AM, Rosseel Y, Kindt R, Van Bocxlaer J, De Kimpe N, Van Peteghem C and De Saeger S (2009) Fungi, mycotoxins and microbial volatile organic compounds in moldy interiors from water-damaged buildings. J Env Monit 11: 1849-1858.

Rand TG, Flemming J, David MJ and Womiloju TO (2006) Comparison of inflammatory responses in mouse lungs exposed to atranones a and C from *Stachybotrys chartarum*. J Toxicol Environ Health A 69: 1239-1251.

Rand TG, Giles S, Flemming J, Miller JD and Puiani E (2005) Inflammatory and cytotoxic responses in mouse lungs exposed to purified toxins from building isolated *Penicillium brevicompactum* Dierckx and *P. chrysogenum* Thom. Toxicol Sci 87: 213-222.

Rao CY, Brain JD and Burge HA (2000) Reduction of pulmonary toxicity of *Stachybotrys chartarum* spores by methanol extraction of mycotoxins. Appl Environ Microbiol 66: 2817-2821.

Rao CY, Fink RC, Wolfe LB, Liberman DF and Burge HA (1997) A study of aflatoxin production by *Aspergillus flavus* growing on wallboard. J Am Biolog Safety Assoc 2: 36-42.

Raper KB and Fennell DI (1965) The genus *Aspergillus*. Robert E. Kriger Publishing Company, New York, NY, USA.

Räty K, Raatikainen O, Holmalahti J, Wright A, Joki S, Pitkaenen A, Saano V, Hyvärinen A, Nevalainen A and Buti I (1994) Biological activities of actinomycetes and fungi isolated from indoor air of problem houses. Int Biodet Biodegr 34: 143-154.

Ren P, Ahearn DG and Crow SA (1998) Mycotoxins of *Alternaria alternata* produced on ceiling tiles. J Ind Microb 20: 53-54.

Ren P, Ahearn DG and Crow SA (1999) Comparative study of *Aspergillus* mycotoxin production on enriched media and construction material. J Ind Microbiol Biotechnol 23: 209-213.

Richard JL, Plattner RD, Mary J and Liska SL (1999) The occurrence of ochratoxin A in dust collected from a problem household. Mycopathologia 146: 99-103.

Roggo BE, Petersen F, Sills M, Roesel JL, Moerker T and Peter HH (1996) Novel spirodihydrobenzofuranlactams as antagonists of endothelin and as inhibitors of HIV-1 protease produced by *Stachybotrys* sp. I. Fermentation, isolation and biological activity. J Antibiot 49: 13-19.

Ruotsalainen M, Hirvonen M-R, Nevalainen A, Meklin T and Savolainen K (1998) Cytotoxicity, production of reactive oxygen species and cytokines induced by different strains of *Stachybotrys* sp. from moldy buildings in RAW264.7 macrophages. Env Tox Pharm 6: 193-199.

Samson RA, Hoekstra ES, Frisvad JC and Filtenborg O (2002) Introduction to Food- and Airborne fungi. 6th and 7th ed. Centraalbureau voor Schimmelcultures, Utrecht, the Netherlands.

Samson RA, Houbraken J, Thrane U, Frisvad JC and Andersen B (2010). Food and Indoor Fungi. CBS Laboratory Manual Series 2. Centraalbureau voor Schimmelcultures, Utrecht, the Netherlands.

Schuster E, Dunn-Coleman N, Frisvad JC and Van Dijck PW (2002) On the safety of *Aspergillus niger* – a review. Appl Microbiol Biotechnol 59: 426-435.

Scott PM (1999) Aflatoxins in corn and peanut butter. AOAC official method 990.33. In: Cunniff P (ed.) Official methods of analysis of AOAC international. AOAC International, Gaithersburg, MD, USA.

Seifert KA and Frisvad JC (2000) *Penicillium* on solid wood products. In: Samson RA and Pitt JI (eds.) Integration of modern taxonomic methods from *Penicillium* and *Aspergillus* classification. Harwood Academic Publishers, Amsterdam, the Netherlands, pp. 285-298.

Seifert KA, Samson RA, Dewaard JR, Houbraken J, Levesque CA, Moncalvo JM, Louis-Seize G and Hebert PDN (2007) Prospects for fungus identification using C01 DNA barcodes, with *Penicillium* as a test case. Proc Natl Acad Sci USA 104: 3901-3906.

Sekita S, Yoshihira K, Natori S, Udagawa S, Muroi T, Sugiyama Y, Kurata H and Umeda M (1981) Mycotoxin production by *Chaetomium* spp. and related fungi. Can J Microbiol 27: 766-772.

Shier WT, Abbas H and Mirocha CJ (1991) Toxicity of the mycotoxins fumonisins $B_1$ and $B_2$ and *Alternaria alternata* f.sp. *lycopersici* toxin (AAL) in cultured mammalian cells. Mycopathologia 116: 97-104.

Slack GJ, Puniani E, Frisvad JC, Samson RA and Miller JD (2009) Secondary metabolites from *Eurotium* species, *Aspergillus calidoustus* and *A. insuetus* common in Canadian homes with a review of their chemistry and biological activities. Mycol Res 113: 480-490.

Smith JE and Solomons GL (1994). Mycotoxins in human nutrition and health. EUR16048 EN. 1994. European Commision, Directorate-General XII. Ref Type: Report.

Smith SL and Hill ST (1982) Influence of temperature and water activity on germination and growth of *Aspergillus restrictus* and *A. versicolor*. Trans Brit Mycol Soc 79: 558-559.

Smoragiewicz W, Cossette B, Boutard A and Krzystyniak K (1993) Trichothecene mycotoxins in the dust ventilation system in office buildings. Int Arch Occup Health 65: 113-117.

Sorenson WG (1999) Fungal spores: hazardous to health. Env Health Persp Supp 107: 469-472.

Sorenson WG, Frazer DG, Jarvis BB, Simpson J and Robinson VA (1987) Trichothecene mycotoxins in aerosolized conidia of *Stachybotrys atra*. Appl Environ Microbiol 53: 1370-1375.

Sorenson WG, Tucker JD and Simpson JP (1984) Mutagenicity of the tetramic mycotoxin cyclopiazonic acid. Appl Environ Microbiol 47: 1355-1357.

Steyn PS and Vleggaar R (1974) Austocystins. Six novel dihydrofuro[3',s': 4,5]furo,3,2-b-xanthenones: from *Aspergillus ustus*. J Chem Soc Perkin I 1: 2250-2256.

Stoev SD, Daskalov H, Radic B, Domijan AM and Peraica M (2002) Spontaneous mycotoxic nephropathy in Bulgarian chickens with unclarified mycotoxin aetiology. Vet Res 33: 83-93.

Stolk AC, Samson RA, Frisvad JC and Filtenborg O (1990) The systematics of the terverticillate *Penicillia*. In: Samson RA and Pitt JI (eds.) Modern concepts in *Penicillium* and *Aspergillus* classification. Plenum Press, New York, NY, USA, pp. 121-137.

Sunesson A-L, Nilsson C-A, Carlson R, Blomquist G and Andersson B (1997) Production of volatile metabolites from *Streptomyces albidoflavus* cultivated on gypsum board and tryptone glucose extract agar – influence of temperature, oxygen and carbon dioxide levels. Ann Occup Hyg 41: 393-413.

Tuomi T, Reijula K, Johnsson T, Hemminki K, Hintikka E-L, Lindroos O, Kalso S, Koukila-Kähköla P, Mussalo-Rauhamaa H and Haahtela T (2000) Mycotoxins in crude building materials from water-damaged buildings. Appl Environ Microbiol 66: 1899-1904.

Ueno Y (1985) Toxicology of mycotoxins. Crit Rev Toxicol 14: 99-132.

Van Emon JM, Reed AW, Yike I and Vesper SJ (2003) ELISA measurement of Stachylysin (TM) in serum to quantify human exposures to the indoor mold *Stachybotrys chartarum*. J Occup Env Med 45: 582-591.

Vesper SJ, Dearborn DG, Yike I, Allen T, Sobolewski J, Hinkley SF, Jarvis BB and Haugland RA (2000) Evaluation of *Stachybotrys chartarum* in the house of an infant with pulmonary hemmorrhage: quantitative assessment before, during, and after remediation. J Urb Health 77: 68-85.

Vesper SJ, Dearborn DG, Yike I, Sorenson WG and Haugland RA (1999) Hemolysis, toxicity, and randomly amplified polymorphic DNA analysis of *Stachybotrys chartarum* strains. Appl Environ Microbiol 65: 3175-3181.

Vesper SJ, Vesper MJ (2002) Stachylysin may be a cause of hemorrhaging in humans exposed to *Stachybotrys chartarum*. Infect Immun 70: 2065-2069.

Vinson D, Esswein E and Page E. (2005) NIOSH health hazard evaluation report. Report HETA #2005-0112-2980. Santa Ana, CA, USA.

Vishwanath V, Sulyok M, Labuda R, Bicker W and Krska R (2009) Simultaneous determination of 186 fungal and bacterial metabolites in indoor matrices by liquid chromatography/tandem mass spectrometry. Anal Bioanal Chem 395: 1355-1372.

Vojdani A, Campbell AW, Kashanian A and Vojdani E (2003a) Antibodies against molds and mycotoxins following exposure to toxigenic fungi in a water-damaged building. Arch Environ Health 58: 324-336.

Vojdani A, Kashanian A, Vojdani E and Campbell AW (2003b) Saliva secretory IgA antibodies against molds and mycotoxins in patients exposed to toxigenic fungi. Immunopharmacol Immunotoxicol 25: 595-614.

Williams DH, Stone MJ, Hauck PR and Rahman SK (1989) Why are secondary metabolites (natural products) biosynthesized? J Nat Prod 52: 1189-1208.

Winter CK, Gilchrist DG, Dickman MB and Jones C (1996) Chemistry and biological activity of AAL toxins. Adv Exp Med Biol 392: 307-316.

Woody MA and Chu FS (1992) Toxicology of *Alternaria* mycotoxins. In: Chelkowski J and Visconti A (eds.) *Alternaria*. Biology, plant disease and metabolites. Elsevier, Amsterdam, the Netherlands, pp. 409-434.

Yike I, Distler AM, Ziady AG and Dearborn DG (2006) Mycotoxin adducts on human serum albumin: biomarkers of exposure to *Stachybotrys chartarum*. Env Health Persp 114: 1221-1226.

# 10 WHO guidelines for indoor air quality: dampness and mold

*Otto O. Hänninen*
*National Institute for Health and Welfare, Kuopio, Finland*

## History of who guidelines for air quality

Clean air is a basic requirement for human health and wellbeing. World Health Organization (WHO) has recognized the special role of indoor air quality as a health determinant. In all climates and most cultures globally a majority of population time is spent in various indoor environments, and all population groups, including those vulnerable due to their age or health status, are exposed to the factors present in indoor air (WHO 2000a). Nevertheless, the history of WHO guidelines for air quality started with a focus on the ambient air.

Soon after the foundation of WHO the international community faced the devastating health consequences of air pollution of the famous London air pollution episode. The 1952 episode occurred four years after the WHO constitution became into force and has remained as one of the landmarks in the progress of air pollution epidemiology and health risk assessment. The activities in evaluating these health hazards culminated in 1987 in the publication of the first edition of WHO Guidelines for Air Quality for Europe (WHO 1987). The Guidelines covered 28 pollutants that occur in indoor and outdoor air and defined 32 guideline levels various averaging times. Nine unit risks were determined for pollutants associated with cancer risks for which it was impossible to find a safe level for setting a guideline.

The growing body of knowledge in the early 1990s resulted in a revision and publication of the second edition of the Guidelines with risk characterization of 35 pollutants (Table 10.1) and including for the first time a dedicated section on indoor air pollutants (WHO 2000a). The second edition defined 29 guideline levels and determined 12 unit risks for carcinogens and now also for particulate matter, for which there was no evidence on safe level; as a consequence the previous guidelines for $PM_{10}$ and total suspended particles (TSP) were omitted. Table 10.1 also highlights the revisions in comparison to the first edition (WHO 1987). Specifically, seven new pollutants were added to the evaluation (butadiene, polychlorinated biphenyls (PCBs), polychlorinated dibenzodioxins and dibenzofurans (PCDDs & PCDFs), fluoride, platinum, $PM_{2.5}$ and man-made vitreous fibers), but only for the fibers a unit risk could be presented. The fluoride evaluation concluded that the guideline based on

*Table 10.1. WHO Air Quality Guidelines for Europe (WHO, 2000a) with the changes in comparison with the 1987 edition highlighted by bold font.*

| Group | Compound | Averaging time | Guideline or unit risk | Notes |
|---|---|---|---|---|
| Organic substances | | | | |
| 1 | acrylonitrile | lifetime | UR $2\times10^{-5}$ | cancer risk per 1 µg/m$^3$ exposure |
| 2 | benzene | lifetime | UR $6\times10^{-6}$ | UR 6 instead of 4 ppm. Cancer risk per 1 µg/m$^3$ exposure |
| 3 | **butadiene**[1] | – | – | no guideline was recommended due to large variability of risk estimates |
| 4 | carbon disulfide | 24 h | 100 µg/m$^3$ | sensory value 20 µg/m$^3$ (30 min) |
| 5 | carbon monoxide[2] | 15 min | 100 mg/m$^3$ | multiple longer time responses up to 8 hours (10 mg/m$^3$) |
| 6 | 1,2-dichloroethane | 24 h | 0.7 mg/m$^3$ | |
| 7 | dichloromethane | 24 h | **3 mg/m$^3$** | **additional value 0.45 mg/m$^3$ for 1 week** |
| 8 | formaldehyde | 30 min | 0.1 mg/m$^3$ | |
| 9 | polynuclear aromatic hydrocarbons | lifetime | UR $9\times10^{-2}$ | cancer risk per 1 µg/m$^3$ exposure |
| 10 | **polychlorinated biphenyls (PCBs)**[1] | – | – | no guideline was recommended due to low inhalation intake in comparison to diet |
| 11 | **PCDDs & PCDFs**[1,3] | – | – | no guideline was recommended due to low inhalation intake in comparison to diet |
| 12 | styrene | **1 week** | **0.26 mg/m$^3$** | **weekly instead of 24 h (800 µg/m$^3$), sensory value 70 µg/m$^3$ (30 min)** |
| 13 | tetrachloroethylene | **1 a** | **0.25 mg/m$^3$** | **annual instead of 24 h value of 5 mg/m$^3$, sensory value 8 mg/m$^3$ (30 min)** |
| 14 | toluene | **1 week** | **0.26 mg/m$^3$** | **weekly instead of 24 h value of 8 mg/m$^3$, sensory value 1 mg/m$^3$ (30 min)** |
| 15 | trichloroethylene | **lifetime** | **UR $4.3\times10^{-7}$** | **unit risk instead of 24 h value of 1 mg/m$^3$** |
| 16 | vinyl chloride | lifetime | UR $1\times10^{-6}$ | 24 h value at 1 µg/m$^3$ dropped |
| Inorganic substances | | | | |
| 17 | arsenic | lifetime | **UR $1.5\times10^{-3}$** | **UR 1.5 instead of 4 ‰. Cancer risk per 1 µg/m$^3$ exposure** |
| 18 | asbestos | lifetime | UR $1\times10^{-4}$ | mesothelioma (lower unit risk for lung cancer) |
| 19 | cadmium | 1 a | 5 ng/m$^3$ | **5 instead of 1-5 ng/m$^3$** |
| 20 | chromium (VI) | lifetime | UR $4\times10^{-2}$ | Cancer risk per 1 µg/m$^3$ exposure |

*Table 10.1. Continued.*

| Group | Compound | Averaging time | Guideline or unit risk | Notes |
|---|---|---|---|---|
| Inorganic substances (continued) | | | | |
| 21 | **fluoride**[1] | – | – | **ecological guideline 1 µg/m³ sufficient for health, too** |
| 22 | hydrogen sulfide | 24 h | 150 µg/m³ | sensory value 7 µg/m³ (30 minutes) |
| 23 | lead | 1 a | **0.5 µg/m³** | **0.5 instead of 0.5-1 µg/m³** |
| 24 | manganese | 1 a | **0.15 µg/m³** | **0.15 instead of 1 µg/m³** |
| 25 | mercury | 1 a | 1 µg/m³ | concerning mainly indoor air |
| 26 | nickel | lifetime | UR 4×10⁻⁴ | cancer risk per 1 µg/m³ exposure |
| 27 | platinium[1] | – | – | no guideline was recommended due to low ambient levels |
| 28 | vanadium | 24 h | 1 µg/m³ | |
| Classical pollutants[1] | | | | |
| 29 | nitrogen dioxide | 1 h | **200 µg/m³** | **200 instead of 400 µg/m³. Also annual guideline 40 µg/m³ instead of 24 h** |
| 30 | ozone and photochemical oxidants | **8 h** | **120 µg/m³** | **120 instead of 150-200 µg/m³. 1 hour guideline dropped** |
| 31 | sulphur dioxide | 10 min | 500 µg/m³ | **also instead of 1 h value a 24 h & annual guidelines at 125 µg/m³ and 50 µg/m³** |
| 32 | particulate matter | | | |
| | TSP | – | – | **guideline dropped** |
| | PM₁₀ | – | – | **guideline dropped. Risks observed below 30 µg/m³ are acknowledged** |
| | PM₂.₅ | – | – | **risks observed below 20 µg/m³ are acknowledged** |
| Indoor air pollutants[1] | | | | |
| 33 | tobacco smoking (ETS)[4] | lifetime | **UR 1×10⁻³** | no safe level. UR = lifetime ETS exposure cancer risk in a home with one smoker |
| 34 | man-made vitreous | lifetime | **UR 1×10⁻⁶** | **cancer risk per 1 fiber/l concentration** |
| 35 | radon[2] | lifetime | **UR 3-6×10⁻⁵** | **cancer risk per 1 Bq/m³ exposure; instead of previous value of 2×10⁻⁴** |

[1] New category or compound added in 2000.
[2] In the inorganic substances section of the 1987 edition.
[3] Polychlorinated dibenzodioxins and –dibenzofurans.
[4] ETS=Environmental tobacco smoke (passive smoking).

ecological effects ($1 \mu g/m^3$) is sufficient to protect human health, too. For particulate matter ($PM_{2.5}$ and $PM_{10}$) a summary of the relative mortality and morbidity risks associated with the exposures was given, but no safe level could be determined or a conclusive unit risk presented. For the other new candidates insufficient evidence was available for guidelines due to large variability between estimates from different studies (butadiene), or the inhalation route contributed only marginally to the human risks (PCBs, PCDDs & PCDFs). The full risk characterization and evaluation of the evidence is available in the original guideline document available in the Internet (WHO 2000a).

During the first decade of the new millennium there has been a constantly increasing awareness among scientists and policy-makers of the global nature and magnitude of the public health problems posed by exposure to air pollution both indoors and in the outdoor environment. A systematic review of health aspects of air pollution in Europe was carried out by the WHO Regional Office for Europe to support the development of the European Union's Clean Air for Europe (CAFE) programme in 2002-2004, which concluded that the new evidence warranted revision of the air quality guidelines for particulate matter (PM), ozone and nitrogen dioxide (WHO 2004). Of particular importance in deciding that the guidelines should apply worldwide was the substantial and growing evidence of the health effects of air pollution in the low- and middle-income countries in the developing world, where air pollution levels are the highest. According to a WHO assessment of the burden of disease due to air pollution, more than two million premature deaths each year can be attributed to the effects of urban outdoor air pollution and indoor air pollution from the burning of solid fuels. More than half of this burden is borne by the populations of developing countries.

In response to this real and global threat to public health, WHO released a Global Update of the WHO Air Quality Guidelines (WHO 2005, 2006a,b). The Global Update focused only on four key pollutants, particulate matter, ozone, and sulphur and nitrogen dioxides (Table 10.2), for which the newest scientific evidence was carefully reviewed. The Global Update lead to a modest tightening of $PM_{10}$ and $O_3$ guideline levels from the previous values (WHO, 1987, and WHO, 2000a, respectively), but a substantial tightening of 24-hour $SO_2$ level, and more importantly, the introduction of guidelines for $PM_{2.5}$. The previous guidelines for $NO_2$ were not changed based on the new evidence. A new element was introduced due to the global role of the guidelines. It was acknowledged that in many areas reaching the tightened and new guidelines may not be feasible in the near future, and therefore interim targets were determined with corresponding risk characterizations.

*Table 10.2. Global update of WHO Guidelines for Air Quality (WHO 2006b) and comparison with previous guideline levels.*

| Compound | Averaging time | Interim targets[1] | | | Guideline (µg m³) | Previous guideline (µg m³) |
|---|---|---|---|---|---|---|
| | | IT1 (µg m³) | IT2 (µg m³) | IT3 (µg m³) | | |
| PM$_{2.5}$ | 24 h | 75 | 50 | 37.5 | 25 | – |
| | 1 a | 35 | 25 | 15 | 10 | – |
| PM$_{10}$ | 24 h | 150 | 100 | 75 | 50 | 70[2] |
| | 1 a | 70 | 50 | 30 | 20 | – |
| O$_3$ | 8 h | 160 | – | – | 100 | 120[3] |
| NO$_2$ | 1 h | – | – | – | 200 | 200[3] |
| | 1 a | – | – | – | 40 | 40[3] |
| SO$_2$ | 10 min | – | – | – | 500 | 500[3] |
| | 24 h | 125 | 50 | – | 20 | 125[3] |

[1] Interim targets introduced first time in the Global Update (WHO 2006).

[2] WHO 1987.

[3] WHO 2000a.

While the Global Update (WHO 2006b) concerned particulate matter, sulphur and nitrogen dioxides and ozone, traditionally considered as ambient air contaminants, the role of indoor air quality was given a separate emphasis in the background review of scientific evidence that included a specific chapter for indoor air quality, and the development of guidelines specific to indoor environments was straightforwardly recommended by the working group for the World Health Organization. As a response to these recommendations, WHO Regional Office for Europe convened a planning meeting for this work in Bonn, Germany, in October 2006 (WHO 2006c) for the development of WHO Guidelines for Indoor Air Quality. The current chapter gives an overview of the status of the ongoing development process with specific focus on the dampness and molds related indoor air quality issues and health.

## Focusing on indoor air

The basic right for, and importance of, healthy indoor air was postulated by a WHO working group already in the 1990s (WHO 2000b, see Table 10.3) parallel to the inclusion of the first indoor air specific section in the second edition of the

*Table 10.3. Nine Principles on everyone's right to healthy indoor air (WHO 2000b).*

| | |
|---|---|
| Principle 1 | Under the principle of the human right to health, everyone has the right to breathe healthy indoor air. |
| Principle 2 | Under the principle of respect for autonomy ("self-determination"), everyone has the right to adequate information about potentially harmful exposures, and to be provided with effective means for controlling at least part of their indoor exposures. |
| Principle 3 | Under the principle of non-maleficence ("doing no harm"), no agent at a concentration that exposes any occupant to an unnecessary health risk should be introduced into indoor air. |
| Principle 4 | Under the principle of beneficence ("doing good"), all individuals, groups and organizations associated with a building, whether private, public, or governmental, bear responsibility to advocate or work for acceptable air quality for the occupants. |
| Principle 5 | Under the principle of social justice, the socioeconomic status of occupants should have no bearing on their access to healthy indoor air, but health status may determine special needs for some groups. |
| Principle 6 | Under the principle of accountability, all relevant organizations should establish explicit criteria for evaluating and assessing building air quality and its impact on the health of the population and on the environment. |
| Principle 7 | Under the precautionary principle, where there is a risk of harmful indoor air exposure, the presence of uncertainty shall not be used as a reason for postponing cost-effective measures to prevent such exposure. |
| Principle 8 | Under the "polluter pays" principle, the polluter is accountable for any harm to health and/or welfare resulting from unhealthy indoor air exposure(s). In addition, the polluter is responsible for mitigation and remediation. |
| Principle 9 | Under the principle of sustainability, health and environmental concerns cannot be separated, and the provision of healthy indoor air should not compromise global or local ecological integrity, or the rights of future generations. |

guidelines (WHO 2000a). This section covered pollutants for which the health risks are specifically confined to indoor spaces, namely environmental tobacco smoke, man-made vitreous fibers and radon. However, the role of indoor air is covered for numerous pollutants for which indoor sources play a significant role in the exposure patterns, including mercury, benzene and butadiene. Moreover, indoor levels of dichloromethane were typically several times higher than outdoor levels, and the most severe health risks from carbon monoxide take place in indoor spaces. Indoor levels of formaldehyde were also more than an order of magnitude higher than outdoors. Levels of styrene may be elevated in newly built houses containing styrene-based materials and dry cleaning operations may raise tetrachloroethylene levels more than 2 orders of magnitude above the background concentrations indoors. Thus for a large number of pollutants of health relevance, indoor environment plays an important role.

The importance of indoor air is magnified by the substantial fraction of time populations spend within buildings. Across the European climates the average time spent indoors varies only between 85% and 90% (Figure 10.1). Due to the time activity patterns, the indoor spaces are the most important exposure environment for outdoor pollutants, even when the buildings partly protect people from them (e.g. Hänninen *et al.* 2004a, Hoek *et al.* 2008).

### Development of WHO Guidelines for Indoor Air Quality

As a reaction to the recommendation by the Global Update of WHO Guidelines for Air Quality the development of WHO Guidelines for Indoor Air Quality was started in 2006. A planning meeting with participants from 17 countries in South- and North America, Asia, Australia and Europe convened in Bonn, Germany, in

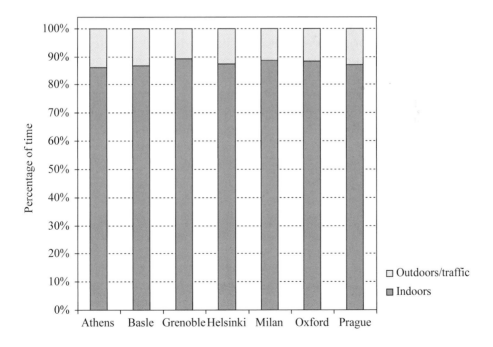

*Figure 10.1. Average annual time spent indoors in seven European cities (data from working age populations EXPOLIS-study; Hänninen et al. 2004b, 2005).*

*Otto O. Hänninen*

October 2006. The working group outlined the development process for indoor air quality and identified three sections that were recommended to be addressed separately: (A) pollutant specific guidelines, (B) biological agents, and (C) indoor combustion of fuels (WHO 2006c).

In 2007 the development work started on the biological agents in the indoor air with special focus on the effects of excess moisture and molds and as of writing this book the guidelines for dampness and mold are passing through the final process before publication later in 2009. The development of chemical specific guidelines for indoor air started in 2008 and continues the systematic review of the scientific evidence and developing the background material for the guideline setting. After having received the background material from the authors invited external reviewers will evaluate the drafts. Based on this reviewed draft, a WHO Working Group (the authors, selected reviewers and members of the Steering Group) will evaluate the risk of each pollutant and recommend the Guidelines in 2009-2010 followed by the development of the third section on indoor combustion.

### Dampness and molds

First section of the recommended indoor air guidelines concerns dampness and molds. Availability of water is a major determinant of microbiological growth. Agents of biological origin prevalent in indoor air include hundreds of species of bacteria, fungi, algae, protozoa, pollens from outdoor air, house dust components from textiles, microbial organisms like house dust mites and allergens originating from pets. Particular attention is paid to the filamentous fungi often called molds that are commonly found growing on indoor surfaces globally. To provide sufficient focus for the working group, WHO started the guideline development with dampness and molds, leaving allergens and other factors associated less with the prevalence of moisture to be addressed later.

Biological growth on indoor surfaces is in principle controlled by the availability of nutrients and moisture. Spores and other life forms of microbes capable for airborne transport are commonly present in the air almost everywhere emitted from existing colonies and ready to launch new colonies where ever the suitable conditions for their proliferation exist. Numerous species of bacteria and molds can extract sufficient nutrients from the normal house dust and the household surface materials themselves, leaving the availability of moisture the main growth limiting factor. Therefore, while temporary dampness itself may be considered harmless, it has been repeatedly shown to rapidly lead to manifestation of various microbial problems, especially molds.

Dampness is observed regularly in residences and other buildings. It is estimated that more than half of buildings in Europe and in North America are affected by moisture problems at some time during their life cycle. In European countries the prevalence of self-reported moisture problems in houses varies from 5% to 40% (Finland and Poland, respectively; WHO 2009a, Figure 10.2).

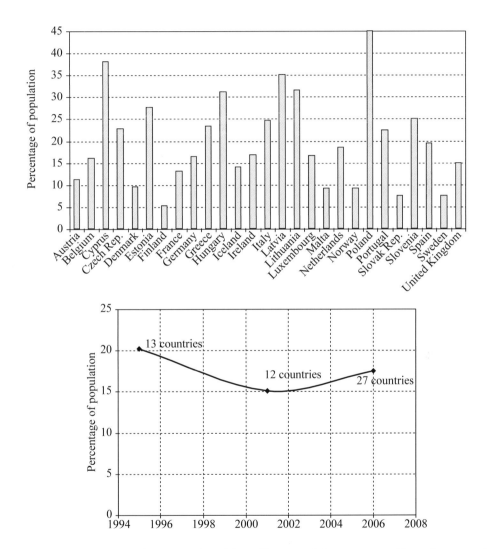

*Figure 10.2. Percentage of homes affected by self-reported dampness problems by countries (top) and European development for 1995-2006 showing an increasing data coverage but no systematic trend (WHO 2009a).*

Health problems associated with moisture and biological agents are for example affecting asthmatic patients, especially children for whom the prevalence of asthma has been increasing during the last decades and approaches or exceeds 20% of the children in some countries. Comparison of asthma prevalence between studies is problematic due to the heterogeneous definition of key symptoms, but even when using standard methods across countries, asthma prevalence varies significantly between countries: e.g. Robertson *et al.* (2005) reported wheezing in the last 12 months in 7-year olds to vary from 7% to 23% and 26% in Switzerland, Melbourne and Chile, respectively.

In 2000-2003 the World Health Organization's Regional Office for Europe undertook a large study to evaluate housing and health in eight European cities (LARES-study in cities Angers, Bratislava, Bonn, Budapest, Ferreira, Forli, Geneva, Vilnius). Survey tools were used to obtain information about housing and living conditions, health perception and health status from a representative sample of the general population in each city (Bonnefoy *et al.* 2003, WHO 2007). In the housing inspection by trained surveyors, visible mold growth was detected in at least one room of almost 25% of the visited dwellings with highest mold occurrence rates in kitchens and bathrooms, the spaces most regularly affected by water use.

The causal chain of events involved in the formation of the health hazard and linking events from sources of water through existence of excessive moisture to biological growth and physical and chemical degradation and further to emission of hazardous biological and chemical agents is quite complex and involves numerous health endpoints affected including various effects on respiratory system, dermal responses and allergies, nervous system and immunology. Epidemiological studies reviewed in WHO (2009b; the reader is referred to the original review for the references) have been able to show the association between excess moisture and a number of these effects on population level. These epidemiological findings have not been able to confirm all suggested outcomes, but cannot exclude the possibility of causal links due to problems in study designs and exposure assessment.

## State of the scientific evidence

WHO guidelines are based on systematic evaluation of the scientific evidence (WHO 2000c). To facilitate the guideline development process, the review of scientific evidence related to dampness and mold was launched in 2006 and the first draft of the background material was prepared for and evaluated at the working group meeting in October. The full background material has now been published together with the WHO Guidelines for Indoor Air Quality: Dampness and Mold (WHO

2009b). The following subsections highlight the main results (see also Table 10.4 for used definitions).

*Table 10.4. Definition of dampness and mold related terms (WHO 2009b).*

| | |
|---|---|
| Air-conditioning system | Assembly of equipment for treating air to control simultaneously its temperature, humidity, cleanliness and distribution to meet the requirements of a conditioned space. |
| Dampness | Any visible, measurable or perceived outcome of excess moisture that causes problems in buildings, such as mold, leaks or material degradation, mold odor or directly measured excess moisture (in terms of relative humidity or moisture content) or microbial growth. |
| Excess moisture | Moisture state variable that is higher than a design criterion, usually represented as moisture content or relative humidity in building material or the air. Design criteria can be simple indicators (e.g. no condensation or relative humidity value) or more complicated representations that take into account continuous fluctuation of moisture (i.e. mold growth index). |
| Moisture | (1) Water vapor; (2) water in a medium, such as soil or insulation, but not bulk water or flowing water. |
| Moisture problem or damage; water damage | Any visible, measurable or perceived outcome caused by excess moisture indicating indoor climate problems or problems of durability in building assemblies. Moisture damage is a particular problem of building assembly durability. Water damage is a moisture problem caused by various leaks of water. |
| Moisture transport | Moisture can be transported in both the vapour and the liquid phase by diffusion, convection, capillary suction, wind pressure and gravity (water pressure). |
| Molds | All species of microscopic fungi that grow in the form of multicellular filaments, called hyphae. In contrast, microscopic fungi that grow as single cells are called "yeasts". A connected network of tubular branching hyphae has multiple, genetically identical nuclei and is considered a single organism, referred to as a "colony" (Madigan and Martinko 2005). |
| Ventilation | Process of supplying or removing air by natural or mechanical means to or from any space; the air may or may not have been conditioned. |

*Otto O. Hänninen*

## Sources of exposures

It has been estimated that approximately one out of four or five buildings in Europe and North America displays current signs of dampness problems (IOM 2004) and that approximately 50% of buildings experience such problems during their lifetime (Mudarri and Fisk 2007). Dampness problems may originate from outdoor air, e.g. due to floods and rainfall, or from anthropogenic sources inherent to the buildings themselves or the use of the building like plumbing and use of water in the household equipment and by the occupants. However, even the driest climates are not fully protected against dampness problems; for example air conditioning units are prone to water condensation.

When prevalent, moisture leads rapidly to the growth of micro-organisms, including bacteria, fungi and mites, but also increases the probability of e.g. termite intrusion. Moisture is frequently also associated with chemical and physical degradation of building materials leading to direct and indirect emissions of organic and inorganic pollutants while at the same time causes physical damage to the structures. The minimum water activity required for fungal growth ranges from 0.7 to almost 1 for some species. Primary colonisers are capable of growth at water activities below 0.8 and include e.g. *Penicillium brevicompactum, Eurotium* spp., *Wallemia sebi*, and *Aspergillus versicolor*. Besides the micro-organisms and their fractions, the health hazards may be caused by biochemical factors emitted by various micro-organisms. These include allergens, endotoxins, mycotoxins, betaglucans and microbial volatile organic compounds (MVOCs).

The heterogeneity of the chemical, biochemical and organism level factors challenge the identification of causal agents for the health hazards. Culture methods have been used to identify viable organisms, but relatively new measurement techniques especially related to the genetic sequences, (e.g. polymerase chain reaction, PCR) are promising for accurate species recognition of viable and nonviable components of microbes.

Micro-organisms in general are ubiquitous in all general environments. Microbes propagate rapidly; the proliferation begins within hours and significant growth can take place in favourable conditions in few days whenever water or moisture is continuously is available. The dust and dirt normally present in most indoor spaces provide sufficient nutrients to support extensive microbial growth. While mold growth is possible on all materials, appropriate material selection is nevertheless important to prevent dirt accumulation, moisture penetration and retention and mold growth. The water potential available on/in materials is the most important factor triggering the growth of micro-organisms, including fungi, actinomycetes and other bacteria.

Moisture availability depends on five main areas; protection of the indoor environment from the natural water outdoors; proper installation of the water systems of the buildings themselves; use of water by the occupants; proper ventilation to provide drying capacity for the normal water use and emissions; and prevention of condensation of moisture from the air. Moreover, many building materials have a capability of storing moisture, which helps the microbes to remain in active state over periods of blocking water availability form these sources.

While mechanical ventilation systems allow for most accurate control of the ventilation rates and moisture in principle, natural ventilation systems are still used in an overwhelming majority of buildings globally. In the eight cities participating the WHO LARES study 40% of the buildings had no ventilation system and in additional 20% of the houses the ventilation system could not be adjusted by the occupants (WHO, 2007). Mechanical ventilation is mainly used in modern high-rise buildings and in buildings applying air conditioning systems (cooling), as well as in colder climates in even small houses applying heat recovery techniques. However, in vast majority of older buildings and mild and warmer climates rely merely on natural ventilation.

## Health hazards

The review of scientific evidence on health effects of microbial contamination in indoor spaces was conducted in three sections: (1) epidemiological, (2) toxicological, and (3) clinical studies. The review conducted by the WHO working group contains hundreds of references, which cannot be replicated here; therefore only a concise summary of the review is given here and the reader is referred to the WHO publication for the full list of references.
1. While epidemiological studies are based on large human populations and realistic, existing exposure levels, the accurate characterization of exposures remains challenging. Therefore the most consistent associations are reported for the existence of moisture problems and health (see Table 10.5 for health association categories).
2. On the other hand, toxicological research allows for a very accurate control of exposures, including both species of micro-organisms involved as well as quantification of the exposure levels, but the results are limited to *in vitro* studies using commonly non-human target organisms and typically high exposure levels in comparison to normal indoor spaces.

*Table 10.5. Categories for classifying the strength of the scientific evidence for associations with specific health outcomes (WHO 2009b).*

The categories refer to the association between exposure to an agent and a health outcome and not to the likelihood that any individual's health problem is associated with or caused by the exposure.

Sufficient evidence of a causal relation

The evidence is sufficient to conclude that a causal relationship exists between the agent and the outcome. That is, the evidence fulfils the criteria for "sufficient evidence of an association" and, in addition, satisfies the following evaluation criteria: strength of association, biological gradient, consistency of association, biological plausibility and coherence and temporally correct association. The finding of sufficient evidence of a causal relationship between an exposure and a health outcome does not mean that the exposure inevitably leads to that outcome. Rather, it means that the exposure can cause the outcome, at least in some people under some circumstances.

Sufficient evidence of an association

The evidence is sufficient to conclude that there is an association. That is, an association between the agent and the outcome has been observed in studies in which chance, bias and confounding could be ruled out with reasonable confidence. For example, if several small studies that are free from bias and confounding show an association that is consistent in magnitude and direction, there may be sufficient evidence of an association.

Limited or suggestive evidence of an association

The evidence is suggestive of an association between the agent and the outcome but is limited because chance, bias and confounding could not be ruled out with confidence. For example, at least one high-quality study shows a positive association, but the results of other studies are inconsistent.

Inadequate or insufficient evidence to determine whether an association exists

The available studies are of insufficient quality, consistency or statistical power to permit a conclusion regarding the presence or absence of an association. Alternatively, no studies of the relation exist.

Limited or suggestive evidence of no association

Several adequate studies are consistent in not showing an association between the agent and the outcome. A conclusion of "no association" is inevitably limited to the conditions, magnitude of exposure and length of observation covered by the available studies.

3. Therefore also review of the clinical evidence where data on human patients is studied in relationship with their exposures and, more importantly, exposure reductions achieved by renovation is needed.

Prior to the WHO work on dampness and mold the work by the Institute of Medicine, USA, (IOM 2004) remained the most comprehensive review of scientific evidence on the association of molds and health. The update conducted by the WHO working group for the guideline development did not provide substantial new information. The IOM review concluded that there is sufficient evidence on associations between upper respiratory tract symptoms, cough, wheeze and asthma symptoms and damp indoor environments on the other hand, existence of excess moisture indoors, as well as existence of visible mold damage indoors. Additionally, existence of mold was associated with hypersensitivity pneumonitis in susceptible persons. The evidence was slightly stronger for the former association, supporting guidance and guidelines targeting moisture control. (IOM 2004).

Additionally, the clinical evidence collected by the WHO working group shows that exposures to molds and other dampness-related microbial agents increase also the risk of rare conditions, such as hypersensitivity pneumonitis/allergic alveolitis, chronic rhinosinusitis and allergic fungal sinusitis (WHO 2009b). Toxicological evidence *in vivo* and *in vitro* supports the epidemiological and clinical findings by showing diverse inflammatory and toxic responses after exposure to specific micro-organisms isolated from damp buildings, including their spores, metabolites and components.

## Conclusions on the health effects

Based on the systematic review of the scientific evidence, the WHO working group concluded in October 2007 in Bonn that there is sufficient epidemiological evidence from studies in different countries and climatic conditions showing that occupants of buildings affected by dampness or mold, both homes and public buildings, are at increased risk of experiencing respiratory symptoms, respiratory infections and exacerbations of asthma (WHO, 2008a). Non-comprehensive evidence suggests an increased risk of developing allergic rhinitis and asthma. Although not many well-documented intervention studies are available, their results show that remediation of dampness problems leads to a reduction in adverse health outcomes.

Population groups such as atopic and allergic individuals are particularly susceptible to exposures to biological and chemical agents in damp indoor environments, but adverse health effects have also been widely demonstrated in non-atopic populations. (e.g. Rönmark *et al.* 1999). The increased prevalence of asthma and allergies in many countries increases the number of people susceptible to the effects of dampness and mold in buildings.

*Otto O. Hänninen*

## Potential global guideline targets

The working group identified a number of factors that are indicators of the health risk and that could be considered by the guidelines (WHO 2008a). Global guidelines aiming at the protection of public health have to address various levels of economic development and different climates, cover all relevant population groups, and enable the design of feasible approaches to reduce health risks from dampness and microbial contamination. In many parts of the world sophisticated techniques in characterizing the biological exposures in indoor environments are not feasible or necessary for elimination of the hazards.

Currently, the relationships between dampness, microbial exposure and various health effects cannot be precisely quantified, so no quantitative health-based guideline values or thresholds can be recommended for acceptable levels of specific micro-organism contamination. Instead, the working group recommended that dampness and mold-related problems be prevented. When they occur, they should be remediated because of the increased risk of hazardous microbial and chemical exposures.

The WHO working group agreed that persistent dampness and microbial growth on interior surfaces and in building structures should be avoided or minimized, as they may lead to adverse health effects. It is important that even materials and surfaces that frequently become in contact with water or moisture dry thoroughly and rapidly after the contact.

## Indicators of dampness and microbial growth

Indicators of dampness and microbial growth include the presence of condensation on surfaces or in structures, visible mold, perceived mold odour and a history of water damage, leakage or penetration of rain or flooding water outdoors. Water or moisture on surfaces and in materials is often visible, but this may not always be the case. Mold growth begins on a microscopic non visible scale, but when the growth is sustained, the mold layer often becomes visible. Mold may be difficult to be distinguished from non-organic dust in some cases, but the prevalence of moisture and visible progression are indicators of the problem.

Thorough inspection and – if needed – appropriate measurements may be used to confirm indoor problems related to moisture and microbial growth. Surface moisture content measurement and regular follow-up of relative humidity are indicators of the problem development, but the measurements may also include microbial sampling and cell cultivation or DNA-based techniques for species identification.

The working group discussed that the existing building standards and regulations on comfort and health do not sufficiently emphasize the requirements to prevent and control excess moisture and dampness. Besides occasional events such as water leaks, excess rain and floods most moisture enters buildings through incoming air, including that infiltrating though the envelope, or is due to occupants' activities. Allowing surfaces to become cooler than the surrounding air may result in unwanted condensation. Thermal bridges in for example metal window frames, inadequate insulation and unplanned air pathways, or cold water plumbing and cool parts of air conditioning units can result in surface temperatures below the dew point of the air that create dampness. The problem of excess moisture and dampness (i.e. presence of sufficient water activity to sustain microbial growth) can be tackled by controlling the quality of the building envelope regarding air infiltration, exfiltration, and pathways of water intrusion, by ensuring adequate thermal insulation and by avoiding condensation indoors through the control of moisture sources and of temperature, humidity and velocity of the air in the proximity of the surfaces, as well as a targeted choice of surface materials on surfaces becoming in contact with moisture.

Well-designed, -constructed and -maintained building envelopes are critical to the prevention and control of excess moisture and microbial growth by avoiding thermal bridges and preventing intrusion by liquid or vapour-phase water. Frequent and rapid drying of surfaces that become in contact with moisture regularly must be ensured by proper ventilation, and the materials of such surfaces have to be water resistant and easy to clean. Management of moisture requires proper control of temperatures and ventilation to avoid high humidity, condensation on surfaces and excess moisture in materials. Ventilation should be distributed effectively in spaces, and stagnant air zones should be avoided.

Local recommendations in different climatic regions should be updated to control dampness mediated microbial growth in buildings and to ensure the achievement of desirable indoor air quality. Dampness and mold may be particularly prevalent in poorly maintained housing for low income people with insufficient heating required for drying and having other priorities combined with limited economical resources. Remediation of conditions related to adverse exposures should be given priority to prevent additional contributions to poor health in populations already living with an increased burden of disease.

*Otto O. Hänninen*

## Chemicals with specific role indoors

Many chemicals, including formaldehyde, volatile organic compounds (VOCs), dioxins, heavy metals, phthalates, and perfluorochemicals are released indoors from building materials, household equipment and consumer products, and air quality may be deteriorated also by microbial pollution from hundreds of species of bacteria, fungi, especially filamentous fungi known as molds. Fungi are capable of producing soluble and volatile metabolites (microbial VOCs, MVOCs) comprising a mixture of compounds that can be common to many species and similar to some common industrial chemicals. More than 200 MVOC compounds have been identified from different fungi (Wilkins *et al.* 2003) including various alcohols, aldehydes, ketones, terpenes, esters, aromatic compounds, amines and sulphur-containing compounds.

A number of international organizations and studies have considered chemicals that would warrant guideline setting, including the previously summarized World Health Organization guidelines as well as the International Agency for Research on Cancer (IARC), the European DG Sanco and EC Joint Research Centre funded project Critical Appraisal of the Setting and Implementation of Indoor Exposure Limits in the EU (INDEX), US Environmental Protection Agency (US-EPA), UK Institute for Environment and Health (IEH) and German Kommission Innenraumlufthygiene (IRK) (WHO 2006c). The INDEX project reviewed exposure and dose-response information on chemicals with indoor significance and prioritized them into two categories in terms of European public health relevance (Kotzias *et al.* 2005).

First priority compounds identified by the INDEX project were formaldehyde, carbon monoxide, citrogen dioxide, benzene and naphthalene (Table 10.6). Formaldehyde is the most important sensory irritant among the chemicals assessed and is frequently present in indoor environments. Carbon monoxide exposures at the normal levels are of moderate concern, but in contrast with all other chemicals, carbon monoxide causes a considerable number of deaths and acute poisonings in the general population. Nitrogen dioxide originates from gas appliances used in homes and frequently reaches levels affecting the pulmonary function of asthmatics, besides the long-term effects of increased respiratory symptoms. Benzene is a genotoxic carcinogen and hence there is no safe level. Naphtalene exposures have generally an order of magnitude margin of safety, but in many countries, mainly in Southern Europe, the use of naphthalene in insecticides increases the indoor levels well above the exposure limits (Kotzias *et al.* 2005). The second priority group of chemicals as well as the pollutants that require more research for more conclusive risk assessment are listed in Table 10.6. The reader is directed to the INDEX report for the full characterization of the chemicals and their health risks in Europe.

*Table 10.6. The three priority groups of chemicals of indoor relevance as identified in the INDEX-project (Kotzias et al. 2005).*

| First priority | Second priority | Third priority |
|---|---|---|
| formaldehyde | acetaldehyde | ammonia |
| carbon monoxide | toluene | limonene |
| nitrogen dioxide | xylenes | α-pinene |
| benzene | styrene | |
| naphtalene | | |

The chemicals selected for the ongoing development of WHO Guidelines for Indoor Air Quality are listed in Table 10.7. The background reviews are under finalization and review in 2009 after which the guidelines will be recommended by an expert meeting.

*Table 10.7. Indoor air pollutants selected for the pollutant specific guideline development (WHO 2006c).*

| Compounds | Risk characterization[1] |
|---|---|
| Benzene | Mild toxicant causing main effects in blood; main effect leukemia (Group 1) |
| Carbon monoxide (CO) | Acute toxicity causes significant number of deaths annually in all countries |
| Formaldehyde | Highly reactive gas causing sensory irritation and cancer (Group 1) |
| Naphtalene | Common exposure, especially in areas where insecticides are used; acutely toxic at high doses, potential carcinogen (Group 2B) |
| Nitrogen dioxide (NO$_2$) | Gaseous pollutant originating from combustion processes; associated with respiratory symptoms |
| PAH (BaP) | Large group of gaseous and particle bound organic substances associated with cancer (Class 1), endocrine system and developmental toxicity |
| Radon (Rn) | A significant source of radiation, causing lung cancer (Group 1) |
| Tetrachloroethylene | Volatile industrial solvent affecting central nervous system and a probable human carcinogen (Group 2B) |
| Trichloroethylene | Volatile industrial solvent affecting central nervous system and a probable human carcinogen (Group 2A) |

[1] Including IARC classification of carcinogens.

*Otto O. Hänninen*

## Relationship of guidelines to health impact assessment and management

Indoor air quality management is made difficult not only by the large number and variation of indoor spaces themselves, but also the complex relationships of the indoor air quality and building design, materials, operation and maintenance, ventilation, as well as behaviour of the building users. While the air quality guidelines and standards which put health in the focus are widely used in outdoor air quality management, no such science-based systematic comprehensive approaches were available for indoor air quality before the start-up of the WHO process.

Although the WHO Guidelines are neither standards nor legally binding criteria, they are designed to offer guidance in reducing the health impacts of air pollution based on expert evaluation of current scientific evidence. They are intended to be relevant to the diverse conditions of all WHO regions, and to support a broad range of policy options for air quality management. Knowledge about (1) the health hazards and (2) indication of the risk related to exposures, summarized by the guidelines, provide an essential scientific contribution to the development of strategies for indoor air quality management. Authorities preparing national strategies, especially in those countries that lack the necessary scientific infrastructure and resources to conduct their own assessments in support of public policy, often find the guidelines an essential resource. The synthesis of the research results that underlie the guidelines has been conducted by outstanding scientists and is subjected to scrupulous peer review.

The guidelines are intended to inform policy-makers and to provide appropriate targets for a broad range of policy options for air quality management in different parts of the world. The guidelines are developed for worldwide use but support actions to protect public health in different contexts. Air quality and other standards set by each country to protect the public health of their citizens are an important component of national risk management and environmental policies. National standards will vary according to the approach adopted for balancing health risks, technological feasibility, economic considerations and various other political and social factors, which in turn will depend on, among other things, the level of development and national capability in air quality management. The guideline values recommended by WHO (e.g. Table 10.1, 10.2 and 10.7) acknowledge this heterogeneity and, in particular, recognize that when formulating policy targets, governments should consider their own local circumstances carefully before adopting the guidelines directly as legally based standards.

## Interventions

### Evidence on the effectiveness of interventions: WHO evaluation of case studies

Practical implementation of actions to reduce, mitigate or prevent dampness and mold has remained a difficult area for public health policies due to the diversification of indoor spaces, the fragmentation of responsibilities and, in the case of homes, the limited mandate of public authorities for interventions. To support the adaptation of the WHO guidelines for indoor air quality and to review the effectiveness of practical interventions in reducing exposure to damp and mold, WHO conducted an intervention case study assessment in 2007-2008.

Thirty case studies on actions and interventions against damp and mold in indoor settings from developed countries with modern building stocks were collected to assess the practical actions and interventions undertaken to address or prevent health-relevant problems of dampness and mold and seventeen case studies were selected for detailed discussion at the WHO expert meeting in Bonn, Germany in 28-29 February, 2008 (WHO, 2008b), identifying elements of good practice based on the actions and technical measures.

The evidence base collected on risk management actions being carried out in different countries in the case of damp and mold in the indoor built environment included all kinds of projects targeted at the prevention, reduction or mitigation of damp and mold problems in residential buildings and dwellings, schools and centres for children and elderly people with focus on non-care elements; workplaces and health care facilities were excluded from the studies. The addressed actions included specifically:
- reduction of relative humidity in indoor air;
- reduction of damp/condensation in/on building structures;
- removal/prevention of mold growth; and
- improvement of ventilation/air exchange rates.

The technical problems addressed by the seventeen in detail evaluated case studies are summarized in Table 10.8. All studies included mold growth and substantial number focused on various problems associated with water, moisture and humidity, followed by studies addressing ventilation system performance and problems. Six case studies included heating and five damaged construction materials.

A variety of technical problems may additionally be associated with mold problems, but they are not reflected in the case studies selected. Unaddressed problems were

Table 10.8. Summary of problems addressed in the intervention case studies (WHO 2008b).

| Technical problem | Case studies | |
|---|---|---|
| | n | % |
| Mold growth | 17 | 100 |
| Problems with water, moisture and humidity | | |
| High levels of indoor relative humidity | 10 | 59 |
| Leakage due to rain water | 7 | 41 |
| Problems with rising damp | 3 | 18 |
| Internal moisture sources in the building | 2 | 12 |
| Leakage due to plumbing problems | 1 | 6 |
| Building affected by major flooding | 1 | 6 |
| Problems related to ventilation | | |
| Lack of indoor ventilation | 7 | 41 |
| Lack of ventilation of crawl spaces | 2 | 12 |
| Spread of contaminants by HVAC system | 1 | 6 |
| Other | | |
| Lack of adequate heating | 6 | 35 |
| Damaged construction material | 5 | 29 |

moisture inside building structures and interstitial condensation, the problem of ageing materials, and the exposure to high outdoor humidity levels.

Nine of the studies measured indoor mold concentration before and after the intervention, mostly in the air and/or on surfaces of building materials. Other compounds or metabolites of microorganisms that were measured in the studies were endotoxins, beta-glucan, ergosterol and volatile organic compounds. All studies that analysed the differences between concentrations of mold and compounds of microorganisms found a reduction or complete disappearance after the intervention.

Eight studies measured moisture in the air and/or on building surfaces. In the limited number of studies where the changes in humidity after the intervention were reported, it resulted in at least a small reduction of moisture. Interventions conducted to reduce humidity were mainly changes in ventilation, reduction of water infiltration and thermal insulation.

Fundamentals of mold growth in indoor environments

Each of the single methods to reduce humidity seems to be successful, but it is not clear which of these measures is the best to resolve the problem and if a combination of improved ventilation and thermal insulation is much better than applying only one of these measures.

Analysing questions on health status, all studies found an improvement of symptoms after the intervention (it is not possible to assess publication or "contribution" bias, which would result from the selective publication/contribution of "successful" studies compared to studies that show negative results). In some studies, however, questionnaires were handed out to the participants both before and after the intervention and differences in health outcomes were assessed, whereas some case studies only asked if symptoms had improved after remediation.

Medical examinations were less frequently implemented than questionnaires on health status. In six studies, the authors stated that they conducted medical examinations. Generally, if effect sizes for the improvement of health status were measured, they were greater for self-reported data than for the results of medical examinations.

### Intervention recommendations

Constructional faults and building damages are the main reason originating from the building itself for damp and mold occurrence. Adequate building construction and maintenance is therefore a key issue to avoid the origin of damp and mold problems from the beginning. High-occupancy buildings are more prone to health problems from damp and mold, and the geographical location and orientation to the north may be a factor for damp and mold problems. In the evaluation the relevance of thermal conditions and indoor surface temperatures for the prevention of excessive moisture was highlighted as well as the necessity of adequate ventilation rates in relation to the climate. Ventilation systems should be quiet to make their use acceptable to building occupants (WHO 2008b).

Addressing the occupant perspective, the group acknowledged the key relevance of occupant behaviour, which – depending on the building – may be able to compensate for minor constructional failures or building faults, but can also trigger damp and mold problems in well-constructed buildings. It was considered necessary to educate building owners as well as occupants about the impact of their behaviour on damp and mold problems, and to provide adequate information on the consequences of building changes and renovation activities. The main objectives of any kind of information or capacity building campaign should be to help occupants to avoid technical problems and to empower them to provide their part of the solution (such as avoiding drying laundry indoors, open flue heaters, excessive water use during cleaning, excessively low temperature indoors or misuse of ventilation). The case study review suggested that measures giving tailored advice to occupants of problem buildings should especially be supported. (WHO 2008b)

From the problem perspective, a number of open questions still remain, such as the unstructured and vague qualification requirements of housing professionals in general (architects, engineers and construction workers) and remediation workers in particular; the lack of evaluations of the costs and benefits of remediation activities; the uncertainty as to whether building occupants need to be relocated (and when); and the frequent conflict between ventilation and energy-saving requirements. Evidence-based answers to these questions will help to inform and improve both the actions undertaken against damp and mold, and increase the health benefits derived from these interventions (Table 10.9).

The working group suggested a need for policy guidelines on technical issues such as remediation work requirements, but acknowledged that such guidelines can only be very broad and need to be supplemented by specific technical norms and codes of practice on national or regional levels.

*Table 10.9. Elements and practices associated with successful intervention (WHO 2008b).*

- Buildings investigated by experts/with proper (tools for) diagnosis.
- Solving the causes of moisture problems.
- Removing mold as well as damaged material (mechanical or chemical cleaning).
- Drying remaining structures.
- Improving ventilation and thermal insulation.
- Using or replacing with proper materials that do not promote mold growth.
- Protecting the workers.
- Preventing cross-contamination/separating clean-up areas from non-infected areas within a building.
- Initiating remediation work as soon as possible after problem identification and minimizing the remediation period to prevent occupant discomfort/complaints.
- Relocating affected building occupants as an effective strategy next to remediation to reduce exposure and health complaints.
- Performing follow-up of remediation/evaluation and quality control of work.

## Acknowledgements

The chapter text is based on the cited WHO documents and the proceedings of the WHO working group meetings in Bonn, Germany, in 2006-2008. The meeting participants, authors and reviewers are acknowledged for their substantial contribution to this work.

## References

Bonnefoy X, Braubach M, Boissonnier B, Monolbaev K and Röbbel N (2003) Housing and Health in Europe: preliminary results of a pan-European study. Am J Public Health 93: 1559-1563.

Bruce NG, Perez-Padilla R and Albalak R (2002) The health effects of indoor air pollution exposure in developing countries. WHO/SDE/OEH/02.05, World Health Organization, Geneva, Switzerland. Available at: http://whqlibdoc.who.int/hq/2002/WHO_SDE_OEH_02.05.pdf.

Ezzati M and Kammen DM (2002) The health impacts of exposure to indoor air pollution from solid fuels in developing countries: knowledge, gaps, and data needs. Environ Health Perspect 110: 1057-1068.

Hänninen OO, Alm S, Katsouyanni K, Künzli N, Maroni M, Nieuwenhuijsen MJ, Saarela K, Srám RJ, Zmirou D and Jantunen MJ (2004b). The EXPOLIS study: implications for exposure research and environmental policy in Europe. J Expo Anal Environ Epidemiol 14: 440-456.

Hänninen OO, Lebret E, Ilacqua V, Katsouyanni K, Künzli N, Srám RJ and Jantunen MJ (2004a) Infiltration of ambient $PM_{2.5}$ and levels of indoor generated non-ETS $PM_{2.5}$ in residences of four European cities. Atmos Environ 38: 6411-6423.

Hänninen OO, Palonen J, Tuomisto J, Yli-Tuomi T, Seppänen O and Jantunen MJ (2005) Reduction potential of urban $PM_{2.5}$ mortality risk using modern ventilation systems in buildings. Indoor Air 15:246-256.

Hoek G, Kos G, Harrison R, De Hartog J, Meliefste K, Ten Brink H, Katsouyanni K, Karakatsani A, Lianou M, Kotronarou A, Kavouras I, Pekkanen J, Vallius M, Kulmala M, Puustinen A, Thomas S, Meddings C, Ayres J, Van Wijnen J and Hameri K (2008) Indoor-outdoor relationships of particle number and mass in four European cities. Atmos Environ 42: 156-169.

IOM (2004) Damp indoor spaces and health. Institute of Medicine, National Academies Press, Washington, DC, USA, 370 pp.

Kotzias D, Koistinen K, Kephalopoulos S, Schlitt C, Carrer P, Maroni M, Jantunen M, Cochet C, Kirchner S, Lindvall T, McLaughlin J, Mølhave L, De Oliveira Fernandes E and Seifert B (2005) The INDEX project: critical appraisal of the setting and implementation of indoor exposure limits in the EU. Final report. Available at: http://ec.europa.eu/health/ph_projects/2002/pollution/fp_pollution_2002_frep_02.pdf.

Mudarri D and Fisk WJ (2007) Public health and economic impact of dampness and mold. Indoor Air 17: 226-235.

Robertson C, Bishop J, Sennhauser F and Mallol J (2005). International comparison of asthma prevalence in children: Australia, Switzerland, Chile. Pediatr Pulmonol 16: 219-226.

Rönmark E, Jönsson E, Platts-Mills T, Lundbäck B (1999) Different pattern of risk factors for atopic and nonatopic asthma among children – report from the Obstructive Lung Disease in Northern Sweden study. Allergy 54: 926-935.

WHO (1987). Air quality guidelines for Europe, World Health Organization, Regional Office for Europe, European series, No 23, Copenhagen, Denmark.

WHO (2000a) Air quality guidelines for Europe; Second edition, World Health Organization, Regional Office for Europe, European series No 91, Copenhagen, Denmark, Available at: http://www.euro.who.int/document/e71922.pdf.

WHO (2000b) The right to healthy indoor air. Report on a WHO meeting, Bilthoven, the Netherlands, 15-17. May, 2000. Available at: http://www.euro.who.int/document/e69828.pdf.

WHO (2000c) Guideline evaluation and use of epidemiological evidence for environmental health risk assessment. Guideline document, World Health Organization, Regional Office for Europe, Copenhagen, Denmark. Available at: http://www.euro.who.int/document/e68940.pdf.

WHO (2004) Health aspects of air pollution. Results from the WHO project "Systematic review of health aspects of air pollution in Europe. Copenhagen. Available at: http://www.euro.who.int/document/E83080.pdf.

WHO (2005) Air quality guidelines global update. Report on a working group meeting Bonn, Germany, 18-20 October 2005. Available at: http://www.euro.who.int/Document/E87950.pdf.

WHO (2006a) Air quality guidelines global update, Executive summary. World Health Organization, Geneva, Switzerland. Available at: http://www.who.int/phe/air/aqg2006execsum.pdf.

WHO (2006b) World Health Organisation Air Quality Guidelines, Global Update 2005. Copenhagen. Available at: http://www.euro.who.int/Document/E90038.pdf.

WHO (2006c) Development of WHO guidelines for indoor air quality. Report on a working group meeting Bonn, Germany, 23-24 October 2006. 22 pp. Available at: http://www.euro.who.int/Document/AIQ/IAQ_mtgrep_Bonn_Oct06.pdf.

WHO (2008a) Development of WHO guidelines for indoor air quality: Dampness and mould. Report on a working group meeting, Bonn, Germany, 17-18. October 2007. Available at: http://www.euro.who.int/Document/E91146.pdf.

WHO (2008b) Interventions and actions against damp and mould Report on a WHO working group meeting 28-29 February 2008, Bonn, Germany. Available at: http://www.euro.who.int/Document/E91664.pdf.

WHO (2009b) WHO Guidelines for indoor air quality: dampness and mould. WHO Regional Office for Europe, Copenhagen O, Denmark.

WHO (2007) Large analysis and review of european housing and health status (LARES). Copenhagen, Denmark. Available at: http://www.euro.who.int/Document/HOH/lares_result.pdf.

WHO (2009a) On-line european environment and health information system (ENHIS): Housing. Children living in homes with problems of dampness. Available at: http://www.enhis.org.

Wilkins K, Larsen K and Simkus M (2003) Volatile metabolites from indoor molds grown on media containing wood constituents. Environ Sci Pollut Res Int 10: 206-208.

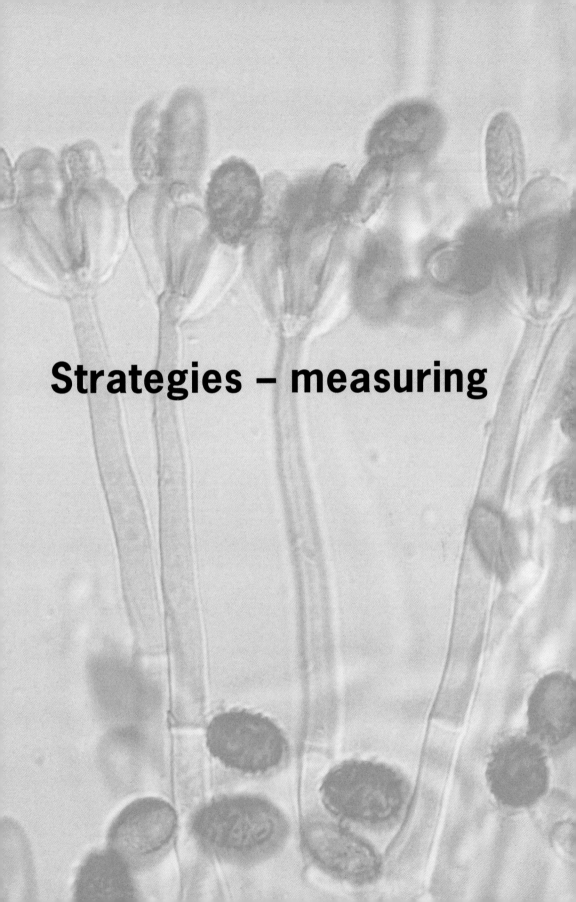

# Strategies – measuring

# 11 Moisture content measurement

*Bart J.F. Erich and Leo Pel*
*Eindhoven University of Technology, Faculty of Applied Physics, Eindhoven, the Netherlands*

## Introduction

Many deterioration processes in buildings are linked to the presence of moisture, i.e. water in liquid and/or vapor form. Changes in moisture levels can cause salts to crystallize or deliquesce, which may result in visual damage such as discoloration and/or mechanical damage such as cracking, chipping and/or disintegration (Pel *et al.* 2001, 2003). Additionally, when materials are moist for longer periods growth of micro-organisms, like algae and/or fungi, is often observed (Adan 1994). In these cases the presence of moisture in the micro-climate of the organisms is crucial for its metabolic activity. Consequently, adequate and accurate determination of moisture or humidity conditions plays a pivotal role both in control and analysis of fungal growth in building practice. Despite this crucial role, awareness of the limitations, restrictions and differences in measurement techniques is often lacking, leading to incorrect use and misinterpretation of the result. Incorrect use and/or misinterpretation may finally lead to inadequate remedies, resulting in considerable higher costs.

The present chapter aims to increase the knowledge on the available moisture measurement techniques, along with the working principles, advantages and disadvantages. Moreover, we will discuss some more advanced techniques to measure the spatial distribution of moisture in materials non-destructively. Such moisture profiles are crucial to understand and describe the material response to transient moisture loads adequately. Before elaborating on the various techniques the basic principles and definitions concerning moisture measurements are introduced.

## Basics of moisture measurements

### Moisture in equilibrium

The amount of moisture, i.e. water in liquid and/or vapor phase, in a porous material is often described in terms of volumetric moisture content. The amount of moisture in air is generally described in terms of relative humidity. These two definitions will be explained in the following subsections, respectively. A relation exists between the relative humidity and the moisture content in a porous material, which will be considered in the last subsection.

## Moisture content

The volumetric moisture content of a material is defined as the volume of water divided by the material volume that contains the water and can be expressed as a percentage:

$$\theta = \frac{V_w}{V_{material}} \cdot 100\%, \tag{1}$$

in which $V_w$ represents the volume of water and $V_{material}$ the volume of the material. The moisture content $\theta$ often represents an average value over a so-called Representative Elementary Volume (REV), or sometimes even represents a bulk value. In fact the moisture content may vary as a function of position, which is the result of several processes that drive moisture transport inside porous materials, such as capillary suction and diffusion. Likewise, also the moisture content by mass is used sometimes. The moisture content by mass is defined as:

$$\theta_{mass} = \frac{m_w}{m_{material}} \cdot 100\% = \frac{\rho_w}{\rho_{material}} \cdot \frac{V_w}{V_{material}} \cdot 100\% = \frac{\rho_w}{\rho_{material}} \theta, \tag{2}$$

where $m_w$ represents the mass of water and $m_{material}$ the mass of the dry material.

## Relative humidity

In air, the absolute amount of water in a certain volume is directly related to the partial vapor pressure of water. In a mixture of ideal gases, each gas has a partial pressure, being the pressure of the gas only. Dalton's law says that the total pressure of a gas mixture is the sum of the partial pressures of each individual gas in the mixture. Relative humidity is the most commonly used parameter used in practice to describe the moisture content in air. The relative humidity is defined as the ratio of the partial pressure of water vapor ($p_{vapor}$) in the mixture and the saturated vapor pressure of water ($p_{sat}$) at a given temperature. It is usually expressed as a percentage:

$$RH = \frac{p_{vapor}}{p_{sat}} \cdot 100\% = \frac{x_{vapor}}{x_{sat}} \cdot 100\%, \tag{3}$$

where $x_{vapor}$ and $x_{sat}$ represent (in case of an ideal gas) the actual and saturated amount of water present per unit of volume, respectively. The saturated vapor pressure is the maximum amount of vapor that air can contain at a certain temperature. The saturated vapor pressure depends on the temperature, and consequently the relative humidity. In case that water and water vapor are in equilibrium, the change in chemical potential for the transition is zero: $d\mu=-s{\cdot}dT+V{\cdot}dp=0$. This equation finally leads to the Clausius-Clapeyron equation (Greiner *et al.* 1995, Schroeder 2000):

$$p_{sat}(T) = p_0 \cdot e^{-\frac{L}{R}\left(\frac{1}{T} - \frac{1}{T_{ref}}\right)},\qquad(4)$$

where $p_0$=101,325 Pa is the water vapor pressure at $T_{ref}$=373 K, $L$=40.7 kJ/mol the enthalpy of vaporization, $p_{sat}$ is the maximum water vapor pressure at a certain temperature $T$ and $R$=8.314 J/Kmol the gas constant. In Figure 11.1 the maximum humidity in air is given as a function of temperature. From this figure the saturated vapor pressure can be calculated. This equation is only valid if the vapor pressure is much smaller than the critical vapor pressure (22 MPa).

Besides relative humidity the term water activity is often used. In 1953 the term water activity was introduced by W.J. Scott (1953). Water activity describes the energy status or escaping tendency of the water in a sample. It indicates how tightly water is "bound" to the substrate or liquid, either structurally or chemically. To measure the water activity, the system has to be in equilibrium, e.g. the temperature should be constant and no net flow of water occurs from substrate or liquid to air or *vice versa*.

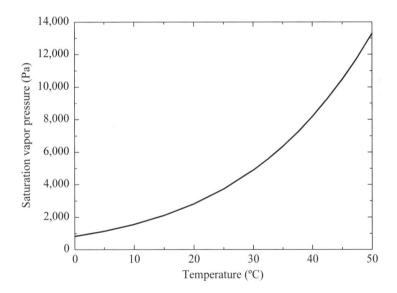

*Figure 11.1. Saturation vapor pressure as a function of temperature, calculated using the Clausius-Clapeyron equation.*

In case of such equilibrium, the water activity is defined as:

$$a_w = \frac{p_{vapor}}{p_{sat}} = \frac{RH}{100\%},$$  (5)

Pure water is taken as the reference or standard state, meaning that in that case the water activity is 1.0. In many cases specific humidity conditions are required for experiments with respect to fungal growth on materials, e.g. as in ISO tests. To achieve this, often humidity chambers are used, which contain saturated salt solutions (see Table 11.1) to obtain the desired humidity. Saturated salt solutions generate a specific water vapor pressure above the solution when placed in a closed environment at a constant temperature. Since salts bind water, the relative humidity is always lower than 100%, because the maximum vapor pressure of the salt solution is reduced compared to pure water. Another possibility is to use water glycerol mixtures to generate a specific relative humidity. Glycerol has a strong interaction with water i.e. retains water, and consequently, reduces the water vapor pressure.

## Sorption isotherm/suction curve

Capillary action is the movement of a liquid along the surface of a solid caused by the attraction of molecules of the liquid to the molecules of the solid. In case of porous materials, capillary action reflects the ability of the materials to take up water. It occurs when the adhesive intermolecular forces between the liquid and a substance are stronger than the cohesive intermolecular forces inside the liquid. In case of a small tubular pore with a radius $r$, water will rise. The equilibrium height is given by

$$h = \frac{2\gamma \cos \alpha}{\rho g r},$$  (6)

Table 11.1. Equilibrium RH (%) of saturated salt solutions at different temperatures.

| Salt | 5.0 °C | 10.0 °C | 15.0 °C | 20.0 °C | 25.0 °C |
|---|---|---|---|---|---|
| Lithium chloride (LiCl) | 11.3 | 11.3 | 11.3 | 11.3 | 11.3 |
| Magnesium chloride (MgCl) | 33.6 | 33.5 | 33.3 | 33.1 | 32.8 |
| Potassium carbonate (K$_2$CO$_3$) | 43.1 | 43.1 | 43.1 | 43.2 | 43.2 |
| Sodium bromide (NaBr) | 63.5 | 62.2 | 60.7 | 59.1 | 57.6 |
| Sodium chloride (NaCl) | 75.7 | 75.7 | 75.6 | 75.7 | 75.3 |
| Potassium chloride (KCl) | 87.7 | 86.8 | 85.9 | 85.1 | 84.3 |
| Potassium sulphate (K$_2$SO$_4$) | 98.5 | 98.2 | 97.9 | 97.6 | 97.3 |

where $\gamma$ is the liquid-air surface tension (N/m), $\alpha$ is the contact angle of a water droplet on the substrate, $\rho$ is the density of liquid (kg/m³), $g$ is acceleration due to gravity (m/s²), and $r$ is radius (m). Depending on the radius, each tube has its own capillary pressure. In case of a capillary with a radius of 1 μm the water rise would be 14 m.

Over the liquid air surface in a capillary, a pressure difference occurs that can be estimated by

$$\Delta p = \frac{2\gamma \cos \alpha}{r},$$ (7)

where $r$ is now the radius of curvature of water-air interface inside the capillary. For a capillary it has been shown by Thomson – Lord Kelvin – that it reduces the maximum saturation pressure and therefore the RH (Thomson 1871, Powles 1985):

$$p = p_0 \exp\left(-\frac{2\gamma \cos \alpha}{r\rho RT}\right),$$ (8)

where $p$ and $p_0$ are the actual partial pressure (Pa) and partial pressure for a free surface (Pa), respectively, $\gamma$ is the liquid-air surface tension (N/m), $\rho$ is the density of liquid (kg/m³), $T$ is the temperature (K), and $r$ is the radius of curvature of the water-air interface inside the capillary (m), and $R$ the ideal gas constant, equal to 8.314 J/kgK. Note, that in many cases the effective surface tension ($\sigma$) of water with the material is used: $\sigma=\gamma\cos\alpha$, where $\gamma$ is the liquid-air surface tension and $\alpha$ the contact angle of water on the (pore) material. Additionally for small pore sizes, instead of using the radius of curvature for $r$, the radius of the capillary tube is used. A porous material can be considered a network of capillary tubes with different radii. As a result a porous material has a complex relation between $RH$ and the moisture:

$$RH = \frac{p}{p_0} = \exp\left(-\frac{g\Psi(\theta)}{RT}\right),$$ (9)

where $g$ is the gravity (m/s²), $T$ the temperature (K), $R$ the ideal gas constant, and $\Psi$ (m) is the suction, which is a function of the actual volumetric liquid moisture content $\theta$ (m³/m³), representing the capillary forces of the whole network of a porous material. Hence for porous materials, there is relation between the relative humidity and the moisture content. This relation is often referred to as sorption isotherm. In Figure 11.2 an example is given.

As can be seen from this figure a different branch of the sorption curve usually exists for wetting and drying. This effect is referred to as "hysteresis", which is often explained by the so-called ink-bottle effect and raindrop effect (see Figure 11.3). Consequently,

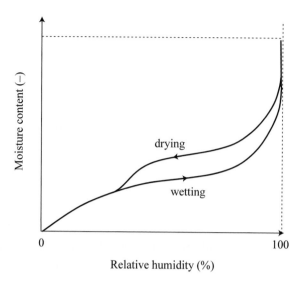

*Figure 11.2. Schematic presentation of a typical sorption isotherm, including a hysteresis loop.*

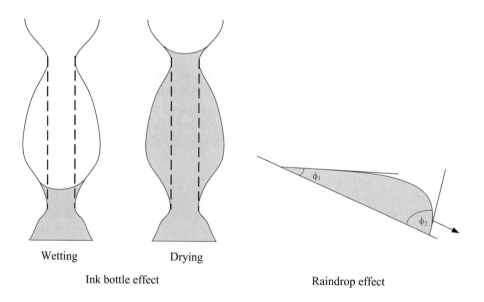

*Figure 11.3. The ink bottle effect: two conditions with the same capillary pressure, explaining the hysteresis in a porous material.*

for many materials there is no one-to-one relation between relative humidity and moisture content. Upon wetting, starting at a low RH, the capillary at the bottom is easily filled, only at high RH the rest will be filled. Upon drying, going from high RH to low RH, only at relatively low RH the water will be evaporating because of the high capillary pressure, when past that point the rest will easily be evaporating. This process is shown schematically in Figure 11.2. Another effect is the raindrop effect. The contact angle can have different values for advancing and receding cases. This can be easily seen if you tilt a plate; the droplet will only start to move after a certain angle. More energy is required to move the water than the sum of the gravitational potential and the energy related to the wetting of the plate. This difference is often explained on the basis of the roughness and/or contamination of the surface.

## Moisture transport

In practice, moisture content of materials may change in time due to varying boundary condition such as rain. Understanding the transport in materials is a primary requirement to describe the surface conditions, i.e. to predict the real surface humidity conditions for fungal growth.

> As each measuring device at the surface affects the surface conditions, transport modeling and measuring of transport to and through the surface should be used to predict the actual surface conditions. The surface conditions usually will differ from the conditions in the adjacent air, and are influenced by the transport in the material, as well as by the boundary layer conditions of the air close to the surface. Especially in highly transient indoor environments, the effects of such climate dynamics may be rather significant.

### Moisture diffusivity

In porous material, moisture can be transported in liquid and/or in vapor form. In the case of moisture transport in a porous material there are many interactions between liquid and vapor transport at the pore level. As it is very hard to model all these transport mechanisms the moisture transport is often simplified by:

$$q = -D(\theta) \nabla \Psi(\theta), \tag{10}$$

where $q$ is the moisture flux, and $D(\theta)$ $(m^2/s)$ is the so-called isothermal moisture diffusivity. In this "lumped" model all mechanisms for moisture transport, i.e. liquid flow and vapor diffusion, are combined into a single moisture diffusivity which depends on the actual moisture content.

For a porous material the conservation of mass (in volumetric quantities) can be written as:

$$\frac{\partial \theta}{\partial t} + \nabla \cdot q = 0. \tag{11}$$

By introducing Equation 10 the transport in a porous material can be described by

$$\frac{\partial \theta}{\partial t} = \frac{\partial}{\partial x}\left(D(\theta)\,\frac{\partial \theta}{\partial x}\right), \tag{12}$$

where in $\theta$ (m³/m³) is the volumetric liquid moisture content and $D(\theta)$ (m²/s) is the so-called isothermal moisture diffusivity. This is a non-linear diffusion equation as $D$ is a function of moisture content and cannot be solved analytically. In general, the moisture diffusivity has to be determined experimentally. By measuring transient moisture profiles, during various transport processes like drying and absorption, the diffusion coefficient can be determined directly (Pel *et al.* 1993, 1996), see Table 11.2. Note that the diffusion is in many cases an effective diffusion coefficient.

### Boundary conditions

Two major boundary conditions can be identified at material/material and material/air interfaces, respectively. Over a material/material boundary, the macroscopic capillary pressure and hence the relative humidity will be continuous:

$$\Psi_l(\theta_l) = \Psi_\gamma(\theta_\gamma) \quad \wedge \quad RH_l = RH_\gamma. \tag{13}$$

For each material the moisture content is a different function of the capillary pressure, so in general this condition will result in a jump of the moisture content at a boundary. This can in fact be measured by several measurement techniques, e.g. NMR. Generally, boundaries conditions are not sharp and in fact an interfacial layer exists, depending on for example air speed.

*Table 11.2. Typical diffusion coefficients D for water absorption (Pel 1995).*

| Type of material | D (m²/s) |
|---|---|
| Fired-clay brick | $3 \times 10^{-9}$ |
| Sand-lime brick | $8 \times 10^{-11}$ |
| Mortar | $1 \times 10^{-9}$ |
| Gypsum | $1 \times 10^{-12}$ |
| Wood (meranti) | $5 \times 10^{-9}$ |

A large change in the boundary conditions can have dramatic effects in drying or wetting behavior. For instance, during drying air flow may change the drying behavior from an externally limited (limited by the air flow) to an internally limited process (limited by the transport behavior of the material). In the latter case, a dry surface layer may be created, which can have a large influence on growth of organisms.

Under isothermal conditions the flux across the boundary material/air is given by:

$$q = \beta(RH_{air} - RH_{material}),\tag{14}$$

where $\beta$ the mass transfer coefficient, $RH_{air}$ the relative humidity of the air, and $RH_{material}$ the relative humidity of the material at the interface. The mass transfer coefficient depends on many parameters, such as air velocity, porosity, and surface roughness. In general, this coefficient has to be determined experimentally.

## Measuring the moisture content in materials

The moisture in the material originates in general from three sources, external water (e.g. rain, rising ground water, melting snow), internal liquid water (e.g. leaking pipes), and water vapor present in the surrounding air. Determining the moisture content as a function of position holds valuable information with respect to the source of water. However, measuring moisture in building materials is challenging even with high-tech laboratory equipment.

Ideally, measuring moisture in building materials should meet the following requirements:
- *High spatial resolution.* The technique should scan large surface areas as well as assess the profile in depth. A typical response depth related to daily indoor air to clarify humidity fluctuations is in the order of tenths of millimeters.
- *Fast response.* To measure transport processes and interaction of fungi with materials, the technique should be able to measure the processes and interactions with sufficiently short time intervals. Typical response times are in the order of one minute.
- *High accuracy, especially at high humidity's.* The technique should be sensitive to small changes in moisture content over time and allow a sound and indisputable interpretation.
- *Robustness.* High humidity's often lead to degradation phenomena, such as corrosion. Therefore, the sensor should be able to withstand high relative humidity (99% or even higher) and temperature extremes, which may depend on the application area.

- *High reproducibility.* A high reproducibility is required. No time shifts and no hysteresis should occur. After cleaning a contaminated sensor, it should still indicate the same value.
- *Portability.* An ideal instrument should be portable, so that it can be used anywhere and any time.
- *Low cost.* The technique should be affordable enough for regular and prolonged use.
- *Non-destructive.* The technique should be non destructive allowing an *in situ* investigation without damaging the material.

Most current techniques are "indirect", as the moisture content is derived from a material property that is sensitive. For such methods, a calibrated relationship between the moisture content and the material property is required. In other words, each measurement requires a material specific calibration.

For scientific studies it is more feasible to measure the moisture spatially, dynamically and quantitatively. Some of the techniques that are able to do that are discussed hereafter, i.e. neutron, gamma-ray transmission and Nuclear Magnetic Resonance (NMR).

### Direct methods

### Gravimetric determination

The best and most absolute method for determining the average moisture content is gravimetric determination, i.e. to measure to sample weight before and after drying. In contrast to vacuum drying, high temperature drying will not always be allowed, as at higher temperatures chemically bound water, rather than free water, may be given off and the nature of the material may be altered. For this reason it is recommended that materials like gypsum should be dried at 45±3 °C or lower. From the difference in weight between wet and dry sample, the absolute moisture content ($\theta_m$) in kg/kg can be determined, i.e.

$$\theta_m = \frac{m_{wet} - m_{dry}}{m_{dry}}, \tag{15}$$

which is often expressed as a percentage. In order to determine the moisture content in volume the specific density $\rho$ (kg/m$^3$) of the material should be known. Thus:

$$\theta_v = \theta_m \frac{\rho_{water}}{\rho_{material}}. \tag{16}$$

The gravimetric test method is *the benchmark technique* for all materials and methods. However, it is destructive, labor intensive and often also time consuming. Furthermore, the spatial resolution is rather low, and is determined by the sample thickness, often in the range of mm's to cm's.

On the basis of the gravimetric method moisture profiles can also be (roughly) determined. A widely applied method in practice is based on drilling and collecting crushed material at incremental depths. The sampling is repeated at various heights. Samples are sealed in airtight containers and are subsequently weighed. Naturally, the drilling process will affect the measurement as during the drilling heat will be generated. Therefore the drilling should be carried out at low speed.

### Indirect methods of determining moisture content

### Electrical properties

The majority of indirect methods are based on the water-induced change of electrical material such as electrical resistance and dielectric permittivity.
- *The electrical resistance.* Probably the best know method is the resistance technique, which measures the material resistance. Since moisture is a conductor, a raise in the moisture content lowers the electrical resistance and increases the conductance. The resistance varies in several orders of magnitude, ranging from some $k\Omega$ for wet materials compared to some hundreds of $M\Omega$ for dry materials. The resistance technique is prone to error, because "water" does not act as a consistent conductor of electricity, i.e. pure water is a very poor conductor whereas salt water is a very efficient conductor. Since salt is almost always present in building materials, the concentration of salt usually determines the actual conductivity. Therefore, a calibration curve should always be made for each material. Additionally, such measurements only result in a volume-based average value between the points of electrical contact. This volume can vary at different sampling locations and, consequently, hand-held resistance meters only give an indication. Despite this feature, hand-held resistance meters (Figure 11.4) are used on site for surface screening in order to determine the appropriate sampling locations for precise gravimetric measurements can be selected.
- *The dielectric permittivity.* Dielectric permittivity is a physical quantity that describes how an electric field interacts with a dielectric medium. The ability of a material to polarize in response to an externally applied field determines the dielectric permittivity. Thus, permittivity relates to the material's ability to transmit (or "permit") an electric field. Using this technique, the moisture content in water can be determined, as the dielectric permittivity for water is much higher

*Figure 11.4. A hand-held resistance moisture meter.*

than for most materials. Table 11.3 summarizes the dielectric permittivity for several materials in case of a static electric field.

In case of an alternating electric field, the dielectric permittivity is a function of the frequency. A moist porous material can be characterized by the effective complex permittivity of the material, relative to free space, $\varepsilon = \varepsilon' - j\varepsilon''$, where $\varepsilon'$ is the dielectric constant and $\varepsilon''$ is the dielectric loss factor. In general both are a function of frequency and temperature. Usually, the dielectric permittivity is determined by measuring the capacitance.

In practice hand held capacitance meters are used, which apply frequencies up to some MHz. Meters have contact and receiver electrodes. Capacitance meters are frequently used to screen the surface differences in moisture content to select the most appropriate locations for destructive testing. Surface contamination and salts do not affect the reading as strongly as they do with resistance measurements. However, once again, the meter can only be used as an indicator for moisture

*Table 11.3. The static dielectric permittivity $\varepsilon_r$ for various materials.*

| Material | $\varepsilon_r$ (-) |
|---|---|
| Water | 80 |
| Air | 1.0 |
| Paper | 2.1 |
| Wood | 3-7 |
| Soil | 2-4 |
| NaCl | 7.1 |

content differences. In case of absolute measurements capacitance should be calibrated specifically for the examined material.

- *Microwave absorption.* The microwave absorption resembles the dielectric principle, except for the fact that the applied frequency is much higher, in the order of some GHz, to realize a higher moisture sensitivity. At such high frequency, water will absorb more electromagnetic energy due to its polar nature. The effect of salt conductivity at these high frequencies is also minimized. The material is placed between the transmitting and receiving antennae. Furthermore, no hand held devices are available.

- *Radar.* Radar systems are based on the transmission of a short pulse of electromagnetic (EM) energy. Radar frequency ranges from 50 MHz to 40 GHz. An antenna receives the signal of energy traveling directly from the source to the receiver, referred to as the direct wave. Furthermore, waves are reflected at interfaces between materials of different permittivity. Radar waves propagation is governed by the permittivity of the material. A typical example of concrete showed a good correlation between moisture content and radar signals (Sbartai *et al.* 2006, Klysz *et al.* 2007).

- *Time-Domain Reflectometry (TDR).* TDR is used as an alternative technique to determine dielectric permittivity of a material. The principle is based on an electric pulse that is guided through the TDR probe, which is inserted in the material under investigation. When the travelling TDR pulse encounters changes in the electrical impedance surrounding the probe, part of the signal is reflected back towards the source. These reflections (expressed in a voltage) can be plotted as a function of time, (see Figure 11.5). The time of arrival and the magnitude of the reflected pulse are used to calculate the dielectric permittivity. This technique is applied mostly for measuring the water content in soil (Pettinelli *et al.* 2002, Wraith *et al.* 2005).

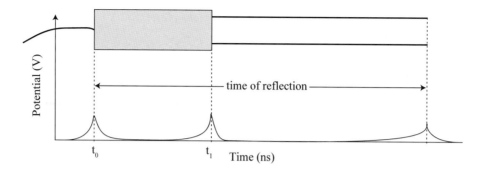

Figure 11.5. Basic principle of Time Domain Reflectometry.

The apparent dielectric permittivity, $\varepsilon_m$, can be determined using:

$$\sqrt{\varepsilon_m} = \frac{c}{2l}(t_{probe} - t_{ref}),\qquad(17)$$

where $c$ is the velocity of the electro-magnetic wave in free space, $3.0 \times 10^8$ m/s, and $l$ is the length of the embedded metal rods (m). $t_1$ and $t_2$ are the points of reflection (s), identifying the entrance and the end of the rods. Since the characteristic length, $l$, and $t_1$ of the probe are known by calibration using liquids of known dielectric constants, $\sqrt{\varepsilon_m}$ can be determined. (The typical frequency ranges between 50 MHz and 3 GHz). Due to the short pulse, ions have hardly any influence as they cannot move within this timeframe. This method can be used with great success with non-consolidated materials like soils. As electrodes require direct contact with the material, its use in consolidated materials like building materials as fired-clay brick is limited.

**Chemical reactions**

*Calcium carbide moisture meter.* The mass of a small amount of sampled material is determined gravimetrically and then placed in a vessel. A defined quantitiy of calcium carbide is then added and the vessel sealed. This reagent reacts with free moisture in the sample and produces acetylene gas, resulting in a pressure increase within the vessel. The reaction is:

$$CaC_2 + 2H_2O \rightarrow C_2H_2 + Ca(OH)_2.\qquad(18)$$

The pressure is directly related to the moisture content of the sample. It is stressed that the sample should be crushed into sufficiently small particles to ensure that the calcium carbide reacts with all the moisture in the sample. On the basis of samples at different locations and depth, spatial information can obtained.

**Acoustic properties**

As water conducts sound waves better than air or solid materials, (ultra)sound can be used to measure moisture. In that case a sound generator and receiver should be placed on opposite sides of a wall. The method is valuable as a screening tool, as it may point out differences in moisture content while scanning different cross-sections. It does not provide any information on moisture distribution as a function of depth, but gives an average value only. The practical application is only limited, as the translation of the transferred signal into moisture content requires a specific calibration for each situation.

**Thermal properties**

The thermal properties of moist porous material, more particular the thermal conductivity, strongly depend on the moisture content. In case of multilayer materials, inhomogeneity in materials or moisture content gradients, the relation between conductivity and moisture content becomes complicated.

- *Thermal conductivity.* Each material has a specific heat conductivity $\lambda$ However, when a porous material is moistened or dries the heat conductivity changes. From the measured value of $\lambda$ the actual moisture content can be calculated. Although thermal conductivity measurements provide some indication of the presence of moisture, generally this technique does not measure moisture content very accurately. Especially when the heat conductivity of the material is equal to the heat conductivity of water the moisture determination content becomes difficult or even impossible.
- *Heat capacity.* Several methods exist to determine the heat capacity and moisture content relationship. One is based on a heated wire inserted into the material. The temperature attained will depend on the thermal conductivity of the material, which on its turn depends on the moisture conductivity. Once again, specific calibration for each material is required.
- *Infrared radiation.* Each material emits infrared radiation, but when the temperature rises the intensity of the radiation increases. Infrared radiation has a wavelength that is larger than the wavelength of normal light. At low temperatures till about 50 °C, most materials do not exhibit a color change that is visible to the naked eye. However, unnoticed changes in the infrared take place, an infrared camera visualizes such IR changes, as it images the emitted infra red electromagnetic radiation. Thermographic data from an IR camera correlate

well with actual surface temperature measurements (e.g. by thermocouples) and are frequently used in building surveys. When moisture is present in building material the heat conductivity is increased. Consequently, the material will have a higher heat loss to the environment, decreasing the surface temperature. Also thermal bridges in building envelopes will decrease the surface temperature, which may cause condensation, thereby increasing the moisture content of the material. Both effects can be observed on the infrared heat camera, and it is therefore used frequently to create visuals of moisture anomalies in (historic) building envelopes (Figure 11.6). Note that exact calculation of the amount of moisture present is almost impossible, since only the surface temperature is measured. Additionally, interpretation is difficult when different materials with different thermal properties are present, also shadows and features of the building can interfere.

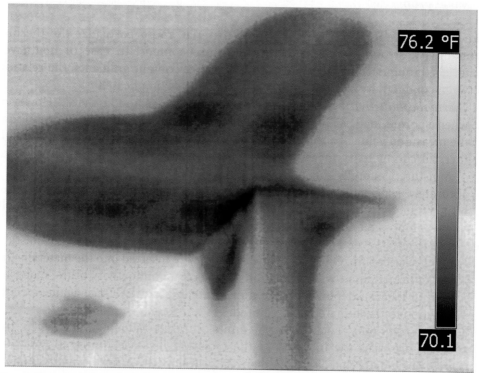

*Figure 11.6. An Infrared Radiation image showing colder (and/or wetter) area's (i.e. the blue-purple areas).*

## Scanning methods

In most cases spatial variations in the moisture content in materials occur due to transport processes. The spatial variation provides information on wetting and drying properties of the material under investigation. Additionally, it may give information on the moisture source and degradation processes. In this subsection, we concisely describe several scanning methods to determine transient moisture content profiles.

### Neutron scattering

When a beam of neutrons passes through a material, the neutrons will interact with the nuclei of this material (unlike γ-rays, which interact with the electrons). The attenuation of the neutron beam is determined by the cross-section for scattering and absorption of the nuclei present in the sample. Because of the relative large scattering cross-section of hydrogen, the intensity of the transmitted beam strongly depends on the amount of water.

The intensity $I$ of a neutron beam after passing a sample of thickness $d$ is

$$I_{wet} = I_0 \exp(-\mu_m d - \mu_w d_w), \tag{19}$$

where $I_0$ is the initial intensity of the neutron beam, $\mu_m$ the macroscopic attenuation coefficient of the material and $\mu_w$ of water, and $d_w$ represent the effective distance the neutrons have interacted with water. In general $\mu_d \ll \mu_w$ and hence the attenuation only depends on water. The major disadvantage of this transmission method is that it requires access to a neutron source, which is generally quite expensive. On the other hand, this method is very sensitive to water and samples can be scanned in 3D.

### X-ray absorption

Gamma rays interact with orbital electrons by three mechanisms:
1. Photoelectric effect, in which the energy of the gamma ray is used to eject an electron from the atom.
2. Compton effect, in which the gamma ray scatters from the atom, while loosing energy and ejecting also an electron.
3. Positron-electron pairs (at high energies >1.02 MeV) are produced at these energies. The excess in energy is used to give the positron and electron a certain amount of energy.

The intensity $I$ of a X-ray beam after passing a sample of thickness $d$ is

$$I_{wet} = I_0 \exp(-\mu_m d - \mu_w d_w), \tag{20}$$

where $I_0$ is the initial intensity of the neutron beam, $\mu_m$ the macroscopic attenuation coefficient of the material and $\mu_w$ of water, and $d_w$ represent the effective distance the x-rays have interacted with water. For X-rays $\mu_d \gg \mu_w$ and hence first a background measurement has to be performed of a dry material in order to determine the moisture content. When using a CT-scanner, the moisture content can be resolved spatially (Wu 2007).

## Nuclear Magnetic Resonance (NMR)

Almost all nuclei have a magnetic dipole moment, resulting from their spin-angular momentum. One may consider a nucleus as a charged sphere spinning around its axis, which corresponds to a current loop, generating a magnetic moment. In an NMR experiment the magnetic moments of the nuclei are manipulated by suitably chosen electromagnetic radio frequency (RF) fields. The frequency of the resonance conditions is given by,

$$f_l = \frac{\gamma}{2\pi} B_0, \tag{21}$$

where $f_l$ is the so-called Larmor frequency (Hz), $B_0$ (T) the externally applied static magnetic field and $g$ is the gyro magnetic ratio, which is dependent on the type of nucleus (for hydrogen at 1 T the frequency is 42.58 MHz). Because of this condition, the method can be tuned to only hydrogen and therefore to water. When a known magnetic field gradient is applied, the constant magnetic field $B_o$ in the resonance condition (Equation 21) has to be replaced by the spatially varying magnetic field,

$$B(x) = B_0 + Gx, \tag{22}$$

where $G$ (T/m) is the magnetic field gradient and $x$ is the position of the precessing magnetic moment. The resonance condition is then spatially dependent. Therefore the moisture content at different positions $x$ in the sample can be measured by varying the resonance frequency without moving the sample.

In a pulsed NMR experiment the orientation of the moments of the spins are manipulated by short RF pulses at the resonance frequency. The amplitude of the resulting signal emitted by the nuclear spins, the so-called "Hahn spin-echo" (Hahn 1950) signal is proportional to the number of nuclei taking part in the experiment.

The spin-echo signal also gives information about the rate at which this magnetic excitation of the spins decays. The system will return to its magnetic equilibrium by two mechanisms: interactions between the nuclei themselves, causing the so-called spin-spin relaxation (characterized by a $T_2$ value), and interactions between the nuclei and their environment, causing the so-called spin-lattice relaxation (characterized by a $T_1$ value). This method can determine the moisture content spatially and quantitatively. For example, in coatings a resolution as high as 5 μm and a profiling time of 10 minutes is achievable.

NMR imaging showing features as previously described are only available in a lab (Figure 11.7). However, in principle also a single sided portable NMR setup is available, which is called the NMR mouse (Eidmann *et al.* 1996, Casanova *et al.* 2003). Although portable, it is limited with respect to other properties, such as sensitivity and resolution, because the magnetic fields can only penetrate partly into the material.

### Comparing methods

Clearly each method has is own advantages and disadvantages. In Table 11.4 these advantages and disadvantages are presented.

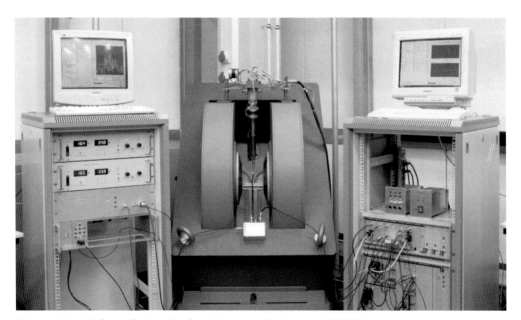

*Figure 11.7. High resolution Nuclear Magnetic Resonance (NMR) setup.*

*Bart J.F. Erich and Leo Pel*

*Table 11.4. Overall comparison of measuring methods.*

|  | Spatial | Destructive | Portable | Process (speed)[1] | Accuracy[2] | Cost[3] |
|---|---|---|---|---|---|---|
| Gravimetry | no | yes | yes | fast | high | low |
| Calcium carbide method | no | yes[4] | yes | fast | medium | low |
| Acoustical method | no | no | yes | fast | low | medium |
| Heat conductivity | no | no | yes | fast | low | medium |
| Resistance | no | no | yes | fast | low | low |
| Capacitance | no | no | yes | fast | low | low |
| TDR | yes | yes / no[5] | yes | fast | high | medium |
| Neutron method | yes | no | no | fast | high | very high |
| X-ray method | yes, CT | no | no | fast/medium | medium | medium |
| NMR | yes | no | possibly | fast/medium | high | medium |

[1] Process speed means time needed for determining the moisture content of the material.
[2] Accuracy of the method.
[3] Indicative initial costs: low: <5,000 €, medium <200,000 €, very high >200,000 €.
[4] Replacement required after each measurement.
[5] Yes for solid materials, no for soil.

## Measuring moisture content in air

In the previous paragraphs, we discussed methods to measure the moisture content in materials. The moisture content may also be deduced on the basis of the relative humidity under equilibrium conditions (see paragraph "Moisture in equilibrium"). Not only the relationship is often unknown, hysteresis is inherently part of the relationship. This implies that precise determination of the material moisture content requires that the materials history should be known. Consequently, this indirect method is not opportune for practical purposes. Nevertheless, accurate determination of the relative humidity is crucial to explain or understand the surface degradation phenomena that are primarily related to this parameter, like mold growth and corrosion.

An ideal technique to measure relative humidity should meet the following performance requirements:
- *High accuracy.* As fungal growth usually takes place in the RH range above 80%, with fast growth in the high RH range (>90%, sometimes even above 95%), the

sensor should be highly sensitive in this range. A change of 1% RH may have huge consequences in growth rate. Since the speed of fungal growth is very sensitive for small changes in relative humidity at high humidity levels (over 90% RH) and well-defined conditions are required for laboratory experiments, the accuracy of the sensor should be at least 0.1% RH. For many of the current RH sensors is a very challenging issue.

- *Quick response.* Since RH may change rapidly, to control the humidity a short response time of the sensor is required to regulate the relative humidity of the environment.
- *Durability and robustness.* High humidity often leads to all kinds of degradation phenomena such as corrosion. The sensor should be able to withstand high relative humidity's (e.g. condensation), chemical contaminants and temperatures.
- *High reproducibility.* A high reproducibility is required, over time the values should remain the same, no hysteresis should be present. After cleaning the sensor it should still indicate the proper value.
- *Independent of temperature/temperature corrected.* When temperature changes the sensor should still indicate the correct relative humidity.
- *Low cost.* The sensor should have a structure that allows cheap and simple production, as a basis for an affordable price.

## Measurement principles and sensors

Like moisture measurements most techniques for determining the relative humidity are indirect, i.e. they determine the moisture from a material property that is affected by water. Consequently, a calibrated relationship between the relative humidity and the responsive material property is always required.

## Dew point principle

The dew point is defined as the temperature when condensation starts, or when the relative humidity is 100%. This is easily understood from the water vapor curve in Figure 11.8. The saturated water vapor pressure depends on temperature, and rises with increasing temperature; in other words warm air can contain more water vapor than cold air. In the dewpoint principle, the RH is determined on the basis of temperature measurements, which can be done accurately. From this partial vapor pressure, the actual relative humidity at the starting temperature is calculated.

The three most common methods to determine the dew point are the chilled mirror, metal oxide and polymer sensor. As the chilled mirror offers the highest accuracy, it is discussed next.

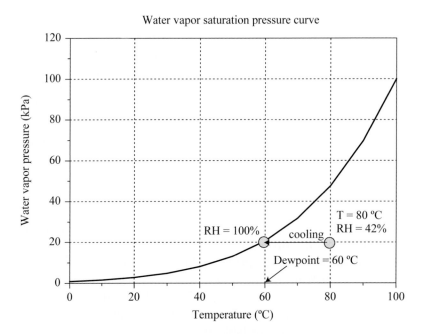

*Figure 11.8. Saturated water vapor pressure and dew point.*

*Chilled mirror.* The chilled mirror method determines the mirror temperature at which condensation occurs during cooling down. This transition is observed on the basis of change in reflection of a light beam. Via an analog or digital control system, the mirror temperature is maintained at the dew point. A precision thermometer embedded in the mirror monitors the mirror temperature at the dew point. This technology can offer the highest accuracy over a wide range of dew points. However, due to its optical measurement principle, it is very sensitive to dirt or dust on the sensor. Accurate chilled mirror devices are expensive and therefore mostly used when an absolute accuracy is needed and frequent maintenance can be done, e.g. in laboratories.

### The wet-bulb temperature principle

Wet-bulb thermometers or psychrometers are based on a dual and simultaneous temperature measurement. The first thermometer measures the so-called dry temperature, whereas the second is wrapped in a wet socket. At relative humidities below 100%, water evaporates from the wet-bulb thermometer, which cools the bulb below ambient temperature, the dry-bulb thermometer. At any given ambient temperature, a lower relative humidity results in a higher evaporation rate and therefore in a bigger difference between the dry-bulb and wet-bulb temperatures. The precise relative humidity is calculated using a psychrometric chart, see Figure 11.9. When the wet-bulb and dry-bulb temperatures are known the relative humidity can be determined from the chart. The accuracy of a simple wet-bulb thermometer depends on how fast air passes over the bulb and how well the thermometer is shielded from the radiant temperature of its surroundings.

### Principles based on mechanical performance

Water may also affect mechanical performance of some materials, such as length and tension. The well-known hair hygrometer is based on this principle.
- *Hair hygrometer.* The hair hygrometer uses a human or animal hair under tension (Figure 11.10). The traditional folk art device known as a "weather house" works on this principle. In order to see changes that occur in time, several hygrometers record the value of humidity on a piece of graduated paper so that the values can be read off the printed scale.

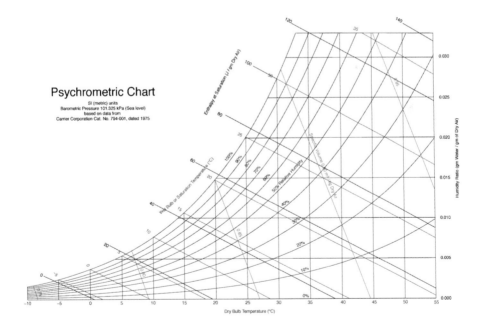

*Figure 11.9. Psychometric chart (SI metric units, barometric pressure 101.325 KPa (sea level), based on data from Carrier Corporation Cat. No. 794-001, 1975).*

**Fundamentals of mold growth in indoor environments**

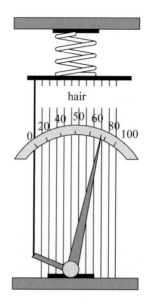

*Figure 11.10. Schematic presentation of a hair hygrometer.*

## The changing electrical impedance principle

The relative humidity can also be measured on the basis of change in electrical impedance. Typically, three categories of materials can be distinguished: ceramics, polymers and electrolytes. Because of the conductivity of water the impedance generally decreases with increasing relative humidity.

- *Resistive relative humidity sensors.* In resistive sensors, the absorbed water vapor leads to an increase in electrical conductivity. The response time for most resistive sensors ranges from 10 to 30 s. Most resistive sensors use alternating voltage (30 Hz to 10 kHz) without a DC component to prevent polarization of the sensor. A distinct advantage of resistive RH sensors is their interchangeability, usually within ±2% RH. A drawback of some resistive sensors is their tendency to have hysteresis. Other drawbacks are the slow response and the fact that long-term stability can only be guaranteed by a continuous operation and by attentive maintenance (Lee *et al.* 2005). In general the small size (down to a few mm), low cost, and interchangeability make these resistive sensors suitable for use in control and display products for industrial, commercial, and residential applications.
- *Capacitive relative humidity sensors.* Approximately 75% of the humidity sensors on the market today are based on the capacitive technique (Rittersma 2002). Capacitive humidity sensors detect moisture induced changes in the dielectric

constant of an applied hygroscopic layer. They consist of a substrate with two conductive electrodes, with a humidity sensitive dielectric layer in between. The substrate is typically glass, ceramic, or silicon. The humidity sensitive materials used are PMMA, porous ceramics, porous silicon, porous silicon carbide and hygroscopic polymers (Lee *et al.* 2005). Capacitive sensors are characterized by low temperature dependence, and its ability to function at high temperatures (up to 200 °C), full recovery from condensation, and reasonable resistance to chemical vapors. The typical response time is in the range of 30 s. The accuracy is around ±2% RH.

The change in the dielectric constant of a capacitive humidity sensor demonstrates a complex nonlinear relationship with the relative humidity. Consequently signal conditioning is required.

State-of-the-art techniques for producing capacitive sensors take advantage of many of the principles used in semiconductor manufacturing to yield sensors with minimal long-term drift and hysteresis.

A commonly used type of capacitive based relative humidity sensors is based on a polymer dielectric layer (Figure 11.11). The advantage of using a polymer is the fact that it is immune to condensed water and therefore has a wide humidity range including condensing environments. Polymer sensors are used in a wide variety of applications, e.g. in industry and meteorology. The main advantage of the technology is the long-term stability.

*Figure 11.11. Example of a capacitive sensor (Honeywell).*

*Bart J.F. Erich and Leo Pel*

## Colorimetric principles

Fiber-optic humidity sensors are mainly based on colorimetry or changes in the refractive index of materials applied on top of the surface of a fiber. Upon humidity changes, the color or refractive index of the material changes, which is subsequently measured and converted into appropriate output signals (Lee *et al.* 2005). For example an optical fiber sensing system exists based on the colorimetric interaction of cobalt chloride with water molecules. Another example is to apply cellulose that swells under different humid conditions, and thereby changes the refractive index. This change in refractive index is then used after calibration to determine the relative humidity. Many other principles exist, however, several drawbacks hinder their application, such as instability, high costs, temperature compensation, large differences in inaccuracies. At this time development in this area is increasing rapidly, and more wide applications of those types of sensors are expected.

## Thermal conductivity principle

Thermal conductivity humidity sensors measure the absolute humidity by quantifying the difference between the thermal conductivity of dry air and that of air containing water vapor. When thermistors are energized, the heat dissipated from a sealed thermistor is greater than that from an exposed thermistor due to the difference in thermal conductivity. The difference in resistance between the thermistors is directly proportional to the absolute humidity. This type of sensor is very durable. It operates at high temperature and is resistant to chemical vapors due to the use of inert materials; i.e. glass, semiconductor material for thermistors, high temperature plastic, and aluminum. The typical accuracy of an absolute humidity sensor is about 5% RH.

### Comparing methods

Clearly each type of sensor has is its specific advantages and disadvantages, depending on the application different factors are more relevant. Many new technologies have facilitated the development of micro humidity sensors. However, the challenge is still to develop a sensor which has excellent characteristics on all aspects, e.g. linearity, high sensitivity, low hysteresis, low cost, and rapid response time. In Table 11.5 the advantages and disadvantages are presented.

*Table 11.5. Comparing sensors.*

| Sensor type | Process (speed)[1] | Stability | Typical size (m) | Accuracy[2] | Cost[3] | Durable[4] | Hysteresis |
|---|---|---|---|---|---|---|---|
| Dew point | medium | high | 0.1 | high (<1%) | medium | high | no |
| Wet bulb | fast | high | 0.5 | high (<1%) | low | high | no |
| Impedance based | | | | | | | |
|    capacitive | medium | medium | $10^{-3}$ | ±2% | medium | medium | yes |
|    resistive | medium | medium | $10^{-3}$ | ±2% | medium | medium | yes |
| Thermal conductivity | medium | medium | $10^{-2}$ | ±2% | medium | high | yes |
| Optical | varying | medium | $10^{-1}$ | varying | high | high | yes |
| Hair hygrometer | medium | medium | $10^{-1}$ | ±2% | low | high | no |

[1] Process speed means time needed for determining the relative humidity.
[2] If not exactly specified the accuracy may vary.
[3] Indicative costs: low: <1 k€, medium <5 k€, very high >5 k€.
[4] Durable is the feature to withstand aggressive environments.

# Measuring the moisture content on the scale level of the fungus

## Measuring moisture in the micro-climate

Zooming into the environment of a fungus and measuring the moisture in its environment is very complex as each sensor contact will affect the micro-climate by definition. At this moment no techniques exist that can accurately determine the moisture content in the environment of a fungus as a function of space and time.

## High resolution nuclear magnetic resonance

Some recent developments in the area of NMR create new perspectives to measure water on small scale levels without any interference with the micro-climate. Current scanners are available with a high spatial resolution. This high resolution can be obtained by introducing a high magnetic field gradient in the direction perpendicular to the sample (GARField approach; Erich *et al.* 2005). As a result the resonance frequency of the nuclei will depend linearly on the position $f(y)=\gamma (B_0 + Gy)$, where $y$ is the position, G the gradient, $\gamma$ is the gyromagnetic ratio (for hydrogen 42.58

MHz/T), and $B_0$ the main magnetic field. Using this approach a 1D spatial resolution of approximately 5 μm can be realised, which is of the same order of magnitude of the smallest size of a fungal fragment. Additionally, the gradient enables measurements of low molecular diffusion constants, e.g. to measure bound water.

## Scanning electron microscopy

To investigate the micro-environment scanning electron microscopy can be used. In traditional scanning electron microscopy microorganisms need to be fixed, frozen or dehydrated, and coated with a conductive film before observation in a high vacuum. In several studies environmental scanning electron microscopes (ESEM) has been applied to image hydrated or dehydrated biological samples in their natural state while using minimal manipulation and without the need for conductive coatings (Collins *et al.* 1993).

## Micro-electrode

Micro electrodes are usually used for measuring oxygen, and can be used to measure oxygen profiles in a microbial mat (Revsbech *et al.* 1983). Such techniques might also allow measuring the moisture content in the direct environment of the fungus.

### Measuring water (transport) in molds

Only a few techniques are available that may be used to study water (transport) in growing fungi:
• Confocal scanning laser microscopy
  Confocal laser scanning microscopy (CLSM) is a technique for obtaining high-resolution optical images. The key feature of confocal microscopy is its ability to produce in-focus images of thick specimens, a process known as optical sectioning. Images are acquired point-by-point and reconstructed with a computer, allowing three-dimensional reconstructions of topologically-complex objects. One can use confocal scanning laser microscopy (CSLM) to measure the concentration profiles of aerial and penetrative biomass during the growth of fungal species on agar (Nopharatana *et al.* 2003a,b). While following this biomass production it might be possible to follow the water transport processes as well, possibly via a direct or indirect method. Researchers already have shown that the water content in the growth environment strongly influences the water content of the biomass (Nagel *et al.* 2001).

- Micro-nuclear magnetic resonance
  New technologies might be developed using micro-coils, which allow analysis of very tiny structures and small volumes of water. At this moment micro-coils are most often used for spectroscopic analysis of materials in an NMR spectrometer. Combining such a tool with a high gradient setup, allows visualization of moisture flow through hyphae. By changing the humidity conditions at both ends of the hyphae the water flow can be monitored.

# References

Adan OCG (1994) On the fungal defacement of interior finishes. PhD Thesis, Eindhoven University of Technology, the Netherlands.

Casanova F and Blumich B (2003) Two-dimensional imaging with a single-sided NMR probe. J Magn Reson 163: 38-45.

Collins SP, Pope RK, Scheetz RW, Ray RI and Wagner PA (1993) Advantages of environmental scanning electron-microscopy in studies of microorganisms. Microsc Res Tech 25: 398-405.

Eidmann G, Savelsberg R, Blumler P and Blumich B (1996) The NMR MOUSE, a mobile universal surface explorer. J Magn Reson 122: 104-109.

Erich SJF, Laven J, Pel L, Huinink HP and Kopinga K (2005) Comparison of NMR and confocal Raman microscopy as coatings research tools. Prog Org Coating 52: 210-216.

Greiner W, Neise L and Stöcker H (1995) Thermodynamics and statistical mechanics. Springer-Verlag, New York, NY, USA.

Hahn EL (1950) Spin echoes. Phys Rev 80: 580-594.

Klysz G and Balayssac JP (2007) Determination of volumetric water content of concrete using ground-penetrating radar. Cement Concr Res 37: 1164-1171.

Lee CY and Lee GB (2005) Humidity sensors: a review. Sensor Letters 3: 1-15.

Nagel FJJI, Tramper J, Bakker MSN and Rinzema A (2001) Model for on-line moisture-content control during solid-state fermentation. Biotechnol Bioeng 72: 231-243.

Nopharatana M, Mitchell DA and Howes T (2003a) Use of confocal microscopy to follow the development of penetrative hyphae during growth of *Rhizopus oligosporus* in an artificial solid-state fermentation system. Biotechnol Bioeng 81: 438-447.

Nopharatana M, Mitchell DA and Howes T (2003b) Use of confocal scanning laser microscopy to measure the concentrations of aerial and penetrative hyphae during growth of *Rhizopus oligosporus* on a solid surface. Biotechnol Bioeng 84: 71-77.

Pel L (1995) Moisture transport in porous building materials. PhD thesis, Eindhoven University of Technology, the Netherlands.

Pel L and Brocken H (1996) Determination of moisture diffusivity in porous media using moisture concentration profiles. Int J Heat Mass Trans 39: 1273-1280.

Pel L, Huinink H and Kopinga K (2003) Salt transport and crystallization in porous building materials. Magn Reson Imag 21: 317-320.

Pel L, Huinink HP, Kopinga K, Rijniers LA and Kaasschieter EF (2001) Ion transport in porous media studied by NMR. Magn Reson Imag 19: 549-550.

Pel L, Ketelaars AAJ, Adan OCG and Van Well AA (1993) Determination of moisture diffusivity in porous media using scanning neutron radiography. Int J Heat Mass Trans 36: 1261-1267.

Pettinelli E, Cereti A, Galli A and Bella F (2002) Time domain reflectometry: calibration techniques for accurate measurement of the dielectric properties of various materials. Rev Sci Instrum 73: 3553-3562.

Powles JG (1985) On the validity of the Kelvin equation. J Phys Math Gen 18: 1551-1560.

Revsbech NP and Ward DM (1983) Oxygen microelectrode that is insensitive to medium chemical composition: use in an acid microbial mat dominated by *Cyanidium caldarium*. Appl Environ Microbiol 45: 755-759.

Rittersma ZM (2002) Recent achievements in miniaturised humidity sensors – a review of transduction techniques. Sensor Actuator Phys 96: 196-210.

Sbartai ZM, Laurens S, Balayssac JP, Ballivy G and Arliguie G (2006) Effect of concrete moisture on radar signal amplitude. ACI Mat J 103: 419-426.

Schroeder DV (2000) An introduction to thermal physics. Addison Wesley Longman, New York, NY, USA.

Scott WJ (1953) Water relations of *Staphylococcus aureus* at 30 °C. Austral J Biol Sci 6: 549-564.

Thomson W (1871) On the equilibrium of vapour at a curved surface of liquid. Philosophical Magazine Series 4 – 42: 448-452.

Wraith JA, Robinson DA, Jones SB and Long DS (2005) Spatially characterizing apparent electrical conductivity and water content of surface soils with time domain reflectometry. Comput Electron Agr 46: 239-261.

Wu XFP (2007) Investigation of water migration in porous material using micro-CT during wetting. Heat Tran Asian Res 36: 198-207.

# 12 The fungal resistance of interior finishing materials

Olaf C.G. Adan
*Eindhoven University of Technology, Faculty of Applied Physics, Eindhoven, the Netherlands; TNO, Delft, the Netherlands*

## Introduction

The interior finish plays a pivotal role in the risk management of fungal growth in the indoor environment. As has been shown in the previous chapter on water relations of indoor fungi, its impact not only stems from material constituents, but moreover from the ability of moisture storage that leads to prolongation of favorable humidity conditions at the material surface.

We have become even more aware of the material's role considering that:
• the industrial trend towards more eco-friendlier products (e.g. waterborne coatings) appears to go hand in hand with an overall increase in product's bio-susceptibility;
• whereas at the same time, the application of bio-inhibitors in finishes is facing stricter environmental legislation and is under prohibitive rules (e.g. the European Biocides Directive 98/8/EC that entered into force May 2000).

Both obviously enlarged the pressure on performance requirements of materials, and finishes in particular.

The recognition of such crucial role calls for an approved and internationally accepted method for assessing the interior finish performance. Such a method lays the foundation for definition of material requirements in codes and may eventually lead to labeling of commercial products. After all, the decorative finish is often an end-users choice.

The early interest in fungal growth on interior products was most obvious in the case of paint products. Biodeterioration of paint and paint films provided a considerable challenge to microbiologists and chemists in the paint industry to develop paint preservatives. Their effort originated from fungal problems during paint manufacturing and in-can spoilage, as well as from defacement of the dried film after application onto the substrate. In the late 70s, the latter gave rise to development of methods for testing the fungal resistance of paint films (e.g. Bravery *et al.* 1978, 1983, 1984, Smith 1978). Since that time, several methods for assessing the fungal resistance of other manufactured materials for interior applications have been put forward as "Standards" or "Recommendations". At present, some of them are internationally accepted.

In this Chapter, current standards to determine fungal resistance of finishing materials for interior applications are reviewed. Deterioration caused by wood-rotting fungi is not considered within this context. Differences exist with respect to the targeted materials domain, but this review particularly addresses the definition of the humidity conditions during test, considering the potential effect of water retention in the material. As a consequence, a new concept for testing is proposed, which has been validated for a wide range of finishing materials during the past decade.

## Present standards to assess fungal resistance of materials for interior applications

The most relevant test methods for assessing the fungal resistance of finishing materials for indoor applications (including coatings, plasters, plastics such as silicon caulking and wood panel products) are summarized in Table 12.1. Except for those methods, many other methods exist with respect to other types of material, such as textiles and wood. These methods have not been included, as our primary interest is on wall and ceiling finishes.

Basically, each test is designed to comparatively evaluate different finishing materials for their relative performance under a given set of conditions. It does not imply that a finish that resists growth under these conditions will necessarily resist growth in the actual application. Most standards cover a specific range of material only. The most important aspects relevant to testing of interior finishes are now commented on briefly.

### Focus

*British Standard BS 1982* (1990) focuses on the fungal resistance of panel-products made of or containing material of organic origin. It prescribes three types of tests, of which only tests in Part 2 and 3 are relevant in the context of the present review. Part 2 concerns testing of the resistance to cellulose attacking microfungi (commonly known as soft rot fungi, e.g. *Chaetomium globosum*), and can be used on all sheet materials such as felt and paper. It can be carried out without the resources of a

*Table 12.1. Current standards to assess fungal resistance of interior finishes.*

| Year | Document number | Title |
|------|-----------------|-------|
| **British Standards** | | |
| 1989 | BS 3900-G6 | Methods of test for paints – assessment of resistance to fungal growth. |
| 1990 | BS 1982-3 | Fungal resistance of panel products made of or containing materials of organic origin. Methods for determination of resistance to mould or mildew. |
| **European Committee for Standardization CEN** | | |
| 2005 | EN-IEC 60068-2-10 | Environmental testing – tests – test J and guidance: mould growth. This standard has several national translations in EU member states, e.g. as a BS, DIN, NEN or IEC standard with the same document number. |
| **American Society for Testing and Materials** | | |
| 2002 | ASTM G21-96 | Standard practice for determining resistance of synthetic polymeric materials to fungi. |
| 2005 | ASTM D3273-00 | Standard test method for resistance to growth of mold on the surface of interior coatings in an environmental chamber. |
| 2005 | ASTM D 5590 | Standard test method for determining the resistance of paint films and related coatings to fungal defacement by accelerated four-week agar plate assay. |
| **International Organization for Standardisation ISO** | | |
| 1997 | ISO 846 | Plastics – evaluation of the action of microorganisms. |
| **Other** | | |
| 2000 | MIL-STD-810F Method 508.5 | Department of defense test method standard for environmental engineering considerations and laboratory tests. |

mycological laboratory and is intended to give a quick general indication of the susceptibility under damp conditions.

Part 3 assesses the mold resistance of materials that are required to present a decorative finish or appearance, such as plasterboard and sheets made of plastics. It should not be used normally to paints and distempers.

*British Standard BS 3900*, part G6 (1989) is one of a series of standards in group G of BS3900, which is concerned with environmental testing of paint films. It describes a method to assess the fungal resistance of paints, varnishes and lacquers, applied in the laboratory to specified panels, either as a part of a multi-coat system or separately. The method has been elaborated after extensive international collaborative work in the late 70s and early 80s, during which variations in technique have been tested both within and between laboratories (Bravery *et al.* 1978, 1983, 1984).

*ASTM G21-96* (2002a) covers determination of the effect of fungi on the properties of synthetic polymeric materials in the form of molded and fabricated articles, tubes, rods, sheets, and film materials. Changes in optical, mechanical, and electrical properties may be determined, for which one is referred to other applicable ASTM methods.

*ASTM D 3273* (2005a) aims to evaluate the comparative mold growth resistance of coatings designed for use in interior environments. It may be useful in evaluating compounds that may inhibit mold growth and the aggregate levels for their use. The test method describes a small environmental chamber and the conditions to determine the relative resistance of paint films in a severe interior environment. For testing, coatings are applied on test panels made of wood (ponderosa pine sapwood) or gypsum board (or other if needed).

*ASTM D5590-00* (2005b) covers an accelerated method for determining the relative resistance of two or more paints or coating films to fungal growth. Basically, it is an agar-plate test without specific climate requirements. This test may be conducted on nutritive agar plates (either PDA – potato dextrose agar or Malt agar) alone or on nutrient salts agar plates, if all samples fail completely. For testing, paints are applied on prescribed substrates i.e. filter paper (glass fiber, grade 391) or drawdown paper (unlaquered chart paper).

*ISO 846* (1997) specifies methods for determining the deterioration of plastics due to the action of fungi, bacteria and soil microorganisms. It includes two methods for assessment of fungal growth. Method A is suitable for assessment of the inherent resistance of plastics to fungal attack in the absence of other organic matter. Therefore, test specimens are exposed in the presence of an incomplete nutritive medium, without a carbon source. Method B focuses on the determination of fungistatic effects. In that case, test specimens are place on a nutritive medium.

Two standards have been reviewed that have another focus than finishes for interior applications. Both, however, are sometimes used in that context.

The military standard *MIL-STD-810F Method 508.5* of the US Department of Defense (2000) aims to assess the extent to which materiel will support fungal growth and how any fungal growth will affect performance – its mission and safety – of the material. The standard focuses on detrimental effects on both natural material, including cellulosic materials (e.g. wood, paper, natural fiber textiles and cordage), animal- and vegetable-based adhesives and leather, and on synthetic materials of which PVC formulations, certain polyurethanes (e.g. polyesters and some polyethers), plastics that contain organic fillers of laminating materials and paints and varnishes are mentioned explicitly. Although its purpose concerns materiel in military applications primarily, the standard is also used to assess materials in civil applications.

The European standard *EN-IEC 60068-2-10* (2005) aims to investigate fungal deterioration of assembled electrical products. IEC 60068 provides a test method for determining the extent to which electrotechnical products support mold growth and how any mold growth may affect the performance and other relevant properties of the product. The test is applicable to products intended for transportation, storage and use under humid conditions over a period of some days at least. Generally fungal growth may have two effects: on the one hand the wet mycelium may cause formation of an electrically conducting path across the surface or cause a serious variation in the frequency impedance characteristics of the circuit, on the other fungi can yield acid products which may cause ageing effects. The method refers to both ISO 846 and MIL-STD-810F.

### Inoculation

All standards considered use an aqueous suspension of a mixture of fungal species, however showing major differences in the selection. Most mixtures have been chosen such that the nature of attack may vary widely, covering a range from paints and plastics to textiles, rubbers and wood panel products. Remarkably, except for *Aspergillus versicolor* and *A. niger* that have minimum water activity $a_w$ for growth of 0.78 and 0.77, respectively, common indoor xerophilic fungi, such as some *Eurotium* species with minimum $a_w$ near 0.7, are not yet considered. As has been observed by Adan (1991), *E. herbariorum* grows well on interior finishes within a 12-week test period.

When a mixture of fungal species intends to reflect the fungal flora in the considered field of indoor application, species with a tolerance to low water activity should be included in fungal resistance tests of interior finishes.

Except for the BS1982 and the ASTM D3273, the considered standards include quantitative requirements on spore concentration and/or volume to be sprayed on the samples. The lowest value is set in ASTM D5590 with $0.8\text{-}1.2\times10^4$ spores/ml, whereas all other require a minimum concentration of about $10^6$ spores/ml.

### Incubation temperature and humidity

The principles of testing are fairly similar. In most tests the relative humidity is maintained at a more or less steady state level below saturation (BS1982, ASTM G21, ASTM D3273, ISO 846, MIL-STD-810F and EN-IEC 60068). In these cases, the relative humidity level is ambiguously defined, and should be set in a range above a threshold value, which varies from 70 to 95%.

Only in the BS3900 intermittent surface condensation is applied (British Standards Institution 1989), which is introduced with simultaneous alterations in temperature. ASTM D5590 does not include specific climate requirements, as it is an agar-plate test.

With respect to temperature there are marked differences. Three standards use temperatures around 24 °C (BS1982, BS3900 and ASTM D5590 with 24±1 °C, 23±2 °C and room temperature, respectively), whereas the others have higher temperatures around 30 °C.

There is a striking similarity in the minimum duration of test. All standards considered have a minimum experimental duration of 4 weeks, which might be extended to 8 weeks (EN-IEC 60068-2-10, in order to assess the effect on electrical properties specifically), or 12 weeks (BS3900 and MIL-STD-810F).

### Control of experimental conditions

All standards include controls to check experimental conditions, including viability of spores. ASTM D5590 includes a positive and a negative growth control, i.e. one that is known to support fungal growth and one that is known to inhibit growth completely.

### Assessment of growth

Generally, growth rates are assessed visually either by the naked eye or microscopically at magnifications up to ×100, mostly using numerical scales for area covered. Typically, these scales are non-linear. Only few of the test methods considered

(BS1982, ASTM D3273 and EN-IEC 60068-2-10) recommend examination by both naked eye and microscope. ASTM D3273 refers a photographic reference standard with a 0 to 10 rating scale in ASTM D3274-95(2002b). That standard provides a numerical basis for rating the degree of fungal and algal growth or soil and dirt accumulation on paint films.

The frequency of inspection of test samples differs strongly. In the majority of tests, samples are assessed only once at the end of the test period. The BS3900 prescribes a first assessment at four weeks and further assessment at 6 and 12 weeks if necessary. The highest frequency is found in the ASTM D3273, wherein a weekly inspection is demanded.

### Preconditioning of materials

None of the standards includes requirements with respect to sterilization of samples prior to inoculation, except for test A in ISO 846 which prescribes cleaning of plastics by dipping in ethanol. Aseptic conditions form the starting point of all. Some standards demand sample preconditioning, such as the BS3900 and the ASTM D3273, requiring drying of paints. The latter prescribes a maximum for the initial water content of wooden panels, which are used as base substrates. For materials with interior applications no artificial weathering is required.

## A new concept to assess fungal growth

The need for a standard on fungal resistance of interior products has been identified as a priority in the frame of the European Products Directive, aiming for harmonization of the European Market for products and their active substances. However, the proposed method in this paragraph not only anticipates the envisaged activities on the European level, but also gives follow-up to the recommendations of the first International Workshop on fungi in indoor environments (Samson *et al.* 1994). Considering the preceding review of current standards, improvement is needed with respect to two pivotal issues at least:
- *Humidity conditions.* Firstly, current testing does not explicitly take account of the potential effect of the material's reservoir function that introduces inertia effects during humidity dynamics. A new standard should have unambiguous moisture loads – differing between steady state and dynamic climatic conditions – to come to a more reliable and reproducible prediction of material differences in fungal resistance. Both climatic extremes actually occur in the indoor environment.
- *Assessment of growth.* Secondly, the majority of tests consider fungal growth (i.e. mostly coverage area) at specific moments in time, in most cases at the end of

the test period only. Such approach may lead to misinterpretation of material performance (Adan *et al.* 1999) as it does not take account of growth *pattern* differences. Apparent differences may disappear or equal trends in growth may show divergent behavior if predicted progress beyond the actual period of test is considered.

### Basic principles of a new concept

The leading principles for a new concept are:

• *Principle 1:* **ambiguous** *humidity conditions, including steady-state* **and** *transient conditions separately*

   The following conditions are suggested:
   – An unsaturated (97%) relative humidity at 22 °C. This is usually near optimum growth conditions of indoor fungi found in unsaturated conditions.
   – An intermittent pattern of condensation at 35 °C and drying (RH <60%) at 22 °C ambient temperature, respectively. This mirrors a worst-case condition in which the reservoir function of the substrate is involved.

   Furthermore, single fungal species are suggested in both respective conditions, being predominant indicator organisms in most domestic environments: *Penicillium chrysogenum* Thom (CBS 401.92) in the former and *Cladosporium sphaerospermum* Penzig (CBS 797.97) in the latter. The use of single fungal species is preferred since it allows an easy check on adventitious contaminants in the experiments. The species may be varied according to the geographical location or according to known selectivity of the substrate.

   Defining such humidity conditions is a matter of great concern, since small deviations in RH may have a huge impact on growth rate in this humidity range. Adan *et al.* (1999) outline a possible solution for an experimental set-up, setting a steady-state 97±0.5% RH. Their experimental arrangement consists of a closed re-circulating system, using counter flow humidification of the air. Furthermore, a steady average air velocity below 5 cm/s is set, being typical for boundary flow along surfaces in the indoor environment.

• *Principle 2: Assessment on the basis of* **growth pattern modeling**

   The analysis of macroscopic growth is based on the entire growth pattern as a function of time, including all stages of growth. It includes the following successive steps.

   First of all, assessment of fungal growth with the naked eye at repeated intervals during the period of test using the BS3900 (British Standards Institution 1989) numerical scale (Table 12.2). During the assessment specimens remain in the incubator.

*Table 12.2. Numerical scale to assess coverage area (from BS 3900).*

| Rating | Appearance |
| --- | --- |
| 0 | no growth |
| 1 | coverage ≤ 1% |
| 2 | 1% < coverage ≤ 10% |
| 3 | 10% < coverage ≤ 30% |
| 4 | 30% < coverage ≤ 70% |
| 5 | 70% < coverage |

Secondly, mathematical modeling of the growth pattern is applied. The vegetative growth of both fungi considered on batch cultures produces sigmoidal curves. Adan (1994) introduced a non-linear regression technique to analyse the sigmoidal curve starting from the mathematical model $y_i = f(t_i, \upsilon) + \xi_i$, where $y_i$ is the rating for the coverage area at day i, $f$ is a deterministic function depending on a vector $\upsilon$ of parameters to be estimated and $\xi$ is a stochastic term that is assumed to be identically and independently normally distributed.

Numerous mathematical functions have been proposed for modeling sigmoidal curves, many of which are claimed to have some theoretical basis (Patten 1971, Ratkowsky 1983). Adan (1994) recommended a reparameterized logistic model reading:

$$y = \frac{\alpha}{1 + e^{4 \cdot \varepsilon \, (\delta - t)/\alpha}} \tag{1}$$

where $\alpha$ is the upper asymptotic value of $y$, $\delta$ the time coordinate of the inflection point and $\varepsilon$ the first derivative with respect to time in the inflection point. a actually represents the estimated final coverage area; $\delta$ can be considered a measure for the growth rate in the exponential stage, i.e. the highest growth rate occurring at the moment of time e (Figure 12.1). The logistic model is a nonlinear model, because the coefficients $\delta$ and $\varepsilon$ to be estimated appear non-linearly.

Given the assumption of independent and identically distributed normal variables, the criterion of least squares in linear models provides the best available estimates in practice. Nonlinear models tend to do so only as the sample size becomes very large. In addition, the distributional properties of the stochastic term are often unknown. In simulations of Ratkowski (1983) the coefficient estimates in the logistic model,

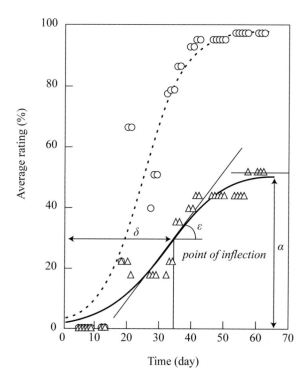

*Figure 12.1. The logistic growth curve of a sensitive (bullets) and less sentive (triangles) material and definition of the response variables used in the statistical analysis. a=the upper asymptotic value; δ=time coordinate of the inflection point; ε=the derivative with respect to time in the inflection point, i.e. the maximum growth rate. In this case, the y-axis refers to the surface coverage area. Another option is to use the non-linear BS3900 rating.*

however, remained stable to various assumptions about the error term. Furthermore, the nonlinear behavior of the coefficients is not serious in practical terms, as simulations showed that the distribution of the coefficient estimators approached that expected for the normal distribution.

## Next steps for improvement

Using such growth pattern modeling allows a distinct evaluation of growth in terms of response variables. Compared to the approach in current standards the statistical reliability increases, thus laying the foundation for an improved discrimination between material performances.

Although this is a step forward, still some features leave room for improvement:
*   The human factor. The critical point concerns the data acquisition, as frequency and continuity depend on the observer availability. It is emphasized that the human factor does not refer to the observer variability, since extensive analysis in previous work revealed high inter-observer agreement (Adan 1994).
*   The non-linear numerical scales that are commonly used for assessment with the naked eye. Although these scales are typically in line with the human judgment and discrimination, the non-linearity confuses the data analysis and obscures a clear interpretation.

Real-time recording at higher magnification using digital microscopy in combination with digital analysis shows prospects to tackle these drawbacks and to improve resolution even more. In recent years, compact low-price digital microscopic tools became available on the market. In some preliminary experiments, their potentials for monitoring fungal growth have been screened. The experimental set-up and technical details are not discussed in this context, but some typical results are included in Figures 12.2 and 12.3.

As with common standards, growth was related to assessment of the coverage area. In this case, a linear relationship between pixel changes and coverage area forms the basis, instead of a non-linear scale for observation. Considering that digital recording at a high sample rate delivers a vast amount of data points, the reliability of the growth characteristics deduced from logistic modeling increases strongly.

The enlarged image of the surface, however, delivers the most interesting feature. The magnified picture provides information on the early stages in growth, which cannot be assessed with the naked eye. The screening experiments indicate that the differences found between materials on the basis of such initial growth observations are identical to those found on the basis of macroscopic observation with the naked eye. Therefore, it is believed that the actual behavior can be predicted in an *earlier* stage of growth, which may drastically reduce the time needed to perform a reliable test. These initial tests indicate that the duration of test can be reduced by a factor up to 10. Further investigation is needed to support these findings. Such faster assessment of materials performance will contribute to a shorter time-to-market of new products in that respect.

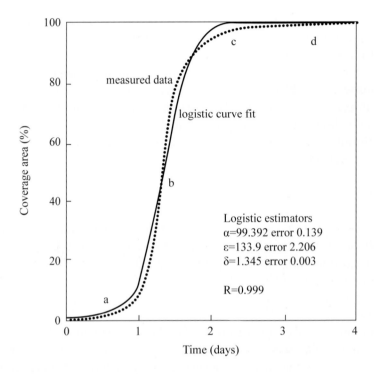

*Figure 12.2. Example of a growth curve, using digital microscopy for data acquisition. The data points are depicted as dots, the logistic curve-fit as a continuous line. Magnification 60×.* Penicillium chrysogenum *on a gypsum substrate, incubated at 97% RH and 22 °C. Note the increased observation speed compared to Figure 12.1.*

## Pilot application on a wide range of materials

During the past decade, the fungal resistance of a wide variety of materials has been assessed using the proposed method of test. Main objectives of this pilot application were to:
- determine whether the test allows sufficient discrimination between materials;
- check the reproducibility in various application areas;
- optimize the test with respect to these issues;
- make the step towards an approved product qualification system.

The following materials and products were subjected to this pilot research: silicon caulking, typically applied in sanitary rooms (Adan and Lurkin 1997a); a wide range of coating types, including waterborne interior paints (Adan *et al.* 1999); specialties,

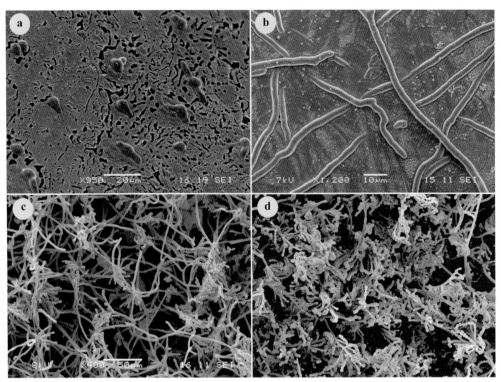

*Figure 12.3. Cryo-SEM observations of developmental stages of* Penicillium chrysogenum *incubated at 97% RH and 22 °C. (a), (b), (c) and (d) refer to the corresponding areas in Figure 12.2. Samples are taken from the experiment in Figure 12.2.*

such as high-absorbing claddings (Adan and Lurkin 1997b) and ceramic coatings (Sanders 2002a); fiber products; gypsum-based plasters and wall papers, including glues (Adan *et al.* 1999); and cement-based panels (Sanders 2002b).

For some products, affected spots in practice obviously show selectivity in terms of fungal species. Therefore, in case of silicon caulking, the test included *Aureobasidium pullulans* var. *melanogenum* (CBS 621.80), *Phoma herbarum* (CBS 366.61) and the yeast *Rhodotorula glutinis* (CBS 2890), mirroring the common microflora predominantly occurring on this type of material in bathrooms (Van Reenen-Hoekstra *et al.* 1991).

In all pilot tests, experimental design was based on a balanced set-up (Figure 12.4), i.e. replicate samples are included and positioned such that variance caused by position (local climatic distortions) can be determined. Although aseptic conditions formed

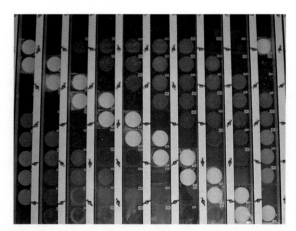

*Figure 12.4. Typical balanced set-up in fungal resistance expriments. View from above, showing fungal growth on circular samples. Each diagonal includes identical samples, so that the position dependency (row, column) can be deduced. In this set-up, air is flowing along the surface. The air flow has fixed climatic conditions and low air velocity (below 5 cm/s, which is typical for indoor boundary flows along walls and ceiling).*

a starting point, it has been shown that adventitious contamination is of no or minor effect (Adan 1994, Adan *et al.* 1999). In all experiments, 10 replicates have been used as a standard. Considering statistics, a minimum of 6 replicates was recommended.

Generally, the logistic model provided a satisfactory fit to the data, as the coefficient of determination $R^2 > 0.96$. The estimator a proved to be the most appropriate tool to compare fungal resistance, with a satisfactory reproducibility (Adan and Lurkin 1997a, Adan *et al.* 1999).

Generally, the pilot application during the past decade yielded a highly reproducible and discriminating picture of material performance in terms of fungal resistance. Tests revealed a significant range of a (varying between 0 and 5) within a test period of 3 months. Furthermore, the experiments showed that growth might be clearly dependent on moisture loads, i.e. material performance under steady state and transient conditions may highly differ. Assessment of growth at a distinct moment in time only, as with most of the current standards, would have obscured the observed differences originating from the present growth pattern analysis.

*Table 12.3. Classification system for fungal resistance of interior finishes ($\alpha_{con}$ and $\alpha_{steady}$ represent the estimators for the transient and steady-state conditions, respectively).*

| Class | Quality | Recommended application | Definition |
|---|---|---|---|
| I | resistant | indoor environments with transient moisture loads such as bathrooms, kitchens, production processes, swimming pools | $\alpha_{con} \leq 1.25$ |
| II | fairly resistant | all other indoor areas, with a more or less steady state indoor humidity, such as living rooms, attics, storage rooms or depots | $\alpha_{con} > 1.25$ $\alpha_{steady} \leq 1.25$ |
| III | sensitive | only on *inner* constructions not being part of the building envelope in environments other than class I | $\alpha_{con} > 1.25$ $\alpha_{steady} > 1.25$ |

## Towards performance requirements in building codes

These findings laid the foundation for a qualification system in the Netherlands with respect to fungal resistance of finishes. This system introduces the classes "resistant" and "sensitive", and an intermediate level "fairly resistant" (see Table 12.3). Labeling is taking place accordingly.

Class definition is based on threshold values of the estimator a, since this variable proved to be the most appropriate and discriminating tool to compare product performance. The estimator of d appeared less suitable, as it usually shows higher variance originating from the logistic fit. d certainly showed to be valuable for refining assessment. The threshold value of a is founded on the overall picture provided by the data set of the pilot study. These data, covering a wide range of materials, showed a ranging between 0-1 or 3-5.

The basic principle underlying this classification is the potential of most products to exhibit a widely divergent behavior as a function of the moisture load. In the past decade, in about 50% of the tested products, the estimators $\alpha_{steady}$ and $\alpha_{con}$ for the steady-state and transient (i.e. condensation) conditions, respectively, were on different sides of the threshold values, underlining the relevance of both tests to assess finishes' performance. Consequently, labeling should be directly connected to the overall characteristics of the indoor humidity load, i.e. to a recommended application.

*Olaf C.G. Adan*

The best quality (I) in terms of resistance reflects the fact that the majority of mold problems occurs in indoor areas with a distinct vapor production (e.g. bathrooms and kitchens in 60% and 40% of cases in the Netherlands, respectively; Anonymous 1993). Logically, it is principally based on $\alpha_{con}$. Under these circumstances, the finishing product is truly dominating the risks for fungal growth, as has been shown fundamentally (Adan 1994). In all other indoor areas, with a more or less steady state indoor humidity, risks of surface growth are a consequence of interaction of finishing product, building construction – thermal bridging in particular –, and average humidity or ventilation. In that case, labeling is based on the $\alpha_{steady}$. It discriminates between fairly resistant finishes that can be applied on thermal bridges, and sensitive finishes that should be applied on inner constructions in dry environments only.

The pilot application clearly revealed that fungal resistance is not a generic characteristic of a type of material, but a product characteristic. Nonetheless, until now some types appear to show uniformity, e.g. waterborne paints tested were all labeled III.

In summary, fungal resistance is a product-based feature, and application-oriented, underlining that consideration of indoor climate dynamics is of major concern. A classification system as suggested earlier is a step towards performance requirements in building codes, making it compulsory. But moreover, labeling of consumer products provides simple decision-support to end-users, i.e. tenants or building owners, the actual occupants. In fact, the finishing layer is mostly the choice of the real end-user.

## References

Adan OCG (1991) Subsoil effects on superficial growth. In: Minutes and Proceedings of the CIB W40 Meeting on "Heat and moisture transfer in buildings", Lund, Sweden, September 10-12 1991, pp. 1-6.

Adan OCG (1994) On the fungal defacement of interior finishes. PhD thesis, Eindhoven University of Technology, Eindhoven, the Netherlands.

Adan OCG (1995) Response of fungi to transient relative humidities. In: Proceedings International symposium on moisture problems in building walls, 11-13 September 1995, Porto, Portugal, pp. 62-74.

Adan OCG and Lurkin JHM (1997a) The fungal resistance of Dutch silicon caulking. In: Proceedings of the CIB-W40 Meeting, 7-10 October, Kyoto, Japan, 1997, pp. 1-10.

Adan OCG and Lurkin JHM (1997b) The fungal sensitivity of three types of condensate absorbent cladding "Firet CondenStop". Report 97-BT-R1222, TNO Building and Construction Research, Delft, the Netherlands.

**Fundamentals of mold growth in indoor environments**

Adan OCG, Lurkin JHM and Van der Wel GK (1999) De schimmelgevoeligheid van afwerkmaterialen. Report 1999-BT-MK-R0205-02, TNO Building and Construction Research, Delft, the Netherlands [in Dutch].

American Society for Testing and Materials (2002a) ASTM G21-96 standard practice for determining resistance of synthetic polymeric materials to fungi.

American Society for Testing and Materials (2002b) ASTM D 3274-95 standard test method for evaluating degree of surface disfigurement of paint films by microbial (fungal or algal) growth or soil and dirt accumulation.

American Society for Testing and Materials (2005a) ASTM D3273 standard test method for evaluating degree of surface disfigurement of paint films by microbial (fungal or algal) accumulation.

American Society for Testing and Materials (2005b) ASTM D5590-00 standard test method for determining the resistance of paint films and related coatings to fungal defacement by accelerated four-week agar plate assay.

Anonymous (1993) Kwalitatieve woningregistratie 1989-1991. Resultaten Landelijke Steekproef, Ministry of Housing, Physical Planning and the Environment, the Hague, the Netherlands [in Dutch].

Bravery AF, Barry S and Coleman LJ (1978) Collaborative experiments on testing the mould resistance of paint films. Int Biodeterior Bull 14: 1-10.

Bravery AF, Barry S and Worley W (1983) An alternative method for testing the mould resistance of paint films. J Oil Colour Chem Ass 2: 39-43.

Bravery AF, Barry S, Pantke M and Worley W (1984) Further collaborative experiments on testing the mould resistance of paint films. J Oil Colour Chem Ass 1: 2-8.

British Standards Institution (1989) BS 3900-G6 Methods of test for paints – assessment of resistance to fungal growth.

British Standards Institution (1990) BS 1982-3 Fungal resistance of panel products made of or containing materials of organic origin. Methods for determination of resistance to mould or mildew.

European Committee for Standardization (2005) EN IEC 60068-2-10 Environmental testing – tests – test J and guidance: mould growth.

International Standardisation Organisation (1997) ISO 846 Plastics – evaluation of the action of microorganisms.

Patten BC (1971) System analysis and simulation in ecology, Vol 1. Academic Press, New York, NY, USA.

Ratkowsky DA (1983) Nonlinear regression modeling – a unified practical approach. Marcel Dekker Inc., New York, NY, USA.

Samson RA, Flannigan B, Flannigan M, Verhoeff AP, Adan OCG and Hoekstra ES (1994) Health implications of fungi in indoor environments. Air Quality Monographs, Vol. 2, Elsevier, Amsterdam, the Netherlands, 602 pp.

Sanders MM (2002a) Fungal resistance of an interior coating. Report 2002-BS-R0012, TNO Building and Construction Research, Delft, the Netherlands [in Dutch].

Sanders MM (2002b) Fungal resistance of a cement based panels. Report 2002-BS-R0020, TNO Building and Construction Research, Delft, the Netherlands [in Dutch].

Van Reenen-Hoekstra ES, Samson RA, Verhoeff AP, Van Wijnen JH and Brunekreef B (1991) Detection and identification of moulds in Dutch houses and non-industrial working environments. Grana 30: 418-423.

US Department of Defence (2000) MIL-STD-810F, Method 508.5 Department of Defence, Test method standard for environmental engineering considerations and laboratory tests.

# 13 Detection of indoor fungi bioaerosols

James A. Scott, Richard C. Summerbell and Brett J. Green*
Sporometrics Inc., Toronto, Canada*

## Introduction

The detection of microorganisms in environmental samples, particularly from air, has a long history, greatly pre-dating the analysis of the chemical contaminants that are now generally more familiar to occupational hygiene. Antonie van Leeuwenhoek (1632-1723) was the first to demonstrate the presence of microbial cells in indoor dust, but it was not until two centuries later that Louis Pasteur (1822-1895) famously demonstrated lactic acid fermentation by airborne microbes by introducing air into sterile broth in swan-necked flasks, thereby defeating the spontaneous generation hypothesis of microbial life (Pasteur 1857). Based on the recognition that living microorganisms could travel through the air, Pasteur's methods were rapidly enlisted in the search for agents of human diseases, such as cholera and typhoid. Another early air sampler, the "aeroconioscope" of Maddox (1870), relied on wind pressure to propel air through a cone tapered to a narrow point positioned above an adhesive-coated slide onto which particles were impacted (Cunningham 1873, MacKenzie 1961). Although this device was effective in capturing particles from the air, airborne concentrations could not be determined because the volume of air drawn through the instrument could not be measured.

One of the earliest studies using volumetric air sampling was conducted by Pierre Miquel (1850-1922). Miquel's device collected particles by impaction on the surface of a sticky slide that was situated beneath a narrow orifice through which air was drawn by suction. The suction was derived from the descent of boli of water separated by long bubbles of air in a vertical tube (Miquel 1883, Comtois 1997). Remarkably, Miquel's technique was able to sample a cubic metre of air using only 40 litres of water (Miquel 1883). Even though much of this early work sought to clarify the epidemiology of notorious human diseases like cholera and typhoid, this work actually furthered our understanding of the aerobiology of relatively large bioaerosols, primarily fungal spores and plant pollen.

During the last 50 years, a great number of volumetric air sampling devices have been developed to suit various purposes. These devices along with their uses and limitations have been reviewed in detail (Cox *et al.* 1995, Macher 1999, American Industrial Hygiene Association (AIHA) 2005). As well, a number of excellent references have appeared dealing with approaches to building investigations for mold growth and the interpretation of sampling results (AIHA 2005, Flannigan *et al.* 2001, Macher 1999).

Within the last two decades, numerous cognizant bodies and governmental jurisdictions have advised that exposure to environmental molds represents a controllable health risk, and that indoor mold growth should not be tolerated irrespective of its contribution to the indoor airborne spore-load (see Chapter 14). This approach has been favored over the establishment of exposure limits based on air sampling for a number of reasons including the following: (1) widely used air sampling methods typically express levels in units such as colony forming units (cfu) or spores per cubic meter, which are not relevant to dose (typically milligrams of contaminant per cubic meter); (2) mold samples contain complex mixtures of chemical contaminants whose constituents vary qualitatively and quantitatively, and cannot readily be defined a priori; (3) no single sampling approach can adequately evaluate the broad spectrum of known and probably health-relevant contaminants expressed by molds; and (4) much of the current knowledge implicating molds as health effectors is derived from proxy indicators of exposure (Douwes *et al.* 2003). Nevertheless, indoor air sampling remains a useful tool in the search for indoor growth sites, in conjunction with a visual inspection by an experienced investigator.

This chapter is by no means an exhaustive review of fungal air sampling techniques, of which there are many. It is intended as a review the uses and limitations of some of the air sampling methods for molds that have been used commonly for occupational hygiene purposes. We have not included sampling approaches that have only been used in research studies, or investigational techniques to assess personal exposure. Our discussions of new technologies, however, include some recent studies introducing or validating techniques that appear to have considerable practical potential.

## Environmental sampling

Virtually all indoor materials, ranging from construction products to furnishings and foods, are susceptible to fungal spoilage under conditions of dampness. A range of methods have been developed to detect the presence of indoor fungal growth. Fundamentally, these methods can be classified into: (1) those that monitor fungal particles in the air by air sampling; and (2) those that characterize fungal growth directly on bulk material specimens.

## Fungi normally present in indoor environments

Airborne fungal particles represent a complex mixture consisting of elements that differ in biology, chemistry and morphology. Under normal circumstances where indoor fungal contamination is absent, the fungal particles that predominate in indoor environments arise from three primary reservoirs: (1) the phylloplane; (2) the pedosphere; and (3) the dermatoplane.

Phylloplane fungi comprise an ecologically similar grouping of unrelated, mostly dry-spored taxa that inhabit plant leaf surfaces. A proportion of them are biotrophic to necrotropic parasites, while many others are saprotrophs. In northern temperate regions, phylloplane taxa predominate in the aerospora during the growing season. Owing to their abundance in the outdoor air, aerosols derived from these fungi, including spores, hyphal fragments and cellular debris, filter into indoor environments through normal activity. Characteristic taxa in this category include certain species of the anamorph genera *Alternaria*, *Cladosporium* and *Epicoccum*, as well as powdery mildews, lignicolous, foliicolous, coprophilous and lichenized ascomycetes, most myxomycetes, and most basidiomycetes including rusts, smuts, agarics, brackets and corticioid fungi (excluding *Poria* and *Serpula*).

Fungi originating from soil reservoirs also commonly occur in indoor environments even when indigenous fungal amplifiers are absent. These mostly wet-spored fungi enter buildings on footwear: this is particularly common in North America, and other places where footwear is not removed upon entering buildings. Secondarily, soil containing fungal material may become aerosolized during construction or excavation activities, and these aerosols may be carried indoors in the same way as phylloplane fungal materials are, that is, passively in air currents or via human movements or actively through the actions of mechanical systems. Strictly speaking, fungi associated with soil as a reservoir are not necessarily indigenous to soil as a habitat. They may originate elsewhere as saprotrophs or necrotrophs and accumulate over time in the soil "spore bank". Taxa common in this category include species of the anamorph genera *Acremonium*, *Beauveria*, *Chromelosporium*, *Clonostachys*, *Fusarium*, *Gliocladium*, *Myrothecium*, *Phoma*, *Trichoderma*, and *Penicillium* subgenera *Biverticillium*, *Furcatum* and *Aspergilloides*.

Human skin is an underestimated but a rich source of fungal material in the indoor environment. In a study by Pitkäranta *et al.* (2008) of indoor dusts analysed via sequence analysis of rDNA clone libraries 14% of all clones recovered represented *Malassezia* yeasts (*Exobasidiomycetidae*), fungi growing on human skin where they are often associated with tinea versicolor, dandruff and seborrhoeic dermatitis (Gupta *et al.* 2001). The abundance of these fungi in indoor environments is not surprising in light of their high prevalence in human populations. The results of this report suggest that *Malassezia* species from human sources may be non-trivial contributors to dust-borne beta-(1,3)-D-glucan. Dermatophytic fungi, by comparison, have been reported from indoor air and dust but are uncommonly recovered in microbiological investigations of indoor air and dust (MacKenzie 1961, Gupta and Summerbell 2000, Summerbell 2000).

Lastly, a number of fungi are known to proliferate in dry indoor habitats, such as settled dust. They do so under normal ambient conditions without superfluous moisture, but their growth is enhanced by dampness. Typical indoor xerophilic taxa include members of the *Aspergillus versicolor* group, *Aspergillus penicillioides* and *Wallemia sebi* (AIHA 2005, Samson *et al.* 2010, Miller 2011). Examination of normal household floor dust often reveals mite fecal pellets that consist largely of comminuted fungal hyphae and conidia that, when cultured, yield *A. penicillioides* and other xerophiles (Van Asselt 1999). In light of such findings, Van Bronswijk and others have suggested that xerophilic molds are important elements of the indoor dust ecosystem, providing nutritionally enriched grazing materials for dust microarthropods such as pyroglyphid mites (Van Bronswijk and Sinha 1973, Van de Lustgraaf 1978, Hay *et al.* 1993, Van Asselt 1999). The observation of low background levels of these and other xerophilic taxa in indoor air and dust is of ambiguous significance and does not necessarily reflect abnormal indoor fungal growth.

## Bioaerosols

The atmosphere is replete with fine particles of different size, shape, composition and origin, which are constantly sent aloft by disturbance and sedimented out by gravity. This suspension of airborne particles, known as an aerosol, contains considerable biologically-derived material. Fungal particles and pollen together account for up to 10% of the total mass (Womiloju *et al.* 2003). The term "bioaerosol" refers more narrowly to the fraction of total aerosol comprising particles immediately originating from biological materials or processes of biological origin. Particulates arising from the combustion of fossil fuels are not considered bioaerosols despite the biological origin of the parent fuel. Bioaerosols are generally interpreted to include whole cells, cell fragments, and non-volatile chemicals arising directly from biological processes, such as proteins and carbohydrates.

## Origin and release

The liberation of fungal bioaerosols is accomplished by a range of mechanisms including both active and passive release. Active release refers to adaptive means of propagule aerosolization, via forces arising from a burst of energy. Ballistic spore discharge, for example, is seen in basidiospore release in the gilled mushrooms, bracket fungi and corticioid fungi, while forcible ascospore discharge is observed in the perithecial and pseudothecial ascomycetes. Ballistoconidium release occurs in the mirror yeasts. In conidial filamentous fungi, an important means of active spore discharge involves the disruption of conidia by hygroscopic torsion of conidiophores and conidial chains (Meredith 1973).

Passive release is accomplished by mechanical disturbance in activities such as construction and excavation, in maintenance practices like sweeping and grass mowing as well as in personal activities including walking and sitting. Adhesion of fungal propagules to mist droplets has also been identified as an important aerial dispersal mechanism. Nebulization of water containing fungal debris results in the generation of a mist containing cells and other materials. Subsequent desiccation of mist droplets can cause the accretion of all non-volatile constituents into a composite solid particle, in a process known as "droplet nucleation" (Wells 1934).

The release of fungal propagules from the structures producing them often demonstrates pronounced temporal patterns, ranging in scale from seasonal to diurnal. Phylloplane taxa, for example, show peak airborne concentrations seasonally in accordance with the life cycles of their plant hosts. In northern temperate climates, airborne levels of these taxa substantially decrease during the winter. Many of the same phylloplane species demonstrate diurnal periodicity of propagule release, with peak concentrations detected during the daylight hours and with lower concentrations present during the night-time hours (Gregory 1971). By contrast, the cooler, often damp night-time conditions are associated with the enhanced release of ascospores and ballistosporic yeasts that otherwise only dominate the aerospora following precipitation (Gregory 1971).

### Factors affecting bioaerosol dispersion, deposition, collection and recovery

#### Aerodyamics

The aerodynamic behavior of bioaerosols is governed by intrinsic properties, such as size, morphology, hygroscopy and density under the influence of prevailing environmental conditions. From a kinetic standpoint, the most important property of a bioaerosol particle is its terminal settling velocity (TSV); this refers to the maximum velocity attained by a particle at the point that gravitational acceleration is offset by viscous resistance of air. In still air, in the absence of thermal or electrostatic disturbances, a 1 μm diameter sphere of unit density has a TSV of 12 cm/hr, whereas the TSV of similar sphere of 10 μm diameter is 100 times greater, 1.2 m/hr (Cox *et al.* 1995).

Particle morphology and density are also important determinants of TSV. For the purposes of estimation of "aerodynamic equivalent diameter" (AED), fungal spores are assumed to have a density slightly greater than that of water, 1.1 g/cm. Non-spherical or morphologically irregular particles, or those of different density are said to have an "aerodynamic equivalent diameter" (AED) equal to the diameter of a hypothetical unit-density spherical particle of equivalent TSV (Hinds 1999). Generally, bioaerosol particles with an aspect ratio (i.e. length : width) of less than 3 can be assumed to have an AED roughly equivalent to the diameter of a spheroid of equivalent volume (Hinds 2005). Particles with a larger aspect ratio than 3, and particularly greater than 10, exhibit variable TSV depending on their orientation, and require a more complex approach to predict their behaviour (Hinds 2005). The AED of a particle is also related to the efficiency with which that particle can be captured and retained by certain kinds of air sampling instruments, particularly those based on impaction.

**Moisture**

Most bioaerosol particles are hygroscopic. This, in combination with their small physical size and relatively large surface area-to-volume ratio, makes them increasingly susceptible to small changes in relative humidity, resulting in rapid changes in particle density and physical size. These changes ultimately cause variation in aerodynamic properties (Reponen *et al.* 1996). As such, the mechanical collection efficiency of a given bioaerosol sampler for the same aerospora will be greater when sampling is carried out under conditions of higher relative humidity.

Liquid impingement is a collection technique that circumvents the problem of desiccation-related cell death by drawing air through a liquid collection medium, often a nutrient broth, water or oil. However, the collection of viable cells in liquid media can result in premature cell germination and unintentional amplification, dilution of analytes used biochemical quantification or cell death. Culturable taxa may also continue to grow and amplify in the collection medium, resulting in an artifactual increase in reflected biomass and changes in cell morphology. Liquid impingement accepts this as a trade-off against the greater ability of liquid-based collection media to protect cell viability by preventing desiccation.

## Electrostatic effects

Most bioaerosol particles acquire a small electrostatic charge through passive equilibration with normally occurring atmospheric positive and negative ions. Electrostatic charges can result in the deposition of particles on vertical surfaces, the agglomeration of particles of opposite charge, and other behaviors that are not predicted by gravitational effects alone. Electrostatic charges can also be an important source of sample loss in the collection head. The action of drawing air at a high flow rate through an orifice or slit can cause a static electric charge to develop on the sampler, typically in the region at and around the outlet of the sampling nozzle. This charge can be dispersed if the sampler is composed of conductive material; however, components of the sampling head composed of non-conductive materials, such as plastic and glass, can build up static electric charges to a sufficient degree that particles become diverted from the air stream during sampling (Cox *et al.* 1995: 18). The loss of collected particles by re-entrainment in the air stream during sampling is well-known (Hinds 1999).

## Electromagnetic radiation

The strong influence of electromagnetic radiation on bioaerosol particles relates mainly to their size. A majority of indoor environmental bioaerosols are less than 10 μm in physical diameter, and therefore are close in magnitude to the wavelength of visible light. The energy imparted by light is readily propagated to aerosol particles. In incident light, opaque particles become heated on the aspect of the particle directly facing the light source and are propelled from the light source by convection. The opposite effect occurs with transparent particles, which function like lenses. Energy is concentrated on the far side of the particle, thereby causing the particle to be propelled under convective force towards the heat source. This is the principle underlying the tendency of incandescent light bulbs in kitchens to acquire a film of grease on their surfaces (the grease originates as an oil mist comprising transparent particles that are generated during cooking, depositing by thermal attraction on the surface of the bulb). Exposure to light in the visible range also increases Brownian motion, which can result in particle movement or deposition behaviors that are not predicted by gravity effects.

Convection effects predominate with longer wavelength, thermal radiation. Convective effects are mostly responsible for keeping fine particles suspended in the air for long periods in absence of persistent mechanical disruption. However, these effects may also result in particle deposition on vertical surfaces adjacent to heat sources, or horizontal surfaces above heat sources.

*James A. Scott, Richard C. Summerbell and Brett J. Green*

# Air sampling

## Volumetric air sampling

A large range of sampling instruments is available for the volume-based measurement of biological particles in air. All air sampling instruments work by separating particles from air, and then collecting and concentrating the particles to permit their measurement and characterization. There are several methods of collection that are classified under the headings impaction, filtration and impingement. The devices most commonly used for bioaerosol detection are those relying on inertial properties, such as impaction samplers.

## Inertial samplers

Most inertial samplers are based on the principle of impaction. The most common impaction sampler is the jet-to-agar impactor. Air is drawn through a slit or round orifice, accelerating any suspended particles to the velocity of the air flow. The jet is directed against a surface producing a deflection in the air stream. Typically this surface is a glass or plastic impaction plate or a growth medium contained in a Petri plate. The trajectory of particles in the deflected air stream is determined by their momentum. Above a certain threshold momentum, particles become impacted on the plate, whereas particles with lower momentum remain in the air stream. In most impactors, the air sampling pump is situated downstream of the collection head and capture media, and the device operates under negative pressure.

Other samplers that utilize inertia as mechanism to separate particles from an air stream include centrifugal impactors and liquid impingers. Centrifugal impactors draw air through an impeller containing an annular collection plate, typically filled with an agar growth medium or coated with an adhesive. Particles are accelerated centrifugally and captured by impaction on the surface of the collection medium. The centrifugal impactors most commonly used in bioaerosol sampling are the RCS Standard and High-flow devices manufactured by BioTest (Dreieich, Germany). The RCS Standard has the strong disadvantage of exhausting air from the intake orifice, making it impossible to measure the flow rate of the device. This limitation was addressed in later versions of the sampler, such as the Plus and High Flow models, for which air flow rate can be calibrated.

Liquid impingers collect particles by passing air through a liquid medium. With these instruments, air is drawn through a series of orifices, producing a stream of bubbles in the liquid medium. The bubbles move through the medium much more slowly than the speed of the air exiting the orifices, causing the inertial impaction of particles at the periphery of air bubbles. Liquid impingers are less prone to kill cells by desiccation than most other particle collection methods. However, this collection method is impractical for field use because of spillage or contamination during handling. Losses may arise from the re-entrainment of collected particles as bubbles burst at the medium surface. Foam accumulation during collection is common.

## Mechanical collection efficiency

The effectiveness with which inertial samplers mechanically collect particles from the air is most commonly described by the $d_{50}$ cut-point, which is the AED of particles that are collected with 50% efficiency. In general, particles of greater AED tend to be collected with high efficiency in keeping with the steep morphology of the collection efficiency curves (Hinds 1999: 125). It should be noted; however, that mechanical collection efficiency is contingent on the consistency of sampling flow rate, which determines the momentum of particles. A reduction in flow rate causes a simultaneous, non-proportional reduction in the $d_{50}$ cut-point. Therefore, $d_{50}$ cut-points are always given relative to a standard flow rate that must be maintained throughout the sampling period. The flow rate of the sampling device is usually directly controlled by the flow rate of the sampling pump used. The pressure drop contributed by the sampling head is low for most of the commonly used impaction samplers. Furthermore, sample loading during sample collection does not cause an increased pressure drop. Some air samplers are equipped with a "critical orifice", a precision narrowing of the air outlet that limits the maximum flow rate of air through the device. For these devices, the flow rate through the device will remain constant as long as the flow rate of the sampling pump does not fall below a certain minimum. Many older Andersen-style samplers are equipped with this flow restricting device; however, it is absent in most recent- model samplers. In these samplers, a careful calibration of the sampling pump is required.

## Collection losses

Even with the foregoing factors optimized, several sampling phenomena remain that, can modify the capture characteristics of the sampler resulting in sample loss. The first of these relates to the modification or redistribution of particles by the air stream itself. At the flow rates used in most inertial impaction samplers, the process of entrainment and passage through the collection orifice creates sheer forces that can comminute aggregated particles, altering their deposition characteristics (Andersen *et al.* 1967). In our experience with Hirst-type slit samplers, cells that become comminuted during collection typically co-deposit in a spray pattern spanning several cell diameters. Depending on the number of non-contiguous cells, the pattern of spread, and the overall loading of the sample, the microscopic interpretation of these cells as representing a single progenitor can be highly subjective.

In addition to modifying the deposition characteristics, the sheer forces produced during collection can also cause cell damage and reducing viability (Andersen *et al.* 1967). Sample loss can also arise from the redistribution of sampled particles by the air stream, depending on the ability of these particles to adhere to the collection plate. Poor adhesion to the capture medium can cause particles of AED greater than the $d_{50}$ cut-point of the sampler to bounce off of the surface of the medium at the moment of impaction. Also, the persistent force of the air stream during sampling can cause captured particles to become re-entrained in the air stream and to exhaust from the sampler. With samplers that use direct microscopy to detect, quantify and characterize captured cells, redistribution of cells on the collection medium can artificially increase the number of apparent capture events.

### Filtration samplers

The use of filter membranes for the capture and collection of bioaerosols has a long history, but, nonetheless, the method is not widely used in routine air sampling. As a collection method, filtration has several advantages over inertial sampling. A primary benefit is that the mechanical collection efficiency of the sampler is governed by characteristics of the membrane. In relatively still air, this method is minimally dependent on face velocity (i.e. the speed of the air moving into the medium) above a certain minimum threshold dependent on the medium type; this velocity is usually about 100 cm/min for mixed cellulose ester membrane filters with a 0.45 μm-pore size (NIOSH 1994). Membranes are available in a wide variety of compositions and collection efficiencies. Membrane samples also have an advantage of being suitable for bioaerosol analysis in conjunction with microscopic or culture-based methods, or with other modes of characterization such as biochemical, immunochemical, genetic or spectral analyses.

Filtration samples are used widely in the characterization of integrated personal exposures to bioaerosols and other aerosols. This is because they can function with lower flow rates than inertial samplers, using relatively small pumps, and also because this sample format is minimally susceptible to overloading. These techniques have been underexploited outside of research-related assessments of bioaerosol exposure.

*Passive air sampling*

### Settle plates

The use of gravity settle plates has a long history in environmental mycology and bacteriology, and these plates continue to be used in some specific applications. This method consists of exposing the agar medium in a Petri dish to the ambient air for a period of time, often 0.5-1 hr, incubating the plate and counting and identifying the resulting colonies. Because the effective volume of air sampled cannot be determined, the colony counts on settle plates cannot be expressed in volumetric units; hence, these data are semi-quantitative. Due to the different settling velocities of airborne particles according to their aerodynamic features, this method has long been known to under-represent taxa with small airborne propagules, such as *Aspergillus* and *Penicillium*, relative to those taxa with larger propagules (Sayer *et al.* 1969, Soloman 1975). This matter can be subjectively taken into consideration when results are evaluated (that is, the analyst can determine that the number of *Penicillium* colonies seen is probably a drastic under-representation), but there are no formulas available to allow formal transformation of the data to accommodate the differences in propagule size and shape. Despite these shortcomings, the persistent recovery of taxa commonly associated with indoor contamination is strongly suggestive of an indoor growth site. These taxa include: *Aspergillus niger, A. fumigatus, A. versicolor, Chaetomium* spp., *Cladosporium sphaerospermum, Paecilomyces variotii, Penicillium* subgen. *Penicillium*, and *Stachybotrys* spp. (Federal-Provincial Committee on Environmental and Occupational Health (Canada) 1995, ACGIH 2001, Flannigan *et al.* 2001, AIHA 2005).

### Dusts

Studies comparing static area sampling with integrated personal sampling for aerosols have shown very poor agreement between the two methods (Lange *et al.* 2000, Niven *et al.* 1992). This inability to use area sampling to predict personal exposure is well recognized in industrial hygiene (AIHA 2005). Attempts to evaluate exposures in the non-industrial indoor environment face similar challenges. For example, analysis of fungal content in indoor air samples is of limited use in the evaluation of the contamination status of the affected environment. The fungal burden in indoor air varies episodically, related to factors including the total mass of settled dust present, the amount of activity, and the various features of the building that in some way moderate the ventilation (AIHA 2005, Ferro *et al.* 2004a). In most indoor environments, settled dust is the primary source of biological particulate contaminants (AIHA 2005: 95). Dust mass and disturbance measurements are well correlated with measured levels of airborne fine particles (Afshari *et al.* 2005) including fine biological particles (Buttner and Stetzenbach 1993, Cole *et al.* 1996, Ferro *et al.* 2004b). Exposure to indoor biological contaminants mainly arises from activity-related disturbance of fine dusts on flooring and furnishings, contributing to a "personal cloud" of fine particulate matter (Burge 1995, Ferro *et al.* 2004a,b). The ability of indoor dust to act as a reservoir for seasonal outdoor aeroallergens, leading to perennial exposures in the indoor environment, has been well demonstrated (Arbes *et al.* 2005, Salo *et al.* 2006). Dust analysis provides a more robust proxy measure for personal exposure than short-term air sampling provides, and it is less influenced than air sampling by temporal and spatial variation. These measures are more reproducible than air sampling results, and provide an integrated picture of exposure, literally capturing transient bursts of indoor bioaerosols (Ren *et al.* 2001, Portnoy *et al.* 2004).

While settled dust indoors may influence respiratory status independently of the exposure to any specific chemical constituent (Elliott *et al.* 2007), most commonly assayed contaminants such as allergens, endotoxins and glucans can be measured in a highly reproducible manner from dust (Antens *et al.* 2006, Heinrich *et al.* 2003). These analytes remain stable in dust during long-term storage (Morgenstern *et al.* 2006). The mass of sieved dust collected by standard methods from a pre-defined area (e.g. 2 m$^2$) provides an index of exposure both to total dust and to a series of health-relevant contaminants in the home (Elliott *et al.* 2007).

### Analytical methods

The most commonly used bioaerosols samplers are grouped into two general categories, spore trap samplers and culturable samplers. Spore trap samplers are analysed by direct microscopy whereas samples collected on growth media must be cultured before identification and enumeration can be done.

### Direct microscopic methods

The earliest methods of characterizing airborne fungi featured light microscopic analysis of liquid-impinged samples; this technique was devised by Louis Pasteur (Ariatti and Comtois 1993). The same approach continues to be a front-line technique for the identification and characterization of fungal bioaerosols. Direct microscopy is used in combination with culture-based techniques for the primary identification of taxa. As well, light microscopy is used increasingly in the visualization of fungal cellular material in methods excluding prior amplification via culture. These methods, while useful and rapid, are poorly able to distinguish many morphologically similar common taxa such as *Acremonium, Aspergillus, Clonostachys, Paecilomyces, Penicillium, Phoma,* etc. Indeed, many common indoor contaminant genera of ascomycetes and basidiomycetes cannot be reliably identified via microscopic examination of their spores. Hyphal fragments from all sorts of fungi are similarly minimally distinguishable by light microscopy.

A range of methods and approaches have been applied to the analysis of spore trap samples. There are various types of counting procedures as well as conventions for identifying spores or other cells. Counting procedures vary according to the type of sampler used. Where the collected material is randomly deposited, as is the case on a filter membrane sample, the deposit may be enumerated by counting random microscopic fields. By contrast, sampling approaches that result in the non-random deposition of the catch must be analyzed using a systematic approach. For example, slit-type impaction samplers have long been used in aerobiological studies (Hirst 1953). These devices draw air through a slit-like orifice below which is situated an impaction plate, often a glass microscope slide, a coverslip or a piece of film coated with grease or adhesive. In sampling, the catch is deposited in a linear trace, shaped by the morphology of the orifice. Counts of captured cells are typically taken along microscopic transects perpendicular to the deposit. Transects must be counted completely in order to avoid bias based on the AEDs of the various particles. Nevertheless, incorrect counting methods have been applied in some large studies comparing sampler collection efficiencies under field conditions; the methodological correctness of published studies must be carefully evaluated (e.g. Lee *et al.* 2004).

It has been recommended that a minimum of 10-15% of the deposit be examined in order to obtain an adequate representation of the taxa collected (Foto *et al.* 2004). However, given the relatively short duration of sampling in all common methods, a number of authors have suggested that the quantitative data derived from bioaerosol sampling procedures are of little value, and that much more useful information is obtained from the qualitative composition of the sample. Thus, the expenditure of analytical time to establish a highly accurate count of spores or other cellular elements is often discouraged in favor of providing a thorough characterization of the taxonomic composition of the sample. The formal count is often followed by an exhaustive examination of the sample to ascertain the total number of distinguishable taxa that are present (Thorn *et al.* 2007).

Another important factor to consider in the microscope examination of spore-trap samples is microscopic magnification and resolving power. Many commonly occurring indoor fungal spores are small, often below 10 μm in diam. The discrimination of fine details of cell morphology and ornamentation necessary to differentiate cells,

*James A. Scott, Richard C. Summerbell and Brett J. Green*

such as the conidia of *Aspergillus* and *Penicillium*, from morphologically similar, hyaline basidiospores typical of many common outdoor *Polyporaceae*, can rarely be accomplished reliably at magnifications below 1000×, and cannot be done using a microscope with poor resolving power. Furthermore, the distinction of these and other small, hyaline fungal cells from background organic and inorganic debris can be exceedingly difficult at low or medium power. Therefore, the routine use of high-power magnification, often in addition to resolution-enhancing optical techniques such as Nomarski differential interference (DIC) or phase contrast microscopy is increasingly recommended for aerobiological analyses (AIHA 2005).

The enumeration of spore clumps or chains is another area where methodology varies among researchers. Many aerobiological studies have taken the approach of counting (or estimating the count) of individual cells or spores in an aggregated particle. This practice, strongly influenced by early pollen aerobiologists investigating outdoor exposures and allergic sensitization, is rooted in the assumption that numerical counts of individual particles, even when aggregated in clumps, may correlate with increased inhalation exposure. While there is some validity to this assumption for long-term, integrative pollen sampling (Van Leeuwen 1924), the relevance of this approach to commonly-used, short-duration fungal spore sampling is unclear. First of all, the purpose for using these methods often relates to the evaluation of whether or not building interiors harbor fungal growth sites. In this type of evaluation, the "dose" is irrelevant; instead, the community composition is of primary importance, since indoor fungal growth generally features distinctive assemblages of species. Secondly, the individual counting of aggregated cells treats the process of collection for each of these cells as if it were a statistically independent event. Actually, the clumps or chains often arise as masses from individual conidiophores or microcolonies, and the individual cells thus collectively represent statistically *dependent* "capture events". This is an important distinction in air-sampling data in practice and research. For example, a coherent chain of 30 *Cladosporium* conidia, if interpreted as adding 30 units to the count per cubic meter of air, artificially implies a higher airborne prevalence of this fungus than is statistically realistic. The potential for surface contamination by spore deposition appears to be 30-fold greater than it actually is. If the goal of air sampling is to express the frequency of occurrence of airborne cells or propagules, counts are much more informative when they document "collection events" rather than the total number of cells collected. Depending on the planned applications of air-sampling data, it may be useful to document details of the morphology of bioaerosol particles as they occur in the air, particularly in regard to whether they are unitary or composite.

## Culture-based methods

Most of the commonly used air sampling methods rely on cultivation of fungi on agar as a means of detection. Bioaerosol sampling methods that rely on culture are subject to recovery losses that arise from mechanical as well as biological effects:

- *Jet-to-plate distance.* The stability of the $d_{50}$ cut-point from sample to sample relies on a consistent distance between the terminus of the sample jet and the surface of the collection plate. In culturable samplers, like the Andersen samplers and other jet-to-agar impactors, this distance is governed by the medium volume of the Petri plates used for collection. In theory, the air stream should not diffuse appreciably between the nozzle exit and a location not more than one jet diameter away from the surface of the collection plate (Hinds 1999: 125). However, the use of a consistent jet-to-plate distance through adherence to the recommended medium volume is important in the maintenance of collection efficiency (Macher *et al.* 2001). Although user guides that accompany sampling devices typically provide these requirements, in practice they are often ignored and it is worthwhile here to review several common standards.

The most commonly used jet-to-agar impactor is the single stage Andersen samplers, also known as the N6 because it derives from the sixth stage of larger, size-partitioning instrument, the 6-stage Anderson cascade impactor. The N6 has an empirically determined $d_{50}$ cut-point of 0.65 µm (Andersen 1958) using a sampling flow rate of 28.3 l/min with standard 100 mm Petri dishes filled with 41 ml of medium (Thermo Electron Corporation 2003). Actually, plate volumes ranging from 40-50 ml are suitable (Macher *et al.* 2001: 687). In contrast, the 2-stage Andersen uses 20 ml per plate, and the 6-stage uses 27 ml per plate (Thermo Electron Corporation 2003). Substitution of 90 mm Petri plates for 100 mm plates is acceptable for all devices. This may create a problem if no adjustment is made for the difference in interior volume. For example, the critical height of the medium surface in a 100 ml diam Petri dish with straight sides containing 41 ml of medium is 5.2 mm. However, many researchers now commonly use 90 mm polystyrene Petri plates with tapered sides. As a case in point, our laboratory ordinarily uses "100×15 mm" Petri plates manufactured by Fischer Scientific Inc., Canada, catalogue no. 08-757-13, which have an inner diameter at the bottom of the plate of 84 mm, and an inner diameter at the top of the lower plate of 87 mm. In order to achieve the predicted $d_{50}$ cut-point with an Andersen N6 sampler using these Petri plates, a 30 ml volume of medium is required. It should be noted that the critical medium height of 5.2 mm is a minimum required to achieve the predicted $d_{50}$ cut-point of the device, and that further increasing the medium height (i.e. increasing the plate volume) would reduce the $d_{50}$ cut-point.

*James A. Scott, Richard C. Summerbell and Brett J. Green*

- *Biological recovery efficiency.* The recovery of propagules or "colony-forming units" requires that the captured cells be amplified by culture. However, the process of sampling itself can dramatically modify cell culturability or viability. These effects become pronounced where solute-rich, semi-solid growth media are used for collection. In the case of a jet-to-agar impactor, for example, air passing through an orifice strikes a highly localized area on the medium surface. Over the course of sampling, this may result in localized water loss accompanied by transient hyperconcentration of solutes on the upper surface of the agar (Andersen 1958, Blomquist *et al.* 1984, Burge 1995, Morris 1995). As the spot impacted by the jet is the locus of cell deposition, this region becomes osmotically ever less hospitable to incident cells as sampling continues. At the conclusion of the sampling period, there is some rebound rehydration of the impaction loci from water provided by the underlying medium; however, the osmotic damage that arises in the deposited propagules may be irreversible. This is particularly true for propagules impacted early in the course of sampling or those of taxa that exhibit low osmotolerance. Depending on the prevailing relative humidity, the duration of sampling, and the degree of osmotic tension that develops while the medium is exposed to the sampling air stream, varying degrees of irreversible cell damage may result. Similar desiccation-related loss effects have been reported with air sampling and vacuum-collection of surface dust on filter membranes (Collett *et al.* 1999, Hinds 1999).

  Another important source of biological loss arises from intraspecies and interspecies competition at sites of co-impaction (Cox and Wathes 1995). This phenomenon is predominantly associated with jet-to-agar impactors like the Andersen N6, where deposition on the surface of the collection medium is non-random, mirroring the arrangement of the jets. The likelihood of co-impaction increases logarithmically with the number of propagules collected, expressed as the number of "positive holes". A statistical correction procedure known as "positive hole correction" has been applied to resolve this problem. This method, however, is widely misapplied and its utility is poorly understood. To be statistically valid, the application of a positive hole correction factor requires the user to determine the number of jets through which at least one particle has impacted. This can only be determined by examination of each jet impaction site by direct microscopy immediately following sampling. The enumeration of colonies following incubation cannot be used as a proxy measure for several reasons. Firstly, the deleterious effects of competition between co-impacted propagules may result in failure to detect any of the co-impacted propagules by culture. Secondly, the turbulence resulting from the deflection of air streams immediately beneath jets can lead to tangential impaction of free particles in the air stream, and can deposit particles arising from secondary redistribution or from comminution of previously collected particles. Thus, only particles located immediately below jets can be counted and all other particles present on the plate must be ignored. After incubation, however, it is impossible to distinguish colonies coinciding with jets from those of interstitial origin. Given these limitations, the proper use of the positive hole correction method for multi-orifice jet-to-agar samplers is impractical for routine bioaerosol sampling.

- *Additional confounders.* The collection of culturable samples necessitates careful handling of sample media to avoid unintended contamination during handling. Carry-over from location to location on the exterior of the sampler, pump or ancillary equipment must also be avoided. Handling-related contaminants are often recognizable by their patterns of distribution, typically concentrated at the margins of agar plates or strips, and also by their consistent taxonomic composition from sample to sample. The use of so-called "field blanks", sample media that are taken to the field, loaded in the sampler and repackaged without sampling, handled always in the same manner as actual samples. The presence of contaminants on field blanks is confirmatory of handing-related contamination, and warrants cautious interpretation of data from the sample set.

Another potential confounder in culturable sampling on agar media arises from the presence of free water in the form of condensation in the Petri dish or sample container. It is widely recommended to cool agar plate or strip samples between field collection sites and the laboratory. This is usually accomplished by placing samples in an insulated box cooled with ice packs. However, rapid temperature changes may cause condensation to form in the sample containers. Free water is problematic on semi-solid growth media because collected cells tend to be redistributed on the medium surface. For yeasts, this redistribution can result in the formation of satellite colonies. Conversely, free water can wash the propagules of filamentous taxa from the surface of the medium and deposit them on the plastic walls of the container. Semi-solid growth media used for culturable sampling should be free of standing water at the time of sampling.

The risk of condensation forming on the interior of growth medium containers is greatly reduced when sample plates or strips are over-packed in strongly insulative materials, such as expanded polystyrene foam or paper, prior to transport. The use of cooling packs during transport should be carefully considered as this practice risks the formation of interior condensation which may lead to spurious results. During incubation, sampling media should be kept inverted to prevent the flow of any free water that may form to the medium surface.

*James A. Scott, Richard C. Summerbell and Brett J. Green*

## Biochemical detection of fungal extrolites

Quantification of fungal bioaerosols using analytical biochemical techniques has recently complemented traditional viable and non-viable sampling methodologies. These bioanalytical approaches encompass various separations, analytical, immunological and biophysical methods that have enabled the isolation, detection, structural identification and quantitative determination of biologically inactive and active compounds. For fungi, the analysis of fungal antigens, allergens, secondary metabolites, macromolecules, ergosterol and microbiological volatile organic compounds (mVOCs) has been made possible using various immunological and biophysical techniques.

## Antigens

The development of monoclonal antibodies (mAbs) and polyclonal antibodies (pAbs) has been a significant analytical biochemical advancement. Antibodies are specialized immune proteins that are produced by B lymphocytes following immunization with native or reduced antigens. Specifically, pAbs are mixtures of immunoglobulin G (IgG) molecules secreted by different B cell lines against multiple epitopes of a specific antigen. In contrast to pAbs, mAbs are very specific Abs produced by fusing single antibody-forming B cells to tumor cells grown in culture. The resulting cell is called a hybridoma; it produces relatively large quantities of identical antibody molecules. Antibodies have been extensively used in combination with enzyme-linked immunosorbent assays (ELISA) for the qualitative or quantitative analysis of fungal allergens, antigens and mycotoxins in indoor, outdoor and occupational environments.

Important considerations when utilizing mAbs or pAbs in fungal exposure assessment studies include the extent of antibody cross reactivity (Trout *et al.* 2004, Schmechel *et al.* 2006) and metabolic spore activity (Green *et al.* 2003, Schmechel *et al.* 2008). This concern was recently addressed following studies that used an enzyme immunoassay-based assay for *A. alternata* (Barnes *et al.* 2006, Salo *et al.* 2005). In the National Survey of Lead and Allergens in Housing study, a pAb-based inhibition ELISA was used to measure the prevalence of *A. alternata* antigen in house dust samples from different locations throughout the United States (Salo *et al.* 2005). Salo *et al.* (2005) concluded that *A. alternata* antigens were detectable in 95-99% of American homes. However, the extent to which this pAb cross reacted with other fungi remained unclear. Recently, Schmechel *et al.* (2008) demonstrated that over 50% of all tested fungi (n=24) inhibited the binding of the same *A. alternata* pAb when tested by ELISA. These findings were confirmed by other immunoassays including inhibition immunoblotting and Halogen Immunoassay (HIA) (Schmechel *et al.* 2008). The strongest inhibition was observed in pleosporalean fungi, including *Alternaria, Drechslera, Exserohilum, Stemphylium* and *Ulocladium* species. Several of these fungi may be common in indoor environments (Verhoeff *et al.* 1994). They have been shown to extensively cross react with *A. alternata* and some medically important fungi (Hong *et al.* 2005, Bowyer *et al.* 2006, Sa´enz-de-Santamaria *et al.* 2006).

The interpretation of antigen quantification data obtained with mAbs or pAbs should include consideration of the extent of spore and hyphal metabolic activity. This has recently been demonstrated by Schmechel *et al.* (2008) using the HIA. In this study, no pAb reactivity to *Curvularia lunata* ungerminated spores was identified; however, following spore germination elevated concentrations of antibody binding was recorded. Similar increases in antigen concentrations have been reported following spore germination for Asp f 1 (Sporik *et al.* 1993), Alt a 1 (Mitakakis *et al.* 2001) and other fungal aeroallergens (Green *et al.* 2003). Although the rate of germination varies according to the fungus involved as well as the incubation time, buffer type and presence of other fungi (Green *et al.* 2006), it is recommended to limit ELISA incubation times to avoid analytical variability associated with spore germination.

Continuing development of species specific antibodies for an extending range of common indoor fungi will allow comprehensive assessment of fungal exposure in the future. Until then, investigators undertaking fungal exposure assessment studies should thoroughly characterize the specificity of the antibody used for the quantification of airborne fungi, and they should limit incubation times. These actions will eliminate analytical bias and ambiguities in the interpretation of epidemiological results.

## Ergosterol

Ergosterol is a sterol contained within fungal cell membranes. The presence of ergosterol in fungi and some protists has made this constituent a useful biomarker of fungal biomass in environmental assessments. The determination of ergosterol content in environmental samples is biochemically complex and involves a series of extractions and hydrolysis steps. Ergosterol is quantified using either high-performance liquid chromatography (HPLC) or gas chromatography-tandem mass spectrometry (GC-MS). Recent studies have demonstrated the utility of detecting this biomarker to provide a proxy measure of fungal biomass in contaminated indoor environments (Park *et al.* 2008). However, these methods are generally expensive and require a large amount of infrastructure and time to process and analyze samples.

Accordingly, these associated limitations have prevented the widespread use of this biomarker to quantify fungal biomass.

## Beta-(1,3)-D-glucan

Glucan has been shown to be a useful measurement. An excellent review of the biochemical measurement of this airborne burden is given by Miller (2011).

## Molecular genetic methods

The identification of indoor fungi by means of diagnostic sequences has been a well established and commonly used technique since the late 1990s (Haugland *et al.* 2004). Though initially developed for identifying *in vitro* cultures, the techniques can readily be modified to analyse the fungal contents of filtered air samples, dust samples or contaminated indoor materials. In an early example, Haugland and Heckman (1998) introduced specific primers for the important indoor fungus *Stachybotrys chartarum*. This study was shortly followed by the first of a series of species-specific quantitative PCR (qPCR) studies for indoor fungi, based on use of the TaqManä fluorogenic probe system combined with the ABI Prismâ Model 7700 Sequence Detector (Haugland *et al.* 1999). In this study, *S. chartarum* was again the object of study. qPCR counts of *S. chartarum* conidia were found to be highly comparable to counts obtained with a haemocytometer. The method was further developed by Roe *et al.* (2001) for direct quantitative analysis of *S. chartarum* in household dust samples.

This Taqman-based qPCR methodology was extended over subsequent years into a broad ranging methodology encompassing many major indoor air fungal groups (Haugland *et al.* 2004) and a variety of applications. In conjunction, key studies considered how best to extract DNA for qPCR and related PCR-based analyses of indoor fungi (Williams *et al.* 2001; Haugland *et al.* 2002; Kabir *et al.* 2003). Meklin *et al.* (2004) employed qPCR to evaluate indoor dust from the presence of 82 mold species or species complexes, including members of the genera *Aspergillus, Cladosporium, Penicillium, Trichoderma* and *Ulocladium*, in addition to *Stachybotrys* and the closely related *Memnoniella*. Comparisons among techniques showed that culturing underestimated numbers of key *Aspergillus* species by 2 to 3 orders of magnitude. "Moldy homes" could be distinguished from putatively uncontaminated "reference homes" using qPCR-based quantification. More medically oriented environmental studies developed Taqman qPCR for *Candida* yeasts in water samples (Brinkman *et al.* 2003) and *Aspergillus fumigatus* conidia in filtered air samples (McDevitt *et al.* 2004). More recently, the accuracy of qPCR for *A. fumigatus* detection in hospitals has been very acutely tested using a comparison with green fluorescent protein (GFP)-expressing conidia of this species (McDevitt *et al.* 2005).

qPCR was used in conjunction with other modern techniques such as a quantitative protein translation assay for trichothecene toxicity in an evaluation of *Stachybotrys* from a house where a case of idiopathic pulmonary hemosiderosis had occurred (Vesper *et al.* 2000). However, haemocytometer counts of the relatively large and conspicuous *Stachybotrys* conidia were heavily relied on for quantitation in the data used in the subsequent analysis. A much more detailed, later qPCR analysis of homes where pulmonary hemosiderosis had occurred showed that *S. chartarum* was part of a group of species, also including *A. fumigatus* and several other *Aspergillus* species, that was significantly elevated in quantity in dust samples in affected homes (Vesper *et al.* 2004). Species abundant in affected homes tended to be hemolytic in *in vitro* testing, whereas the common species associated with reference homes were generally not hemolytic. Another significant application of the qPCR technique was to sensitively monitor *Aspergillus* contamination during hospital renovation (Morrison *et al.* 2004) and related infection control applications.

Relatively recent developments have included detailed studies of the fungal contents of dust from various sources, including studies optimizing qPCR to overcome chemical PCR inhibitors in dust (Keswani *et al.* 2005, Vesper *et al.* 2005). Multi-species qPCR has been applied to compare outdoor and indoor air (Meklin *et al.* 2007) and to analyse both fungi and bacteria in building materials such as chipboard, paper materials and insulation (Pietarinen *et al.* 2008). An elegant recent study has used propidium monoazide to inactivate DNA from dead cells prior to using qPCR to quantify viable conidia (Vesper *et al.* 2008). An online information page is available about the now widely used technology developed by R.A. Haugland, S.J. Vesper and other members of the US Environmental Protection Agency group (http://www.epa.gov/nerlcwww/moldtech.htm). Primers are given for 130 target species. For a more detailed account on molecular methods for bioaerosols characterization see Summerbell *et al.* (2011).

## References

Afshari A, Matson U and Ekberg LE (2005) Characterization of indoor sources of fine and ultrafine particles: a study conducted in a full-scale chamber. Indoor Air 15: 141-150.

American Conference of Governmental Industrial Hygienists (ACGIH) (2001) Air sampling instruments for evaluation of atmospheric contaminants, 9th edition. Cohen BS and Hering SV (eds.). ACGIH Press Cincinnati, OH, USA.

*James A. Scott, Richard C. Summerbell and Brett J. Green*

American Industrial Hygiene Association (AIHA) (2005) Field guide for the determination of biological contaminants in environmental samples, 2nd edition. Dillon HK, Heinsohn PA and Miller JD (eds.) AIHA Press, Fairfax, VA, USA, 284 pp.

Andersen AA (1958) New sampler for the collection, sizing and enumeration of viable bioaerosol sampling. J Bacteriol 76: 471-484.

Andersen JD and Cox CS (1967). Microbial survival. In: Gregory PH and Monteith JL (eds.) Airborne microbes, Cambridge University Press, Cambridge, UK, pp. 203-226.

Antens CJ, Oldenwening M, Wolse A, Gehring U, Smit HA, Aalberse RC, Kerkhof M, Gerritsen J, De Jongste JC and Brunekreef B (2006) Repeated measurements of mite and pet allergen levels in house dust over a time period of 8 years. Clin Exp Allergy 36: 1525-1531.

Arbes S, Sever M, Mehta J, Collette N, Thomas B and Zeldin D (2005) Exposure to indoor allergens in day-care facilities: results from two North Carolina counties. JACI 116: 133-139.

Ariatti A and Comtois P (1993). Louis Pasteur, the first experimental aerobiologist. Aerobiology 9: 5-14.

Barnes C, Portnoy J, Sever M, Arbes S, Vaughn B and Zeldin DC (2006) Comparison of enzyme immunoassay-based assays for environmental *Alternaria alternata*. Ann Allergy Asthma Immunol 97: 350-356.

Blomquist G, Stroem G and Stroemquist L (1984) Sampling of high concentrations of airborne fungi. Scand J Work Environ Health 10: 109-113.

Bowyer P, Fraczek M and Denning DW (2006) Comparative genomics of fungal allergens and epitopes show widespread distribution of closely related allergen and epitope orthologues. BMC Genomics 7: 251.

Brinkman NE, Haugland RA, Wymer LJ, Byappanahalli M, Whitman RL and Vesper SJ (2003) Evaluation of a rapid, quantitative real-time PCR method for cellular enumeration of pathogenic *Candida* species in water. Appl Environ Microbiol 69: 1775-1782.

Burge HA (1995) Bioaerosols in the residential environment. In: Cox CS and Wathes CM (eds.) Bioaerosols handbook. Lewis Publishers, Boca Raton, FL, USA, pp. 579-597.

Buttner M and Stetzenbach L (1993) Monitoring airborne fungal spores in an experimental indoor environment to evaluate sampling methods and the effects of human activity on air sampling. Appl Environ Microbiol 59: 219-226.

Cole EC, Dulaney PD, Leese KE, Hall RM, Foarde KK, Franke DL, Myers EM and Berry MA (1996) Biopollutant sampling and analysis of indoor surface dusts: characterization of potential sources and sinks. In: Tichenor, BA (ed.) Characterizing sources of indoor air pollution and related sink effects, ASTM STP 1287. American Society for Testing and Materials, West Conshohocken, PA, USA, pp. 153-165.

Collett CW, Nathanson T, Scott JA, Baer K and Waddington J (1999) The impact of H.V.A.C system cleaning on levels of surface dust and viable fungi in ductwork. Proceedings of Indoor Air '99, Edinburgh 3, pp. 56-61.

Comtois P (1997) Pierre Miquel: the first professional aerobiologist. Aerobiologia 13: 75-82.

Cox CS and Wathes CM (eds.) (1995) Bioaerosols handbook. Lewis Publishers, Boca Raton, FL, USA.

Cunningham DD (1873) Microscopic examinations of air. Government Printer, Calcutta, India, 58 pp.

Douwes J, Thorne P, Pearce N and Heederik D (2003) Bioaerosol health effects and exposure assessment: progress and prospects. Ann Occup Hyg 47: 187-200.

Elliott L, Arbes SJ, Harvey ES, Lee RC, Salo PM, Cohn RD, London SJ and Zeldin DC (2007) Dust weight and asthma prevalence in the national survey of lead and allergens in housing (NSLAH). Environ Health Perspect 115: 215-220.

Federal-Provincial Committee on Environmental and Occupational Health (Canada) (1995) Fungal contamination in public buildings: a guide to recognition and management. Environmental Health Directorate, Ottawa, Ontario, Canada.

Ferro AR, Kopperud RJ and Hildemann LM (2004a) Elevated personal exposure to particulate matter from human activities in a residence. J Expo Anal Environ Epidemiol 14 (Suppl. 1): S34-S40.

Ferro AR, Kopperud RJ and Hildemann LM (2004b) Source strengths for indoor human activities that resuspend particulate matter. Environ Sci Technol 38: 1759-1764.

Flannigan B, Samson RA and Miller JD (eds.) (2001) Microorganisms in home and indoor work environments: diversity, health impacts, investigation and control. Taylor Francis, London, UK.

Foto M, Plett J, Berghout J and Miller JD (2004) Modification of the Limulus amebocyte lysate assay for the analysis of glucan in indoor environments. Anal Bioanal Chem 379: 156-162.

Green BJ, Mitakakis TZ and Tovey ER (2003) Allergen detection from 11 fungal species before and after germination. J Allergy Clin Immunol 111: 285-289.

Green BJ, Tovey ER, Sercombe JK, Blachere FM, Beezhold DH and Schmechel D (2006) Airborne fungal fragments and allergenicity. Med Mycol 44: S245-S255.

Gregory PH (1971) The Leeuwenhoek Lecture, 1970: airborne microbes: their significance and distribution. Proc Roy Soc Lond B 177: 469-483.

Gupta AK and Summerbell RC (2000) Tinea capitis. Med Mycol 38: 255-287.

Gupta AK, Kohli Y, Summerbell RC and Faergemann J (2001) Quantitative culture of *Malassezia* species from different body sites of individuals with or without dermatoses. Med Mycol 39: 243-251.

Haugland RA and Heckman JL (1998) Identification of putative sequence specific PCR primers for detection of the toxigenic fungal species *Stachybotrys chartarum*. Molec Cell Probes 12: 387-396.

Haugland RA, Brinkman NE and Vesper SJ (2002) Evaluation of rapid DNA extraction methods for the quantitative detection of fungal cells using real time PCR analysis. J Microbiol Methods 50: 319-323.

Haugland RA, Varma M, Wymer LJ and Vesper SJ (2004) Quantitative PCR of selected *Aspergillus, Penicillium* and *Paecilomyces* species. Syst Appl Microbiol 27: 198-210.

Haugland RA, Vesper SJ and Wymer LJ (1999) Quantitative measurement of *Stachybotrys chartarum* conidia using real time detection of PCR products with the TaqManÔ fluorogenic probe system. Molec Cell Probes 13: 329-340.

Hay DB, Hart BJ and Douglas AE (1993) Effects of the fungus *Aspergillus penicillioides* on the house-dust mite *Dermatophagoides pteronyssinus* – an experimental reevaluation. Med Vet Entomol 7: 271-274.

Heinrich J, Hölscher HB, Douw JB, Richter K, Koch A, Bischof W, Fahlbusch B, Kinne RW and Wichmann H-E (2003) Reproducibility of allergen, endotoxin and fungi measurements in the indoor environment. J Expo Anal Environ Epidemiol 13: 152-160.

Hinds WC (1999) Aerosol technology: properties, behavior, and measurement of airborne particles, 2nd edition. Wiley, New York, NY, USA.

Hinds WC (2005) Aerosol properties. In: Ruzer LS and Harley NH (eds.) Aerosols handbook: measurement, dosimetry, and health effects. CRC Press, Boca Raton, FL, USA, pp. 19-33.

Hirst JM (1953) Changes in atmospheric spore content: diurnal periodicity and the effects of weather. Trans Brit Mycol Soc 36: 375-395.

Hong SG, Cramer RA, Lawrence CB and Pryor BM (2005) Alt a 1 allergen homologs from *Alternaria* and related taxa: analysis of phylogenetic content and secondary structure. Fungal Genet Biol 42: 119-129.

Kabir N, Rajendran N, Amemiya T and Itoh K (2003) Quantitative measurement of fungal DNA extracted by three different methods using real-time polymerase chain reaction. J Biosci Bioeng 96: 337-343.

Keswani J, Kashon ML and Chen BT (2005) Evaluation of interference to conventional and real-time PCR for detection and quantification of fungi in dust. J Environ Monit 7: 311-318.

Lange JH, Kuhn BD, Thomulka KW and Sites SLM (2000) A study of area and personal airborne asbestos samples during abatement in a crawl space. Indoor Built Environ 9: 192-200.

Lee KS, Bartlett KH, Brauer M, Stephens GM, Black WA and Teschke K (2004) A field comparison of four samplers for enumerating fungal aerosols I. Sampling characteristics. Indoor Air 14: 360-366.

Macher JM (ed.) (1999) Bioaerosols: assessment and control. ACGIH Press, Cincinnati, OH, USA.

Macher JM and Burge HA (2001) Sampling biological aerosols. In: Air sampling instruments for evaluation of atmospheric contaminants, 9th edition, pp. 661-701.

MacKenzie DWR (1961) The extra-human occurrence of *Trichophyton tonsurans* var. *sulfureum* in a residential school. Sabouraudia 1: 58-64.

Maddox RL (1870) On an apparatus for collecting atmospheric particles. Monthly Microscopy J 3: 286-290.

McDevitt JJ, Lees PSJ, Merz WG and Schwab KJ (2004) Development of a method to detect and quantify *Aspergillus fumigatus* conidia by quantitative PCR for environmental air samples. Mycopathologia 158: 325-335.

McDevitt JJ, Lees PSJ, Merz WG and Schwab KJ (2005) Use of green fluorescent protein-expressing *Aspergillus fumigatus* conidia to validate quantitative PCR analysis of air samples collected on filters. J Occup Environ Hyg 2: 633-640.

Meklin T, Haugland RA, Reponen T, Varma M, Lummus Z, Bernstein D, Wymer LJ and Vesper SJ ( 2004) Quantitative PCR analysis of house dust can reveal abnormal mold conditions. J Environ Monit 6: 615-620.

Meklin T, Reponen T, McKinstry C, Cho SH, Grinshpun SA, Nevalainen A, Vepsalainen A, Haugland RA, Lemasters G and Vesper SJ (2007) Comparison of mold concentrations quantified by MSQPCR in indoor and outdoor air sampled simultaneously. Sci Total Environ 382: 130-134.

Meredith DS (1973) Significance of spore release and dispersal mechanisms in plant disease epidemiology. Ann Rev Phytopathol 11: 313-342.

Miller JD (2001) Mycological investigations of indoor environments. In: Microorganisms in home and work environments. Flannigan B, Samson RA and Miller JD (eds). CRC Press, Boca Raton, FL, USA, pp. 231-246.

Miquel P (1883) Les organismes vivants dans l'atmosphère. Gauthier-Villars, Paris, France.

Mitakakis TZ, Barnes C and Tovey ER (2001) Spore germination increases allergen release from *Alternaria*. J Allergy Clin Immunol 107: 388-390.

Miller JD (2011) Mycological investigations of indoor environments. In: Flannigan B, Samson RA and Miller JD (eds.) microorganisms in home and indoor work environments: diversity, health impacts, investigations and control. Taylor and Francis, Abingdon, UK.

Morgenstern V, Bischof W, Koch A and Heinrich J (2006) Measurements of endotoxin on ambient loaded PM filters after long-term storage. Sci Tot Environ 370: 574-579.

Morris KJ (1995) Modern microscopic methods of bioaerosol analysis. In: Cox C and Wathes CM (eds.) Bioaerosol handbook. CRC Press, Boca Raton, FL, USA, pp. 285-316.

Morrison CJ, Yang C, Lin KT, Haugland RA, Neely AN and Vesper SJ (2004) Monitoring *Aspergillus* species by quantitative PCR during construction of a multi-storey hospital building. J Hosp Infect 57: 85-87.

NIOSH (1994) Method 7400, Issue 2: Asbestos and other fibers by PCM. NIOSH Manual of Analytical Methods (NMAM), 4th edition. US Department of Health and Human Services, Cincinnati, OH, USA.

Niven RM, Fishwick D, Pickering CAC, Fletcher AM, Warburton CJ and Crank P (1992) A study of the performance and comparability of the sampling response to cotton dust of work area and personal sampling techniques. Ann Occ Hygiene 36: 349-362.

Park JH, Cox-Ganser JM, Kreiss K, White SK and Rao CY (2008) Hydrophilic fungi and ergosterol associated with respiratory illness in a water-damaged building. Environ Health Perspec 116: 45-50.

Pasteur L (1857) Mémoire sur la fermentation appelée lactique. C R Acad Sci Hebd Seances Acad Sci 45: 913-916.

Pietarinen VM, Rintala H, Hyvärinen A, Lignell U, Kärkkäinen P and Nevalainen A (2008) Quantitative PCR analysis of fungi and bacteria in building materials and comparison to culture-based analysis. J Environ Monit 10: 655-663.

Pitkäranta M, Meklin T, Hyvärinen A, Paulin L, Auvinen P, Nevalainen A and Rintala H (2008) Analysis of fungal flora in indoor dust by ribosomal DNA sequence analysis, quantitative PCR, and culture. Appl Env Microbiol 74: 233-244.

Portnoy JM, Barnes CS and Kennedy K (2004) Sampling for indoor fungi. JACI 113: 189-198.

Ren P, Jankun TM, Belanger K, Bracken MB and Leadereret BP (2001) The relation between fungal propagules in indoor air and home characteristics. Allergy 56: 419-424.

Reponen T, Willeke K, Ulevicius V, Reponen A and Grinshpun SA (1996) Effect of relative humidity on the aerodynamic diameter and respiratory deposition of fungal spores. Atmos Environ 30: 3967-3974.

Roe J, Haugland RA, Vesper SJ and Wymer LJ (2001) Quantification of *Stachybotrys chartarum* conidia in indoor dust using real time, fluorescent probe-based detection of PCR products. J Expo Anal Environ Epidemiol 11: 12-20.

Sa´enz-de-Santamaria M, Postigo I, Gutierrez-Rodriguez A, Cardona G, Guisantes JA, Asturias J and Martínez J (2006) The major allergen of *Alternaria alternata* (Alt a 1) is expressed in other members of the *Pleosporaceae* family. Mycoses 49: 91-95.

Salo PM, Arbes S, Sever M, Jaramillo R, Cohn R, London S and Zeldin D (2006) Exposure to *Alternaria alternata* in US homes is associated with asthma symptoms. JACI 118: 892-898.

Salo PM, Yin M, Arbes SJ, Cohn RD, Sever M, Muilenberg M, Burge HA, London SJ and Zeldin DC (2005) Dustborne *Alternaria alternata* antigens in US homes: results from the National Survey of Lead and Allergens in Housing. J Allergy Clin Immunol 116: 623-629.

Samson RA, Houbraken J, Thrane U, Frisvad JC and Andersen B (2010) Food and indoor fungi. CBS laboratory manual series 2. Centraalbureau voor Schimmelcultures, Utrecht, the Netherlands.

Sayer W, Shean DB and Ghosseiri J (1969) Estimation of airborne fungal flora by the Andersen sampler versus the gravity settling culture plate. J Allergy 44: 214-227.

Schmechel D, Green BJ, Blachere FM, Janotka E and Beezhold DH (2008) Analytical bias of cross-reactive polyclonal antibodies for environmental immunoassays of *Alternaria alternata*. J Allergy Clin Immunol 121: 763-768.

Schmechel D, Simpson JP, Beezhold D and Lewis DM (2006) The development of species specific immunodiagnostics for *Stachybotrys chartarum*: the role of cross-reactivity. J Immunol Methods 309: 150-159.

Soloman WR (1975) Assessing fungus prevalence in domestic interiors. J Allergy Clin Immunol 56: 235-242.

Sporik RS, Arruda LK, Woodfolk J, Chapman MD, Platts-Mills TAE (1993) Environmental exposure to *Aspergillus fumigatus* allergen (Asp f I). Clin Exp Allergy 23: 326-331.

Summerbell RC (2000) Form and function in the evolution of dermatophytes. In: Kushwaha RKS and Guarro J (eds.) Biology of dermatophytes and other keratinophilic fungi. Revista Iberoamericana de Micología, Bilbao, Spain, pp. 30-43.

Summerbell RC, Green BJ, Corr D and Scott JA (2011). Molecular methods for bioaerosols characterization. In Flannigan B, Samson RA and Miller JD (eds.) Microorganisms in home and indoor work environments: diversity, health impacts, investigations and control. Taylor and Francis, Abingdon, UK, in press.

Thermo Electron Corporation (2003) Series 10-800: single stage viable sampler instruction manual, p/n 100074-00. Thermo Electron Corporation, Franklin, MA, USA.

Thorn RG, Scott JA and Lachance MA (2007) Methods for studying terrestrial fungal ecology and diversity. In: Reddy CA (ed.) Methods for general and molecular microbiology, 3rd edition. ASM Press, Washington, DC, USA, pp. 929-951.

Trout DB, Seltzer JM, Page EH, Biagini RE, Schmechel D, Lewis DM and Boudreau AY (2004) Clinical use of immunoassays in assessing exposure to fungi and potential health effects related to fungal exposure. Ann Allergy Asthma Immunol 92: 483-491.

Van Asselt L (1999) Interactions between domestic mites and fungi. Indoor Built Environ 8: 216-220.

Van Bronswijk JEMH and Sinha RN (1973) Role of fungi in the survival of Dermatophagoides (Acarina: Pyroglyphidae) in house dust environment. Environ Entomol 2: 142-145.

Van de Lustgraaf B (1978) Ecological relationships between xerophilic fungi and house-dust mites (Acari: Pyroglyphidae). Oecologia (Berl) 33: 351-359.

Van Leeuwen WS (1924) Bronchial asthma in relation to climate. Proc R Soc Med (Ther Pharmacol Sect) 17: 19-26.

Verhoeff AP, Van Wijnen JH, Van Reenen-Hoekstra ES, Samson RA, Van Strien RT and Brunekreef B (1994) Fungal propagules in house dust. II. Relation with residential characteristics and respiratory symptoms. Allergy 49: 540-547.

Vesper SJ, Dearborn DG, Yike I, Allan T, Sobolewski J, Hinkley SF, Jarvis BB and Haugland RA (2000) Evaluation of *Stachybotrys chartarum* in the house of an infant with pulmonary hemorrhage: quantitative assessment before, during and after remediation. J Urban Health 77: 68-85.

Vesper SJ, McKinstry C, Hartmann C, Neace M, Yoder S and Vesper A (2008) Quantifying fungal viability in air and water samples using quantitative PCR after treatment with propidium monoazide (PMA). J Microbiol Methods 72: 180-184.

Vesper SJ, Varma M, Wymer LJ, Dearborn DG, Sobolewski J and Haugland RA (2004) Quantitative PCR analysis of fungi in dust from homes of infants who developed idiopathic pulmonary hemorrhaging. J Occup Environ Med 46: 596-601.

Vesper SJ, Wymer LJ, Meklin T, Varma M, Stott R, Richardson M and Haugland RA (2005) Comparison of populations of mould species in homes in the UK and USA using mould-specific quantitative PCR. Lett Appl Microbiol 41: 367-373.

Wells W (1934) On air-borne infection. II. Droplets and droplet nuclei. Am J Hyg 20: 611-618.

Williams RH, Ward E and McCartney HA (2001) Methods for integrated air sampling and DNA analysis for detection of airborne fungal spores. Appl Environ Microbiol 67: 2453-2459.

Womiloju TO, Miller JD, Mayer PM and Brook JR (2003) Methods to determine the biological composition of particulate matter collected from outdoor air. Atmos Environ 37: 4335-4344.

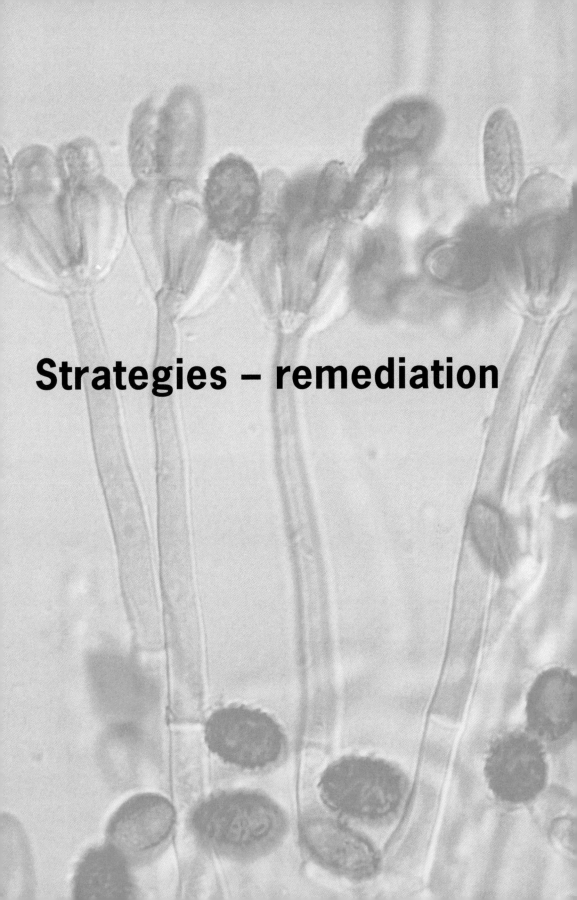

# Strategies – remediation

# 14 Mold remediation in North American buildings

*Philip R. Morey*
*ENVIRON International Corporation, Gettysburg, PA, USA*

## Introduction

The association of microbial growth within buildings with adverse health effects was recognized in Europe and North America more than 25 years ago (Arnow *et al.* 1978, Morey *et al.* 1984, Samson 1985, Hunter *et al.* 1988). Many of these early studies focused attention on mechanical ventilation and air conditioning systems as a source of the moisture and microbial problems associated with adverse health effects. Emphasis in most early studies was placed on identifying the types and concentrations of microorganisms including molds found in the air and on surfaces of water damaged building materials. It was also shown that disturbance of moldy materials resulted in elevated counts of airborne spores (Hunter *et al.* 1988).

In the early 1990s, in North America, investigators, especially in Florida and New York City, began to quantify mold growth in buildings based on the extent (e.g. square meters) of visible colonization on interior surfaces (Jarvis and Morey 2001). A panel discussion held on May 7, 1993 and sponsored by the New York City Department of Health and the New York City Human Resources Administration recommended a semi-quantitative approach for assessing and remediating moldy buildings based on the extent (e.g. $<0.2m^2$; 0.2 to $3m^2$; $>3m^2$ in one area or one zone of a building) of visible colonization on interior surfaces (NYC 1993). Coincident with the development of the New York City Guidelines, investigations in two new courthouses in Martin (NIOSH 1993) and Polk Counties, Florida (Morey 1992, Jarvis and Morey 2001) showed that extensive mold growth had occurred on biodegradable construction materials especially paper faced wallboard (Figures 14.1 and 14.2). In both courthouses, the restoration costs (approximately \$45 and \$20 million dollars for Polk and Martin County Courthouses, respectively), which included mold remediation plus the temporary relocation of occupants, exceeded the initial capital construction costs of the buildings. The mold remediation process used in both courthouses and in other buildings in New York City at that time involved physical removal of colonized building materials and dust suppression by methods similar to those used in asbestos abatement. The publication of the 1993 New York City guidelines began the process of development of specific protocols for mold remediation, which are now evolving as distinct from asbestos abatement procedures.

*Figure 14.1. Large courthouse undergoing building restoration which included removal of extensive visible mold growth primarily on and within exterior and nearby demising wall systems.*

It should be recognized that the concept of removal of visible mold from interior surfaces was described more than 3 millennia ago in Leviticus 13 and 14 of the Hebrew Bible (Heller *et al.* 2003). According to the interpretation of Leviticus 14: 39-40 (Heller *et al.* 2003: 589), mold on walls in blighted houses was physically removed and discarded into an unclean place outside of the city. Mold remediation today in North America is a more complex process than in ancient times due in part to the architectural complexity of modern buildings and to the increasing use of construction and finishing materials that are moisture intolerant or sensitive to biodeterioration.

*Figure 14.2. Concealed mold growth caused by condensation occurred within many wall cavities in the building shown in Figure 14.1. Unconditioned warm-humid outdoor air entered this wall cavity and condensation occurred on the paper facer of the relatively cool wallboard (occupied side of wallboard is air-conditioned). Mold growth (primarily* Aspergillus versicolor*) occurred on these chronically moist surfaces. About 20,000 m² of moldy wallboard was removed from this building.*

*Philip R. Morey*

This chapter reviews representative mold remediation guidelines that have been published in the USA and Canada during the past decade. The needs for containment and dust suppression during mold remediation as well as appropriate quality assurance to document the effectiveness of clean up are discussed in the context of North American guidelines.

## Examples of mold remediation strategies

General principles of mold remediation include the following: (a) physical removal of mold colonized porous finishes and construction materials and physical removal of mold colonization from surfaces of non-porous or smooth materials, (b) prevention of cross-contamination of occupied or clean areas by dusts generated by mold remediation elsewhere in the building, and (c) fixing the water or moisture problem that caused mold growth on interior surfaces. Factors that influence mold remediation include the location of mold growth in building construction, the amount ($m^2$) of visible mold on finishes and construction materials, and air pathways that may be present to transport microbial contaminants to building occupants. General approaches to mold remediation are illustrated in the following six case studies.

### Case study 1. Biodeterioration of fiberboard in the outer portion of the building envelope

Mushroom-like growths were visually apparent on the painted surface of oriented strand board (OSB) siding in a new apartment building shortly after occupancy (See Figures 14.3-14.5). Destructive inspection of the siding revealed the presence of massive mold growth on the backside of OSB (Figure 14.4) and considerable visible mold growth on underlying exterior sheathing (plywood and particle board) (Figure 14.5). Mold growth had not occurred on the paper-faced wallboard on the occupied side of the building envelope. The main cause of biodeterioration in the exterior portion of the building envelope was water penetration.

The mold remediation strategy in case study 1 involved careful removal of OSB and other moldy porous materials in the exterior portion of the envelope by workers on scaffolds. Shrouds were used to minimize the dissemination of dusts into the vicinity of nearby buildings. Workers carrying out the demolition of the exterior portion of the envelope were provided with appropriate personal protective equipment (e.g. gloves, eye protection, respirators with P-100 filters, and disposable coveralls). Occupied apartments were positively pressurized (air filtration device [AFD] with high efficiency particulate air [HEPA] filter) so as to keep fugitive demolition dusts

*Figure 14.3. Macroscopic mushroom-like growth (arrow) protrudes from moist-oriented strand board in new construction.*

*Figure 14.4. Back surface of oriented strand board showing mushroom-like growth seen in Figure 14.3.*

**Fundamentals of mold growth in indoor environments**       *387*

*Figure 14.5. Mold growth on exterior sheathing beneath location when moldy oriented strand board had been removed. Note mushroom-like growth from oriented strand board locations above window (arrows).*

from entering interior spaces through loose construction. Occupants were required to keep windows and balcony doors closed during remediation activities.

### Case study 2. Chronic water leaks around window

Extensive, visible mold growth was present on the wall cavity surface of paper-faced wallboard and the Kraft paper face of batt insulation in a newly constructed school building (Figures 14.6 and 14.7). This mold growth was caused by chronic water leaks in the building envelope especially around windows. Mold remediation involved removal of the paper faced wallboard along the envelope wall contiguous with occupied spaces. In order to protect adjacent clean and occupied spaces, a depressurized (-5 Pascals) containment extending about 2 meters into the interior was built so that dusts and debris could be confined to the demolition work area. Air containing microbial contaminants from the containment work area was exhausted through HEPA filters to the outdoor atmosphere. Demolition debris including moldy wallboard and insulation was bagged and preferably discarded by passage to an outside dumpster via a chute to the outdoors. Some demolition debris was transported in bags through the interior of the building for disposal outdoors. The outside surfaces of the bags were cleaned so as to avoid cross contamination of interior spaces.

*Figure 14.6. Extensive mold growth occurred on the wall cavity surface of paper faced gypsum wallboard and on the Kraft paper face of a bad insulation. Note the location where electrical socket was located (arrow).*

*Figure 14.7. Enlargement of a portion of Figure 14.6 showing mold growth on the wall cavity surface of the wallboard. Rust (arrows) from the metal stud that had been in physical contact with the wallboard provides evidence (in addition to the mold growth) that the envelope wall was chronically wet.*

**Fundamentals of mold growth in indoor environments**

### Case study 3. Plumbing leaks in wall cavities

Water leaks from plumbing, especially in stacked bathrooms in high-rise buildings, can cause mold growth on nearby construction materials, especially paper faced wallboard (Figures 14.8 and 14.9). In Case study 3, mold growth was vertically extensive in risers and shafts because the water damage occurred on many floors below the actual source of the leakage.

Remediation of extensive mold growth associated with water leaks from interior plumbing was similar to that described in Case study 2 with some added precautions. One additional precaution involved making certain (by negative pressurization and sealing) that dusts generated during removal of moldy materials were not dispersed inadvertently into other zones of the building by passage through risers and shafts, which also contain plumbing lines. Since mold remediation was occurring in interior or core zones in the building, care was taken that air being exhausted from demolition work areas was HEPA filtered and exited the envelope through temporary openings such as at windows. In order to avoid a repetition of the mold problem, it was necessary that the retrofitted plumbing be tested for leakage before piping was enclosed in newly constructed walls.

*Figure 14.8. Mold growth (arrows) and water damage on a wallboard in close proximity to a pipe leak.*

*Figure 14.9. Mold growth on a paper faced wallboard in a wall cavity. The plumbing leak is behind the moldy wallboard.* Stachybotrys *and* Chaetomium *are the prominent isolates from the moldy wallboard.*

### Case study 4. High indoor relative humidity, i.e. consistently in the 70-85% range

Mold growth was observed on furniture, upholstery, and textiles in a newly constructed building located in a warm humid climate. An average of about 3 to 5m$^2$ of visible mold growth was detected in each room especially on fiberboard surfaces in cabinet drawers (Figure 14.10) and on the under surface of tables. A lesser amount of visible mold was found on upholstery, linens and drapes. The relative humidity in rooms at the time when mold growth was detected was consistently in the 70 to 85% range. All porous materials present in these rooms were discarded with a few exceptions for valuable items.

### Case study 5. Mold growth on airstream HVAC system surfaces

Growth of mold within a HVAC system (Figure 14.11) can be problematic because the air being transported within the mechanical system is designed to be delivered directly to the breathing zone of room occupants. Care must be exercised to

*Figure 14.10. Mold growing on the inside of fiberboard surfaces of a dresser drawer. Prominent species growing on the fiberboard included* Aspergillus penicillioides *and* Eurotium herbariorum.

decommission the HVAC system when mold remediation of airstream surfaces occurs. Mold that is present on smooth or non-porous surfaces can be removed simply by wiping with water and detergent. Porous surfaces that are moldy (Figure 14.11) in HVAC equipment are best remediated by discarding the moldy material. In some cases, cleaning the moldy surface with HEPA vacuum, followed by treatment

*Figure 14.11. The porous insulation on the inside (airstream) surface in a large air handling (conditioning) unit is visually moldy (arrows). The insulation was removed from the air handling unit.*

of the surface with an fungistatic encapsulant has been used as an alternative to discarding the moldy porous material (Yang and Ellringer 2004).

### Case study 6. Water disasters, waterlogging of porous materials

Extensive moldy growth occurs throughout buildings subject to water disasters such as from hurricanes, floods, and firefighting activities. In the example in Figures 14.12 and 14.13, most porous materials were chronically waterlogged (note plants growing out of the carpet) and moldy. Porous materials, as in this example, are discarded. Containments during remediation are generally unnecessary, unless the clean-up area is contiguous to clean or occupied zones unaffected by the water disaster. Personal protective equipment is required for workers involved in clean-up activities. Drying biodegradable materials quickly after a water disaster is required to minimize mold growth. This may not always be possible as electrical power to operate dehumidification equipment may be unavailable in the building after the water disaster and the availability of portable generators may be limited. In buildings, such as in Figures 14.12 and 14.13, a thorough HEPA vacuum cleaning of salvageable non-porous and semi-porous materials (metal studs, concrete, glass, etc.) is required prior to considering the work area to be a normal construction zone.

Three of the previously described case studies illustrate the occurrence of concealed mold growth either in the building envelope (Case studies 1 and 2; associated with water leakage into the envelope) or in interior wall systems (Case study 3; associated

*Figure 14.12. View of a guestroom in a hotel, severely damaged several years earlier by a hurricane. Note the presence of plants growing on the floor.*

*Figure 14.13. Small seedlings (arrows) growing on the carpet in another room. The growth of plants indicates a high water activity and biodeterioration of wooden flooring beneath the carpet.*

with leakage from plumbing). Forensic investigations involving destructive inspection of construction materials were required to determine the extent of mold growth and water damage in these three case study buildings.

A very common kind of moisture/mold problem (not illustrated in the six case studies) occurs in air-conditioned buildings, especially in the Southeast USA when warm humid outdoor air infiltrates the envelope resulting in condensation and mold growth usually on wall coverings and other surfaces along the interior surface of the external wall (Morey and Cornwell, 2006). This kind of problem is exacerbated by depressurization of the indoor air and use of low perm wall coverings that trap water molecules in the wall system. The following general actions are recommended to prevent condensation and mold growth in air-conditioned buildings in warm humid climates or seasons in USA buildings (Morey 1995, see ASHRAE 2005a,b, Lstiburek and Carmody 1994 for detailed discussions):

- Minimize infiltration of warm humid air by installation of air and vapor retarders in the exterior portion of the envelope.
- Maintain a slight positive pressurization of the indoor air relative to the ambient air.
- Use water vapor permeable materials (permeance greater than 5 perms) on interior surfaces.

- Avoid cooling the indoor air below the mean monthly outdoor dew point temperature.

In North American buildings, In cold climate areas, condensation and mold growth can occur on interior surfaces of envelope walls where insulation is deficient (thermal bridges). Some of the recommendations to minimize condensation in buildings located in cold climate zones include (Morey 1995, see ASHRAE 2005a;b, Lstiburek and Carmody 1994 for detailed discussions):

- Install the vapor retarder on the side of the insulation facing the occupied space.
- Prevent leakage of moist indoor air into the building envelope.

## New York City guidelines

In the United States, the three editions of the New York City guidelines on assessment and remediation of visible mold growth in buildings have had a significant impact on mold remediation technology. These guidelines are briefly reviewed below.

The 1993 Guidelines were written because at that time there were significant gaps with regard to evaluating and remediating mold growth in buildings. The 1993 Guidelines focused on *Stachybotrys atra* (*chartarum*) because of widely publicized growth of this mold species in wet New York City buildings.

Visual inspection involving the categorization of the extent (square meters) of mold growth on interior surfaces was recognized as the most important component of the assessment process. Air sampling for *Stachybotrys* was not viewed as an essential component of the assessment process. Remediation strategies could according to the 1993 document be based primarily on the amount of mold found during the inspection process. Four types (levels) of remediation based on the extent and location of visible mold growth were recognized namely: *Level I*, small (<0.1 m$^2$) areas of growth on interior surfaces; *Level II*, large (0.1-3.0 m$^2$) isolated interior surfaces; *Level III*, large scale (>3 m$^2$) interior surfaces; and *Level IV*, for visible mold growth in HVAC systems. The general philosophy of the 1993 document recommended increasingly complex actions with regard to dust suppression and use of personal protective equipment as the scale of mold growth on interior surfaces increased. For example, complete isolation of the remediation work area (critical barriers, depressurization via AFDs with HEPA filters) was recommended when the extent of visible mold colonization on interior surfaces attained the Level III status. It was mentioned that air monitoring could be used to determine if *Stachybotrys* spores were escaping into building zones exterior to the mold remediation work area.

The second edition of the New York City Guidelines released in 2000 clarified some of the general principles of the 1993 guidelines and provided additional technical details on appropriate mold remediation activities. The 2000 edition of the NYC Guidelines recognized that all molds (not just *Stachybotrys*) that grow on interior surfaces should be remediated. It was noted that during clean-up activities, remediation personnel may be exposed to levels of microbial particulate sufficient to be associated with onset of organic dust toxicity syndrome and hypersensitivity pneumonitis. The concept was reaffirmed that visual inspection of the building and its HVAC system was the most important step in assessing a mold problem. Additionally, bulk and surface sampling were recognized as not being required to initiate the mold remediation process.

For purposes of remediation, the 2000 edition of the NYC Guidelines categorized the extent of visible mold found on interior surfaces as follows: *Level I*, small areas <1.0 m$^2$; *Level II*, mid-sized isolated areas 1-3 m$^2$; *Level III*, large isolated areas, 3-10m$^2$; *Level IV*, extensive contamination, >10 m$^2$; *Level V*, small scale (<1 m$^2$) HVAC system contamination of air stream surfaces and large scale (>1 m$^2$) HVAC system contamination of air stream surfaces. Professional judgment was recognized as being important in mold remediation (e.g. more strenuous dust control and containment measures are needed if clean-up activities generated excessive amounts of aerosolized dust).

The 2000 edition of the NYC Guidelines provides some limited advice on the clean-up of moldy non-porous (e.g. sheet metal), moldy semi-porous (e.g. concrete), and moldy porous (e.g. ceiling tiles) materials. In general, porous materials with more than a small area of visible mold growth should be discarded. Visually, moldy non-porous and semi-porous materials that are structurally sound can be cleaned and reused (NYC 2000). The 2008 edition of the IICRC Standard and Reference Guide for Professional Mold Remediation (IICRC 2008: 52) provides practitioners with definitions of porous, semi-porous and non-porous materials based on ability to absorb and adsorb moisture and potential for biodeterioration. Thus, non-porous materials do not readily take up moisture and are resistant to biodeterioration (e.g. glass, hard plastic, finished wood products). Semi-porous materials (e.g. concrete) take up moisture slowly, but can support mold growth if organically based (e.g. unfinished wood products). Porous materials which are organically based can take up moisture readily and are highly susceptible to biodeterioration (e.g. paper-faced wallboard, textiles, and clothing).

The use of antimicrobial agents such as ozone and chlorine dioxide in place of physical removal of visibly moldy materials was discouraged. Attention to appropriate hazard communication was recommended for buildings in which large-scale mold remediation was occurring. For example, informing occupants about mold clean-up activities and the schedule for completion including actions to fix the moisture problem that led to mold growth were considered necessary steps in the remediation process.

The third edition of the NYC Guidelines (2008) reaffirmed general principles in earlier guidelines namely (a) visual inspection remained the most important step in the mold assessment process, (b) air monitoring for mold spores was not considered a routine component of the assessment process for mold (with the exception of hidden mold), (c) an important goal of the mold remediation process was the removal of visible colonization in a manner that does not result in cross contamination of adjacent occupied and clean zones, and (d) the use of antimicrobials was not recommended as a substitute for physical removal of mold growth. For interior spaces excluding HVAC systems, the Third Edition of the NYC Guidelines recognized only three levels of contamination, namely small *Level I*, (<1.0 m$^2$), medium *Level II* (1-10 m$^2$); and large *Level III* (>10 m$^2$) isolated areas with visible mold growth. As with the 2000 edition, professional judgment was recognized as being important in specifying dust suppression and containment methods and appropriate personal protective measures for remediation workers.

Three of the case studies described earlier (i.e. Case study 1, 2 and 3) involved extensive visible mold growth concealed within the building envelope. While the first NYC Guidelines focused on mold growth on interior surfaces visible to the human eye, a consensus has developed that concealed mold growth in building construction should be included in estimating if the extent of growth is small, medium, or large scale (Prezant *et al.* 2008, see also Chapter 15). The rationale for including hidden growth during mold remediation assessment includes the likelihood that spores from concealed growth will eventually enter and degrade the indoor air (Miller *et al.* 2000, Morey *et al.* 2003) plus the difficulty in management of hidden contamination for the life of the building (See discussion in Prezant *et al.* 2008).

## Containment and dust suppression during mold remediation

The degree to which the mold remediation work area should be isolated or contained from the surrounding occupied or clean areas is dependent on several factors including the extent (m$^2$) of visible mold growth present on interior surfaces. In addition, the ease with which moldy materials can be physically removed from the building is important with regard to the degree to which the remediation work area needs to be isolated. According to the NYC 2000 Guidelines, the remediation of small isolated moldy areas (Level I) does not need containment. Examples of small isolated areas include "...ceiling tiles and small areas on walls" (NYC 2000: 7). Careful handling, bagging, and removal of a moldy ceiling tile (or removal of small isolated wall areas) can be carried without a formal containment. Using a HEPA vacuum cleaner to capture dust generated at its source during the removal process should be adequate for these kinds of small-scale removals.

When mid-sized or large isolated areas (NYC 2000 Guidelines, e.g. one or several wallboard panels) are undergoing mold remediation, more vigorous removal techniques are used and dispersion of dusts including microbial particulate becomes more likely.

Table 1 presents spore trap sampling data collected within a mold remediation work area when approximately 3 to 4 m$^2$ of visibly moldy wallboard was being removed. The mold remediation work area was contained and depressurized (approximately -5 Pascals). Basidiospores dominated the airborne mold spores collected in the outdoor air as well as in the room air prior to the start of mold remediation. The concentration of *Penicillium/Aspergillus* spores within the remediation work area was approximately three orders of magnitude greater than the levels in the outdoor air or in the room prior to the onset of the clean-up process. *Stachybotrys* spore levels increased in the remediation work area to about three orders of magnitude over concentrations measured in room air prior to onset of remediation activities. A limitation to interpretation of sampling results in Table 14.1 was the accuracy of the collected data because of the short duration of spore trap sampling time (<30 seconds) within the remediation work area (caused by elevated dust and spore levels associated with wall removal).

Additional longer-term air samples were also collected in the same remediation work area described for Table 14.1, and analyzed by PCR methodology (Table 14.2). *Stachybotrys chartarum* and *Penicillium chrysogenum* were detected in the remediation work area at levels of about 50,000 and 4,000 spore equivalents/m$^3$, respectively, over the time period (>1 hour) when the moldy wallboard and debris/dust were being removed from the work area. *Stachybotrys chartarum* and *Penicillium chrysogenum* were not detected in long-term outdoor air samples (Table 14.2).

Previous work by Rautiala *et al.* (1996, 1998) and Morey and Hunt (1995) showed that mold remediation activities resulted in two to four order of magnitude increases in airborne mold spore (e.g. *Penicillium/Aspergillus*) and actinomycete levels. Level II and III remediation procedures as recommended by the NYC 2000 Guidelines do not specifically call for construction of a fully depressurized containment. However,

*Table 14.1. Rank order taxa of mold spores outdoors and indoors within containment prior to and during removal of moldy wallboard (spore trap analysis).*

| Fungal taxa | Spores/m$^3$ |
|---|---|
| Outdoor air on roof | |
| Basidiospores | 370 |
| *Penicillium/Aspergillus* | 90 |
| *Cladosporium* | 70 |
| Ascospores | 55 |
| Indoors in room prior to removal | |
| *Basidiospores* | 150 |
| *Penicillium/Aspergillus* | 25 |
| *Stachybotrys* | 25 |
| *Cladosporium* | 10 |
| Ascospores | 10 |
| Indoors within containment during removal | |
| *Penicillium/Aspergillus* | 220,000 |
| *Stachybotrys* | 21,500 |
| *Cladosporium* | 12,000 |
| Ascospores | 4,500 |
| Basidiospores | 4,250 |

Note: Sampling period 5-10 minutes except for short term (<0.5 minute) samples collected during demolition; n=4-8.

the examples in Tables 14.1 and 14.2 plus the literature (Rautiala *et al.* 1996, 1998, Morey and Hunt 1995) suggest that it is prudent to use a depressurized containment to protect occupants in zones external to the remediation work area.

The data in Tables 14.1 and 14.2 also call attention for the environmental health professional to use cautious professional judgment in recommending dust suppression and control methodologies during mold remediations. In addition to the square meters of mold colonization found in an interior space, the following factors are important in forming professional judgment with regard to appropriate dust suppression and containment during mold remediation:
- Degree of disturbance associated with the removal of moldy building materials.
- Proximity of occupants to the mold remediation work area.

*Table 14.2. Rank order taxa of mold spore equivalents (se) in containment during mold remediation and in the outdoor air.*

| Fungal taxa | se/m³ |
|---|---|
| Outdoor air on roof | |
|     *Aspergillus niger* | 2 |
|     *Aspergillus fumigatus* | 1 |
|     *Cladosporium cladosporioides* | 1 |
| Indoors within containment during wall removal | |
|     *Stachybotrys chartarum* | 50,000 |
|     *Penicillium chrysogenum* | 4,000 |
|     *Aspergillus sydowii* | 2,900 |
|     *Chaetomium globosum* | 400 |
|     *Aspergillus versicolor* | 100 |

Note: Air sample collected on fiberglass filter for >1.0 hr.; PCR analysis (Hung *et al.* 2005).

- Presence of pathways (e.g. elevator shafts; building risers) through which aerosolized dusts can be transported to occupants.
- Density of mold growth (e.g. light, moderate, heavy; see Miller 2000) on interior surfaces.

The data in Tables 14.1 and 14.2, in addition to other published data on spore concentrations in mold remediation work areas (Rautiala *et al.* 1996, 1998, Morey and Hunt 1995, IICRC S520, 2008, Chapter 6) also call attention to the necessity of providing excellent respiratory protection for workers involved in the restoration of moldy buildings.

## Quality assurance during mold remediation

An important aspect of North American guidelines on mold remediation is documentation that the steps taken to remove visible mold growth were effective. There is consensus among guidelines that the water problem that led to mold growth should have been fixed and that visible growth and associated debris and dust on interior surfaces should have been physically removed (NYC 2000, EPA 2001, CCA 2004, Health Canada 2004). However, unless all surfaces, including those hidden within building infrastructure, are inspected for visible mold and dampness, it is difficult to be certain that all residual contamination was removed (IOM 2004: 300). Consequently, some investigators in North America use air and surface sampling for mold to ascertain remediation effectiveness.

A difficulty associated with using sampling as a quality assurance indicator is that even if all visible colonization is removed during cleaning some mold spores will still be present on the remediated surfaces (IOM 2004: 300). Furthermore, there is a total absence of health-based guidelines describing acceptable levels or kinds of molds present in the air or on interior surfaces.

The Institute for Inspection Cleaning and Restoration Certification (IICRC 2008) attempted to define acceptable cleanliness after remediation in terms of "normal fungal ecology". Normal fungal ecology (IICRC, 2008: 17) is described as "...an indoor environment that may have settled spores, fungal fragments or traces of actual growth whose identity, location and quantity are reflective of a normal fungal ecology for a similar indoor environment". It is left up to the professional judgment of the investigator to determine if normal fungal ecology occurs when a specified amount of phylloplane spores are present in the air or in dusts on interior surfaces (Horner *et al.* 2004).

Tables 14.3 and 14.4 illustrate the interpretation difficulty of relying on sampling as a quality assurance measure in mold remediation. Preremediation air sampling was carried out in rooms where Level II and Level III (NYC 2000) amounts of visible mold growth occurred on interior surfaces (Table 14.3). *Wallemia sebi, Penicillium brevicompactum*, and *Eurotium amstelodami* dominated the indoor taxa in moldy rooms. *Cladosporium herbarum* was the dominant taxon in outdoor samples (Table 14.3).

Air sampling was carried out in rooms several weeks after completion of mold remediation. Mold remediation in rooms included fixing the moisture problem that originally caused the mold growth, removal of all visibly moldy materials, removal of visually evident surface dusts, and operation of the HVAC system for at least a week. Only trace levels of spores (e.g. *Emericella rugulosa, Cladosporium cladosporioides*) were found in room air (Table 14.4). *Wallemia sebi*, the dominant taxon prior to clean-up, was not detected in post-remediation air samples.

Comparison for similarity of rank order taxa indoors and outdoors is usually used in determining if air in the remediated indoor environment was normal or typical. The low numbers of culturable taxa indoors in Table 14.4 can be interpreted as an indication of successful mold remediation. However, some investigators might

Table 14.3. Rank order taxa concentrations of culturable fungi. indoors and outdoors before mold remediation.

| Fungal taxa | Average conc. (cfu/m³) |
|---|---|
| Outdoor air on roof | |
| Cladosporium herbarum | 298 |
| Aspergillus niger | 27 |
| Eurotium amstelodami | 22 |
| Cladosporium sphaerospermum | 15 |
| Penicillium citrinum | 12 |
| Penicillium purpurogenum | 12 |
| Penicillium chrysogenum | 10 |
| Aspergillus flavus | 10 |
| Outdoor air at grade level | |
| Cladosporium herbarum | 261 |
| Alternaria alternata | 14 |
| Aspergillus niger | 10 |
| Cladosporium sphaerospermum | 8 |
| Yeasts, Sporobolomyces | 6 |
| Indoor air in moldy rooms | |
| Wallemia sebi | 640 |
| Penicillium brevicompactum | 146 |
| Eurotium amstelodami | 136 |
| Yeasts, Sporobolomyces | 89 |
| Cladosporium sphaerospermum | 80 |
| Aspergillus ochraceus | 42 |
| Emericella | 22 |
| Penicillium citrinum | 25 |

Note: n=4 for outdoor air; n=7 for indoor air; Laboratory #1 accredited by the American Industrial Hygiene Association, Environmental Microbiology Proficiency Analytical Testing Program; culture conditions: Malt extract agar; culture plate impactor operating at a flow rate of about 0.2 m³/minute.

find the slightly higher concentration of *Emericella rugulosa* indoors during post-remediation sampling as an indication of need for additional cleaning. The dominance of *Cladosporium herbarum* in outdoor samples in Table 14.3 and the absence of the same taxon in Table 14.4 might also be considered problematic by

*Table 14.4. Rank order taxa concentrations of culturable fungi indoors and outdoors after mold remediation.*

| Fungal taxa | Average conc. (cfu/m³) |
|---|---|
| Outdoor air on roof | |
| *Cladosporium cladosporioides* | 15 |
| *Curvularia lunata* | 11 |
| *Penicillium implicatum* | 5 |
| *Penicillium brevicompactum* | 3 |
| *Aspergillus fumigatus* | 3 |
| *Drechslera hawaiiensis* | 2 |
| *Aspergillus japonicus* | 2 |
| Outdoor air at grade level | |
| *Penicillium brevicompactum* | 36 |
| *Curvularia lunata* | 18 |
| *Alternaria alternata* | 8 |
| *Penicillium implicatum* | 7 |
| *Aspergillus alliaceus* | 4 |
| *Aspergillus japonicus* | 3 |
| *Emericella rugulosa* | 3 |
| Indoor air in formerly moldy rooms | |
| *Emericella rugulosa* | 5 |
| *Cladosporium cladosporioides* | 2 |
| *Cladosporium sphaerospermum* | 2 |
| *Penicillium citreonigrum* | 2 |
| *Curvularia lunata* | 2 |
| *Penicillium sclerotiorum* | 1 |
| *Penicillium brevicompactum* | 1 |

Note: n=4 for outdoor air; n=7 for indoor air. Laboratory #2 accredited by the American Industrial Hygiene Association, Environmental Microbiology Proficiency Analytical Testing Program; culture conditions: Malt extract agar; culture plate impactor operating at a flow rate of about 0.2 m³/minute.

some investigators. It should be noted that *Cladosporium* speciation in Tables 14.3 and 14.4 were carried out by different laboratories, both of which were accredited by the American Industrial Hygiene Association (AIHA) Environmental Microbiology Proficiency Analytical Testing Program.

The example in Table 14.4 shows that reliance on sampling data, as a primary indicator of successful mold remediation can be problematic. In addition, reliance on sampling data for quality assurance in mold remediation will continue to be difficult until health-based interpretation guidelines are promulgated by cognizant authorities.

The AIHA (2001) recommended use of key inspection actions as appropriate indicators that mold remediation has been carried out successfully. These actions include:

- Fixing the moisture problem that caused mold growth.
- Use of professional judgment to specify containment and dust suppression methods to prevent cross-contamination in the building.
- Documenting that an appropriate remediation was specified and that mold removal was actually carried out according to that plan.
- Verifying by inspection that (a) moldy materials were removed, (b) the remediated areas were HEPA-vacuumed, and (c) the amount of dust left on interior surfaces is sufficiently low so as to preclude the need for recleaning.

Among the AIHA quality assurance actions, fixing the moisture problem which caused the mold growth and documenting that visible contamination was removed are the most important. Both of these key quality assurance indicators are achieved by physical inspection throughout the remediation process. While simple tools such as moisture and relative humidity meters are useful in evaluating whether or not the moisture problem has been fixed, more complex evaluation tools such as water testing of the integrity of the building envelope are useful in more complex water entry issues. The ultimate quality assurance for documenting successful remediation in buildings with severe mold and moisture problems is successful reoccupancy (see Jarvis and Morey 2001 for the case study of the building in Figures 14.1-14.3).

While reliance on sampling in North American buildings is not recommended as a key quality assurance indicator of successful mold remediation, it is important to recognize sampling is an important assessment tool for determining if a building has a mold problem. Sampling can also be useful in detecting the presence of hidden mold (Miller *et al.* 2000, Morey *et al.* 2003) or mold growth overlooked during the inspection process (Morey 2007).

## North American guidelines on mold remediation

In addition to the 1993, 2000, and 2008 editions of the New York City Guidelines, a number of North American professional society and cognizant authorities have published important documents on mold remediation. These documents, which are not reviewed in this chapter, include:

- U.S. Environmental Protection Agency, Mold remediation in schools and commercial buildings, EPA 402-K-01-001, 2001.
- Occupational Safety and Health Administration, A Brief guide to mold in the workplace, *www.osha.gov/dts/shib/shib101003.html*, 2003.
- AIHA, Report of microbial growth task force, AIHA 458-EQ-01, 2001.
- AIHA, Assessment, remediation, and post-remediation verification of mold in buildings, AIHA Guideline 3-2004.
- Institute of Medicine (IOM), Damp indoor spaces and health, Chapter 6, Prevention and remediation of damp indoor environments, 2004.
- American Conference of Governmental Industrial Hygienists, Bioaerosols, assessment and control, Chapter 15, Remediation of microbial contamination (Shaughnessy and Morey, 1999).
- IICRC, Standard and reference guide for professional mold remediation, IICRC S520, 2008.

Four additional documents, two from Canada (Canadian Construction Association 2004, Health Canada 2004) and two from the United States (California Research Bureau – Umbach and Davis 2006, University of Connecticut Health Center – Storey *et al.* 2004) are reviewed in the following paragraphs.

### Health Canada (2004). Fungal contamination in public buildings: health effects and investigation methods

Chapter 3 of Health Canada (2004) discusses the "Investigation of fungal contamination of the non-industrial workplace". Section 3.2 on "General Principles" mentions that investigations in buildings with mold problems should be guided by methods such as those published by the AIHA (Hung *et al.* 2005) and the ACGIH (Macher 1999). These AIHA and ACGIH publications contain detailed information on selection and operation of microbial sampling instruments, development of a sampling plan, analytical procedures used in sample evaluation, and advice on data interpretation. Furthermore, the investigation should be carried out by "... appropriately trained individuals entering the building to conduct inspection, sampling documentation and production of reports" (Health Canada 2004: 37). Health Canada's emphasis on "... appropriately trained individuals" is timely and important because modern buildings are complex structurally and the evaluation of microbial ecology in damp buildings is often not straightforward.

On page 38, Health Canada (2004) states that actions taken by investigators during assessment and remediation must not result in increased contamination of the building or further risk to occupants. The term "remediation" is also defined to include "...both the thorough cleaning of any mold growing in the building and correcting the building defect that led to mold growth..." (Health Canada 2004: 38).

An important new concept found on page 40 of Health Canada (2004) is the recommendation that for buildings where mold growth is extensive (e.g. Levels III and IV of NYC 2000, ACGIH 1999a) the amount of dust remaining on surfaces after cleaning be less than 100 mg/m$^2$. The intent of this recommendation is to objectively document that dust removal was effective. Furthermore, when mold remediation activities including dust removal are complete "... the building can be treated like a normal construction site for build-back" (Health Canada 2004: 40).

Air sampling is mentioned as a useful quality assurance indicator of successful mold remediation when carried out one or two weeks after repairs to the affected areas have been completed and the HVAC system has become operational (Health Canada 2004: 40). The logic of this recommendation is that in large buildings it may take several days for phylloplane mold spores to infiltrate into the former remediation work area. Additionally, air sampling carried out one or two weeks after completion of build back can also provide an indication of a return to a normal mold exposure condition (phylloplane molds dominate) in the remediated area. A similar rank order profile of airborne mold spores would then be expected indoors and outdoors, so long as hidden or unremediated mold growth sites are absent.

### Standard Construction Document CCA 82-2004. Mould guidelines for the Canadian construction industry

This comprehensive document by the Canadian Construction Association (CCA, 2004) contains information on mold problems in buildings including (a) legal, insurance, and health risk considerations; (b) operational and construction practices to reduce moisture intrusion; and (c) recommendations on mold assessment and mold remediation. Section 9.1 of the document mentions three factors that should be considered during the development of a mold remediation plan namely:
• the location of the mold growth;
• the extent (m$^2$) of the growth;
• the sensitivity of building occupants to mold exposure.

According to CCA 82-2004, it should always be assumed that the presence of visible mold growth on interior surfaces means that an exposure risk exists. Because of this exposure risk, "universal precautions" such as the use of respiratory protection by clean-up workers is necessary. In addition, "controlled conditions" should be used to avoid aerosolization of microbial particulate into building areas external to the mold remediation work zone. Controlled conditions imply that dust suppression and containment techniques should be used in the area undergoing mold remediation.

Section 9.2 of CCA 82-2004 presents detailed actions that should be taken to remediate small (<1.0 m$^2$), medium (1-10 m$^2$), and large (>10 m$^2$) scale areas with visible mold growth. Important mold remediation recommendations include the following:

- Containment and depressurization of the mold remediation work area is recommended for both medium and large-scale clean-ups.
- An occupational health and safety professional should provide project oversight for large-scale remediations. Included in project oversight is the monitoring of remediation work practices.
- The use of disinfectants and antimicrobial agents during mold remediation is discouraged because the objective of the clean-up process is not to sterilize interior surfaces. After restoration, interior surfaces should be returned to normal conditions implying that some spores will still be present.
- Any remaining colonization (e.g. conidiophores, mycelium) on interior surfaces after clean up is an indication of an unsuccessful mold remediation.

### CRB 06-001. Indoor mold, a general guide to health effects, prevention, and remediation

This guide was prepared by the California Research Bureau (Umbach and Davis 2006) for the California Legislature because of moldy building problems widely publicized in that State in the 1990s. The guide was written to be understandable to policy and public health officials, building managers and maintenance personnel and the general public. It contains sections on health risk from mold exposure, methods for assessing mold problems, methods for avoiding mold growth, policy options, and mold remediation techniques.

Chapter 5 of the CRB 2006 guide on "remediating mold" states that when the area of mold growth is more than "trivial", it is necessary to contain or suppress the aerolization of mold spores and dusts that will be generated during remediation. Determining the amount (m$^2$) of visible growth on interior surfaces for some buildings is described as the result of a simple inspection. In other buildings where mold growth may be hidden, the use of boroscopes or the destructive removal of building components (e.g. walls) may be necessary to determine that extent of the mold problem. These destructive inspection techniques should, according to the CRB 2006 guide, be accompanied by use of personal protective equipment and containment. In a discussion on the isolation of "affected areas", it is noted that mold spores are buoyant and easily dispersed by passive air currents. In this context, small isolated areas with "very heavy "growth should be contained during remediation.

Use of a respirator is recommended by the CRB 2006 guide for any construction project generating dust. A higher, unspecified level of respiratory protection is advised when moldy construction materials are being remediated (Umbach and Davis 2006: 41). Household cleaners and disinfectants, according to the CRB 2006 guide, are said to be used appropriately to remove the trivial amount of mold that grows on the grime found on mostly non-porous surfaces as in shower enclosures. However, household cleaners and disinfectants are described as being "not suitable" for the remediation of moldy building materials, which are porous (Umbach and Davis 2006: 43). It is mentioned that spraying moldy surfaces with disinfectants and cleaners can result in the aerosolization of hydrophobic mold spores.

Encapsulation of mold-affected surfaces is described as not being an effective mold remediation method. Thus, painting over moldy surfaces and attempting to seal-off a contaminated space are considered inappropriate mold remediation methods (Umbach and Davis 2006: 44).

### UCONN (2004). Guidance for clinicians on the recognition and management of health effects related to mold exposure and moisture indoors

This publication was developed by the Center for Indoor Environments and Health, University of Connecticut Health Center (UCONN; Storey *et al.* 2004) under a cooperative agreement with the U.S. Environmental Protection Agency. It is important because the portions of the guideline on mold remediation and environmental assessment are discussed in relation to mold and moisture related illness.

Chapter 6 of the UCONN guide (Storey *et al.* 2004) on "Environmental Assessment" describes the importance of the qualitative walk-through inspection for moisture problems and microbial growth sites. The conclusion is made that the walk-through inspection, together with recommendations to correct moisture/mold problems, provide more useful information for the healthcare provider than spore counts (Storey *et al.* 2004: 47). Limitations to the development of numerical guidelines for mold exposure are summarized as follows: "There is an allure to establishing a fungal concentration standard for indoor air to guide decisions. However, threshold levels for fungal concentrations in the indoor air have not been established and with our current knowledge would not be helpful in understanding exposure risk to patients" (Storey *et al.* 2004: 55). The reasons for the current absence of numerical guidelines for fungal exposure have been discussed extensively elsewhere (ACGIH 1999b, IOM 2004).

It is noted on page 55 of the UCONN guide (Storey *et al.* 2004) that the presence of species of dampness/water indicator molds such as *Stachybotrys*, *Aspergillus*, *Penicillium*, and *Fusarium* indoors over and above the levels present outdoors is an indicator of an indoor moisture problem. However, *mold remediation procedures should not be based solely on air sampling results* showing the presence of these species. Section 7 of the UCONN guide (Storey *et al.* 2004: 58) on "Environmental Remediation Guidance" contains recommendations on mold remediation as follows:

• Porous building materials that have been chronically wetted and damp should be discarded.
• Cross contamination of clean indoor areas during mold, remediation should be avoided.
• Mold remediation activities will result in the aerosolization of spores. It is important that patients affected moisture and mold-related illness not participate in the cleanup.

## References

AIHA (2001) Report of Microbial Growth Task Force. American Industrial Hygiene Association, Fairfax, VA, USA.

AIHA (2004) Assessment, remediation, and post-remediation verification of mold in buildings. American Industrial Hygiene Association, Fairfax, VA, USA.

Arnow P, Fink J, Schlueter D, Barboriak J, Mallison G, Said S, Martin S, Unger G, Scanlon G and Kurup V (1978) Early detection of hypersensitivity pneumonitis in office workers. Am J Med 64: 236-242.

ASHRAE (2005a) Thermal and moisture control in insulated assemblies – fundamentals. In: 2005 ASHRAE handbook: fundamentals. American Society of Heating, Refrigerating and Air-Conditioning Engineers, Atlanta, GA, USA, Chapter 23.

ASHRAE (2005b) Thermal and moisture control in insulated assemblies – applications. In: 2005 ASHRAE handbook: fundamentals. American Society of Heating, Refrigerating and Air-Conditioning Engineers, Atlanta, GA, USA, Chapter 24.

CCA (2004) Mould guidelines for the Canadian construction industry. Standard Construction Document CCA 82-2004. Canadian Construction Association, Ottawa, Canada.

EPA (2001) Mold remediation in schools and commercial buildings. U.S. Environmental Protection Agency, EPA 402-K-01-001, Washington, DC, USA.

Health Canada (2004) Fungal contamination in public buildings: health effects and investigation methods, H46-2/04-358E. Health Canada, Ottawa, Canada.

Heller RM, Heller T and Sasson J (2003) Mold: "Tsara' At," Leviticus, and the history of a confusion. Perspect Biol Med 46: 588-591.

Horner WE, Worthan AW and Morey PR (2004) Air and dustborne mycoflora in houses free of water damage and fungal growth. Appl Environ Microbiol 70: 6394-6400.

Hung L-L, Miller JD and Dillon HK (eds.) (2005) Field guide for the determination of biological contaminants in environmental samples, 2nd ed. American Industrial Hygiene Association (AIHA), Fairfax, VA, USA.

Hunter CA, Grant C, Flannigan B and Bravery AF (1988) Mould in buildings: the air spora of domestic dwellings. Int Biodeterior Biodegr 24: 84-101.

IICRC (2008) Standard and reference guide for professional mold remediation, S520, Institute of Inspection Cleaning and Restoration Certification, Vancouver, Canada.

IOM (2004) Damp indoor spaces and health. Institute of Medicine, National Academics Press, Washington, DC, USA.

Jarvis JQ and Morey PR (2001) Allergic respiratory disease and fungal remediation in a building in a subtropical climate. Appl Occup Hyg 16: 380-388.

Listiburek J and Carmody J (1994) Moisture control for new residential buildings. In: Trechsel HR (ed.), Moisture control in buildings, American Society for Testing and Materials, ASTM Manual Series, MNL 18, West Conshohocken, PA, USA, pp. 321-347.

Macher J (ed.) (1999) Bioaerosols assessment and control. American Conference of Governmental Industrial Hygienists (ACGIH), Cincinnati, OH, USA.

Miller JD (2001) Mycological investigations of indoor environments. In: Flannigan B, Samson RA and Miller JD (eds.), Microorganisms in home and indoor work environments, Taylor & Francis, London, UK, pp. 231-246.

Miller JD, Haisley PD and Reinhardt JH (2000) Air sampling results in relation to extent of fungal colonization of building materials in some water-damaged buildings. Indoor Air 10: 146-151.

Morey PR (1992) Microbiological contamination in buildings: precautions during remediation activities. In: IAQ92, Environments for People, American Society of Heating, Refrigerating and Air-Conditioning Engineers, Atlanta, USA, pp. 171-177.

Morey PR (1995) Control of indoor air pollution. In: Harber P, Schenker MB and Balmes JR (eds.) Occupational and environmental respiratory disease. Mosby, St. Louis, MO, USA, pp. 981-1003,

Morey PR (2007) Microbial remediation in non-industrial indoor environments. In: Yang C and Heinsohn P (eds.) Sampling and analysis of indoor microorganisms, John Wiley & Sons, New York, NY, USA, pp. 231-242.

Morey PR and Cornwell M (2006) Moisture in the building envelope: problems continue even though solutions are straightforward. Proceedings of Healthy Buildings 2006, Vol 3: 17-20.

Morey PR and Hunt S (1995) Mold contamination in an earthquake damaged building. Proc. Healthy Buildings 1995, Vol 3: 1377-1381.

Morey PR, Hodgson MJ, Sorenson WG, Kulliman GJ, Rhodes WW and Visvesvara GS (1984) Environmental studies in moldy office buildings: biological agents, sources, and preventive measures. Ann Am Conf Gov Indust Hyg 10: 21-35.

Morey PR, Hull MC and Andrew M (2003) El Niño water leaks identify rooms with concealed mould growth and degraded indoor air quality. Int Biodeterior Biodegr 52:197-202.

NIOSH (1993) Health Hazard Evaluation Report HETA 93-1110-2575, Martin County Courthouse and Constitutional Office Building, Stewart, Florida, National Institute for Occupational Safety and Health, Cincinnati, Ohio.

NYC (1993) Guidelines on assessment and remediation of *Stachybotrys atra* in indoor environments. New York City Department of Health, New York City Human Resources Administration, and Mount Sinai – Irving J. Selikoff Occupational Health Clinical Center, New York, NY, USA.

NYC (2000) Guidelines on assessment and remediation of mold in indoor environments. New York City Department of Health, New York, NY, USA.

NYC (2008) Guidelines on assessment and remediation of fungi in indoor environments. New York City Department of Health and Mental Hygiene, NewYork, NY, USA.

OSHA (2003) A brief guide to mold in the workplace, occupational safety and health administration. U.S. Department of Labor, Washington, DC, USA.

Prezant B, Weekes DM and Miller JD (eds.) AIHA (2008) Recognition, evaluation, and control of indoor mold. American Industrial Hygiene Association, Fairfax, VA, USA.

Rautiala S, Reponen T, Hyvarinen A, Nevalainen A, Husman T, Vehvilainen A and Kalliokoski P (1996) Exposure to airborne microbes during the repair of moldy buildings. Am Ind Hyg Assoc J 57: 279-284.

Rautiala S, Reponen T, Nevalainen A, Husman T, and Kalliokoski P (1998) Control of exposure to airborne viable microorganisms during remediation of moldy buildings: report of three case studies. Am Ind Hyg Assoc J 59: 455-460.

Samson RA (1985) Occurrence of moulds in modern living and working environments. Eur J Epidemiol 1: 54-61.

Shaughnessy RJ and Morey PR (1999) Remediation of microbial contamination. In: Macher J (ed.) Bioaerosols assessment and control. Proceedings American Conference of Governmental Industrial Hygienists, Cincinnati, OH, USA, pp. 15.1-15.7.

Storey E, Dangman KH, Schenek P, DeBernardo RL, Yang CS, Bracker A and Hodgson MJ, (2004) Guidance for clinicians on the recognition and management of health effects related to mold exposure and moisture indoors. Univ. Connecticut Health Center (UCONN), Farmington, CT, USA.

Umbach K and Davis P (2006) Indoor mold: a general guide to health effects, prevention, and remediation. CRB 06-001, California Research Bureau, CA, USA.

U.S. Environmental Protection Agency (2001) Mold remediation in schools and commercial buildings, EPA 402-K-01-001. Washington, DC, USA.

Yang CS and Ellringer PJ (2004) Antifungal treatments and their effects on fibrous glass liner. ASHRAE J 46: 35-40.

# 15 Mold remediation in West-European buildings

*Thomas Warscheid*
*LBW-Bioconsult, Wiefelstede, Germany*

## Introduction

The problem of indoor mold infestation is actually of growing concern in the evaluation and sanitation of water damages in buildings. Such damage may be caused by high humidity levels during living, condensation due to insufficient insulation or water penetration by leakages in piping or flooding hazards.

While in the past such microbial induced building damages have been widely neglected or trivialized, today a partly overreaching awareness for the presence and growth of fungal spores and metabolites unsettles many parties with respect to possible health and material related impacts. Parties involved are building enterprises and handcrafts men, technical experts and courts, health authorities and physicians as well as the directly affected owners and users of domestic homes and business offices.

In this respect, it was tried in North America in the early 90s (NYC 1993, 2000, 2008, EPA 2001, 2002, AIHA 2001, 2004, Health Canada 2004) of the last century and in Europe, mainly Germany, at the beginning of the new millennium (LGA 2001, 2004, UBA 2002, 2005, BG Bau 2006, VDI 2008) to standardize the analytical assessment, evaluation and sanitation of indoor mold infestations by the development and formulation of recommendatory guidelines. They were primarily meant as an orientation to evaluate the extent of mold infestations and to define appropriate sanitation measures. Those guidelines were never supposed to give particular evidence on health related aspects of indoor molds.

American guidelines mainly recommend a semi-quantitative approach for assessing and remediating moldy buildings based on the extent of visible colonization of interior surfaces. In Germany, additional numerical evaluation schemes have been developed in order to judge the extent of a mold infestation by viable counts in the air and on building materials as well as non-viable organic particles respectively.

This chapter analyzes the specific conditions and causes which have presumably led to the obviously growing problem of indoor mold infestation. It stresses the essential need for an interdisciplinary approach in the anamnesis of damage factors, exposure conditions and the microbial burden in order to define appropriate, practical and sustainable sanitation measures. The chapter critically deals with the common use and interpretation of referring mold guidelines, with special emphasis on the German (European) situation. It addresses possibilities, limits and gaps of microbiological analysis of mold infestation. Finally, common and different approaches within the remediation of mold infestations are outlined considering national differences in respective building technologies, materials and climate conditions.

## Basic causes of indoor mold infestation

All life on earth is basically dependent on the availability of water and therefore also in the building environment the key to mold control is primarily based on moisture control. There is a multitude of sources for microbially available water indoors and they encompass in principle material moisture, condensation, high air humidity or the immediate impact of water itself. The following causes may play a role:

- *Initial moisture content of materials and structures.* Due to time pressure and economical necessities today, buildings are constructed more rapidly and without considering prevailing climatic conditions. This way, building materials are more often still wet during installation (e.g. wood, mortars) and they are given no chance to desiccate during dry seasons (e.g. winter).
- *Mold growth susceptibility of building materials and finishes.* The increased use of organic building materials (wooden fabrics, paper faced wallboards, insulation materials) and coatings (e.g. dispersion paints, detergents, additives) meets the requirements for a rapid and cost effective construction of buildings, but leaves a moisture intolerant, biodegradable building environment most sensitive to later biodeterioration processes.
- *Dynamics in use and growing energy prices.* The variability in the use of buildings (e.g. increased single households, frequent moving) and an improper maintenance of buildings (e.g. reduced expected life time, shortage of money, energy savings) aggravate the situation for buildings nowadays. High humidity levels within a building could be controlled by regular heating and ventilation, but increased costs for oil, gas and electricity as well as referring national and international energy saving legislations in the course of the Kyoto/Bali/Copenhagen agreements make it often quite difficult to meet those requirements in an optimal way.
- *Growing public awareness of health-related risks.* Finally, the natural attention of people for healthy living, wellness and relaxation leads in consequence also to an increased sensibilization concerning possible health-related impacts due to microbial infestations indoors. However this respective awareness sometimes seems to turn out into an unreasonable hysteria, it has to be carefully considered that allergenic predispositions and immunodeficiency have considerably

increased especially in industrial countries. Nevertheless the common high hygienic standards seem not to be guiltless for this situation.

## Risk-assessment of mold growth

The problem of indoor mold growth is multidisciplinary by definition, Therefore, a proper handling of mold infestations indoors can only be achieved by a strong interdisciplinary approach of different professionals. The key players are:

- The *building expert* (e.g. *engineer, architect*) who is the first to identify, document and evaluate damage and exposure situation. In cooperation with the building representatives he has to review construction plans (e.g. buildings materials, insulation, water proofing) to investigate all possible causes of water damages (e.g. water damage, condensation moisture, heating and ventilation behavior) and to describe the extent of moisture infiltration and type of materials obviously infested by mold (e.g. moisture detection, blower door measurement).
- If the causes for high humidity levels on indoor material surfaces can not be easily clarified, an *expert in building physics* would be needed, to provide additional information on the distribution of moisture (e.g. electrical resistance, electrical capacitance, gamma-ray spectroscopy, building material properties (e.g. adsorption isotherm, diffusivity) possible thermal bridges (e.g. infrared camera) and the conditions of heating and ventilation (e.g. energy balance, blower door test).
- In order to evaluate the spreading, intensity and type of the microbial burden within the building construction an environmentally skilled microbiologist is needed. In relation to the intensity of the respective damage, the *microbiologist* will analyze macro- and microscopically the extent and nature of the microbial infestation (e.g. adhesive tape, stained material samples). He takes cotton swab samples, contact agar and air samples in order to enrich and isolate most prevailing types of microorganisms on selective nutrient media. In addition, he can additional characterize the metabolic activity and distribution of the microorganisms (e.g. respiration, ATP-content); these information will help to differentiate between an actively growing microflora or inactive microbial contamination. This way the microbiological analysis helps to define the extent of the necessary dismantling and to control the quality of sanitation measures.
- With the taxonomical determination of microorganisms present at site, the microbiological assessment will provide further information for *medical experts* in charge for the evaluation of any health-related risks of building users and for the definition of precautionary measures in regard to environmental and personal protection measures during mold removal.

- During and after desiccation and sanitation of the mold damage, the engineer and the microbiologist have to ascertain together with the *sanitation company* the efficacy of the remedial treatment by visual inspection, moisture measurements and microbiological analyses.

## Sanitation of mold infestations – goals and limits

It will hardly be possible to achieve sterility in a living environment, but for hygienic and aesthetical reasons the imperative should be the removal of all actively microbial infested building materials and the minimization of the fungal and bacterial contamination by cleaning and disinfection to a naturally given level especially within the indoor air. An effective mold sanitation should therefore provide:
1. no visible microbial contamination;
2. no microbial infestation on building materials and coatings;
3. no malodor; and
4. no significant air-borne contamination by microbes and cell fragments within the building.

First of all, the causes of the water damage, high air humidity or condensation have to be fixed and a sufficient respectively sustainable desiccation of the building envelope needs to be ensured. In naturally very humid areas (i.e. kitchen, bathrooms) moisture control needs to be achieved by proper ventilation or moisture-adsorbing building materials.

The mode and extent of the multidisciplinary analysis is determined by the prevailing reasons and motivations for the handling of mold infestations. These might be:
1. precautions for health-related impacts (e.g. public and private buildings, schools, kindergartens, offices, archives);
2. special attention to identify causes for individual health-related complaints (e.g. allergies, intoxications, infections);
3. immediate repair of constructional defects respectively rising dampness (e.g. sanitation of old and new buildings); and
4. conflicts between landlords and tenants (e.g. private living space, company housing, official residences).

The growing sensibility for microbial infestations indoors and their possible health related impacts leads actually especially in Germany more and more to "all-over" sanitation measures, meaning the "precautionary" removal of any suspicious (i.e. presumably microbial infested and contaminated) building materials in water damaged buildings. These far-reaching actions happen often without any appropriate, interdisciplinary evaluation and consideration of the constructional, economical and microbiological conditions as well as alternative sanitation options, such as desiccation, disinfection and separation measures.

This trend is all the more critical, since actually sound microbiological scientific data are missing which give proof for a naturally given, and therefore acceptable, background level of the microbial load in various building structures (considering their life time, nature of exposure and basic building materials used).

## Mold guidelines in Germany

### Basic idea behind mold guidelines

In most countries in Europe, guidelines for mold inspection, meaning the prevention, analysis, evaluation and sanitation of mold growth indoors, are widely missing. In Germany many official organizations and professional associations, like the Federal Environmental Agency (Umweltbundesamt, UBA), regional public health departments (e.g. Landesgesundheitsamt Stuttgart, LGA) as well as professional associations of building biologists (e.g. Verband der Baubiologen, VDB), building engineers (e.g. Verein Deutscher Ingenieure, VDI) and employees liability insurances (e.g. Berufsgenossenschaft Bau, BG) have dealt in recent years with the prevention, assessment, evaluation and sanitation of mold damages. Their concerted work result has been formulated in mold guidelines, policy recommendations for mold sanitation measures and standards for the assessment of the microbial burden indoors (LGA 2001, 2004, UBA 2002, 2005, BG Bau 2006, VDI 2008).

The mold guidelines of the UBA and LGA Stuttgart can be regarded as first positioning documents giving useful information and orientation for non-professionals in the complexity of moisture and microbial burden indoors and contain important practical advices with respect to possible remedial interventions.

### Analytical assessment and evaluation criteria

The evaluation of microbial loads in air, dust and material surfaces within the German guidelines are mainly based on a visual assessment of the size of the microbial affected spot and on classical cell respectively particle counts.

*Thomas Warscheid*

## Visible mold growth limits

The "Mold sanitation Guideline" of the German Federal Environmental Agency defines that "massive" mold infestations are given already at 0.5 m² of visible growth (UBA 2005), which is in comparison to North American guidelines a very small spot with major economical consequences for the building industry and insurance companies to take professional action. The "Guidelines on Assessment and Remediation of Fungi in Indoor Environments" of the New York City Department of Health and Mental Hygiene defines four different categories of visible microbial infestation (e.g. Level I: <1 m², Level II: 1-3 m², Level III: 3-10 m² and Level IV: 10 m²), where Levels III and IV result in major sanitation and precautionary measures (see also Chapter 14).

## Limits in viable and particle counts

The definition of "orientation values" concerning viable cell and particle counts within the indoor air in German guidelines refer rather to empirical data of service laboratories than on a profound, scientifically validated research (Trautmann *et al.* 2005a,b).

They are taken for a respective classification of low, medium and high levels of the microbial burden present in the particular indoor environment. Those classes are meant to define whether a (probably also hidden) microbial indoor infestation/ contamination is unlikely, not to be excluded or presumably present.

Over and above that, this quantitative classification is hardly applicable as a proof for successful mold sanitation, since in the course of those sanitation activities the small-sized fungal spores can be widely spread free even when specific containment and preventive dust suppression is used. At the best the classification can help to judge carefully the efficacy of a final HEPA (= High Efficiancy Particulate Air) vacuum fine cleaning (i.e. HEPA-class H 13, meaning 99.95-99.997% removal of particles ≥0.3 μm referring to the German standard DIN 24183).

Besides the evaluation of the quantitative data of cell and particle counts in air or on materials within the German guidelines a special emphasis is given to a sound qualitative, taxonomical differentiation between:
- moisture-related species (e.g. *Acremonium* spp., *Aspergillus versicolor, A. penicillioides, A. restrictus, Chaetomium* spp. *Phialophora* spp., *Scopulariopsis brevicaulis, S. fusca, Stachybotrys chartarum, Engyodontium album, Trichoderma* spp.) indicating a conspicuous building damage; and

- environmental common strains (e.g. *Cladosporium* spp., *Alternaria* spp., *Botrytis* spp., yeast).

Nevertheless, it is also clearly stated within the respective guidelines, that those analytical data are not recommended for an immediate medical evaluation. Over and above that, judging numerical data of the aerial microbial burden, we should consider whether fungal strains present on building materials tend to build morphologically low, medium or low amounts of spores, leading to different concentration of air-borne microorganisms (LGA 2001, 2004, UBA 2002, 2005).

**Microscopic analysis**

In order to complete the microbiological assessment and evaluation of a mold infestation given within a particular object the German guidelines recommend additionally microscopic analyses of adhesive tapes and contact agar sampling from suspicious material surfaces, whereas in this case qualitative instead of quantitative statements can be achieved (VDI 2008).

Within the first German mold guidelines (LGA 2001, UBA 2002, 2005), another additional option for the indirect assessment of a possibly hidden microbial burden is noted, which refers to the analysis of microbial volatile organic compounds ("MVOC") by a personal air sampler. The realization and interpretation of "MVOC"-measurements requires a high expertise, since the analytical data may hardly correlate with an actual microbial infestation and raise the risk of false positive findings. For the evaluation of "MVOC"-data, the natural background level of those volatiles and possible different non-biogenic sources (e.g. building products and related sources) need strictly be considered (Virnisch *et al.* 2003, Keller *et al.* 2004, Schleibinger *et al.* 2004, Fischer *et al.* 2005, Moriske 2005). This applies also to the use of "mold dogs", frequently used to detect those "MVOC", which also holds a large risk at wrongly positive and negative results for the same reason (Oswald 2004); not to mention the ethical objections to expose the dog involuntary and intentionally to molds. For these reasons the chemical assessment of "MVOC" and use of "mold dogs" has nowadays been noticeably reduced referring to recent German mold guidelines (LGA 2004, UBA 2005, VDI 2008).

**Required skills and expertise**

All German mold guidelines commonly stress as condition precedent that people and laboratories, being in charge for the anamnesis, analysis, evaluation and sanitation of mold infestations, have to provide a sound scientific and technical expertise, as well as appropriate microbiological skills and qualifications to interpret microbiological data with respect to the entire damage situation. This, however, is quite often insufficiently defined or based on one-sided judgment. In this respect, an interdisciplinary approach of building engineers, enterprises, environmental microbiologists, health professionals and sanitation companies is of greatest importance. The microbiological assessment of moisture and mold problems can never replace, but has to go along with an entire analysis of the building construction (Oswald 2003).

## Limits and risks of mold guidelines

During the development and definition of mold guidelines in Germany environmental microbiologists (which are notably experienced in the ecology and natural distribution of microorganisms) were only scarcely represented compared with medical microbiologists. As a consequence, next to classical viable cell and particle counts, modern analytical tools for additional proofs on the distribution and metabolic activity of microorganisms on building material profiles (e.g. microscopical staining methods, glucan and protein analysis, ATP-measurements) have scarcely been taken in account so far. However, they may significantly contribute to the differentiation of a considerable active infestation and/or a possibly neglectable inactive contamination within the microbial burden.

### Reproducibility and background data

The analytical approach to which actual mold guidelines in Germany refer is based on a very simple, microbiological methodology. Non-professionals are not aware of the intrinsic problems concerning the reproducibility of viable cell and particle counts depending on time of day, season, climate, way of sampling, or choice of nutrient media. Ecological implications indicated by the composition of the fungal and bacterial microflora present may probably be helpful to understand the progression and significance of the microbial-induced damage process. However, they remain in the individual scope of microbiological laboratory and particular expert and can not be easily judged by non-professionals.

Moreover, reliable natural background levels of viable cell and particle counts for various building constructions (e.g. solid or light weight buildings, size and complexity, insulation systems (Pessi *et al.* 2002), with respect to their life time and nature of exposure (e.g. age of buildings, climatic conditions (Wouters *et al.* 2000,Richardson 2004) and basic building materials (e.g. organic respectively mineralic compounds, new or *in situ* (Hicks *et al.* 2005, Trautmann 2005) are largely missing. This complicates the interpretation and evaluation of microbiological data achieved from air, dust and material samples without long-term experience, or scientific technical expertise. In this regard, we need to define the extent how building materials constructions are already "naturally" contaminated by microorganisms during production, environmental exposure and application (e.g. storage on the construction site), before in the later those infections are misinterpreted as "unacceptable microbial burden".

"Orientation values" of microbiological data for viable cell and particle counts in actual German mold guidelines, are therefore often overestimated, misused and misinterpreted. Semi-professional experts, increasingly push legal authorities and insurances to take those data for granted, as standardized and scientifically based legal thresholds.

In the U.S. and Canada similar numerical approaches have been undertaken twenty years ago, but have rapidly been skipped due to the inconsistency of the analytical assessment of viable counts and particles. They were replaced by a far more simple visual inspection and evaluation of the size of mold infestation. Microbiological measurements are scarcely applied, taking for granted that all microbial infested building materials has to be removed dust-free, no matter which type of mold is present, and the moisture problem that caused mold growth has to be fixed (EPA 2001, AIHA 2004, Health Canada 2004).

In conclusion, neither the German approach with its detailed classification system based on a time-consuming and cost-intensive high analytical effort to determine variable counts of viable cells and particles, nor the American way to achieve the microbial burden indoors simply by visual inspection can sufficiently cope with the requirements needed to evaluate the hygienic situation indoors.

Instead of overestimating or neglecting the presence and distribution of allergenic aerial fungal spores, their assessment and evaluation needs to be restricted. This favors a profound microscopical analysis of the microbial contamination and infestation of building materials, and the determination of the microbial organic burden and metabolic activity (e.g. ergosterol, ATP – adenosine-triphosphate) and the prove of possible health-related microbial metabolites (e.g. mycotoxins, endotoxins, proteins, glucan) depending on the particular situation.

## Gaps and necessary improvements in mold guidelines and their analytical approach

While the actual German mold guidelines are bearing the risk of an increasing overestimation of mold problems indoors, the American approach seems to focus merely on the surface only and leaves probably hidden microbial endangerments unconsidered. Maybe the truth and most reliable approach in the assessment and evaluation of mold indoors can be found in future in between, taking advantage of the rapid development of modern microbiological techniques.

### Modern analytical tools

The main future challenge will be to differentiate clearly between (a) a settled, possibly acceptable microbial contamination (e.g. presence of viable, but metabolic inactive microorganisms) and (b) a meaningful microbial infestation (i.e. metabolic active fungi and bacteria) in and on building materials. This will be the most crucial point within the evaluation of mold damage and the necessity of extensive sanitation measures. The limited information of classical viable cell and particle counts, in the air, dust or on building materials, needs to be improved by additional modern microscopical and biochemical analyses that will describe an indoor microbiological situation more precisely. This is explained in detail hereafter:

Classical light microscopy of adhesive tapes and material surfaces using staining solutions (i.e. lactic acid cotton blue, methylene blue, lactophenole) allows to identify microbial cells and structures. Nevertheless, building materials can be examined in more detail for presence and distribution of microbial biofilms (e.g. staining by PAS – periodic acid Schiffs-reagent), fungal hyphae (e.g. staining by calcofluor white) or metabolic active microbial infestation (e.g. staining by FDA – fluoresceine-diacetate) in depth of a material in order to visualize the penetration and to define the amount of microbial infested building material that has to be removed during sanitation measures.

Furthermore, the non-destructive assessment of metabolic activity (e.g. ATP analysis – standard in hygiene inspection in hospitals and food industry) allows a rapid and semi-quantitative control of a non-visible microbial burden present on material surfaces especially within the definition and control of sanitation measures. Moreover, the significance of cell counts on building materials can be improved by the additional measurement of the protein content as marker for the microbial biomass, supports the differentiation between microbial contamination and infestation as well. Furthermore, it helps to clarify the origin microbial spores, cell fragments and possible toxic metabolites (e.g. mycotoxins, endotoxins, glucan) and clearly indicates the most important targets for the supposed bioremedial interventions (Warscheid 1996; 2000; 2003).

Consequently, clearly microbial infested building materials need to be removed, but contaminated areas can be simply cleaned and disinfected, when the moisture-related causes have been fixed. It should be stressed, that visibly microbial infestation in water damaged building structures (i.e. due to condensation, water leakages or flooding) requires an immediate and careful removal of building materials and is – in cases of doubt – subject for microbiological examinations of intensity and taxonomical composition (e.g. occupational health and safety).

## Improving the assessment and evaluation of air-borne spores and particles

The possible impact and contribution of spore contaminations to the indoor hygiene due to sedimented spores or hidden microbial infestations can be determined by a professional analysis of experienced microbiological laboratories and careful interpretation and critical evaluation of the assessment of air-borne microorganisms, as proposed by the German guidelines (LGA 2001, 2004, UBA 2002, 2005, Haas *et al.* 2005, BG Bau 2006, VDI 2008).

The analytical protocol is based on international accepted standards. The quality assurance of the laboratory has to be proven by a regular and successful participation within round robin tests organized by the LGA Stuttgart.

In contrast to the widely accepted protocol and mostly satisfying results of viable counts of air-borne microorganisms, the significance of measurements of the microbial burden in dust is questionable. Recent round robin tests resulted in such a variation of data that all options of interpretation, could not be excluded (Baudisch and Kramer 2006, Gabrio *et al.* 2007). One should be cautious, if alone by microscopical determination, quantification and evaluation of non-viable, microbial particle counts extensive conclusions with respect to microbial burden indoors are drawn. Dust examinations should be interpreted in the context of the aforementioned microbiological investigations (Trautmann 2006).

## Molecular-biological techniques

Finally, even more frequently molecular-biological techniques enter the assessment and evaluation of the microbial burden indoors, since they are supposed to overcome deficiencies of cultural enrichment of microorganisms on specific growth media (Witt and Krause 2007).

These modern procedures may proof the presence of certain, non-culturable microorganisms. However, once again, no practice-relevant conclusions concerning their metabolic activity and quantitative spreading within the material profile should be made. Nevertheless, ongoing molecular-biological investigations might help to address new microbial strains in the indoor environment with significant health-impairing effects for humans. Thus molecular-biological analysis refers actually to research applications, which is beyond the scope of guidelines and provides no practical perspective at this stage.

## Challenges for future mold guidelines

The major challenges in health related impacts of indoor fungi concern the ecological implications of mycotoxin production by molds and the occurrence of actinomycetes.

### Mycotoxins

Actually, the detection of potentially toxic indoor fungi is often misleadingly set equal to the presence of mycotoxins and probably overrated in regard to their possible health-related impacts (Fischer and Dott 2005). Recent scientific research has shown that the fungal production of significant amounts of these metabolic compounds requires certain nutrient support (i.e. specific building materials) and, as a general rule, extremely high moisture levels (>90% rh) (Tuomi *et al.* 2000, Nielsen *et al.* 2004).

For example, the species *Stachybotrys chartarum*, a mycotoxin producing mold in the indoor environment is commonly found on water-damaged gypsum-boards and is subdivided into different subspecies. However, only one of these tends to produce considerable high amounts of trichothecene, which is difficult to address during the daily analytical practice (Andersen *et al.* 2002).

The evaluation of health-related impacts of mycotoxins (including fungal glucane, bacterial endotoxins) within the indoor environment is actually subject for ecological and medical research and should be considered in building practice with the upmost precaution and sufficient expertise.

## Actinomycetes

Actinomycetes are known to cause severe health problems to certain professionals (e.g. garbage collectors). Otherwise, they can be found quite frequently in the natural environment (e.g. soil, compost) and can be addressed as an indicator for a complex microflora.

The possible excretion of toxic metabolites by some indoor species (e.g. *Streptomyces, Nocardiopsis*), should not be neglected, especially with respect to high moisture levels (Lacey 1997, Peltola *et al.* 2001). Under which ecological conditions their health-related impacts receive significant importance, remains unclear so far and needs further scientific research (Lorenz *et al.* 2003).

## Health implications of indoor mold

The environmental medical evaluation of mold infestations indoors is multi-faceted and, considering the actual state of knowledge, so far not conclusive. Health implications depend considerably on the constitution of the individual concerned (e.g. allergic or immune-deficient people, children and elderly) and can be validated only in the framework of environmental medical investigations.

The possible health-relevant reactions essentially comprise (1) the allergenic effect of fungal spores on the respiratory system and skin organs (e.g. cold, cough, asthma, mucous membrane irritations, alveolitis) as well as (2) immunological (fever-) reactions of its cell components (e.g. glucan, endotoxine) on the human immune system (e.g. organic dust syndrome). The resulting and continuous stress to the immune system can cause nonspecific symptoms, such as indisposition, tiredness and listlessness.

Mold infestations also tend to release different alcohols and ketons (e.g. "MVOC" = microbial organic volatile compounds) causing moldy odors, which with continuous admission into the room atmosphere can cause headache, comparable to fusel oil in alcoholic beverages of minor quality.

The evaluation of health-related impacts of mycotoxins (including glucans, bacterial endotoxins) within the indoor environment, needs further ecological and medical research and should be considered in the building practice with caution and sufficient expertise. Immediate infections by molds indoors seem to occur seldom and are restricted to immune-deficient patients affected by systemic mycosis (e.g. aspergillosis).

*Thomas Warscheid*

Therefore, a possible correlation cannot be excluded between mold infestations indoors and immediate respectively delayed, allergenic as well as respiratory illnesses. In the context of a sustainable health precaution, the avoidance and/or minimization of microbial active indoor infestations needs to be aimed for as much as possible.

## Intervention steps in sanitation

The following and successive steps are essentially needed to develop a profound strategy for successful sanitation measures.

### Basic information

The definition of range and goal of remedial interventions needs first of all collection and evaluation of relevant information concerning:
1. type and definition of damage;
2. cause of damage;
3. spreading of damage (i.e. microbial infestation in the building);
4. biosusceptibility of the building materials present;
5. intensity and taxonomical composition of the microbial infestation; as well as
6. possible health effects.

### Precautionary measures for sanitation

Microbiological investigations of the air-borne and material-related microbial burden can provide a basis for risk-assessment to specify necessary protective measures for assuring the occupational health and safety of employees and to prevent the contamination of the surrounding object area.

Risk assessment is based on "classes for endangerment" depending on the expected concentration of air-borne fungal spores (e.g. low, medium, high). The exposure level for workers should not exceed an empirical defined level of 50.000 cfu/m$^3$ during sanitation activities. Above that level, the time of exposure (i.e. less or more than 2 hours) and the availability of technical facilities to reduce the spreading of fungal spores and dust determine the need for further personal protection equipment and appropriate operating instructions (BG Bau 2006).

Besides common hygienic measures (e.g. cleaning, care of skin, avoidance of eating, drinking and smoking), the personal and environmental protective measures within these operating instructions cover basically:

1. the minimization of dust- and aerosols (e.g. wetting of infested material surfaces, machinery with integrated exhaustion, HEPA vacuum cleaner);
2. prevention of cross-contamination (e.g. coverage, encapsulation); and
3. personal protection (i.e. protective clothing, filter masks, gloves, eye protection, gloves, clothes).

With increasing "classes of endangerment" (i.e. >50.000 cfu/m$^3$ during sanitation activities, time of exposure >2 hours) additional measures are claimed regarding:
4. a clear separation of black and white areas (e.g. with and without sluice); and
5. a sufficient ventilation and air-exchange within the contaminated area (e.g. containment with negative pressure).

A respective skilled operating manager, supported by medical consultancy (Lorenz *et al.* 2005, BG BAU 2006), needs to assure the compliance and organization of such protective measured.

## Decontamination

The main objectives for successful sanitation of mold indoors comprise the dust-free removal of microbial infested building materials by immediate disposal and a solid HEPA vacuum cleaning of the formerly infested area. Disinfection treatments (i.e. hydrogen peroxide, ethanol, peracetic acid) are optional, depending on constructive conditions, type of building material and the intensity of remaining microbial contamination. They also might help to eliminate remaining moldy odors (LGA 2001, LGA 2004, UBA 2005).

## Moisture control and preventive measures

The technical feasibility of a sufficient drying process and its control is an essential requirement to the building engineer or the sanitation enterprise in charge.

Nevertheless, in order to achieve a sustainable effect of the remedial interventions, the highest priority has to be given to the removal/remediation of the moisture problem that caused the mold growth (e.g. high humidity levels, condensation moisture, free water penetration by leakages in piping or flooding hazards), appropriate house-keeping and constructional improvements (e.g. occupancy, insulation, water proofing, moisture protection).

The extensive substitution of microbial sensible building materials (e.g. gypsum, gypsum board, derived timber products, wallpaper and glue, emulsion paint) by less susceptible building materials (e.g. lime plaster, lime sludge, lime cement plaster, silicate boards, brick, natural stone, silicate paint, thermoplastic foil) (Warscheid 2007; 2008) should support such measures.

### Microbiological clearance and after-care

All steps of interdisciplinary damage analysis and wide-spread sanitation measures presented here should be accompanied by means of an appropriate microbiological investigation (e.g. monitoring the air-borne and material-related microbial burden by microscopical, biochemical and cultural techniques) and professional evaluation of the hygienic situation by an experienced microbiologist within a recognized qualified laboratory, capable to differentiate between ordinary building dirt, microbial contamination at site or a still immanent infestation of building materials. The evaluation of the microbial endangerment determines the necessity and the range of any remedial intervention.

It is the ultimate ambition that after all sanitation measures (1) the moisture problem that caused mold growth is definitely fixed and a sufficient desiccation is proven (e.g. building anamnesis, moisture measurements), (2) no visible, and so far as possible no hidden, microbial infestation on building materials remains (e.g. visual inspection and entire documentation of sanitation activities), (3) the contamination by fungal spores does not significantly contribute to the microbial burden in the indoor air (e.g. measurement of air-borne microorganisms respectively dust level) and (4) no further olfactory impairments are detectable (Warscheid 2008).

However, frequently both economic obligations and legal decision criteria oppose the ideal remediation. This may limit the necessary investigation and sanitation expenses, so that modification of the strategy becomes necessary.

## Future prospects

The investigation and evaluation of indoor microbial burden as related to water damages implies an interdisciplinary cooperation between building experts, microbiologists and sanitation companies.

In the case of an acute and immediate affection of individuals, health professionals with a background in environmental medicine will be needed to clarify the possible impact of microbial infestation. Moisture problems and mold growth may refer to lack of maintenance by landlords, incorrect heating and ventilation of tenants or structural faults by building enterprises. In that case, lawyers and insurance companies have a legitimate interest of receiving reliable information how to judge the microbial damage and achieve a satisfactory sanitation result for their clients.

The establishment of professional networks can help to improve this multidisciplinary communication, to compile qualified and coordinated strategies and to develop meaningful approaches for the building practice.

In this regard, the actual German mold guidelines (LGA 2001, 2004, UBA 2002, 2005, BG Bau 2006, VDI 2008) give an exploratory basis for prevention, analysis, evaluation and sanitation of mold growth indoors. Nevertheless, they fail to give fair answers for a comprehensive risk assessment in particular case studies. This is due to a limited analytical approach, missing background data in regard to the natural contamination of building constructions and reasonable statements on possible health related risks of indoor mold.

Mold guidelines, as discussed here, cannot be considered as state-of-the-art, either from practical or from scientific point of view, and are much less suitable for legal authorization. They can serve as helpful guides for the building practice, but need further improvements in detail and beyond pure business interests. In that case, they may prevent an aggravated and uncontrolled hysteria of people and an unjustified legal as well as economic abuse of the growing problem.

Practical experiences acquired by multidisciplinary-experienced professionals (e.g. building engineers, microbiologists, physicians, sanitation specialists) within various case studies as well as scientific research (microbiological and medical) need to be considered to improve and differentiate the previously mentioned mold guidelines (e.g. evaluation), policy recommendations for mold sanitation measures (e.g. economical acceptance) and standards for the microbiological assessment of the microbial burden indoors (e.g. methodology).

In this respect, substantial research on the ecological background and health-related impacts of the microbial burden indoors is urgently needed, particularly since the issues of microbial ecology today are frequently obscured by the modern molecular-biological research.

Differences in building technology between or within different countries, have to be considered before at a common consensus for the "back-bone" of an international mold guideline can be achieved. The variety and modifications of analytical methods needed for the assessment of a microbial burden of building materials should not be overlooked.

In this context still open questions remain concerning (1) reliable background levels for the natural microbial burden of various building constructions and materials with respect to their life time and nature of exposure, (2) an approximate estimation of health-related impacts by the release of immune-relevant cell fragments (e.g. glucane), toxic metabolites (e.g. mycotoxins, endotoxins) and bacterial infestations (e.g. actinomycetes) and (3) improvement and validation of microbiological methods (e.g. microbial contamination vs. infestation).

## References

AIHA (2001) Report of microbial growth task force. American Industrial Hygiene Association, Fairfax, VA, USA.

AIHA (2004) Assessment, remediation, and post-remediation verification of mold in buildings. American Industrial Hygiene Association, Fairfax, VA, USA.

Andersen B, Nielsen KF and Jarvis BB (2002) Characterization of *Stachybotrys* from water-damaged buildings based on morphology, growth and metabolic production. Mycoligia 94: 392-403.

Baudisch C and Kramer A (2006) Quantitative Bestimmung von *Aspergillus-* und *Penicillium-* arten in der 63 µm Hausstaubfraktion. In: Keller R, Senkpiel K, Samson RA and Hoekstra ES (eds.) Partikuläre und molekulare Belastungen der Innenraum- und Außenluft, Band 10. Verlag Schmidt Römhild, Lübeck, Germany, pp. 59-85.

BG Bau – Berufsgenossenschaft der Bauwirtschaft (2006) Handlungsanleitung: Gesundheitsgefährdungen durch biologische Arbeitsstoffe bei der Gebäudesanierung (BGI 858). Fachausschuß Tiefbau, Berlin, Germany.

BG Bau – Berufsgenossenschaft der Bauwirtschaft (2006) Richtlinie "Kontaminierte Bereiche" (BGR 128), Fachausschuß Tiefbau, Berlin.

EPA (2001) Mold remediation in schools and commercial buildings. U.S. Environmental Protection Agency, EPA 402-K-01-001, Washington, DC, USA.

EPA (2002) A brief guide to mold, moisture, and your home. U.S. Environmental Protection Agency, EPA 402-K-02-003, Washington, DC, USA.

Fischer G and Dott W (2005) Die Relevanz von sekundären Stoffwechselprodukten (MVOC, Mykotoxine) von Schimmelpilzen für die Innenraumhygiene. Umweltmed Forsch Prax 10: 329.

Fischer G, Möller M, Gabrio T, Palmgren U, Keller R, Richter H, Dott W and Raul R (2005) Vergleich der Messverfahren zur Bestimmung von MVOC in Innenräumen. Bundesgesundheitsbl – Gesundheitsforsch – Gesundheitsschutz 48: 43-53.

Gabrio T, Weidner U and Seidl HP (2007) Schimmelpilze in Innenräumen – Nachweis, Bewertung, Qualitätsmanagement – Auswertung der bisher durchgeführten Ringversuche unter besonderer Berücksichtigung des 10. und 11. Ringversuchs. In: Fortbildung Ringversuch "Schimmelpilze". München, Germany.

Haas D, Habib J, Schlacher R, Unteregger M, Galler H, Buzina W, Marth E and Reinthaler FF (2005) Einfluss des Schimmelbefalls in Innenräumen auf die Sporenzahl der Raumluft. Umweltmed Forsch Prax 10: 328.

Health Canada (2004) Fungal contamination in public buildings: health effects and investigation methods. Health Canada H46-2/04-358E, Ottawa, Canada.

Hicks JB, Lu ET, De Guzman R and Weingart M (2005) Fungal types and concentrations from settled dust in normal residences. J Occup Environ Hyg 2: 481-492.

Keller R, Reinhardt-Benitez S, Döringer K, Eilers J, Laußmann D, Mergener H-J, Ohgke H, Schmidt A, Senkpiel K, Solbach W, Wlaker G, Weiß R and Butte W (2004) Hintergrundwerte von flüchtigen Schimmelpilzmetaboliten in unbelasteten Wohngebäuden. Gefahrstoffe – Reinhaltung der Luft 64: 187-190.

Lacey J (1997) Fungi and actinomycetes as allergens. In: Kay AB (ed.) Allergy and allergenic diseases, Blackwell Scientific Publications, Oxford, UK, pp. 858-887.

LGA – Landesgesundheitsamt Baden-Württemberg (2001) Schimmelpilze in Innenräumen – Nachweis, Bewertung, Qualitätsmanagement. Landesgesundheitsamt Baden-Württemberg, Stuttgart, Germany.

LGA – Landesgesundheitsamt Baden-Württemberg (2004) Handlungsempfehlung für die Sanierung von mit Schimmelpilzen befallenen Innenräumen, Leitfaden, Landesgesundheitsamt Baden-Württemberg, Stuttgart, Germany.

Lorenz W, Hankammer G and Lassl K (2005) Sanierung von Feuchte- und Schimmelpilzschäden – Diagnose, Planung und Ausführung. Rudolf Müller Verlag, Köln, Germany.

Lorenz W, Trautmann C, Dill A, Mehrer (2003) Bakterielle Verunreinigungen in Innenräumen III – 4.3.2. In: Moriske H-J and Turowski E (eds.) Handbuch für Bioklima und Lufthygiene 10 Erg. Lfg. Ecomed, Landsberg, Germany.

Moriske H-J (2005) 12. WaBoLu – Innenraumtage vom 2.-4. Mai 2005 – Mikrobielle und chemische Verunreinigungen. Bundesgesundheitsbl – Gesundheitsforsch – Gesundheitsschutz 48: 1296-1303.

Nielsen KF, Holm G, Uttrup LP and Nielsen PA (2004) Mold growth on building materials under low water activities – influence of humidity and temperatureon fungal growth and secondary metabolism. Int Biodeter Biodeg 54: 325-336.

NYC (1993) Guidelines on assessment and remediation of *Stachybotrys atra* in indoor environments. New York City Department of Health, New York City Human Resources Administration, and Mount Sinai – Irving J. Selikoff Occupational Health Clinical Center, New York, NY, USA.

NYC (2000) Guidelines on assessment and remediation of mold in indoor environments, New York City Department of Health, New York, NY, USA.

NYC (2008), Guidelines on assessment and remediation of fungi in indoor environments, New York City Department of Health and Mental Hygiene, New York, NY, USA.

Oswald R (2003) Schimmelpilzbewertung aus der Sicht des Bausachverständigen. In: 29. Aachener Bausachverständigentage 2003. Leckstellen in Bauteilen; Wärme – Feuchte – Luft – Schall. Vieweg, Wiesbaden, Germany, pp. 120-126.

Oswald R (2004) Schwachstellen – "Die Schimmelspürhund-Unauffälligkeit" – eine wesentliche Eigenschaft von Gebäuden? Deutsch Bauz 9: 85-88.

Peltola J, Anderssen MA, Haahtela T, Mussalo-Rauhamaa H, Rainey FA, Kroppenstedt RM, Samson RA and Salkinoja-Salonen MS (2001) Toxic-metabolite-producing bacteria and fungus in an indoor environment. Appl Environ Microbiol 67: 3269-3274.

Pessi A-M, Suonketo J, Pentti M, Kurkilahti M, Peltola K and Rantio-Lehtimäki (2002) Microbial growth inside insulated external walls as an indoor air biocontamination source. Appl Environ Microbiol 68: 963-967.

Richardson N (2004) Orientierungshilfen zur Bewertung von Schimmelpilzbefall im Fußbodenaufbau. In: 8. Pilztagung des VDB – Berufsverband Deutscher Baubiologen, VDB, Jesteburg, Germany, pp. 125-136.

Schleibinger H, Laußmann D, Eis D, Samwer H and Rüden H (2004) Sind MVOC geeignete Indikatoren für einen verdeckten Schimmelpilzbefall? Umweltmed Forsch Prax 9: 151-162.

Trautmann C (2005) Aussagekraft von Schimmelpilzuntersuchungen. In: 9. Pilztagung des VDB – Berufsverband Deutscher Baubiologen, VDB, Jesteburg, Germany, pp. 17-32.

Trautmann C (2006) Nachweis und Bewertung von non-viable Sporen im Innenraum In: Keller R, Senkpiel K, Samson RA und Hoekstra ES (eds.) Partikuläre und molekulare Belastungen der Innenraum- und Außenluft, Band 10. Verlag Schmidt Römhild, Lübeck, Germany, pp. 315-332.

Trautmann C, Gabrio T, Dill I and Weidner U (2005b) Hintergrundkonzentrationen von Schimmelpilzen in Hausstaub. Bundesgesundheitsbl – Gesundheitsforsch – Gesundheitsschutz 48: 29-35.

Trautmann C, Gabrio T, Dill I, Weidner U and Baudisch C (2005a) Hintergrundkonzentration von Schimmelpilzen in Luft. Bundesgesundheitsbl – Gesundheitsforsch – Gesundheitsschutz 48: 12-20.

Tuomi T, Reijula K, Johnsson T, Hemminki K, Hintikka E-L, Lindroos O, Kalso S, Koukila-Kähkölä P, Mussalo-Rauhamaa H and Haahtela T (2000) mycotoxins in crude building materials from water-damaged buildings. Appl Environ Microbiol 66: 1899-1904.

UBA – Umweltbundesamt (2002) Leitfaden zur Vorbeugung, Untersuchung, Bewertung und Sanierung von Schimmelpilzwachstum in Innenräumen. Umweltbundesamt Berlin, Berlin, Germany.

UBA – Umweltbundesamt (2005) Leitfaden zur Ursachensuche und Sanierung bei Schimmelpilzwachstum in Innenräumen ("Schimmelpilzsanierungsleitfaden"). Umweltbundesamt Berlin, Berlin, Germany.

VDI – Verein Deutscher Ingenieure (2008) Messen von Innenraumluftverunreinigungen – Messstrategien bei der Untersuchung von Schimmelpilzen in Innenräumen VDI 4300 Bl. 10. VDI, Düsseldorf, Germany.

Virnisch L, Lorenz W and Trautmann C (2003) MVOC aus neuen Materialien. Z Umweltmed 4: 180-183.

Warscheid Th (1996) Biodeterioration of stones: analysis, quantification and evaluation. In: Proceedings of the 10th International Biodeterioration and Biodegradation Symposium, Dechema, Monograph 133, Frankfurt, Germany, pp. 115-120.

Warscheid Th (2003) The evaluation of biodeterioration processes on cultural objects and respective approaches for their effective control. In: Koestler RJ, Koestler VH, Charola AE and Nieto-Fernandez FE (eds.) Art, biology, and conservation 2002: Biodeterioration of works of art. The Metropolitan Museum of Art, New York, NY, USA, pp. 14-27.

Warscheid Th (2007) Wandbaustoffe aus Kalk ein natürlicher Schutz gegen Schimmelpilzbildung in Innenräumen. Der Bausachverständige 1: 11-16.

Warscheid Th (2008) Mikrobielle Belastungen in Estrichen im Zusammenhang mit Wasserschäden. In: Osward R (ed.) 33. Aachener Bausachverständigentage 2007. Bauwerksabdichtungen: Feuchteprobleme im Keller und Gebäudeinneren. Vieweg+Teubner, Wiesbaden, Germany.

Warscheid Th and Braams J (2000) Biodeterioration of stones – a review. Int Biodeterior Biodegr 46: 343-368.

Witt A and Krause S (2007) Nachweis von Bauholz- und Schimmelpilzen mittels DNA-Diagnostik. Der Bausachverständige 3: 24-26.

Wouters IM, Douwes J, Doekes G, Thorne PS, Brunekreef B and Heederik DJJ (2000) Increased levels of markers of microbial exposure in homes with indoor storage of organic household waste. Appl Environ Microbiol 66: 627-631.

# 16 Protection of wood

*Michael F. Sailer and Waldemar J. Homan*
*TNO, Delft, the Netherlands*

## Introduction

The use of wood by mankind is as old as mankind itself. The main reason why the Stone Age is not called the Wood Age is the fact that most wooden artefacts from that time have already been taken up in nature's recycling system. Wood has excellent material properties. It is strong in comparison to its weight, it is relatively easy to process – if needed even with flint stone tools – it has good insulation properties and its aesthetics are unrivalled. Finally, as any renewable resource, at the end of the product's life span, the material can be taken up by nature again and with the aid of water, $CO_2$, sunlight and a low quantity of soil nutrients new wood will be produced by the trees.

The degradability of wood with moisture content above fiber saturation, however, can be a disadvantage in those situations where a long life span of the wooden products is desired. Therefore, human beings try to increase the service life of wood in buildings since centuries. Thousands years ago, the use of durable wood species, resins and essential oils in order to avoid wood rot was already known (Moll 1920). The ancient Greeks and Romans (Griffin 1981) recognised the effects of sulphur. The reasons however that caused rot remained unknown over a long period. A big step was made by the recognition that wood rot is caused by microorganisms. Anton de Bary demonstrated in 1850 that some plant diseases are related to microorganisms (Bavendam 1962).

Where in Chapter 6 the actual life and metabolism of a wide variety of fungus species is described, this chapter will deal with what can be done against fungal decay of wood. For that, a short introduction in what wood deterioration actually is will be given and subsequently the traditional, modern and possible future ways of wood protection will be described.

### Decay of wood

The decay of wood is caused by microorganisms that use wood or other lignocellulosic materials and plants as a nutrition source. Depending on their enzymatic potential the microorganism are able to metabolise different components of the wood with negative effects on strengths properties (Huckfeld and Schmidt 2005, Schultz *et al.* 2008).

The details of wood degradation are nowadays better known but are still not completely understood. The wood biomass is consisting mainly of polysaccharides such as hemicelluloses, celluloses and lignin (Table 16.1). It is therefore a potential nutrient for the wood destroying fungi (Rayner and Boddy 1988).

In degraded wood three main rot types are typically distinguished. These degradation patterns are schematically shown in Figure 16.1.

*Table 16.1. Proportion of main wood components (Fengel and Wegener 1984).*

|  | Softwood (mass %) | Hardwood (mass %) |
|---|---|---|
| Cellulose | 42-49 | 42-51 |
| Hemicelluloses | 24-30 | 27-40 |
| Lignin | 25-30 | 18-24 |
| Extractives | 2-9 | 1-10 |

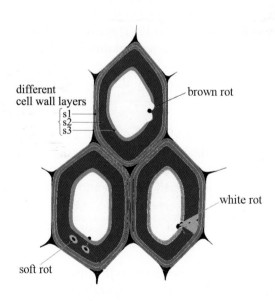

*Figure 16.1. Schematic presentation of the three degradation patterns s (changed according to Montgomery 1982).*

## Brown rot

Brown rot is a typical pattern of hemicellulose or cellulose degradation fungi. In which way that would happen was controversially discussed (Reese 1975). The use of celluloses as a nutrient requires different numbers of enzymes that split the macromolecule (Griffin 1981). A widely accepted pathway of cellulose degradation pattern is based on work with the ascomycete *Trichoderma* spp. The degradation according to this theory is initiated at the amorph areas of the microfibrilles which are first attacked by cellulases. The non-reducing free ends formed are then attacked and cleaved into tri- and bisaccarides which are further divided into glucose (Rayner and Boddy 1988). Koenigs showed (1972) the similarity of wood degraded by a $H_2O_2$/ $Fe_2$+ system and brown rot caused by basidiomycetes. This degradation pathway was considered to be different compared with other degradation systems (Montgomery 1982) and was later on frequently assessed and extended.

Some years ago low-molecular weight compounds were suggested to be involved in the brown rot degradation processes (Xu and Godell 2001, Goodell 2003). Low-molecular weight compounds are much smaller than enzymes. This would explain how the wooden cell walls can be penetrated by active substances, since the enzymes are considered to be too large to penetrate into the wooden cell wall to reach the cellulose (Highly *et al.* 1983).

## White rot

The white rot morphology is characterized by the removal of the darker (brown) lignin leaving white cellulose as the remaining substance. Lignin is the general name for a complex macromolecule in wood with a relatively high biological persistence. Kirk (1975) one of the pioneers in assessment of lignin degradation suggested different degradation steps and showed a decrease in methoxygroups. At this time, the structure of enzymes and degradation pathways involved were not clear. It is now known that lignin peroxidase, mangan-peroxidase and laccases are involved in the lignin degradation processes. In this context, the role of lignin as a sole nutrition for microorganisms is discussed. It is assumed that lignin can only be used in co-metabolism with other substances like celluloses and hemicelluloses (Fritsche 2000).

## Soft rot

Soft rot is a degradation pattern that is usually found in wood exposed to relative high moisture content and with enough oxygen available. Although this type of decay was first described for preserved planks used in cooling towers (Findlay and Savory

1954) it is not a typical decay pattern for buildings. Soft rot is characterised by two types (Anagnost *et al.* 1992, Eaton and Hale 1993). Type 1 causes the formation of cavities or chains; type 2 is similar to white rot. Soft-rot fungi degrade cellulose and hemicelluloses with a typical pattern in the s2 cell wall layer (see Figure 16.1).

## Wood protection by biocides

### Political and environmental aspects of industrial wood protection by application of biocides

The first (non-natural) biocides were based on metal salts like mercury chlorides, copper compounds and arsenic. In the beginning of the 19th century, with the start of the use of wooden sleepers for railways, the use of carbolineum oil got an enormous push.

Although man has always tried to find ways to expand the durability of wood a real boost to wood preservation was given by Bethel in the beginning of the 19th century. Bethel invented a process for pressure impregnation of railway sleepers and poles with creosote and tar oil (Hill 2006), followed by a second boost at the beginning of the 20th century. In 1892, the paint producer Bayer introduced the biocide Antinonnin as a crop protection which was also used in the building sector and later in preservation of wood for mining.

The name Dr. Wolman is strongly associated with the development of fluorine based preservatives and mine wood applications. In the mines, the miners have a preference to use wood as supports. In contrast to for instance steel, wood gives an early warning by cracking before breaking, thus providing the miners some time to get away before a collapse of a corridor occurs. The most common wood species used for these supports was pine (*Pinus sylvestris* L.). In the moist conditions of the mines however, this non-durable wood species was very susceptible to fungal attack thus loosing its original strength prematurely. Wolman tried to solve this problem by experimenting with many different chemical mixtures, which he impregnated in the wood.

One of the most effective wood preservatives is CCA (copper, chromium, arsenic). In the course of many decades, the formulation of the CCA was improved, leading in the end to a mixture that had very good "fixation" properties to the wood. Therefore, modern CCA formulations have a low leaching rate during use. In CCA treated wood that underwent a modern fixation process, the average leaching rate of the preservative during a life span of 25 years is in the order of magnitude of 5% (Homan *et al.* 1999).

Although CCA over many years of development has become very effective, concerns have been raised about ultimate release of arsenic and chromium into the biosphere, if not during the life span of the product than in the waste stage. Especially where human contact with the treated timber is likely, this has become a political issue. In a number of countries, e.g. Sweden and the Netherlands, political questions have been asked about the contact of children with CCA treated wood in playground equipment. Although it was proven that children would need to ingest 10-30 kg of soil from the immediate vicinity of the pole for it to pose a hazard (Henningson and Carlsson 1984), the general reaction was: "better safe than sorry". For that reason, many countries have introduced precautionary approaches into their legislation with regard to the use of CCA treated wood where a high probability of human contact exists.

The total market for treated timber is huge and it is very difficult to obtain accurate statistics on it. One of the reasons for this is the fact that different countries use different classification systems. Another reason is the fact that the industry, which has only a limited number of players, is not willing to release data on this. Jermer (1990) made first a good overview of the European situation. Within the 15 member states of the European Union prior to enlargement, approximately $18 \times 10^6$ m$^3$ of timber was preservative treated. Of this $10.3 \times 10^6$ m$^3$ was for construction timbers. In 1991 the global annual consumption of CCA was estimated to be 118,000 ton, with Europe consuming 39,000 ton. Japan used 95,500 ton of CCA in 1980, but this had declined to 2,000 ton in 1989. In US 1970, the total volume of CCA treated wood was 1.1 million m$^3$. This increased in 1996 to 13.2 million m$^3$ CCA-treated products (Holton 2001).

**US Environmental Protection Agency**

In February 2002, the US Environmental Protection Agency (EPA) announced a voluntary decision by the lumber industry of the United States to replace the sale to consumers of CCA-treated wood with alternative preservative systems by the end of 2003 (Hill 2006). This affected all residential uses including decking, picnic equipment, playground equipment, or residential fencing.

**EC Biocidal Products Directive**

In the EC, Commission Directive 2003/02 (6 January 2003) was published, concerned with restrictions on the use and marketing of arsenic. According to this directive, CCA-treated wood will not be allowed for certain end uses including any residential or domestic constructions. As of 30 June 2004, the net effect of this directive is to severely restrict the use of CCA-treated wood in situations in which there is potential human contact.

In the meantime, the implementation of the Biocidal Products Directive took place. From 2008 all wood preservation products that are based on biocides will require authorization according to the Biocidal Products Directive (BPD). The BPD will regulate the authorization of all biocide applications. Every individual active ingredient needs to be authorized following the system that is laid down in the BPD. Formulated products, that are usually composed of several actives, will be authorized by the member states, but only if all individual actives are authorized under the BPD. For the introduction within the BPD, product type 8 (wood preservatives) was one of the two pilots. The BPD is a highly complex system of judging several aspects of active ingredients in which the environmental aspects play a major role. Of any substance, the toxicity data need to be provided. After this, a tiered approach is followed. In the first step it is assumed that all of the applied product will end up in the receiving environment. This leads to a so-called Predicted Environmental Concentration (PEC) which will be compared with a so called Predicted No Effect Concentration (PNEC) which is based on the toxicity data. If PEC < PNEC the product can be admitted. If not, the second tier will be followed in which a new PEC will be calculated based upon emission scenarios. And so on. It is clear that although the calculations and the used models suggest a high scientific content of this process, many parameters that influence the outcome – for instance the size of receiving compartment or the type of exposition – are merely political decisions.

### Biocides

In discussions during the past decades on the use of biocides (Table 16.2) for material protection in general and on wood protection in particular, there has been very limited emphasis on the merits of these systems. It is important to recognise that a major improvement of the life-time of a (wood) product owing to the use of biocides has a very positive influence on environmental aspects. In several Life Cycle Assessment studies, preservative treated timber has a relatively good environmental profile compared to its alternatives (Künninger and Richter 2001). In almost all cases this is due to the considerable life-time improvement caused by the wood preservative treatments. However, there are limitations to the use of biocides too. A major issue is the persistence of biocides. Many of the more traditional biocides, both in crop protection and in material protection, are relatively persistent. The metal-based wood preservatives such as chromated copper arsenic can not be degraded and will end up in the environment, unless a good recollection system for the products has been set up. Although mapping the fate of biocides is complicated,

*Table 16.2. Development and introduction of preservative agents.*

| Year | Actives | Abbreviation | Specifies |
|------|---------|--------------|-----------|
| 1680 | creosote | | broad spectrum |
| 1920 | acid copper chromate | CC | broad spectrum |
| 1930 | pentachlorphenol | PCP | broad spectrum |
| | chromated copper arsenate | CCA | broad spectrum |
| | ammonical copper arsenate | ACA | broad spectrum |
| 1950 | inorganic borates | | fungi and insects |
| 1960 | LOSP vacuum treatments e.g. TBTO | TBTO | still used in joinery a/o in UK |
| 1980 | alkylammonium compounds | BAC | fungi |
| | | DDAC | fungi |
| 1989 | copper bis (N cyclohexyl diazenium dioxide) | Cu-HDO | fungi also in ground contact |
| 1990 | azoles: | | |
| | tebuconazole | Tebu | fungi and insects |
| | azaconazole | Aza | fungi and insects |
| | cyproconazole | Cypro | fungi and insects |
| 2000 | other non arsenic and chromium free systems (e.g. ammonical copper quat) | ACQ | fungi and insects |
| >2000 | betaines | | fungi (above ground) |

it has become very clear that at least parts of them are introduced into the food chain, ultimately accumulating in several predator species. In crop protection, this has led to the development of biocides that are easier degradable. In the Biocidal Products Directive (BPD) the biodegradability of the biocide is a major item. For material protection this is a bit more complicated. Where in agricultural application a life-time of the biocide in the order of magnitude of half a year is very acceptable, in fact may even be desired, in material protection this would be useless. In wood preservation, a biocide that protects the wood from deterioration should last at least 10 years and preferably 25 years or longer. And yet, at the end of its service life the biocide should be easily degradable. It is obvious that these demands are conflicting for many of the applications. However, in above ground situations, solutions have been found. If wood is not in contact with soil, the influence of soil inhabiting bacteria is limited. Therefore specific fungicides that are vulnerable to degradation by bacteria can be applied. They can be easily broken down by bacteria and are thus meeting the biodegradability requirements of the BPD.

## Classical types

Both creosote and many metal type formulations of wood preservatives are typical broad-spectrum biocides. For many years there have been tens of different formulations types developed and used in wood preservation. These include CC (copper chromium), CCB (copper chromium boron), CCP (copper chromium phosphor), Cu-naphtenate, PCP (penta chlorophenol).

## Newer types

In the last decades the use of creosote and several metal type formulations was increasingly restricted by legislation. Consequently, in many countries a number of "modern" wood preserving substances were assessed as potential substitutes. New generation biocides are more based on organic active ingredients. In many cases they are susceptible to decay by micro-organisms, especially bacteria. This is, in a way, even a prerequisite for the development of new actives. They preferably should be biodegradable. For this reason ground contact applications are limited. Many actives fall into this category. When it comes to actives directed against insects, the possibilities are even more extended. Insecticides, however, are out of the scope of this book.

### Interactions of wood preservatives and wood

The cell structure is characteristic for each wood species, and therefore an important factor for different wood properties. If the pathway from one cell to its adjacent cell are blocked, it is difficult for liquids to penetrate. This also influences the penetration depth of wood preservatives. In Europe, the wood species Spruce (*Picea abies* Karst.) is commercially the most important wood species for wood preservation. In this species, once the wood is dried, the connections from one tracheïd to the next are closed by aspiration of the bordered pits (Figure 16.2, Bauch 1971).

Consequently, the wood species is very difficult to impregnate. Johansson and Nordman-Edberg (1987) among others, describe refractory effects. In practice, this means that there are only two ways to overcome this. Either the wood is treated before it is dried when the pits are still open and the penetration is easy, or the wood will be dried and processed to its final dimensions (including profiles, cross cuts, drilled holes, etc.) and then treated. The latter way of treating leads to a relatively high concentration of the preservative in the outer few millimetres of the wood. If the preservative is meant to be fixed to the woods cell wall then it is clear that by far not all molecules of the product are at the right location in the cell wall. The majority

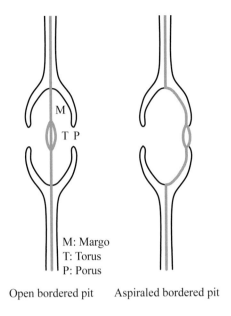

M: Margo
T: Torus
P: Porus

Open bordered pit    Aspiraled bordered pit

*Figure 16.2. Open and aspirated bordered pit in* Picea abies *Karst.*

will simply bulk in the lumina of the fibers (Figures 16.3 and 16.4). Therefore, the fixation of the product to the wood is limited. Consequently, leaching of the product may occur, leading to undesired environmental effects and ultimately, when the level of the preservative concentration drops below the biological reference value, failure of the product.

Depending on the application of the end product, different biocides and treatments are used. Wood in ground contact or water is usually more exposed to moisture and requires a good penetration and fixation of the agents. There are a number of wood preservatives that cannot be applied in ground or water contact, for example the azoles, BAC and DDAC. If wood is used above ground, the treatments vary from non biocidal treatments (oil treatments, durability by design) to pressure treatments with a wide variety of biocides.

### Interactions of biocides with micro-organisms

The interactions between microorganisms and wood preserving agents are not in all cases well understood. The response of microorganisms to biocides is depending on a number of different factors like growth conditions, concentration and availability. The effects of fungicides on fungi can be different depending on the mode of action. They

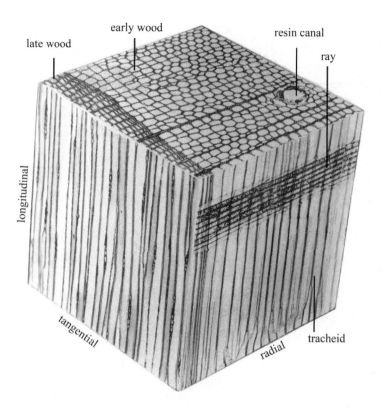

*Figure 16.3. Anatomical structure of softwood.*

can show inhibitory or lethal effects on organism and different effects on different stages of growth. Fungi further respond in different ways to fungicides, not all groups have the same level of sensitivity or resistance to a certain fungicide. The complicated interaction of fungal response and conditions is visible in one example described by McDonough *et al.* (1960). In the yeast phase the sensitive strain of *Blastomyces dermatides* was completely inhibited using a cycloheximine concentration of 90 μM at 37 °C, whereas the mycelium phase at 25 °C was resistant to 50 times higher concentrations. Several microorganisms can tolerate the effects of biocides or reduce their efficiency like the copper tolerant *Antrodia vaillantii, Leucogyrophana pinastri* and copper sensitive *Poria monticola, Gloeophyllum trabeum* (e.g. Leithoff *et al.* 1995, Humar *et al.* 2005).

Most of the examples refer to biocides that are already known for a long time. The mode of action of many biocides is complicated and in many cases not fully clarified

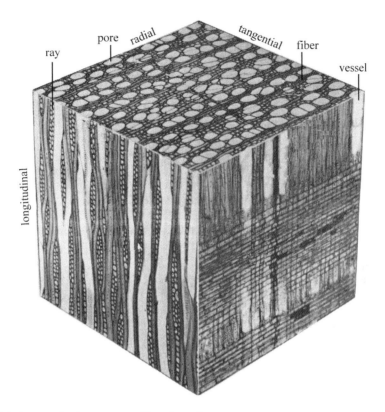

*Figure 16.4. Anatomical structure of hardwood.*

yet. The reasons are the often complicated interactions that can cause more than one effect that is recognized. Some examples of the impact on the biological processes are mentioned in the next paragraphs.

**Influence on proteins**

Typical bioinhibitors are heavy metals like copper, silver or mercury can affect enzymes. They are already effective in low concentrations in metal salts as well as in organic compounds, they react with the SH-groups of proteins, which has a strong influence on the enzyme structure. Treatments with less than 2 kg CCA/m³ wood showed already effects on wood degrading microorganisms (Choi *et al.* 2001).

## Cell membranes

Typical bioinhibitors are surface-active agents and typical disinfectants like phenols and creosols. They are enriched in the lipoprotein membranes and affect the semi-permeability of the cell membranes.

## Inhibition of macromolecule synthesis

Typical bioinhibitors are sulfonamides that are structurally very similar to 4-aminobenzoate, a substance that is part of a co-enzyme (structure analogy). The organism also uses sulfonamides in place of the 4-aminobenzoate to create the co-enzyme, which however leads to a co-enzyme that does not work anymore (Schlegel 1992). Other biocides like cycloheximidine or glutarimide are considered to inhibit DNA and protein synthesis.

## Influence on energy metabolism

Typical bioinhibitors are organic substances that are considered to have effect on the mitochondria of microorganisms like thujaplicine. Thujaplicine, a metabolite of some trees belonging to Cupressacea, is described to block the electron transport process in the cytochrome chain at very low concentrations of $10^{-6}$ to $10^{-7}$ M. The $O_2$ uptake and functioning of the tricarboxylic acid cycle and the formation of ATP from ADP of wood degrading fungi is inhibited (Griffin 1981).

# Other ways of wood protection

In nature, when a tree falls down in the forest it will be broken down by microorganisms and insects. In this way, the constituents are taken up once again into the circle of the ecosystem. The nutrients that the tree used during the growing process will become completely available again. In order to prevent wood from decaying in the living tree, the living cells in the sapwood can actively fight the decaying organisms. In the heartwood, there are not many living cells available. Some tree species protect their heartwood by depositing extractives with fungicidal or insecticidal action, e.g. terpenoids or thujaplicine. Another common protection mechanism is the closure of the wood cells in the heartwood. Next to the application of bio-inhibitors, there are several other options to prevent and control biodegradation of wood. In order to describe them, some understanding of the minimum requirements for wood decaying fungi is needed.

## Effects on degradation activities

The biodegradable properties of wood are obviously in many cases undesirable in building applications. Therefore, it is essential to know the physical conditions in which the decaying microorganisms thrive and use this knowledge to protect the wood. Actually, controlling these conditions is the most desired option for sustainable protection of wood; however, it is not always optimal or feasible in practice. For fungal growth on wood, the following six conditions have to be met.

### 1. Suitable nutrition

For wood decaying fungi this is, obviously, wood. In contrast to wood decaying fungi, many molds and blue-stain fungi merely digest substances that are in (extractives) or on the wood. Although these fungi do not break down the wood's cell wall components itself, many blue staining fungi can digest the substituents of the middle lamellae and of the pit membranes and thereby they do prepare the pathways for their successor fungi, e.g. by opening the wood structure. Many authors describe the increase of low molecular carbohydrates on the wood surface and the occupation by microorganisms repetitively (Theander *et al.* 1993, Terziev *et al.* 1994, Terziev and Nilsson 1999). Sharpe and Dickinson (1992, 1993) showed that sugar monomers and precursors of lignin, that were diffused to the surface, can be used as nutrient by *Aureobasidium pullulans*, surface molds and discoloring fungi. The lignin and the cellulose, however, can not be used (Wolf and Liese 1977, Ritschkoff *et al.* 1997). In nature *Aureobasidium pullulans* is found abundantly on surfaces of leafs and in between leaf cells of several plant species. The fungus probably uses the substances that are present in the intercellular spaces as nutrient (Andrews *et al.* 1994, Carlile and Watkinson 1995). The presence of external food sources, for instance nitrogen in soil, can accelerate the growth of fungi.

### 2. Presence of oxygen

Air contains about 21% oxygen. The solubility of $O_2$ in water is low and in static liquids or waterlogged material it diffuses really slowly. Most wood decaying fungi are aerobic (Rypáček 1966, Bavendamm 1974, Scheffer 1986, Schmidt 1994, Xie *et al.* 1997). For growing and surviving many fungi need only a small amount of oxygen. In the decay area of *Acer sacharum* oxygen levels of 0.8% were measured. Bavendamm (1974) indicated a minimum $O_2$ content of 0.5 vol%. Scheffer (1986) has found in his own work and in literature that 0.3-0.4 vol% suffices. Typical fungi that already die after one week of low $O_2$ levels are *Meruliporia incrassata* and *C. puteana*. Fungi will not grow when wood is completely waterlogged, for instance under water or in soils under the ground water level. There are other fungi that are facultative anaerobe (examples *Aqualinderella, Blastocladia*) and even some that are obligate anaerobes (example *Neocallimastix*). Excessive high oxygen levels may also prove toxic. Although this does not occur in nature, in some process environments it may be relevant. *Penicillium chrysogenum* suffers oxygen toxicity at an air pressure of 1.5 bar.

## 3. Moisture content

Like in other building materials the moisture content of wood is often expressed as the mass percentage of moisture as compared to the dry mass of the material.

MC = (mw – md) / md

Where:
MC = moisture content;
mw = mass of the wet wood;
md = mass of oven dried wood.

If the average moisture content of wood is more than 25-30% during a longer period (weeks), the risk of fungal growth increases drastically. Microorganisms however cannot use all the water that is present in the substrate, but only the water that is not bound by dissolved substances such as salts, sugars, etc. Finally, the absolute moisture content is not the essential factor on fungal activity, but the water activity and matrix potential, which correspond under equilibrium conditions to specific pore dimensions (Table 16.3). They determine to what extend the microorganisms can use the water. The water activity $(a_w)$ is defined as the ratio between the water vapor pressure of the solution (pS) and the water vapor pressure of pure water (pW) under the same (equilibrium) conditions (Griffin 1977, Schmidt 1994, Carlile and Watkinson 1995; see also Chapters 1 and 2).

$a_w$ = pS/pW

Generally it is believed that a fungal spore needs free water in its immediate surrounding environment in order to start growing. In wood in which the water is completely homogeneously distributed, free water will only be available above fiber saturation (f.s.). Depending on wood species, the range is around 25-35% wood moisture content.

*Table 16.3. Water availability in different environments (at equilibrium conditions).*

| Water activity (–) | Water potential (Mpa) | Pore radius (μm) | Examples |
|---|---|---|---|
| 1.0 | 0 | | cell lumina, pure water |
| 0.9993 | -0.1 | 1.5 | fiber saturation |
| 0.98 | -2.8 | 0.5 | sea water |
| 0.97 | -4.2 | 0.035 | lower limit wood destroying fungi |
| 0.96 | -5.6 | 0.026 | leaf litter |
| 0.90 | -14.5 | 0.01 | ham |

In practice, however the moisture is usually not completely homogeneously distributed in the wood. Therefore even at lower general wood moisture contents, very locally the moisture content of the material can be above f.s. At these specific spots, a fungus can start growing. If the moisture content drops, it will stop growing, but will not die. If the moisture content increases again, the growth continues. Dry rot fungi can transport water through their mycelia over long distances. As an example, the dry rot fungus *Serpula lacrymans* can destroy wood on the second floor of a building, transporting the moisture from the cellar.

## 4. Temperature

The sensitivity of fungi to temperature is strongly dependent on the fungal species. When the temperature drops below 4 °C almost all fungal growth stops, although the fungi do not die. Some fungi die with temperatures above 55 °C, others can withstand extreme high temperatures. The temperatures that generally occur in buildings are very suitable for fungal growth. The optimum is different for each fungus. For many wood-destroying fungi the optimum is in the range of 20-30 °C.

## 5. Toxic substances

The principle of biocidal protection systems is based on toxicity. Also the natural durability of some wood species is based on toxic substances that the tree produces itself. Examples of this are thujaplicine (Figure 16.5), a natural fungicide that is produced by Western Red Cedar (*Thuja plicata*) and the tannins that can be found, for example in Oak species. Often these substances can also leach out of the wood and for that reason even durable wood species, when applied wrongly, can leach and therefore become less durable. There have been several attempts (e.g. Grohs and Kunz 1998) to try and extract such chemicals in order to protect less durable wood species. Most of these ideas, however are not economical viable.

## 6. Time

Wood decaying processes are relatively slow compared to for instance the growth of molds. The speed of decay is depending on many factors. Each situation is different and therefore shows a different decay speed. In Figure 16.6 relative speeds of typical decay types are indicated. Brown and white rot fungi can do their work in weeks to months under optimal conditions. Soft rot fungi usually need months to years and bacterial decay of wood can take more than 100 years or more under anaerobic conditions.

## Biotic interactions

A very important factor, affecting many of the previously mentioned 6 factors, is the influence of other organisms. These can work both in a symbiotic and in a competitive way. Many microorganisms excrete substances to create advantages over there competitors. This happens at species level (for example *Penicillium* ssp. excretes the antibiotic penicillin), at strain level and even at individual level (for example the white rot fungus *Coriolus versicolor* creates clearly visible black borderlines between the territoria of individual fungi within the same trunk). Very often, the decay of wood is a story in which succession is very important. Pioneer fungi open the wood structure, making pathways open for successor fungi and water. Often higher organisms like insects also collaborate in the process. In circumstances with very low oxygen conditions bacteria can attack the wood. This is a very lengthy process and can take up to 100 years or more.

Besides fungi there are many other wood decaying organisms. In sea water marine these are organisms like *Limnoria spp.*, and *Teredo navalis*. On land, world wide,

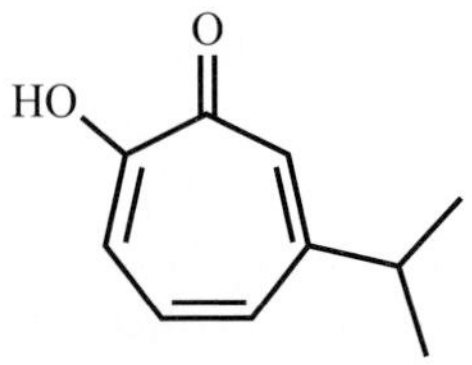

*Figure 16.5. Chemical structure of thujaplicine.*

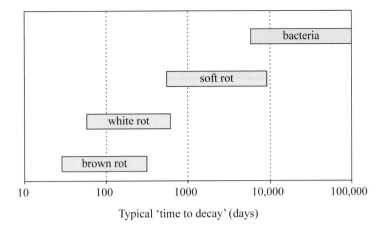

*Figure 16.6. Relative "time to decay" of different decay types.*

insects are the relevant wood decaying organisms. Subterranean termites are dominant but also species like *Hylotrupus baljulus*, *Lyctus* ssp., *Xestobium* ssp., and *Anobium punctatum* are economically of major importance. Many insects live in symbiosis with fungi. *Donkioporia* is an example of a fungus that attracts a/o termites. Some termite species actively breed and maintain fungus gardens in their mounds.

### Extension of service life by non-toxic measures

In the previous paragraph the minimal conditions for fungal growth were given. From this it can be derived how fungal growth and thus wood degradation can be controlled or prevented. "Constructive wood conservation" also known as "durability by design" is built upon this principle. In most of the applications the moisture content of the wood is the parameter that offers the best control possibilities. The basic principle then is simple: "take care that the wood stays dry or – if/when that is not possible – take care that after wetting it dries rapidly". The time that the material stays wet is the dominant factor for growth of fungi. Rather than looking at an overall moisture content of the material the time-of-wetness (TOW) around the spore or the hyphen is decisive (i.e. the very local TOW is important).

### Extended service life by design

Tjeerdsma *et al.* (2003) worked out examples of "durability by design" in exterior applications (e.g. see Figure 16.7). The basic rules for "constructive wood preservation" are:

- Take measures to:
  1. prevent moisture *uptake* or ingression of water by avoiding horizontal surfaces, horizontal holes, capillary seams or surfaces;
  2. prevent moisture *uptake* through the cross-cut sides by avoiding deepened connectors, drilled holes for steel bolts, etc.;
  3. avoid the enclosure of moisture (paint, sand and dirt in seams, multilateral enclosures in constructions);
  4. prevent formation of condense water (especially in climate separating constructions);
  5. avoid ground contact;
- and create:
  7. good outflow of water out of the wooden object (for instance the outer part of a window sill allows for significant faster drying if a slope of 15° is applied, compared to the standard slope of 8°);
  8. good ventilation of the wooden object (for instance ventilation openings should be wider than 8 mm in order to prevent drops of water to be adhered between the surfaces).

### Water-repellent surface treatments

Additionally to the constructive measurements wood can be kept dry by different treatments based on impregnation or surface application of water repellent substances like oils, resins, waxes or coatings (Banks and Voulgaridis 1980, Ritschkoff *et al.* 1999, Sailer and Rapp 2001). These treatments fill the cell lumina (Figure 16.8) but do not change the cell wall properties substantially (Hyvönen *et al.* 2005).

Depending on the definition this type of treatment can be also be considered as wood modification, which will be explained in the next paragraph.

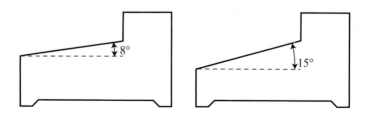

*Figure 16.7. Standard and improved windowsill, for better off flow of rainwater.*

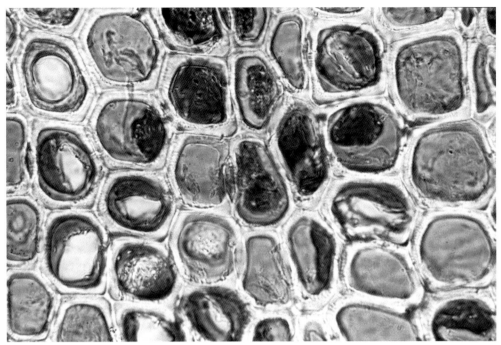

*Figure 16.8.* Pinus sylvestris *with filled cell lumina after impregnation with fatty acids and growth of* Coniophora puteana.

## Wood modification to increase durability

The generic term wood modification is not protected and therefore not well defined. Hill (2006) gave the following definition of wood modification:

> Wood modification involves the action of a chemical, biological of physical agent upon the material, resulting in a desired property enhancement during the service life of the modified wood. The modified wood should itself be *non-toxic* under service conditions and, furthermore, there should be no release of any toxic substances during service, or at end of life following disposal or recycling of the modified wood. If the modification is intended for improved resistance to biological attack, then the mode of action should be non-biocidal.

In short, the general goal of wood modification is to improve properties of the material and even to add new functionalities. Examples of this are adding water repellent properties, UV resistance, fire retardancy, etc. Wood modification is strongly emerging and will be developed to a great extent in the near future. A number of wood modification processes have found their way to the market. The raw material is often fast growing softwood. Here we mention the *thermal* processes such as Plato®, Stellac® and Thermowood®, which have a considerable loss in strength as major drawback and the *chemical* processes Kebony®, Accoya® en Belmadur®.

In general, with chemical modification crucial wood properties like durability or dimensional stability can be improved more than with thermal modification and besides that tailor made property improvement can be achieved (Homan and Jorissen 2004). For all existing processes there are some disadvantages. High costs for production is one of them. Also in the functionalization there is still room for improvement. The ester bond in Accoya (i.e. acetylated wood), for instance, can be hydrolyzed in alkaline soils. Therefore, when acetylated wood is applied in alkaline soils the acetylation can be undone and subsequently the wood can start to be decayed.

Although there is a myriad of possibilities to chemically changes in the cell wall constituents of wood, the most common modification processes are based on replacing the hydrophilic hydroxyl groups with a hydrophobic chemical group. The basic principle is as follows:

$$Wood - OH + X \rightarrow Wood - Y + Z$$

Where:
Wood = cellulose, lignin or hemi-cellulose;
X = a reactive agent;
Y = a hydrophilic group;
Z = a by-product.

By substitution of the hydrophilic OH-groups with hydrophobic groups the hygroscopicity of the wood drops. For instance in acetylated wood the equilibrium moisture content (EMC) of wood in air with a relative humidity of 65%, is reduced to less than half the EMC of untreated wood. This phenomenon on itself is providing a higher durability, but there is a second effect. Many wood decaying fungi first attack the wood on the very same OH-groups. If the enzyme of the digestion system of the fungus does not encounter these groups, because they were substituted, it can no longer break down the material. Both effects make modified wood highly durable. Additionally wood modification reduces the swelling and shrinkage behaviour of wood. In practical applications, this has advantages for crack formation, glue adherence, paint adherence and thereby provides additional life-time to the product.

Durability obtained by wood modification has certain benefits over the wood preservation methods in which biocides are used. The absence of toxic chemicals in the end product is important, but at least as important is the fact that by using very specific chemicals and treatment, tailor made products can be made. In fact, it is possible to produce material with very specific properties "fit for purpose" (Figure 16.9).

## Application of durable wood species

Nature provides several wood species with a high natural durability. Most of these species are hardwoods from the tropics. Natural durability can be based upon several mechanisms. The natural durability of some wood species like Basralocus is based on the *wood density*. Although there are many exceptions, a general rule of thumb is the higher the wood density the *higher* the *lignin content* and the higher the durability.

Besides this density dependent lignin content effect, several tree species deposit *substances* that are *toxic* for wood decaying fungi in the heartwood. The high durability of many tropical hardwoods is based upon this. Also the durability of

*Figure 16.9. Durability and dimensional stability provided by acetylation of radiate pine offers unexpected application possibilities (Photo and sundial design by W.J. Homan).*

the light wood species Western Red Cedar is based on natural toxins, in this case thujaplicine. Another example of is Alaska Yellow Cedar. The *anatomical structure* of this (also light) wood species makes it almost impossible for water in its liquid form to penetrate the wood. In this way, the wood, with the exception of the outer 3-5 cells, is often too dry to be deteriorated by fungi. For almost every wood species, the combination of constitution of the cell wall, chemical deposits, and wood anatomy determines the durability.

Concern about the availability of durable tropical hardwoods is increasing. Primary forests are becoming rare and even on secondary forests there is a high pressure. With increasing populations, all around the world the need of land for other uses is increasing. It is observed world wide, that if wood out of forests provides less income for the local population than alternative land use, the forest disappears. Another consequence of globally growing population and economies is the increasing demand on wood. All these factors lead to decreasing availability and increasing prices of durable hardwood. The price of soft woods for the packaging and pallet industry and for the pulp and paper industry has gone up 50% between 2005 and 2008.

## Protection of wood using other microorganisms

Microorganisms can create different interrelations amongst each other and with plants. The interactions are actually part of a complex system that is developed by co-evolution and that usually has different effects of the partners involved in the interaction.

Research related to microorganisms as a biological control agent (biological control) was rapidly growing in the last 50 years (Schoeman *et al.* 1999). Most of the research was related to medical uses and the control of agricultural pests and pathogens. The growing concern about the environmental impact of biocides and insecticides led to initial ideas of using biological control in the wood conservation. The protection of wood with other organisms is based on the idea that the wood is used as a host for protecting microorganisms, which should prevent growth of degrading organisms, without degradation of the wood. A high cell density of microorganisms using sugars or other substances that can be easily metabolized can influence the growth of wood degrading fungi (pathogens). Investigation of the mechanisms of biological control suggests three different ways of inhibition that probably do not act independent (Hulme and Shields 1975, Schoeman *et al.* 1999):

- occupation of a resource and competition for nutrition's;
- production of antibiotic substances;
- mycoparasitism and production of extracellular enzymes.

A number of case studies have been carried out with the objective to introduce such ideas in the preservation of timber. One of the most assessed microorganisms is *Trichoderma* spp. Inhibition effects of *Trichoderma* and other soil fungi on wood degrading brown fungi were frequently reported and patented, e.g. US4996157 (Smith *et al.* 1991). The practical application however is, up till date, still limited due to persistence problems.

Another well know example with practical application is the non pathogenic *Peniophora gigantea* limiting the growth of the pathogen *Heterobasidion annosum* in the roots of trees. When the surface of trunks is inoculated with spores of *Peniophora gigantea*, *Heterobasidion annosum* stops from growing through the trunks to other non infected trees. Schmidt and Müller (1996) reported the use of *O. piliferum* to prevent stain in pine sapwood. Because of the lack of strong pigmentation ascospore isolates of *Ophiostoma picea* and *O. pluriannulatum* were used to reduce the growth with wild stems of *Ophiostoma* White-McDougall *et al.* (1998).

Most studies are reporting the use of mycelial fungi, fewer reports are known using yeast and bacteria. Initial trails were reported by Greaves (1970) and Kreber and Morell (1993) who assessed 15 bacterial and fungal isolates to inhibit fungal stain. Patent WO9712737 (Poppen and Pallaske 1997) is claiming the use of *Bacillus* and *Streptomyces* as an effective biocontrol measure. Some treatments to prevent post harvest deterioration of fruit were carried out by using *Candida* ssp. (Mc Laughlin *et al.* 1992). The efficiency of the yeast like fungus *Aureobasidium pullulans* as a biocontrol agent of postharvest diseases is mentioned by Schena *et al.* (2002). A different approach is patented in WO0075293 (Poulsen and Kragh 2000), which is claiming the use of enzymes of marine algae from the family of *Gigartinaceae* as a biocontrol agent of wood.

Most of the literature describes the successful application of bio-control under laboratory conditions. Under these conditions, it seems to be easier to establish microorganisms with bio-control properties.

The inevitable influence of changing environmental factors in practical application of the microorganisms is considered as an important factor of their function. Schoeman *et al.* (1999) are distinguishing between physical/chemical factors and biological factors.

The most effective control can therefore be expected by using microorganisms and considering the changing dynamic environmental parameters. One of the most successful applications of organisms is achieved with fungi in the genus *Trichoderma* that have been developed as a biocontrol against fungal diseases of plants (Harman 2006). These fungi grow in their natural habitat on the root surface and can be effective against root diseases, but also improve the yield of plants, e.g. maize. The positive effects of this biocontrol are now leading to increased commercial application world-wide (Ubalua and Oti 2007).

## New developments

The recent developments in the area of biocides have led to ecologically more friendly wood protection products. As a precautionary policy, it is desired to use biocides only when they are strictly needed and even then only under strict conditions. One of the ways to reduce the amount of biocides needed to protect material is controlled release. Preferably, a controlled release system should hold on to the biocide until the moment it is needed. For instance, in crop protection and medicine a large number of examples of controlled release systems exist. This ranges from slow release, extended release to induced or triggered release. Similar developments are at present also being brought into the building material sector (see also Chapter 15).

Already in the 1970's systems like pills and rods containing biocides were developed. These products were put into the joints of window frames and they contained a mantle that became permeable under the influence of water. More modern systems function at a much smaller scale, e.g. with the intercalation of biocides in nano particles.

Some of the disadvantages of wide spread "slow release" systems, like leaking out of the active component and a high dose of active component, can be overcome by a true release on demand system. The Bioswitch technology (De Jong *et al.* 2005) is based on a carrier system that contains an active component that is only released when it is needed. In Bioswitch the entrapped active component becomes only available when an external stimulus triggers the release. For wood protection the trigger for release of a preservative could be the presence of initial fungal contamination, thereby preventing further growth of the fungus. Several release on demand systems have been developed by TNO in the Netherlands based on a carrier system of biodegradable micron sized particles (Homan *et al.* 2007). Thereby the active component is released only when a specific enzyme is present that digests the particles made of a specific biopolymer. Amongst the biopolymers used were starch derivatives, cellulose derivatives and proteins. This resulted in carrier particles that were sensitive to the triggers amylase, cellulase or protease respectively. Amongst the active components were the antimicrobials lysozyme and gentamicin. These charged molecules were entrapped in the Bioswitch particles by electrostatic interactions.

This is only one example of a binding-release mechanism. The Bioswitch technology is versatile and can therefore be adapted to many specific applications.

# References

Anagnost SE, Worrall JJ and Wang CJK (1992) Diffuse cavity formation in soft rot of pine. Doc No IRG/WP 92-154, Int Res Group on Wood Preservation, Stockholm, Sweden, 9 pp.

Andrews JH, Harris RF, Spear RN, Lau GW and Nordheim, EV (1994) Morphogenesis and adhesion of Aureobasidium pullulans. Can J Microbiol 40: 6-17.

Banks WB and Voulgaridis EV (1980) The performance of water repellents in the control of moisture absorbtion by wood exposed to the water. In: British Wood Preservation Association 198, Grange Press, Cambridge, UK, pp. 43-53.

Bauch J (1971) Die Struktur der Hoftüpfelmembranen in Gymnospermen und ihr Einfluss auf die Wegsamkeit des Holzgewebes. Habilitationsschrift Universität Hamburg, Hamburg, Germany.

Bavendam W (1962) Kleine Geschichte des Holzschutzes. Holz-Zentralblatt 83: 982-983.

European Commission (1998). Directive 98/8/EC of the European Parliament and of the Council of 16 February 1998 concerning the placing of biocidal products on the market. OJ Eur Union L 123: 1-63.

Carlile MJ and Watkinson SC (1995) The fungi. Academic press, London, UK, 482 pp.

Choi SM, Ruddick JNR and Morris PI (2001) The possible role of mobile CCA components in preventing spore germination, in checked surfaces, in treated wood exposed above ground. Doc No IRG/WP 01-30263, Int Res Group on Wood Preservation, Stockholm, Sweden, 17 pp.

De Jong AR, Boumans H, Slaghek T, Van Veen J, Rijk R, and Van Zandvoort M (2005). Active and intelligent packaging for food: is it the future. Food Addit Contam 22: 975-979.

Eaton RA and Hale MDC (1993) Wood-decay, Pest and Protection, Chapman & Hall, London, UK, 546 pp.

Findlay WPK and Savory JG (1954) Moderfäule: Die Zersetzung von Holz durch niedere Pilze. Holz, Roh- u. Werkstoff 12: 293-296.

Fritsche W (2000) Mikrobiologie. 2$^e$ Auflage. Spektrum, Akademischer Verlag Gustav Fischer, Stuttgart, Germany, 622 pp.

Greaves H (1970) The effect of selected bacteria and actinomyetes on the decay capacity of some wood rotting fungi. In: Becker G and Liese W (eds.) Material und Organismen 2. Duncker und Humblot, Berlin, Germany, pp. 265-279.

Griffin DM (1977) Water potential and wood-decay fungi. Ann Rev Phytopathol 15: 319-329.

Griffin DH (1981) Fungal physiology. John Wiley & Sons, Chichester, UK, 383 pp.

Grohs M and Kunz B (1998) Studie zur Nutzung von Kernholzextrakten als potentielle biologische Holzschutzmittel. Holz Roh- u. Werkstoff 56:217-220.

Harman GE (2006) Overview of mechanisms and uses of *Trichoderma* spp. Phytopathology 96: 190-194.

Henningson B and Carlson B (1984) Leaching of copper, chrome and arsenic from preservative-treated timber in playground equipment. Doc. No. IRG/WP 3149. Int Res Group on Wood Preservation, Stockholm, Sweden.

Highley TL, Murmanis LL and Palmer JG (1983) Electron microscopy of cellulose decomposition by brown-rot fungi. Holzforsch 37: 271-278.

Hill CAS (2006) Wood modification: chemical, thermal and other processes. John Wiley & Sons Ltd. Chichester, UK.

Holton WC (2001) Special treatment: disposing of CCA-treated wood. Environ Health Perspect 109: A274-A276.

Homan WJ and Jorissen AJM (2004) Wood modification developments. HERON 49: 361-386.

Homan WJ and Van Oosten AL (1999) Statistically stable models for determination of PEC. Doc No IRG/WP 99-50135, Int Res Group on Wood Preservation, Rosenheim, Germany.

Homan WJ, Jetten J, Sailer M, Slaghek T and Timmermans J (2007) Bioswitch: a versatile release on command system for wood protection. In: European Conference on Wood Modification 3, ECWM 2007, Cardiff, Wales, UK.

Huckfeldt T, Schmidt O (2005) Hausfäule- und Bauholzpilze, Diagnose und Sanierung, Rudolf Müller Verlag, Koln, Germany, 378 pp.

Humar M, Šentjurc M, Amartey SA and Pohleven F (2005) Influence of acidification of CCB (Cu/Cr/B) impregnated wood on fungal copper tolerance. Chemosphere 58: 743-749.

Hyvönen A, Piltonen P and Niinimäki J (2005) Biodegradable substances in wood protection. In: Jalkanen A and Nygren P (eds.) Sustainable use of renewable natural resources – from principles to practices. University of Helsinki Department of Forest Ecology Publications 34, Helsinki, Finland.

Hulme MA and Shields JK (1975) Antagonistic and synergistic effects for biological control of decay. In: Liese W (ed.), Biological transformation of wood by micro-organisms. Springer Verlag, Berlin, Germany, pp. 52-63.

Jalkanen A and Nygren P (eds.) (2005) Sustainable use of renewable natural resources – from priciples to practices. University of Helsinki, Department of Forest Ecology, Publication 34.

Jermer J (1990) Production of preservative treated wood in some countries. Doc No IRG/WP 3598, Int Res Group on Wood Preservation, Stockholm, Sweden.

Johansson I and Nordman-Edberg K (1987) Studies on the permeability of Norway spruce. Doc No IRG/WP 2295, Int Res Group on Wood Preservation, Stockholm, Sweden.

Kirk TK (1985) The chemistry and biochemistry of decay. In: Nicholas DD (ed.) Wood deterioration and its prevention by preservative treatments, volume 1, Degradation and Proctection of wood. Syracuse University Press, Syracuse, USA, pp. 149-181.

Koenigs JW (1972) Production of extracellular hydrogen peroxidase and peroxidase by wood-rotting fungi. Phytopathology 62: 100-110.

Kreber B and Morell JJ (1993) Ability of selected bacterial and fungal bioprotectants to limit fungal stain in ponderosa pine sapwood. Wood Fiber Sci 25: 23-34.

Künninger T and Richter K (2001) Ökobilanz von Konstruktionen im Garten- und Landaschaftsbau. EMPA 115: 43.

Leithoff H, Stephan HL, Lenz MT and Peek R-D (1995) Growth of the copper tolerant brown rot fungus *Antrodia vaillantii* on different substrates. Doc No IRG/WP 95-10121, Int. Res. Group on Wood Preservation, Stockholm, Sweden, 10 pp.

McDonough ES, Georg LK, Ajello L and Brinkman S (1960) Growth of dimorphic human pathogenic fungi on media containing cycloheximide and chloramphenicol. Mycopath Mycol Appl 13: 113-120.

McLaughlin RJ, Wilson CL, Droby S, Ben-Arie R and Chatutze E (1992) Biological control of postharvest diseases of grape peach and apple with the yeasts, *Kloekera apiculata* and *Candida guilliermondii*. Plant Dis 76: 470-473.

Moll F (1920) Holzschutz. Seine Entwicklung von der Urzeit bis zur Umwandlung des Handwerks in Fabrikbetrieb. Bd 10 der Beiträge zur Geschichte der Technik und Industrie. VDI und Springer, Berlin, Germany, pp. 66-92.

Montgomery RAP (1982) The role of polysaccaridase encymes in the decay of wood by basidiomycetes. In: Frankland JC, Hedger JN and Swift MJ (eds.) Decomposer basidiomycetes, their biology and ecology, Cambridge University Press, Cambridge, UK.

Poppen H and Pallaske M (1997) Agent based on natural active substances for protecting technical materials against damages and destruction by harmful organisms. Patent nr. WO9712737 (A1) – 1997-04-10.

Poulsen CH and Kragh KM (2000) Anti-fouling composition. Patent nr. WO0075293 (A2) – 2000-12-14.

Rayner ADM and Boddy L (1988) Fungal decomposition of wood. Its biology and ecology. Wiley, Chichester, UK, 587 pp.

Reese ET (1975) Polysaccharides and the hydrolysis of insoluble substrates. In: Liese W (ed.) Biological Transformation of Wood by Microorganisms: Proceedings of the Sessions on Wood Products Pathology at the 2nd International Congress of Plant Pathology. Springer, Berlin, Germany, pp. 165-181.

Ritschkoff A-C, Rättö M and Thomassin F (1997) Influence of the nutritional elements on pigmentation and production of biomass of bluestain fungus *Aureobasidium pullulans*. Doc No IRG/WP 97-10198, Int. Res. Group on Wood Preservation, Stockholm, Sweden, 7 pp.

Ritschkoff A-C, Rättö M, Nurmi A, Kokko H, Rapp A and Militz H (1999) Effect of some resin treatments on fungal degradation reactions. Doc No IRG/WP 99-10318, Int. Res. Group on Wood Preservation, Stockholm, Sweden, 8 pp.

Rypáček V (1966) Biologie holzzerstörender Pilze. VEB Gustav Fischer Verlag, Jena, Germany, 211 pp.

Sailer M and Rapp AO (2001) Use of vegetable oils for wood protection. COST Action E22: Environmental optimization of wood protection. Conference in Reinbek, Germany, 8-9 November 2001.

Scheffer T (1986) $O_2$ requirements for growth and survival of wood-decaying and sapwood-staining fungi. Can J Bot 64: 1957-1963.

Schena L, Sailer MF and Gallitelli D (2002) Molecular detection of strain 147 of *Aureobasidium pullulans*, a biocontrol agent of postharvest diseases. Plant Dis 86: 54-60.

Schlegel HG (1992) Allgemeine Mikrobiologie. 7 Auflage, Thieme Verlag, Stuttgart, Germany, 296 pp.

Schmidt O (1994) Holz- und Baumpilze Biologie, Schäden, Schutz, Nutzen. Springer-Verlag, Berlin, Germany, 246 pp.

Schmidt O and Müller J (1996) Praxisversuche zum biologischen Schutz von Kiefernholz vor Schimmel und Schnittholzbläue. Holzforsch Holzverwend 48: 81-84.

Schoeman M, Webber JF and Dickinson DJ (1999) The development of ideas in biological control applied to forest products. Int Biodeterior Biodegr 43: 109-123.

Schultz TP, Militz H, Freeman MH, Goodell B and Nicholas DD (2008) Development of commercial wood preservatives. efficacy, environmental, and health isssues. ACS Symposium Series 982: 324-336.

Sharpe PR and Dickinson DJ (1992) Bluestain in service on wood surface coatings. Part 1: The nutritional requirements of *Aureobasidium pullulans*. Doc No IRG/WP 92-1556, Int. Res. Group on Wood Preservation, Stockholm, Sweden, 13 pp.

Sharpe PR and Dickinson DJ (1993) Blue stain in service on wood surface coatings. Part 3: The nutritional capability of *Aureobasidium pullulans* compared to other fungi commonly isolated from wood surface coatings. Doc No IRG/WP 93-10035, Int. Res. Group on Wood Preservation, Stockholm, Sweden, 9 pp.

Smith VL, Wilcox WF and Harman GE (1991) Biological control of *Phytophthora* by *Trichoderma*. Patent US4996157.

Terziev N and Nilsson T (1999) Effect of soluble nutrient content in wood on its susceptibility to soft rot and bacterial attack in ground test. Holzforschung 53: 575-579.

Terziev N, Bjurman J and Boutelje JB (1994) Mould growth at lumber surfaces of pine after kiln and air drying. Doc No IRG/WP/94-40033, Int. Res. Group on Wood Preservation, Stockholm, Sweden, 10 pp.

Theander O, Bjurman J and Boutje JB (1993) Increase in the content of low-molecular carbohydrates at the surfaces during drying and correlations with nitrogen content, yellowing and mould growth. Wood Sci Technol 27: 381-389.

Tjeerdsma BF, Homan WJ, Jorissen AJM and Banga J (2003) Hout in de GWW-sector: duurzaam detailleren in hout. Stichting CUR, Gouda, the Netherlands.

Ubalua AO and Oti E (2007) Antagonistic properties of *Trichoderma viride* on postharvest cassava root rot pathogens. Afr J Biotechnol 6: 2447-2450.

Xu G and Goodell B (2001) Mechanisms of wood degradation by brown-rot fungi: chelator-mediated cellulose degradation and binding of iron by cellulose. J Biotechnol 87: 43-57.

White-McDougall WJ, Blanchette RA and Farell RL (1998) Biological control of blue stain fungi on *Populus tremoloides* using selected *Ophiostoma* isolates. Holzforsch 52: 234-240.

Wolf F and Liese W (1977) Zur Bedeutung von Schimmelpilzen auf die Holzqualität. Holz Roh- u. Werkstoff 35: 53-57.

# 17 Coating and surface treatment of wood

*Hannu Viitanen and Anne-Christine Ritschkoff*
*VTT Technical Research Centre of Finland, Espoo, Finland*

## Introduction

Coatings have a twofold basic functionality: (1) to protect the underlying material against deterioration and degradation by the adjacent environment and (2) to decorate or to improve the aesthetic properties of surface. Protection should be given against physical, chemical and biological attack, including water, chemical agents, UV-light, dirt and living organisms, fungi and algae in particular. The aesthetic function refers to characteristics like color performance, gloss and desired surface structure. Both functionalities play a crucial role in health and comfort and exploitation costs due to maintenance. This chapter focuses on wood protection, as related to the latest developments in wood coating technology and surface treatment.

## Microbial activity on coated surfaces

### Typical microbiological flora of coatings on wood

Bacteria, fungi and algae are the most common micro-organisms that contaminate paints and plasters, not only causing disfigurement, but even material degradation and deterioration. Algal surface growth may introduce severe defacement of building and civil construction, which – moreover – may lead to long-lasting moisture retention and consequent moisture-related damage (Gillat 2006). The so-called rot fungi are connected to damage of wood or wood-based materials. In some cases wood coatings may worsen damage, since the drying of wood may be hampered.

### Fungi

The most common fungal species in paint disfigurement are *Alternaria dianthicola*, *A. tenius*, *Aspergillus flavus*, *A. versicolor*, *A. niger*, *Aureobasidium pullulans*, *Cladosporium cladosporioides*, *C. sphaerospermum*, *Fusarium* spp., *Chaetomium* spp., *Paecilomyces* spp., *P. variotii*, *Penicillium* spp., *P. brevicompactum*, *Phoma* spp., *Trichoderma* spp. and *Ulocladium* spp. (Gillat 1991). In tropical climate regions, the so-called primary invaders of paint films are the genera *Alternaria, Aspergillus, Aureobasidium, Bipolaris, Chaetomium, Cladosporium, Curvularia, Fusarium, Mucor, Nigrospora, Paecilomyces, Penicillium, Phoma, Rhizopus* and *Trichoderma* (Joshi *et al.* 1997).

Although fungal damage is usually associated with discoloration, the blue-stain species may also penetrate the material causing structural damage to wood coatings.

## Algae

Common species on wood and wood-based material are *Trentepohlia odorata, Anacystis montana, A. thermale, Chlorococcum* spp. *Scytonema hofmanii, Calotrix parietina, Schizotrix* spp., *Oscillatoria lutella* and *Chlorella* spp. (Gillat 1991).

## Biofilms

Bacteria (e.g. *Pseudomonas*) are often found to be initial colonizers in a succession of growth during discoloration of paint films. A symbiose of fungi and bacteria may grow on organic contamination of the coating surface, forming a so-called biofilm (Morton and Surman 1994). The formation of such biofilm is usually considered to be part of the surface soiling, i.e. the biofouling process of the coatings. Biofilms are complex microbial ecosystems, including organic and inorganic nutrients that are used for growth and colonization, extracellular polysaccharides, enzymes and metabolites secreted by organisms and producing glycocalys. The latter may act as a physical barrier against biocides and cleaning agents. Metabolic products may induce severe corrosion – biocorrosion – or may also function as substrates for other micro-organisms, e.g. sulfate-reducing bacteria. Their products can in turn also cause damage to metals, concrete and organic coatings. Gaylarde and Gaylarde (2000) found biofilms containing algae, cyanobacteria, protozoa, fungi, slime molds, actinomycetes and other bacterial groups on discolored surfaces of painted buildings in five Latin American countries. A total of 1,363 different morphotypes was detected, in which more than half of the population was composed of cyanobacteria in residential locations. This is significantly lower than in urban and rural locations, which is probably related to better building maintenance.

### Conditions of fungal growth on wood coatings

## RH, water content and temperature

The surface microclimate conditions are highly dependent on the physical parameters that define the macro-climate in the surroundings: the ambient relative humidity, ambient air temperature, driving rain, wind, solar radiation. Critical conditions for fungal growth have been reviewed by several researchers (e.g. see references in Chapters 2 and 6). In terms of water activities, the limits for fungal growth are between 0.78 and 0.98 (under equilibrium conditions implying a RH between 78 and

98%) depending on the fungal species, substrate, temperature and exposure time (Figures 17.1A and B).

The first models describing the complex relation between water, temperature and growth rate were based on laboratory studies on agar cultures (Ayerst 1969, Block 1953, Smith and Hill 1982, Grant *et al.* 1989), resulting in so-called isoplets. On coated wood the critical humidity conditions will depend on the properties of the coating, e.g. the water liquid and vapor permeability, pigments, solvents, additives, biocides, etc. Only a few later studies on climate dynamics exist, addressing the time-of-wetness and (Adan 1994, Viitanen 1996).

To initiate wood decay by brown rot fungi, the ambient relative humidity should be above a 95% and the moisture content of pine and spruce sapwood above 25-30% (Viitanen 1996) *for weeks or months.* Such decay will develop fast, when the moisture content exceeds the fiber saturation point (i.e. above 99.9% RH or above a wood moisture content of 30%). Morris *et al.* (2006) modeled decay development in wooden sheathing and found critical ambient humidity condition for decay development to be about RH 98-99%, depending on the temperature and exposure time. In wood structures, decay will develop fastest when the moisture content of untreated wood is between 40-120%, depending on the decay type.

Discoloring fungi will normally grow fast on untreated wood under humid conditions. These fungi, however, will not cause real decay in the wood under normal circumstance. Under high moisture exposure, some blue-stain species, like *Aureobasidium pullulans, Chaetomium globosum,* may cause also soft rot in the wood (see Chapter 6). Water activity, nutrient content of the substrate and biocides are the main ways to control growth and activity of biological organisms. Porous finishing materials may retain moisture over long periods of time, resulting in increased risks of fungal growth (Adan 1994, Van der Wel *et al.* 1998). In the case

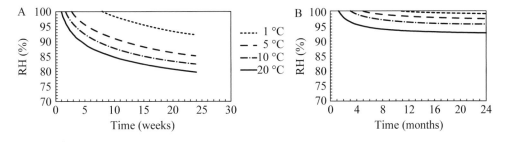

*Figure 17.1. RH and temperature effects on (A) mold growth initiation and (B) the early stage of decay development in pine sapwood (Viitanen 1996, 1997a,b).*

of transient moisture loads, fungal growth risks are therefore related to material properties in particular.

## The outdoor climate

For wood material, microbes play often a key role in material durability. The microbial activity is often highest in the tropical and subtropical climate and lowest in the boreal and arctic climate. Several approaches exist for biological activity as a function of outdoor climate characteristics. Koeppen's climate classification was originally developed for the botanical and agricultural use, but it also gives an overview on the world macro-climate mapping for environmental biological activity. Several climate areas are based on temperature and precipitation. Kottek *et al.* (2006) recently presented a new version of the climate classification.

Different climate indices have been introduced. Dawson *et al.* (2005) used two models to compare the climate of Braunschweig, Germany and Rotorua, New Zealand, which are based on (a) global radiation, days of rainfall and total precipitation, and (b) the average of monthly highest temperature, total sunshine and number of rainy days. Rain fall, sunshine, mean daily global irradiance, and mean daily temperature were higher in Rotorua, but the mean daily relative humidity was higher in Germany. Experiments on coated wood panels showed a striking difference. A solvent-borne stain and a hybrid paint was qualified as stable for end-use in both countries. An acrylic paint performed better in Germany. More stringent fungicide use appeared to be required in New Zealand to suppress fungal growth.

For algal growth on surfaces several criteria have to be met. Algae require day light, high humidity conditions and trace elements, especially fertilizers like nitrogen compounds. Temperature is not a selective factor and the growth medium pH may significantly affect algal growth rates.

## Wood and paint properties

Some ingredients of solvent-based paints, like oils and fatty acids, may be sensitive for fungal growth in relation to biodegradation and which requires biocides in order to protect the paint film. Initial fungal infection, however, is often dependent on nutrient source from the environment, e.g. organic dust (Gillat 1991). For exterior surfaces, the exogenous contamination, like bird and animal droppings, nitrogen compounds from the air, drain water, pollen from the vegetation, will change the original conditions of the coated surfaces more suitable for organisms to grow. Solar UV radiation can also promote fungal growth on exterior surface. Fungal growth is often started on the surface of the coatings (Figure 17.2) and the fungi can penetrate through the porous or cracking of paint film to the substrate.

Fungal growth can also be started under the paint film, in the substrate and the properties of substrate affect on the condition of the coatings of exterior surface. Wood material is normally sensitive for fungal growth, especially when containing low molecular sugar and nitrogen compounds concentrated during kiln drying in the wood surface (Terziev *et al.* 1996, Viitanen and Bjurman 1995). The small molecules can penetrate through the permeable coatings and give higher growth response of the fungi on the surface. The properties of wood are playing an important role for the durability of coated wood against discoloration and decay. Viitanen and Ahola (1998, 1999) found, that the coatings on kiln-dried wood, having higher nutrient content, are more susceptible for discoloring, than these on the wood having lower nutrient content. Bardage and Bjurman (1998) found, that blue-stain fungi, growing in the coated substrate, decreased the adhesion of alkyd emulsion paint. Change in a paint constituent sometimes resulted in a significant change in the adhesion of paint to the wood substrate. On the contrary, the adhesion values of the acrylic dispersion paint tested, became significantly higher after inoculation of blue-stain fungus. In the case of transient moisture loads of porous building materials, specially formulated waterborne paints, may decrease the risk of fungal growth in the substrate regulating the moisture transport between the coating and substrate (Van der Wel *et al.* 1998).

For timber joinery, the controlling the moisture content is essential to prevent swelling, deformation of the component and attack by decay fungi. The moisture absorption characteristics of timber vary with each species, direction of moisture

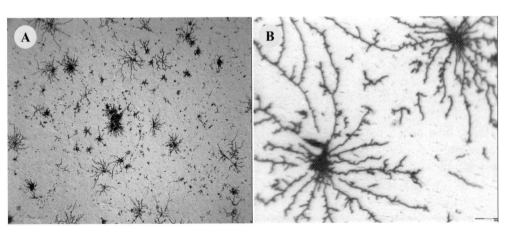

*Figure 17.2. Growth of (A) mold and (B) blue-stain fungi on an acrylate paint. Hyphae are concentrated in the small pores at the paint surface.*

penetration, timber anatomy and other characteristics. The most significant route for moisture penetration into joinery is through open end-grain surfaces (Derbyshire 1999) and this can be prevented through good joint design and end-grain sealing processes. The moisture transmission characteristics of coatings vary according to formulation, film thickness and opacity. The permeability of the coating affects the moisture levels and distribution in the wood substrate. Practical experience has shown that there is a correlation between coating permeability and performance during weathering, and that better weathering performance is achieved with coatings of low permeability to liquid water and permeability high to water damp. Derbyshire (1999) showed that high permeability coatings were not as effective as might have been expected in letting the wood dry out. Coatings of higher permeability allowed faster drying of the wood only during periods of warm dry weather. However, the rapid movement of moisture into and out of the wood encouraged the formation of micro-checks in the coating and precipitated early loss of moisture control.

The protection of ligno-cellulose based material by using only a coating for exposure in ground contact is not a durable solution. Micro-organisms will attack the water-resistant coating layer and then penetrate into the cellulose material to metabolize and produce $CO_2$, $H_2O$, glucose cleaved from cellulose, and small molecules decomposed from the coatings (Zhang *et al.* 1999).

Friebel *et al.* (2004) studied the susceptibility of different coatings to discoloring fungi using experiments in weathering chamber and natural weathering (north facing at a sample angle of 45 degrees) and found a correlation between fungal growth in the test chamber and the natural exposure. The results demonstrated, that the blue-stain species, *Aureobasidium pullulans* and *Sydowia polyspora* are able to metabolize any coating material, if the conditions are suitable for fungal growth. If the coating is intact, the fungi will first grow on the surface, but if the coating has a defect, the blue stain fungi will also grow into the wood. For this reason, a sufficient biocide protection for the wood against blue stain is necessary if a biocide film is omitted.

## Biocidal protection of paints on wood

The protection of material is largely based on the properties of coatings and the substrate. For steel construction, the coatings will protect the substrate until the damage or injuries of the coatings. For wood material, the protection is more based on the moisture and water permeability and biocidal effect of the coating and substrate. For many porous building materials, like wood material, there is interaction between substrate, coating and details of structure. The wood quality has a high impact on the properties of permeable and translucent coatings and lower effect when the coating is thick and low-permeable. For the exterior use of the wood, the coatings are used to protect the wood surface against weathering, water uptake and discoloring. The primers under the top coatings are most important for the performance of the permeable and semi-permeable coating types (e.g. acrylates).

Biocides are needed to protect the coatings and substrate against microbial attack. There are several requirements for active agents for coatings and paints in-can and film preservation (Weber 1999). For film preservation, the requirements for active agents are:
- effect against fungi and algae;
- long lasting effect (preservation);
- extremely low water-solubility;
- low vapor pressure;
- stability in the presence of UV light;
- stability over a wide pH-range;
- good compatibility with all ingredients and materials;
- no discolorations in container and film;
- easy incorporation, dosing and handling;
- no effect on other than biological properties;
- acceptable risk for man, animals and the environment during use and on disposal;
- high degree of cost-effectiveness.

During the last years, the environmental aspects have attracted increasing importance. The requirement to protect the materials and buildings still exists, but there are compromises between effectiveness, environmental or health aspects. Also the conditions for intended use and structural details play important role (Gillatt 2006).

Typical fungicides against decay are propiconazole, zinc oxides, copper components. Normally the wood will be protected against decay using impregnation process, where whole pine sapwood or the high permeable part of wood is treated by pressure treatments. Coating is giving a protection against surface discoloring and water penetration to the surface.

Biocides and fungicides are very important components of coatings used in wet and exterior conditions. Different biocides, like tin-based compounds (tri-n-butyltin methacrylate) and pentachlorophenol acrylate, have been polymerized into acrylics to improve the fungicidal effect. Chlorinated phenols and tin-based compounds like TBTO or TBTN are not anymore used for wood preservatives. As the most common substances for film preservation are urea derivatives, isothiazolinones, dithiocarbamates, benzimidazoles, triazines, benzothiazoles, carbamic acids, thiophthalimides, sulfenic acids, sulfones, and pyridine-N-oxide derivatives (Weber 1999, Gillatt 2003). The protective activity of the biocides of coating is also dependent on other properties of coatings: pigments, binders, opacity, thickness and permeability of the paint film, and properties of the substrate. The capacity of coating protection is also affected by the quality of substrate. For example, high nutrient content of substrate will diminish the resistance of coatings against fungi (Viitanen and Ahola 1999).

Effective exterior coatings biocides are blends of fungicides and algicides (Gillatt 1991, Weber 1999). The classical formulation consists an algicides such as diuron [3-(3,4-dichlorophenyl)-1,1-dimethylurea] or triazine [2-tert-butylamino-4-cyclo-propylamino-6-methylthio-1,3,5-triazine] and fungicides such as carbendazim [methylbenzimidazol-2-ylcarbamate] or OIT [2-n-octyl-4-isothiazolin-3-one] or IPBC [3-iodo-2-proponyl butyl carmabate] (Weber 1999, Gillatt 2006).

In Table 17.1, the spectrum of efficacy of some active ingredients is shown. The inhibitory concentration of fungicides is widely varied depending on the organisms and test type used (Table 17.2). The resistance of coatings against organisms, however, is not only based on the type and the concentration of a biocide. The fungicidal and algicidal efficacy of active ingredients in coatings depended also on pH of the paint and structure of the wall (Gillatt 2006).

Table 17.1. Spectrum and solubility of selected bio-active ingredients (according to Weber 1999).

| Active ingredient | Efficacy | | Solubility organic solvent | | |
|---|---|---|---|---|---|
| | Fungi | Algae | Water | Polar | Non-polar |
| Organic metal compounds | + | + | – | + | + |
| Urea derivatives | – | + | – | + | + |
| Isothiazolines | + | + | – | + | (+) |
| Dithiocarbamate derivatives | + | – | – | (+) | (–) |
| Benzimidazole derivatives | + | – | – | + | |
| Triazine derivatives | – | + | – | + | + |
| Benzothiazole derivatives | + | (+) | – | (+) | (+) |
| Carbamic acid derivatives | + | (+) | – | + | (+) |
| Thiophtalimide derivatives | + | (+) | – | (+) | (+) |
| Sulfenic acid derivatives | + | (+) | – | + | (+) |
| Sulfone derivatives | + | – | – | + | (+) |
| Pyridine-N-oxide derivatives | + | (+) | – | – | – |

*Table 17.2. Minimum inhibition concentration of four fungicides for various fungi (Gillatt 2006).*

| Organism[1] | Minimum inhibitory concentration (ppm) | | | |
|---|---|---|---|---|
| | Carbendazim | Chlorothalonil | IPBC | OIT |
| Alternaria alternata | >1000 | 0.75 | 2 | 1.5 |
| Aspergillus niger | 5–10 | 1000 | 2 | 5 |
| Aureobasidium pullullans | 0.1–0.5 | 1 | 1 | 0.5 |
| Candida albicans | >1000 | 1000 | – | 2.5 |
| Ceratocystis pilifera | 0.5 | – | 1 | – |
| Chaetomium globosum | 0.5 | 5 | 5 | 10 |
| Cladosporium cladosporioides | 0.5 | – | 2 | – |
| Gliocladium virens | 1.0 | – | 5 | – |
| Lentinus tigrinis | >1000 | 2 | 2 | – |
| Penicillium funiculosum | – | 5,000 | – | 5 |
| Penicillium glaucum | 0.5 | 2 | 1 | 2.5 |
| Rhodotorula rubra | 5 | – | 20 | 5 |
| Saccharomyces cerevisiae | – | 350 | – | 1.5 |
| Sporobolomyces roseus | 1–2 | – | 7.5 | – |
| Trichoderma viride | 1–2 | 5,000 | 100 | – |
| Ulocladium atrum | >1000 | – | – | – |

[1] The use of the fungal names in this list does not concur with the current nomenclature.

Viitanen and Ahola (1999) found, that the biocidal effect of coatings on wood surfaces performed better against discoloration, when biocides were included in the all layers of coating systems, e.g. preservative, primer and topcoat, and the protection of coatings were effective in more susceptible wood types (Figure 17.3). The content of fungicides in different layers can be optimized, and the total content of fungicides can be on the acceptable level regarding the environmental requirements. The fungicides used in the coating formulations were propiconazole, isothiazolone, and IPBC, single or in binary mixtures. Samples included: (1) non-treated (control), (2) dipping in water-born alkyd preservative containing fungicide prior to painting, (3) no dipping prior to painting, (4) primer only (higher fungicide content), (5) topcoat paint only (lower fungicide content), and (6) primer/topcoat. The IPBC fungicide showed the best short-term efficacy, but the efficiency of all fungicides decreased

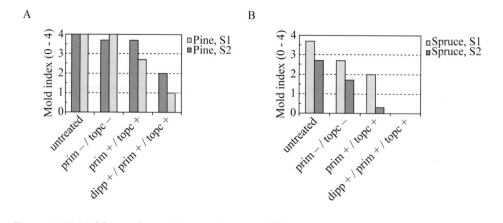

*Figure 17.3. Mold growth on (A) coated pine and (B) spruce sapwood after surface treatment and 26 weeks incubation at RH 100% and 20 °C (Figure after Viitanen and Ahola 1998, 1999). S1=kiln dried surface; S2=resawn surface, 10 mm from original surface; prim=acrylate primer; topc=top coat; −=nof fungicide; +=fungicide added.*

with time. Both the alkyd and acrylate formulations required fungicide additives to endure biological degradation, weathering, and the thicker the coating, the better the resistance to fungal attack.

The performance of fungicides in coating is more active, when used during application of the paint and coatings, than during the cleaning or disinfection of discolored surfaces. Shirakawa *et al.* (2004) studied the disinfection of masonry facades attacked by fungi and algae using hypochlorite and high-pressure water jet cleaning. The disinfection gave a reduction of at least 85% in the microbial population. Paint was applied with or without a biocide formulation (0.25% weight/weight) of – carbendazim, N-octyl-2H-isothiazolin-3-one and N-(3,4-dichlorophenyl)N,N-dimethyl urea. These biocides reduced the fungal colonization up to 10 months after painting on one building. However after 12 months the biocides showed no significant difference. The major fungal contaminant was the genus *Cladosporium*. After 10 months, *Aureobasidium* was also associated with black discoloration. On both the re-painted facades, fungi were detected before algae. SEM analyses showed that fungal contamination (living cells?) was present not only on the surface, but also between the old and new paint films and between old paint and rendering mortar. Under these conditions the properties of the building facade and the micro-climate conditions seem to be more important than the effectivity of the biocide controlling fungal growth.

## Novel trends in coating technology

In the last decades paint industry has given considerable attention to functional and bioactive coatings. Modern coating techniques based on nanotechnology enable development, tailoring and characterization of new coating systems, which provide improved properties to different substrate materials. By using very thin coatings, the surface properties can be modified without changing the substrate properties or appearance. Recent developments in surface science and nanotechnology offer new opportunities for modern engineering concepts for the fabrication of functional surfaces with "passive" and active structures (Kallio *et al.* 2005, Shchukin and Möhlwald 2007).

### *Functional coatings based on sol-gel technology*

Recently, it has been observed that one of the most promising methods to improve surface properties of various materials as well as to provide new properties into the surface, is to use thin coatings tailored with nanohybride materials made by sol-gel technology.

Sol-gel technologies have developed greatly during the last twenty years. In the sol-gel reaction, homogenous inorganic materials with desirable properties of hardness, optical transparency, chemical durability, tailored porosity and thermal resistance can be formed. At the beginning, sol-gel was mainly used for producing purely ceramic coatings, foams, fibers and powders. The discovery, that the sol-gel processing readily yields both inorganic and a hybrid organic-inorganic materials has greatly widened the scope of applications (Arkles 2001). In the sol-gel process the chemistry plays the most important role. The sol-gel process involves the evolution of inorganic nanoscale networks through the formation of a colloidal suspension (sol) and the gelation of the sol to form a network in a continuous liquid phase (gel). Through controlled hydrolysis and condensation reactions different end products can be obtained.

The raw materials in sol-gel processing are usually silicon or metal alkoxide precursors. The most widely used metal alkoxides are the alkoxysilanes, such as tetramethoxysilane (TMOS) and tetraethoxysilane (TEOS). A mixture of water and alcohol is used as a solvent. The raw materials and by-products of the reactions can be selected so that the process is environmentally sound. (Brinkner and Scherer 1990, Witucki 1993, Arkles 2001, Li *et al.* 2001, Okawa 2002, Vesa *et al.* 2004).

## Improved functionality for wood materials by sol-gel coatings

The sol-gel mediated surface modification of ceramics and metals has been studied intensively during several years. The suitability of sol-gel technology on wood surface has been, however, investigated to a much lesser extent. Multifunctional metal alkoxides (alkoxysilanes) are reported to be used in order to affect the water repellency, dimensional stability, UV-resistance, anti-fouling, anti-scratching and anti-flammable properties of wood materials. The functional properties of wood material obtained by sol-gel mediated technology can vary depending of the product end use demands. For example, the functional properties can be achieved by continuous or intermittent thin films with the thicknesses of 0.1 μm to 20 μm and controlled porous structure. The treatment can also be soaked into the wood surface structure in desired depth to form functional barriers that are durable against surface degradation during outdoor exposure conditions (Saka 1997, Winfield 2001, Allen 2002, Rodriquez *et al.* 2003, Tshabalala 2003, Tshabalala *et al.* 2003, Mai *et al.* 2004, Donath *et al.* 2004, Miyafuji *et al.* 2004, Stojanovic *et al.* 2004, Vesa *et al.* 2004, Ritschkoff *et al.* 2005).

Wood-based substrates are well suited to sol-gel mediated modification. The study by Tshabalala (Tshabalala 2003, Tshabalala *et al.* 2003) shows that during the sol-gel deposition free silanols and alcohol are formed to wood cell wall due to the hydrolysis of alkoxy groups by the water bound to wood cell wall reactive groups. Further on, the silanols undergo polycondensation to form polysilanols attached to wood components by hydrogen bonds. The hydrogen-bonded polysilanols lose water upon heating and covalent bonds between the sol-gel film and wood components are formed (Figure 17.4).

In the studies of Saka *et al.* (2001), Tshabalala (2003), Tshabalala *et al.* (2003), Donath *et al.* (2004) and Mai *et al.* (2004), the effect of functional alkoxysilanes on wood moisture behavior and dimensional stability was studied. The functionalization of wood substrates was carried out either by coating wood surface with sol-gel thin coating or by impregnation of wood material with sol-gel substances. Tshabalala (2003) and Tshabalala *et al.* (2003) showed that the sol-gel deposit on the wood substrates resulted in clearly lowered rates of moisture sorption (Figure 17.5). The net effect of the sol-gel deposit is regarded to decrease the surface concentration of hydrogen-bonding sites and also to hinder the formation of hydrogen bonds between such sites and water molecules. Ritschkoff *et al.* (2005), have also shown that water and moisture behavior of wood can be improved by sol-gel coatings. The effect of the sol-gel coatings on the repellence properties of wooden substrates was assessed by measuring static contact angles of distilled water on the surfaces as a function

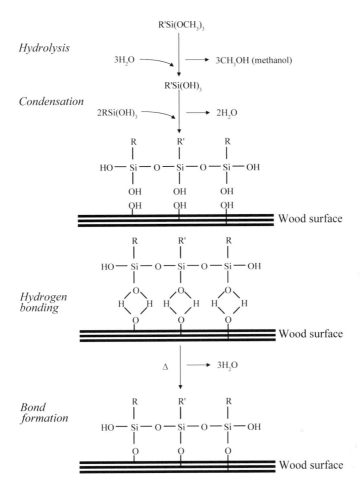

*Figure 17.4. Scheme for sol-gel deposition of alkoxysilane on wood surfaces (Tshabalala* et al. *2003a).*

of time. The contact angles of water showed that the sol-gel thin coatings clearly improved the water-repelling properties of birch (Figure 17.6).

Ultra violet light is one factor that decomposes the lignin fraction of wood quite fast. Wood biodegrades as a result of the interaction of fungi, water and UV-light. Wood also changes its color, which is harmful for its appearance. Ultra violet light resistance can be improved with sol-gel coatings. In recent studies, inorganic and organic UV-stabilizers have been applied into sol-gel matrices with considerable

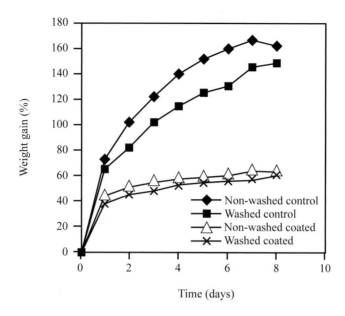

*Figure 17.5. Increase in weight (%) of sol-gel treated wood samples in distilled water at 25 °C (Tshabalala et al. 2003a).*

good effects on the photostability of wood (Pacaud *et al.* 1998, Allen 2002, Miyafuji *et al.* 2004).

In the accelerated weathering tests carried out by Tshabalala (2003) the combination of functional alkoxysilanes in the sol-gel coating on wood substrate exhibited good resistance to photochemical degradation (UV-radiation). Similar results were reported by Donath *et al.* (2004). These researchers however, noticed that the processing and deposition of coatings were crucial to achieve the desired properties, as well as to sustained activity during the ageing of the material.

### Antimicrobial coating technology

Developing antimicrobial coatings is mainly focused on materials that prevent the microbial growth and activity. In order to achieve long-active solutions, research and development has been focused on microbe-repelling, biocide-release, and contact-active antimicrobial solutions (Tiller *et al.* 2005). Most commonly, bulk materials or coatings are impregnated with biocides, such as silver ions, which are then slowly released into the environment to prevent microbial activity (Chainer 2001, Edge *et al.* 2001). Another approach to achieve long-term protective antimicrobiality

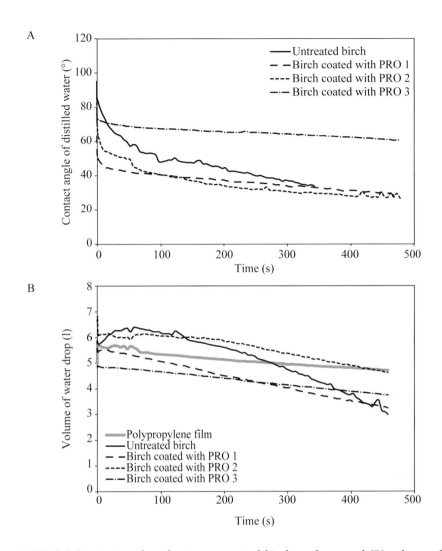

*Figure 17.6. (A) Contact angles of water on coated birch surfaces and (B) volume of water droplets on surfaces (Ritschkoff* et al. *2005).*

is the introduction of either hydrogel-forming non-charged coatings, e.g. grafted poly(ethylene glycol) (PEG) or rendering the surface ultrahydrophobic. In both cases, the adhesion of microbes is strongly reduced (Hyde *et al.* 1997, Bakker *et al.* 2003). A more recent approach is based on surface modification that prevents microbial contact without releasing an exhaustible compound. Such systems are based on grafted antimicrobial polymers, e.g. poly N-alkyl-4-vinylpyridinium salts (Tiller *et al.* 2001). Coatings containing photocatalytic $TiO_2$ have been considered as an active surface that prevents microbial growth by light-induced formation of hydroxyl radicals (Ohko *et al.* 2001a,b).

Recently, coatings with silver and other metal based nanoparticles have been introduced as new potential way to prevent microbial growth (Aymonier *et al.* 2002). Studies on the interaction of silver nanoparticles in coating matrices with various bacteria and viruses have showed high antimicrobial activity (Figovsky *et al.* 2005). The antimicrobial activity of silver is rather selective, being more active against bacteria and viruses than fungi.

**Functional coatings based on photocatalytic $TiO_2$**

Today there is an increasing interest in incorporating photocatalytic functionalities into various material surfaces, coatings and technical devices. Due to the strong oxidation power and superhydrophilic properties of $TiO_2$ under UV illumination, $TiO_2$ coated substrates have shown to be antimicrobial and self-cleaning under conditions, where materials are exposed to sufficient amount of UV irradiation (Nakajima *et al.* 2000, Ohko *et al.* 2001a,b, Gefroy *et al.* 2002, Guan 2004). The photocatalytic process includes chemical steps that produce reactive radicals (i.e. hydroxyl radical, hydrogen peroxide, superoxide) that can cause fatal damage to micro-organisms (Blake *et al.* 1999).

Jacoby *et al.* (1988), Ohko *et al.* (2001a) and Wolfrum *et al.* (2002) have shown the antimicrobial effect of photocatalytic $TiO_2$ against fungi, *Escherichia coli* and other bacterial strains, like MRSA. However, in many cases the efficacy has been mainly shown with the normal plate test (cfu) and no knowledge on the long-term efficiency is available (Ohko *et al.* 2001a). The antimicrobial effect of photocatalytic $TiO_2$ against algal growth has been demonstrated by Linkous *et al.* (2000). Algal growth on exposed surfaces can have great aesthetic and economic consequences. In general, algae are more resistant to the more common chemical oxidants than bacteria and so from a photocatalytic control perspective, they offer a greater challenge. Linkous *et al.* (2000) have shown that algal growth on cement substrate coated with photocatalytic $TiO_2$ (10 wt %) was reduced by 66% in comparison to the unprotected cement surface. The extent of inhibition was shown to be related to the amount of near-UV light emitted from a irradiation source.

## Functional coatings based on controlled release of biocides

Microbial colonization on surfaces is inhibited by the addition of biocides into the coatings and surface finishing products. Biocides must be mobile so that they can migrate from the coating interface to the surfaces in order to be able to prevent the development of micro-organisms. Because biocides are essentially toxic, the environmental impact of their release requires careful monitoring and control (Edge *et al.* 2001). For example, in modern wood protection systems biocides are often excluded. The consequence of this is that the wood remains vulnerable to growth of e.g. algae, molds or blue stain and decay fungi.

Development of a new generation of functional coatings, which have both passive mechanical characteristics originating from the matrix material and active response sensitivity to changes in the local environment, opens a new way for the fabrication of future high-tech functional surfaces. Novel feedback-active surfaces can be composed of a passive coating matrix and active structures for fast response of the coating properties to environmental impacts (Shchukin and Möhlwald 2007).

Encapsulation of bioactive agents (i.e. an active molecule is retained within a protective framework) provides an eco-friendlier route for coating protection technology. Such technology is already applied in other fields, e.g. drug release, fragrance delivery (Edge *et al.* 2001). The active compounds are encapsulated into nano-containers with a shell possessing controlled permeability. These capsules can be introduced into a pre-treatment, primer, or topcoat coating matrix. The release of the biocide from the nano-containers can be free, controlled, or triggered. In the triggered release the agent slowly releases either by interactions with the material, e.g. specific pore size, charge, hydrophobic interactions, or by degradation of the matrix. In the case of triggered release the material contains a recognition structure that controls degradation or swelling of the material in dependence on the specific environmental conditions. Depending on the nature of the sensitive components (e.g. weak polyelectrolytes, metal nano-particles) introduced into the triggered system, reversible and irreversible changes of the shell permeability can be induced by various stimuli: variation of the pH, ionic strength, temperature, ultrasonic treatment, alternating magnetic field, electromagnetic irradiation (Figure 17.7; Tiller *et al.* 2005, Shchukin and Möhlwald 2007).

Homan *et al.* (2007) have introduced novel wood preserving concepts based on "release on demand" principles. The developed concept is based on a carrier system of biodegradable micron sized particles. The release of the bioactive component is triggered by the presence of a specific enzyme derived from fungal colonization. A

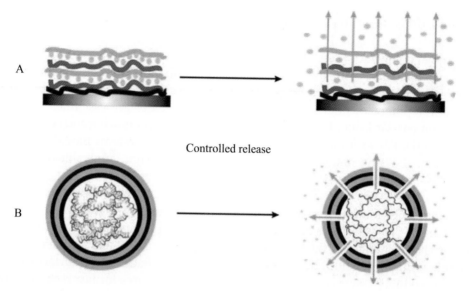

A

Controlled release

B

*Figure 17.7. Schematic presentation of the entrapment/release of active material. (A) Active material is freely dispersed in the "passive" matrix of a coating; (B) active material is encapsulated into nano-containers that possesses controllable shell permeability properties (Shchukin and Möhlwald 2007).*

study of Edge *et al.* (2001) on encapsulated biocides within a modified silica matrix showed, that clear reduction in the levels of biocide delivered to the interface of paints and coatings to prevent microbial spoilage by fungi and blue stain fungi was achieved, by the controlled release of biocide from porous inorganic matrix based on silica. Sol-gel films in controlled release systems promise many advantages as they are cheap and inert. Sol-gel coatings have excellent mechanical properties and are stable to humidity, light heat. Moreover, the coating process is compatible with all common coating technologies and allows control of layer thickness and film quality, and wide spectrum of bioactive compounds can be incorporated into the sol-gel film (Böttcher *et al.* 1999). In the studies of Böttcher *et al.* (1999) the release of benzoic, sorbic and boric acids incorporated to into silica films demonstrate that the sol-gel is a versatile new method for embedding and immobilizing biocidal additives within an inorganic matrix. The release of the activate compounds correlated well to the antimicrobial activity against common and decay fungi.

# Testing the resistance and performance of coatings on wood

## Accelerated tests

Accelerated tests are used most often as screening tests to evaluate the effectiveness of coatings or active agents against different organisms. The role of these tests has been changed due to new legislation and directives when the development of biocides is more difficult and complicated. Exterior surfaces can become infected with several types of organisms and broad-spectrum antifungal/anti-algal biocides have been earlier developed to prevent such growths. However, changing legislation relating to such biocides, pressure from environmental and consumer groups and revised regulations concerning insulation of buildings, have placed greater and greater demands on biocides and have created a need for new products with novel properties. Testing of new coating products is very important and several different types of test systems have been used. The test methods to evaluate efficacy, resistance and durability of wood materials have been reviewed by Viitanen and Gobacken (2005) during the COST E 37 action (COST 2007).

The first step to assess antifungal activity of the coating materials *in vitro* is to embed coating solutions into agar plates and measure the diameters of radial fungal growth after inoculation. The second step may be to treat different material with studied coatings and expose the small samples to organisms on agar plates. Examples of accelerated tests are:

- The ASTM D5590-00: determining the resistance of paint films and related coatings to fungal defacement by accelerated four-week agar plate assay or ASTM D3456-86 (2002): standard practice for determining by exterior exposure tests the susceptibility of paint films to microbiological attack.
- In chambers of high humidity the response of coated materials to fungi and algae can be tested, e.g. ASTM D3273-00: standard test method for resistance to growth of mold on the surface of interior coatings in an environmental chamber. Also other types of chamber tests have been developed, like Mycologg, accelerated laboratory test using programmable climate chambers (Gobacken *et al.* 2004).
- EN 152 is specifically developed for testing efficacy of wood preservatives against blue-stain (CEN 1988). Van den Bulcke *et al.* (2007) studied several coating systems on different wood species using the EN 152 reverse blue-stain method. The reverse method appears to be a more realistic approach of exterior conditions and offers the possibility of resistance calculation. The use of image-processing algorithms and neural networks can be used for rating the results. Furthermore, three-dimensional reconstruction of the fungal development in wood gave more complete implementation of the coating, substrate and fungal growth.

- The standard CEN/TS 839 (CEN 2008) has been developed for determination of the protective effectiveness of surface treatments against wood destroying basidiomycetes. The method is mainly targeted for wood preservatives, not coatings. This method, could however, be used also to test the performance of critical coated details (like end grains) in high exposure conditions.

### Performance tests and service life evaluations

Performance tests are developed to measure the resistance, durability and performance of the coated material in exterior or in intended use conditions. Typical performance tests are tests in exterior conditions in test fields. The performance of the exterior tests, however, is also dependent on local climate conditions, test procedure used, substrate, testing time, etc.

For wooden coatings, the EN 927 (CEN 2006) presents:
- EN 927-1 classification and selection;
- EN 927-3 natural exposure test;
- EN 927-4 and EN 927-5 water vapor/liquid water permeability; and
- EN 927-2 performance specification.

EN 927-5 is shown to give significant differences in water absorption values for different types of coatings in wood (Ekstedt 2002). The combination of a standard procedure for water absorption movement, natural weathering/artificial weathering, fungi and decay resistance tests gives more information regarding expected durability, long term performance and service life evaluation (COST 2007).

The ISO 15686-1 factor method includes several general factors in order to evaluate the service life of building components for different performance requirement levels (ISO 2006). Service life means the period of time after installation during which a building or its parts meets or exceeds the performance requirements. This implies a minimum acceptable level of a critical property, and can be defined as the limit state. The basic concept on building performance and life cycle including performance degree and maintenance is shown in the Figure 17.8. The performance and maintenance of coating is often a key factor for the function and performance of the substrate and the whole construction. Especially the design and coating of details are important.

The long term durability of building structures depends on several factors, but the important stage for evaluation of durability and service life consist of evaluation

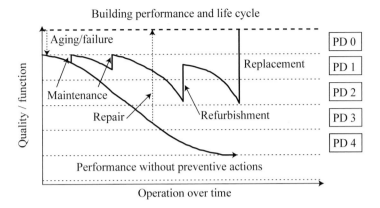

*Figure 17.8. Building performance life cycle as a function of quality, performance degree (PD), failure, maintenance, refurbishment, repair and replacement (based on ISO 15686-7).*

of the exposure conditions. The ISO 15686 identifies a wide range of parameters important to Service Life Prediction:

A = Quality of components, e.g. wood natural durability, treatment and coatings;
B = Design level, e.g. protection by design;
C = Work execution, e.g. joints and details;
D = Indoor environment, e.g. temperature, RH, condensation;
E = Outdoor Environment, e.g. climate, driving rain, shadow;
F = In use conditions, e.g. wear, mechanical impacts;
G = Maintenance level, e.g. repair, revisions, repainting.

For wood materials and components, the factors B, C, E and G are obvious the most important. However, values for these parameters should be determined regionally, taking into account all the local effects of the external factors considered to be important. The quality of components refers to different wood species, including sap and heartwood parts, primers, coating, modification, impregnation, etc. According to Brischke *et al.* (2006), the overall consideration of all possible influences of decay and service life of wood products should take care of reliable evaluation using a reference database of wood durability. When best practices are followed, coated spruce boards and coated birch plywood have been used for long time in exterior conditions without any significant durability problems in Nordic Countries (Viitanen *et al.* 2009).

# References

Adan OCG (1994) On the fungal defacement of interior finishes. Eindhoven University of Technology, PhD Thesis, Eindhoven, the Netherlands.

Allen NS (2002) Behaviour of nanoparticle (ultrafine) titanium dioxide pigments and stabilisers on the photooxidative stability of water based acrylic and isocyanate based acrylic coatings. Polymer Degrad Stabil 78: 467-478.

Arkles B (2001) Commercial applications of sol-gel-derived hybrid materials. MRS Bulletin, May 2001: 402.

ASTM (undated) ASTM D3273-00. Standard test method for resistance to growth of mold on the surface of interior coatings in an environmental chamber. American Society for Testing and Materials, West Conshohocken, PA, USA.

ASTM (undated) ASTM D5590-00. Determining the resistance of paint films and related coatings to fungal defacement by accelerated four-week agar plate assay. American Society for Testing and Materials, West Conshohocken, PA, USA.

ASTM (2002) ASTM D3456-86. Standard practice for determining by exterior exposure tests the susceptibility of paint films to microbiological attack. American Society for Testing and Materials, West Conshohocken, PA, USA.

Ayerst G (1969) The effects of moisture and temperature on growth and spore germination in some fungi. J Stored Prod Res 5:127-141.

Aymonier C, Schlotterbeck U, Antonietti L, Zacharias P, Thomann R, Tiller JC and Mecking S (2002) Hybrids of silver nanoparticles with amphiphilic hyperbranched macromolecules exhibiting antimicrobial properties. Chem Comm 24: 3018-3019.

Bakker DP, Huijs FM, De Vries J, Klinjnstra JW, Bussher HJ and Van der Mei HC (2003) Bacterial deposition to fluoridated and non-fluoridated polyurethane coatings with different elastic modulus and surface tension in parallel plate and a stagnation point flow chamber. Colloid Surface Biointerfaces 32: 179-190.

Bardage S and Bjurman J (1998) Adhesion of waterborne paints to wood. J Coating Tech 70: 39-47.

Blake D, Maness P-C, Huang Z, Wolfrum E and Huang J. (1999) Application of the photocatalytic chemistry of titanium dioxide to disinfection and the killing of cancer cells. Separ Purif Meth 28: 1-50.

Block SS (1953) Humidity requirements for mold growth. Appl Microbiol 1: 287-293.

Böttcher H, Jagota C, Trepte J, Kallies K-H and Haufe H (1999) Sol-gel composite films with controlled release of biocides. J Contr Release 60: 57-65.

Boxall, Carey JK and Miller ER (1992) The effectiveness of end-grain sealers in improving paint performance on soft wood joinery. Part 3. Influence of coating type and wood species on moisture control and fungal colonisation. Holz Roh- u. Werkstoff 50: 227-233.

Brinker CJ and Scherer GW (1990) Sol-gel science. The physics and chemistry of sol-gel processing. Academic Press Inc, San Diego, CA, USA.

Brischke C, Bayerbach R, and Rapp AO (2006) Decay-influencing factors: a basis for service life prediction of wood and wood-based products. Wood Mater Sci Eng 1: 91-107.

CEN (1988) EN 152:1988. Test methods for wood preservatives – Laboratory method for determining the preventive effectiveness of a preservative treatment against blue stain in service. Part 1-2. European Committee for Standardization, Brussels, Belgium.

CEN (2006) EN 927:2006. Paints and varnishes – Coating materials and coating systems for exterior wood. Part 1-5. European Committee for Standardization, Brussels, Belgium.

CEN (2008) CEN/TS 839:2008 Wood preservatives – Determination of the protective effectiveness against wood destroying basidiomycetes – Application by surface treatment. European Committee for Standardization, Brussels, Belgium.

Chainer J (2001) Home steel home. AK steel partners with AgION to build world´s first antimicrobial steel house. AISE Steel Tech 78: 59-60.

COST (2007) COST Action E37 Report. Task Force "Performance Classification".

Dawson BSW, Gottgens A and Hora G (2005) Natural weathering performance of exterior wood coatings on *Pinus sylvestris* and *Pinus radiata* in Germany and New Zealand. JCT Coatings Tech 2: 539-546.

Derbyshire H (1999) Surface coatings: protecting wooden joinery against moisture. In: Turkulin H (ed.) Surface properties and durability of exterior wood building components, International Conference, Zagreb, Croatia, Apr. 30, 1999. University of Zagreb, Faculty of Forestry, Zagreb, Croatia.

Donath H, Militz and Mai C (2004) Wood modification with alkoxysilanes. Wood Sci Tech 38: 555-566.

Edge M, Seal K, Allen NS, Turner D and Robinson J (2001) The enhanced performance of biocidal additives in paints and coatings. Prog Org Coating 43: 10-17.

Ekstedt J (2002) Studies on the barrier properties of exterior wood coatings. PhD Thesis. KTH – Royal Institute of Technology, Stockholm, Sweden, 63 pp.

Figovsky O, Shapalov L and Kydryatzev B (2005) The use of nanotechnology in production of bioactive paints and coatings. PRA´s third international conference dedicated to hygienic coatings & surfaces. 16-17 March 2005, Paris, France.

Friebel S (2004) Fungicide-free topcoats for wood applications. PRA Fourth International Woodcoatings Congress "Developments for a sustainable future", The Hague, the Netherlands, Paper 25.

Gaylarde PM and Gaylarde CC (2000) Algae and cyanobacteria on painted buildings in Latin America. Int Biodeterior Biodeg 46: 93-97.

Gefroy C, Charvet C and Aubay E (2002) Titanium dioxide aqueous dispersion, substrate obtained from said dispersion and self-cleaning method for said substrate. Patent application WO00238682.

Gillat JW (1991) Developments in prevention of biodeterioration of emulsion paints, control of film fungal and algal growth. The growth of airborne molds and yeast on surface coatings and its prevention by biocides. Surface Coatings Australia 28: 6-12.

Gillat JW (2003) Breaking the mold. Polymers Paint Colour J 193: 21-22.

Gillat JW (2006) Dry-film biocides – the next generation. Paintindia 2006: 169-176.

Gobakken LR, Mattson J, Jakobsen B and Evans FG (2004) Durability of surface coating systems. Mycologg – an accelerated mycologigal test. IRG/WP/04-20301, Int. Res. Group on Wood Preservation, Stockholm, Sweden.

Grant C, Hunter CA, Flannigan B and Bravery AF (1989) The moisture requirements of molds isolated from domestic dwellings. Int Biodeterior 25: 259-284.

Guan K (2004) Relationship between photocatalytic activity, hydrophilicity and self-cleaning effects of $TiO_2/SiO_2$ films. Surf Coating Tech 191: 155-160.

Homan W, Jetten J, Sailer M, Slaghek T and Timmermans J (2007) Bioswitch: a versatile release on command system for wood protection. In: European Conference on Wood Modification 3, ECWM 2007, Cardiff, Wales, UK.

Hyde FW, Alberg M and Smith K (1997) Comparison of fluorinated polymers against stainless steel, glass and polypropylene in microbial biofilm adherence and removal. J Ind Microbiol Biotechnol 19: 142-149.

ISO (2006) ISO 15686. Building and construction assets – Service life planning. International Organization for Standardization, Geneva, Switzerland.

Jacoby W, Maness P, Wolfrum E, Blake D and Fennell J (1998) Minralization of bacterial cell mass on a photocatalytic surface in air. Environ Sci Tech 32: 2650-2653.

Joshi CD, Mukundan U and Bagool RD (1997) Fungal fouling of architectural paints in India. Paintindia 1997: 29-34.

Kallio M, Mannila J, Vesa A, Mahlberg R, Ritschkoff A-C and Oligschläger T (2005) Modification of surface properties of metals by sol-gel coatings. Available at: https://www.corrdefense.org/Academia%20Government%20and%20Industry/T-59.pdf.

Kottek MJ, Grieser C, Beck BR and Robel F (2006) World map of the Köppen-Geiger climate classification updated. Meteorol Z 15: 259-263.

Li Z-J, Furuno T, and Katoh S (2001) Preparation and properties of acetylated and propionylated wood-silicate composites. Holzforsch 55: 93-96.

Linkous C, Carter G, Locuson D, Ouelette A, Slattery D and Smith L (2000) Photocatalytic inhibition of algae growth using $TiO_2$, $WO_3$ and cocatalyst modifications. Environ Sci Tech 34: 4754-4758.

Mai C and Militz H (2004) Modification of wood with silicon compounds. Treatment systems based on organic silicon compounds – a review. Wood Sci Tech 37: 453-461.

Miyafuji H, Kokaji H and Saka S (2004) Photostale wood-inorganic composites prepared by the sol-gel process with UV absorbent. J Wood Sci 50: 130-135.

Morris P, Symons P, and Clark J (2006) Resistance of wood sheating to decay. Wood protection, March 21-23, 2006. New Orleans, Lousiana, USA.

Morton LHG and Surman SB (1994) Biofilms in biodeterioration – a review. Int Biodeterior Biodegr 34: 203-221.

Ohko Y, Saitoh S, Tatsuma T and Fujishima A (2001b) Photoelectrochemical anticorrosion and self-cleaning effects of a $TiO_2$ coating for Type 304 stainless steel. J Electrochem Soc 148: B24-B28.

Ohko Y, Utsumi Y, Niwa C, Tatsuma T, Kobayakawa K, Satoh Y, Kubota Y and Fujishima A (2001a) Self-sterilizing and self-cleaning of silicone catherers coated with TiO$_2$ photocatalyst thin films: a preclinical work. J Biomed Mater Res (Appl Biomater) 58: 97-101.

Okawa S (2002) Improvement of wood surface by inorganic modification. Trans Mat Res Soc Japan 27: 637-640.

Pacaud B, Bousseau J-N and Lemaire J (1998) Naniitotania as UV-blockers in stains. Eur Coating J 11: 842-848.

Ritschkoff A-C, Mahlberg R, Löija M, Kallio M, Mannila J and Vesa A (2005) Sol-gel hybrid coatings for wood products with improved surface durability and repellence properties. PCI Paint & Coatings Industry 21: 96-101.

Rodriques R, Estevez M, Vargas S and Mondragon M (2003) Hybrid ceramic-polymer material for wood coating with high wearing resistance. Mat Res Innovat 7: 80-84.

Saka S, Miyafuji H and Tanno F (2001) Wood-inorganic composites prepared by sol-gel process. J Sol-Gel Sci Tech 20: 213-217.

Shchukin DG and Möhwald H (2007) Self-repairing coatings containing active nanoreservoirs. Small 3: 926-943.

Shirakawa MA, John VM, Gaylarde CC, Gaylarde P and Gambale W (2004) Mold and phototroph growth on masonry facades after repainting. Mater Struct 37: 472-479.

Smith SL and Hill ST (1982) Influence of temperature and water activity on germination and growth of *Aspergillus restrictus* and *Aspergillus versicolor*. Trans Br Mycol Soc 79: 558-560.

Stojanovic S, Bauer F, Gläsel H-J and Mehnert R (2004) Scratch and abrasion resistant polymeric nanocomposites – preparation, characterisation and applications. Mat Sci Forum 453-454: 473-478.

Terziev N, Bjurman J and Boutelje JB (1996) Effect of planning on mold susceptibility of kiln-dried and air-dried Scots pine (*Pinus sylvestris* L.) lumber. Mat and Org 30: 95-103.

Tiller JC, Liao CJ, Lewis K and Klibanov AM (2001) Designing surfaces that kill bacteria on contact. Proc Natl Acad Sci USA 98: 5981-5985.

Tiller JC, Sprich C and Hartmann L (2005) Amphiphilic conetworks as regenerative controlled releasing antimicrobial coatings. J Contr Release 103: 355-367.

Tshabalala MA (2003) Accelerated weathering of wood surfaces coated with multifunctional alkoxysilanes by sol-gel deposition. J Coating Tech 75: 37-43.

Tshabalala MA, Kingshott P, VanLandingham MR and Plackett D (2003) Surface chemistry and moisture sorption properties of wood coated with multifunctional alkoxysilanes by sol-gel-process. J Appl Polymer Sci 88: 2828-2841.

Van den Bulcke J, Van Acker J and Stevens M (2007) Laboratory testing and computer simulation of blue-stain growth on and in wood coatings. Int Biodeterior Biodegr 59: 137-147.

Van der Wel GK, Adan OCG and Bancken ELJ (1998) Towards an ecofriendlier control of fungal growth on coated plasters? FATICEP Congress 24, Volume C, pp. 15-26.

Vesa A, Kallio M, Mannila J, Ritschkoff A-C and Mahlberg R (2004) Improvement of the abrasion resistance of stainless steel with nanocomposite sol-gel coatings. Fifth Nordic Conference on Surface Science. Tampere 22-25 Sept. 2004.

Viitanen H (1996) Factors affecting the development of mold and brown rot decay in wooden material and wooden structures. PhD Thesis, Dept. of Forest Products, Swedish University of Agricultural Sciences, Uppsala, Sweden.

Viitanen H and Ahola P (1998) Mold growth on low VOC Paints. Advances in Exterior Wood Coatings and CEN Standardisation. Brussels, Belgium, 19-21 October 1998, paper 16.

Viitanen H and Ahola P (1999) Mold growth on low VOC paints. Pitture e Vernici Europe 75: 33-37 and 39-42.

Viitanen H and Bjurman J (1995) Mold growth on wood under fluctuating humidity conditions. Mat and Org 29: 27-46.

Viitanen H and Gobacken L (2005) Inventory of existing test methods for fungi. Cost E 37, 12 p.

Viitanen H, Toratti T, Peuhkuri R, Ojanen T, and Makkonen L (2009) Evaluation of exposure conditions for wooden facades and decking. Doc No IRG/WP 09-20408. International Research Group on Wood Protection, Stockholm, Sweden, 18 p.

Weber K (1999) Application of biocides in waterborne coatings. Royal Soc Chem 243: 61-73.

Winfield PH (2001) The use of flame ionisation technology to improve the wettability and adhesion properties of wood. Int J Adhesion Adhesives 21: 107-114.

Witucki GL (1993 )A silane primer: chemistry and applications of alkoxy silanes. J Coating Tech 65: 57-60.

Zhang L, Zhou J, Huang Jin, Gong P, Zhou Q, Zheng L and Du Y (1999) Biodegradability of regenerated cellulose films coated with polyurethane/natural polymers interpenetrating polymer network. Ind Eng Chem Res 38: 4284-4289.

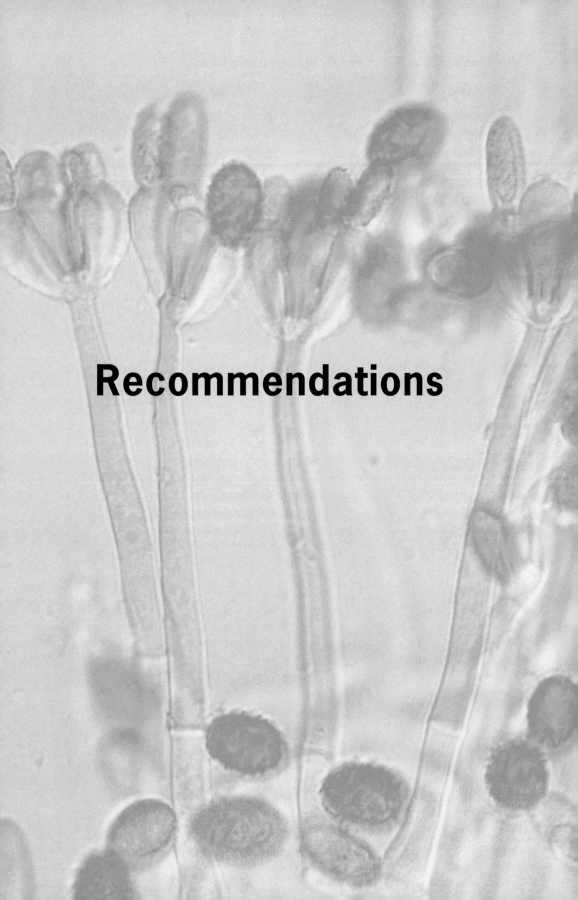

# Recommendations

# 18 Recommendations

*Olaf C.G. Adan[1,2] and Robert A. Samson[3]*
*[1]Eindhoven University of Technology, Faculty of Applied Physics, Eindhoven, the Netherlands; [2]TNO, Delft, the Netherlands; [3]CBS-KNAW Fungal Biodiversity Centre, Utrecht, the Netherlands*

## Introduction

The following recommendations are compiled from Statements and recommendations from the Second International Workshop on fungi in indoor environments: "towards strategies for living in healthy buildings" held in Utrecht, the Netherlands, 17-19 March 2005, together with the data, concepts and visions expressed in this book.

The workshop confirms the general recommendations of the First International Workshop (Samson *et al.* 1994) based on the known health effects of fungi. It is recognized that (visible) fungal growth in non-industrial indoor environments is not acceptable on medical and hygienic grounds.

## Fundamentals

In general, the mycobiota indoors is well studied, but the surveys have been mainly focused on counts and identifications of viable cultures. The importance of non-viable propagules – and in particular fungal fragments – should be more emphasized and molecular methods should be applied to determine the true mycobiota indoors in various climatological regions.

Fungal growth and its relation to water is well understood under steady-state climate conditions, but the concepts of the key physical parameters should be included when designing experiments on fungal growth on building materials or understanding problems of mold contaminations in the built environment. Pragmatically, there is a broad consensus in the scientific community that surfaces can be kept free form mold growth if the relative humidity of the adjacent air is maintained below 80%. The fundamental understanding of the relation between mold growth and indoor climate dynamics, however, is still in its infancy. Experimental evidence described in this book shows that controlling mold growth risks based on the *ambient* relative humidity alone is no guarantee at all for a "mold-free" environment. Short humidity peaks may result in mold growth.

In case of such transient humidity conditions, the properties of the finishing layer on walls, floor and ceiling, and especially their moisture reservoir function, play a pivotal role with respect to surface mold growth. A targeted application of finishing materials then becomes one of the key instruments for controlling indoor mold growth.

## Health

Our current knowledge shows that indoor molds are responsible for allergy, rhinitis, asthma, and few conjunctivitis. In most cases children are affected by indoor molds. Several cases for *Alternaria* are described, while the exposure to allergens can cause asthma attacks. An increase of asthma cases is observed, but the reason for this increase is unknown.

Since all buildings have some mold, and because a large percentage of the existing housing stock has dampness and mold problems, more effort is needed to resolve the thresholds of effect for mold. This cannot be done until true markers of exposure (internal dosemetry) are identified.

In the 2009 WHO report it is stated that: "The most important means for avoiding adverse health effects is the prevention (or minimization) of persistent dampness and microbial growth on interior surfaces and in building structures". Similarly, Krieger *et al.* (2010) concluded in their review for the US Centers for Disease Control/NCHH that one of three interventions ready for implementation in houses was: "Combined elimination of moisture intrusion and leaks and removal of moldy items". We fully concur with these conclusions.

Future studies on the health implications are needed. Immunosuppressive patients should be checked, when they are discharged from the hospital and follow-up of their living conditions in their home environment should be carried out (e.g. through distribution of pamphlets with specific recommendations)

## Strategies

### I. Inspection

The workshop stated:
- Inspection should always include identification of the source of moisture and/or water ingress.

- Clear and visible fungal growth does not to be sampled. Measuring microbiological contamination should only be considered in case of suspected health effects (allergy, infectious organisms, etc.).
- Surface sampling should be documented based on pictures or drawings describing the extend of growth. Destructive sampling may be considered in case of hidden mold. Recommended methods for surface sampling include direct plating, sticky tape, and contact plates. Dust samples should only be applied in special cases.
- Air sampling should be considered only when mold growth or moisture damage is not observed or cannot be detected.

## II. Detection

With respect to current methods and practices, it is stated that:
- Cytotoxicity testing
  - The observed symptoms in mold-infested buildings are *not* in any way suggesting that individuals have been exposed to cytotoxic agents. The rare cases in which persons have been exposed to high levels of the *Stachybotrys* chemotype producing the macrocyclic trichothecenes are an exception.
  - There is no practical use of cytotoxicity testing based on cell-cultures as no validation exists on what is measured.
- Mycotoxin testing
  - No appropriate reference standards for mycotoxin analyses in indoor environments exist. At present, mycotoxin analysis has no practical relevance yet.
  - It is recommended to use specific analytical methods based on LC-MS/MS or GC-MS only for mycotoxins detection

The new detection methods hereafter need validation, before they can be recommended for practical use:
- *PCR methods.* Existing PCR kits are based on a limited number of species. In addition, the taxonomy of the common fungi in indoor environments is still in a state of flux and more molecular date should become available to improve the PCR approach.
- *Structural components*, including ergosterol, glucans, enzymes (Mycometer test) and ATP.

## III. Microbiological aspects in detection

Several methods are available to detect molds in indoor environments and to evaluate their significance for indoor air quality. Concerns exist that in some countries the methods are not following the general recommendations as described in several recent publications and guidelines. Hospitals and other clinical environments are special cases and situations where other principles should be used, because indoor molds could be present as infectious or opportunistic organisms.

**Microbial Volatile Organic Compounds (MVOC)**

Currently, there is no scientific evidence suggesting that MVOC's can be used as an indicator of indoor environmental quality. Furthermore there is no scientific evidence of their health implications.

**Mold concentration in the air**

The presence of both dead and viable spores and fragments in the air is extremely fluctuating and consequently concentrations of fungal propagules can generally not be used to measure mold exposure. The workshop cannot recommend standards of (un)acceptable mold concentrations expressed in numbers of colony forming units.

**Media for detection and isolation**

Two general purpose media for isolation and detection of fungi are recommended:
• 2% malt extract agar.
• Dichloran 18% glycerol agar.

**Quality assurance**

This needs to be applied by the various standard used in the different part of the world. The workshop therefore strongly recommends that laboratories investigating mold problems should perform and participate with the proficiency testing.

*IV. Requirements for building and construction*

In order to tackle mold in indoor environments and their implications to human well-being, the workshop addressed building performance and the requirements for building and construction, usage and maintenance in particular. Essentially, such requirements form the starting point of "source" control in any strategy for healthy indoor environments. Irrespective of climatic and regional differences, we arrive at the following general statements and recommendations:

A robust system to cope with molds and moisture in indoor environments should *always* include coherent requirements with respect to three pillars: thermal performance of the building envelope, ventilation and finishing materials.

## Pillar 1: Thermal performance of the building envelope

- In temperate and cold climates, thermal bridging often determines risks for fungal growth on surfaces of the building envelope. The criterion of a minimal thermal performance (expressed in properties such as thermal resistance, temperature ratio) is commonly used in building codes in various regions, but should be reconsidered in view of recent insights (i.e. mold-temperature-water activity relations) and new incentives from an energy point of view.
- In warm climates and depressurized (air-conditioned) buildings in particular, thermal performance requirements should include requirements dealing with risks of convective moisture transfer ("leakage") in particular. Developing recommendations and guidelines for design according to sound principles of building physics as well as for airtight construction according to best practices and performance control are inevitable follow-up actions to improve robustness.

## Pillar 2: Ventilation

A minimum ventilation rate should be (re)designed from the health point of view. Further justification is needed on how ventilation in airtighter buildings influences health in the context of implications of indoor mold growth. Minimum ventilation rates in that respect should also be adapted to present living and changing occupants demands and wishes.

## Pillar 3: Finishing materials

There is a crucial difference between the living or working area climate and the microclimate. Particularly when indoor climate dynamics are considered, the finish becomes a crucial part of the envelope. At present, this factor is often overlooked.

Performance requirements for fungal resistance of finishing materials should be developed and be considered in building codes as far as high risk situations are concerned (e.g. bathrooms in dwellings). From the health point of view, there is an urgent need for significant tests to assess fungal susceptibility of materials and to search for (communication) tools to address the end-user.

An urgent need exists for assessment methods for mold growth susceptibility that makes sense. Such methods are a prerequisite to tackle health risks related to secondary emissions of fungal growth on finishing and building materials. Development of new and standardized methods should address at least the following insufficiencies in current methods:

• The effect of moisture retention as a crucial factor determining susceptibility. Actually, this concerns the reality of indoor climate dynamics that is often overlooked. Material susceptibility may highly differ under steady state and transient climate conditions.
• Justification of the mycobiota to be used in view of the application area of the material: xerophilic and/or hydrophilic, a mixture or single species, pure cultures.
• Analysis and assessment of growth, reproducibility. Current methods focus on assessment at specific moments in time during test only. Reproducibility has always been poorly addressed in present methods.
• Climatic conditions: unambiguous definition of humidity, temperature and air velocity.
• Preconditioning of material.

Considering the fact that the finish in most cases is a consumers' choice, the introduction of a product classification or labelling system that makes the consumer aware of the consequences of the choice is recommended. Fungal resistance of finishes is a product-based feature.

Susceptible materials should no be used in high risk situations, i.e. where high levels of moisture generation occur such as in bathrooms and kitchens in the domestic environment, or in production processes with local water vapour sources. More research is needed on the vulnerability and effects of new building materials.

## V. Policy

A number of policy recommendations were developed regarding improving indoor environmental quality by reducing mold exposure. One might first question why policies regarding indoor mold are necessary at all. It is because, in a *laissez-faire* scenario in which the free market was allowed to run uninterrupted, no or hardly any stakeholders would be motivated to solve the problems associated with indoor mold. Those who design and build houses are usually not the ones who live in them, and thus have no incentive to erect moisture-proof buildings unless a relevant regulation is in place. Landlords, unless motivated by law, usually do not make moisture prevention a priority in the buildings that house their tenants. At the most basic level, the people living in these damp houses often have no knowledge of the link between mold, dampness, and adverse health effects; thus, they often do not make moisture remediation a priority. For these reasons, public policies are necessary to prevent and reduce mold and moisture in buildings. We arrive at the following five statements and recommendations:

1. Indoor mold and moisture, and their associated health effects, are a *society-wide* problem. In modern societies, people spend about 90% of their time indoors, in one of several environments: homes, workplaces, schools, or public areas such as restaurants and stores. While certain subpopulations are indeed more vulnerable to the adverse health effects associated with mold exposure, enough epidemiological evidence exists to show the link between respiratory symptoms and mold for all humans.

2. The economic consequences of indoor mold and moisture are enormous. While moisture in indoor environments does not lead directly to human disease, it is a precursor to a variety of contaminants, including allergens from house dust mites, cockroaches and mold that cause adverse health effects. On a population level, these contaminants exacerbate asthma symptoms in sensitized individuals. Further, the growth of mold indoors has been linked with increased upper respiratory disease in individuals. These diseases are associated with very large economic burdens. The economic costs stem from both the direct costs of health care and from lost work and school days as well as days of restricted activity. Therefore, it is in the best interest of public health to remediate both residences and workplaces that have excessive moisture, and to improve building standards such that moisture is prevented from entering materials and fabric in the first place and retention in the second.

3. Relevant policies should include specific precautions to protect the most vulnerable subpopulations to indoor mold problems: children, the elderly, and the very poor. There is substantial evidence that children who are exposed to indoor air mold in the first years of their lives have a significantly higher probability of developing chronic respiratory diseases such as asthma. From a health-economic standpoint, the loss of quality of life and productivity from a lifetime of disease is enormous – both to the individual and to the society. The elderly, by virtue of their weakened immune systems, are particularly vulnerable to infectious diseases that could be exacerbated through indoor dampness. In addition, they are often poorer and less able to physically remediate their indoor environments. The very poor are the most likely to be living in housing with severe physical problems, including moisture- and mold-related problems, and the least likely to have the means (money and education) by which to remediate such problems. Oftentimes they also lack the access to information regarding the extent of health problems associated with indoor moisture and appropriate remediation responses.

4. Policymaking needs to occur at two different levels – the level of improving building design and structure, and the level of communicating with the public. Even the most optimally-designed buildings could lead to diseases caused by indoor moisture problems, if the people living in such buildings do not properly maintain their indoor environments. Likewise, public education alone cannot solve the problem if people's houses are too damaged to begin with. A combination of measures is key: (1) to ensure proper building design and construction, and (2) to communicate with the public of risks associated with damp indoor environments and means of remediation. Perhaps the most feasible way to achieve both goals is to delegate responsibility at different levels of government. At the national level, regulations can be put in place regarding basic building codes, quality control (e.g. at construction sites) and maintenance procedures. At the local and community government levels, public education and information dissemination can be more effectively carried out.

5. Information dissemination among all relevant stakeholders, including professional and public education, is necessary to develop strategies for achieving healthy indoor environments. The true problem lies in education and information dissemination: increasing people's awareness that a problem exists in the first place, and then giving them the necessary information to improve their indoor environments. That adverse health effects are still widespread due to indoor dampness and its associated contaminants indicates that much broader efforts are needed to provide information to the public. In the long term, the key to healthier indoor environments is collaboration among the different stakeholders in the problem. In the short term, enhanced public health education is particularly crucial because individuals at risk – those that occupy residences and workplaces – are given the ability to adopt those practices that can improve their health and quality of life.

## References

Krieger J, Jacobs DE, Ashley PJ, Baeder A, Chew GL, Dearborn D, Hynes HP, Miller JD, Morley R, Rabito F and Zeldin DC (2010) Housing interventions and control of asthma-related indoor biologic agents: a review of the evidence. J Public Health Manag Pract 16, E-Supp: S11-S20.

Samson RA, Flannigan B, Flannigan ME, Verhoef AP, Adan OCG and Hoekstra ES (eds.) (1994) Health implications of fungi in indoor environments. Elsevier, Amsterdam, the Netherlands.

World Health Organization (2009) WHO guidelines for indoor air quality: dampness and mould. WHO Regional Office for Europe, Copenhagen, Denmark, 228 pp.

# Contributors

**Olaf (O.C.G.) Adan**
Eindhoven University of Technology
Faculty of Applied Physics
P.O. Box 513
5600 MB Eindhoven
The Netherlands
and
TNO
P.O. Box 49
2600 AA Delft
The Netherlands

**Mirjam (M.) Bekker**
Eindhoven University of Technology
Faculty of Applied Physics
P.O. Box 513
5600 MB Eindhoven
The Netherlands

**Jan (J.) Dijksterhuis**
CBS-KNAW Fungal Biodiversity Centre
Applied and Industrial Mycology
P.O. Box 85167
3508 AD Utrecht, The Netherlands

**Bart (S.J.F.) Erich**
Eindhoven University of Technology
Faculty of Applied Physics
P.O. Box 513
5600 MB Eindhoven
The Netherlands

**Jens (J.C.) Frisvad**
Technical University of Denmark
Department of Systems Biology
Center for Microbial Biotechnology
Søltofts Plads
Building 221, room 204
2800 Kgs. Lyngby
Denmark

**Brett (B.J.) Green**
Allergy and Clinical Immunology Branch
Health Effects Laboratory Division
National Institute for Occupational Safety and Health
Centers for Disease Control and Prevention
1095 Willowdale Road
Morgantown, WV 26505-2888
USA

**Otto (O.O.) Hänninen**
National Institute for Health and Welfare (THL)
Department of Environmental Health
P.O. Box 95
70701 Kuopio
Finland

**Waldemar (W.J.) Homan**
TNO
P.O. Box 49
2600 AA Delft
The Netherlands

**Tobias (T.) Huckfeldt**
Wood Biology Division
Department of Wood Science
University of Hamburg
Leuschnerstraße 91
21031 Hamburg
Germany

**Fundamentals of mold growth in indoor environments**

**Henk (H.P.) Huinink**
Eindhoven University of Technology
Faculty of Applied Physics
P.O. Box 513
5600 MB Eindhoven
The Netherlands

**David (J.D.) Miller**
c/o Department of Chemistry
College of Natural Sciences
228 Steacie Building
Carleton University
Ottawa, Ontario K1S 5B6
Canada

**Philip (P.R.) Morey**
ENVIRON International Corporation
8725 West Higgins Road, Suite 725
Chicago-O'Hare, Illinois 60631
USA

**Kristian (K.) Fog Nielsen**
Technical University of Denmark
Department of Systems Biology
Center for Microbial Biotechnology
Building 221
2800 Kgs. Lyngby
Denmark

**Leo (L.) Pel**
Eindhoven University of Technology
Faculty of Applied Physics
P.O. Box 513
5600 MB Eindhoven
The Netherlands

**Michael (M.F.) Sailer**
TNO
P.O. Box 49
2600 AA Delft
The Netherlands

**Robert (R.A.) Samson**
CBS-KNAW Fungal Biodiversity Centre
Applied and Industrial Mycology
P.O. Box 85167
3508 AD Utrecht
The Netherlands

**Detlef (D.) Schmechel**
Allergy and Clinical Immunology Branch
Health Effects Laboratory Division
National Institute for Occupational Safety and Health
Centers for Disease Control and Prevention
1095 Willowdale Road
Morgantown, WV 26505-2888
USA

**Olaf (O.) Schmidt**
Wood Biology Division
Department of Wood Science
University of Hamburg
Leuschnerstraße 91
21031 Hamburg
Germany

**James (J.A.) Scott**
Sporometrics Inc.
219 Dufferin Street, Suite 20-C
Toronto, Ontario M6K 1Y9
Canada

**Anne-Christine (A-C.) Ritschkoff**
VTT Technical Research Centre of Finland
P.O. Box 1000
02044 VTT Espoo
Finland

**Richard (R.C.) Summerbell,**
Sporometrics Inc.
219 Dufferin St., Suite 20-C
Toronto, ON M6K 1Y9
Canada

**Hannu (H.) Viitanen**
VTT Technical Research Centre of Finland
P.O. Box 1000
02044 VTT Espoo
Finland

**Thomas (Th.) Warscheid**
LBW-Bioconsult
Schwarzer Weg 27
26215 Wiefelstede
Germany

# Index

## A

*Absidia corymbifera* 108
*Acarus siro* 217
*Acer sacharum* 447
acoustical method 324
acoustic properties 319
*Acremonium* 103, 107, 355, 365, 418
  – *murorum* 108
  – *strictum* 108
actinomycetes 425
activation 73, 78, 79, 80, 81
  – limited growth 73, 74, 76, 81
activity 37
adsorption 56
aerial hyphae 87, 91, 92
aeroconioscope 353
aerodyamics 357
aerodynamic equivalent diameter 358
aeromycota 221
aerosolized curdlan 199
aflatoxin 248
  – B$_1$ 109, 257, 258
AFLP – *See*: amplified fragment length polymorphism
*Agrocybe praecox* 121
air
  – conditioning system 287
  – humid 394
  – quality 277, 296
  – sampling 493
  – tightness 29
airborne 383
  – chlamydospores 212
  – fungal conidia 211
  – fungal particles 355
  – fungal spores 426
  – hyphae 222
  – hyphal fragments 212
  – microbes 353
  – spores 423
  – water vapor transport 54
airway inflammation 231
algae 284, 464, 470
algal growth 466
algicides 470
alkoxysilane 474, 475
alkylammonium compounds 441
allergen 184, 195, 288
  – indoor 183
  – perennial 183
allergic
  – alveolitis 291
  – fungal sinusitis 291
  – reactions 20
  – sensitization 231
allergy 291, 492
*Alternaria* 48, 103, 107, 195, 211, 218, 224, 231, 234, 246, 261, 371, 419, 463, 492
  – *alternata* 93, 217, 223, 402, 403, 471
  – *dianthicola* 463
  – *tenius* 463
  – *tenuissima* 108, 261
alternariol 250, 261
altertoxin 261
ammonical copper arsenate 441
amplified fragment length polymorphism (AFLP) 166
  – analysis 166
*Amylocorticiellum*
  – *cremeoisabellinum* 121
  – *subillaqueatum* 121
*Amyloporia* 141
*Amylostereum areolatum* 121
*Amyloxenasma allantosporum* 121
*Anacystis*
  – *montana* 464
  – *thermale* 464

anastomosis 87
anatomical structure 444, 445
annellides 103
*Anobium punctatum* 451
antigens 195, 370
antimicrobial
– agents 407
– coatings 476
*Antrodia* 119, 122, 139, 157, 164
– *crassa* 121
– *gossypium* 121
– *malicola* 121
– *serialis* 119, 129, 140, 142, 145, 146, 163
– *sinuosa* 117, 119, 129, 140, 141, 143, 144, 163, 167
– *sordida* 121
– *vaillantii* 117, 119, 125, 129, 131, 140, 141, 143, 144, 151, 160, 161, 162, 165, 166, 444
– *xantha* 117, 119, 129, 140, 143, 145, 163
*Aqualinderella* 448
*Archegozetes longisetosus* 217
arctic climates 105
*Armillaria* 124
arsenic 278
*Arthrobotrys oligosporus* 91
asbestos 278
ascomycetes 101, 102
ascomycota 101
ascospores 95, 102, 399
aspergillic acid 258
*Aspergillus* 21, 28, 43, 60, 95, 97, 102, 103, 105, 106, 107, 111, 191, 192, 193, 197, 211, 218, 224, 263, 363, 365, 366, 372, 398, 399, 409, 463
– *alliaceus* 403
– *alternata* 222, 233, 261, 370, 371
– *calidoustus* 21, 108, 110, 246, 256, 257
– *candidus* 108

– *clavatus* 108
– *flavus* 108, 111, 257, 402, 463
– *fumigatus* 47, 108, 111, 195, 199, 217, 231, 246, 258, 372, 400, 403
– *glaucus* 259
– *insuetus* 246, 256, 257
– *japonicus* 403
– *niger* 44, 88, 90, 95, 96, 97, 108, 222, 223, 233, 246, 257, 363, 400, 402, 463, 471
– *ochraceus* 110, 111, 258, 402
– *oryzae* 89, 90, 91, 95
– *parasiticus* 231
– *penicillioides* 217, 356, 392, 418
– *restrictus* 48, 108, 418
– *sydowii* 246, 400
– *terreus* 108, 111
– *ustus* 21, 110, 213, 256
– *versicolor* 44, 46, 48, 49, 108, 109, 111, 197, 213, 216, 226, 227, 228, 230, 232, 234, 246, 255, 259, 265, 288, 339, 356, 363, 385, 400, 418, 463
– *westerdijkiae* 108, 110, 111, 258
asperphenamate 250, 260
*Asterostroma*
– *cervicolor* 119, 152, 153
– *laxa* 153
– *laxum* 119, 152
asthma 15, 20, 183, 189, 193, 194, 214, 231, 286, 291, 492
asthmatic attacks 231
*Athelia fibulata* 121
atranone 248, 252
– A 200
– C 200, 201
aurasperone 257
– B 250
*Aureobasidium* 103, 463, 472
– *pullulans* 46, 93, 108, 347, 447, 457, 463, 465, 468, 471
austocystins 256

azaconazole — 441
azoles — 441

## B

*Bacillus* — 457
bacteria — 284, 422
*Ballistoconidium* — 357
basidiomycetes — 101, 102, 114
 – wood-decaying — 117
basidiomycota — 101
*Basidioradulum*
 – *crustosum* — 121
 – *radula* — 121
basidiospore — 398, 399
*Beauveria* — 355
benzene — 278, 295
benzimidazole derivative — 470
benzothiazole derivative — 470
betaine — 441
bioaerosol — 356, 362
 – particles — 358
 – sampler — 364
 – sampling — 367
biocide — 438, 440, 443, 458, 465, 469, 470
 – controlled release — 479
biocorrosion — 464
biodegradation — 466
biodeterioration — 386, 394, 396
biofilms — 464
bioinhibitor — 335, 445, 446
biological
 – agents — 284, 286
 – recovery efficiency — 368
biomass — 71, 80, 81
 – evolution — 71
Bioswitch — 458
biotic interactions — 450
*Bipolaris* — 463
*Bjerkandera adusta* — 119
blastic arthric — 103
blastic single — 103

*Blastocladia* — 448
*Blastomyces dermatides* — 444
*Blattella germanica* — 217
blue-stain — 468, 481
 – fungi — 467
*Boletus* — 119
*Botryobasidium* — 121
botryodiploidin — 260
*Botrytis* — 103, 419
 – *cinerea* — 93, 108
BPD – *See*: EC Biocidal Products Directive
brevianamide — 201, 260
 – A — 200
British Standard — 336, 337
bronchitis — 189
brown rot — 122, 437
 – fungi — 450
building
 – code — 349, 350
 – envelope — 29, 52
butadiene — 278

## C

cadmium — 278
calcium carbide
 – method — 324
 – moisture — 318
*Calotrix parietina* — 464
*Candida* — 457
 – *albicans* — 471
 – *peltata* — 108
capacitance — 324
capillary
 – action — 308
 – condensation — 56
 – pressure — 309
carbamic acid derivative — 470
carbon
 – disulfide — 278
 – monoxide — 278, 295
carcinogen — 277

carpet dust 109, 256
cavity wall principle 29
cellulose 436, 437, 438
cement-rendered brick 43
CEN – *See*: European Committee for Standardization
centrifugal impactor 360
*Ceraceomyces sublaevis* 121
*Ceratocystis pilifera* 471
*Cerinomyces pallidus* 121, 122
chaetoglobosin 254
*Chaetomium* 103, 218, 363, 418, 463
– *aureum* 108
– *globosum* 108, 213, 246, 255, 265, 336, 400, 465, 471
chilled mirror 326
*Chlorella* 464
*Chlorococcum* 464
chromated copper arsenate 441
*Chromelosporium* 355
chromium 278
chromium-free system 441
chronic rhinosinusitis 291
chrysogine 259, 260
*Chrysonillia* 103
– *sitophila* 108
cladosporin 201
*Cladosporium* 21, 22, 28, 43, 46, 47, 48, 95, 97, 98, 103, 107, 111, 195, 211, 224, 231, 246, 262, 263, 366, 372, 399, 403, 419, 463, 472
– *cladosporioides* 46, 60, 93, 94, 108, 112, 216, 217, 262, 400, 401, 403, 463, 471
– *herbarum* 108, 112, 262, 401, 402
– *macrocarpum* 108, 112
– *sphaerospermum* 108, 112, 213, 262, 342, 363, 402, 463
Clausius-Clapeyron equation 23, 307
clay brick 43
climate dynamics 58
*Clitopilus hobsonii* 121

*Clonostachys* 103, 355, 365
coating 465, 467, 468, 482
– technology 473, 476
colorimetric principle 330
commercial identification kit 114
compatible solutes 55
condensation 293, 385
– interstitial 29
confocal scanning laser microscopy 332
conidia 95, 103
conidiogenesis 102
conidiophore 97, 102, 105, 112
conidium 102
– formation 102
conifer mazegill 147
*Coniophora* 122, 135, 157, 165
– *arida* 119, 129, 136, 138, 139, 163
– *fusispora* 121, 136
– *marmorata* 117, 119, 129, 138, 139, 163, 166, 167
– *olivacea* 129, 136, 138, 139, 163
– *puteana* 117, 119, 125, 129, 131, 135, 136, 137, 138, 158, 159, 160, 161, 162, 163, 164, 166, 167, 447, 453
*Coprinus* 119, 122
*Coriolus versicolor* 450
creosote 441, 442
*Crepidotus* 121
*Crustoderma dryinum* 121
cryo-SEM 347
*Cryptococcus* 113
– *laurentii* 108
*Ctenolepisma longicaudata* 217
culturable sampler 364
*Curvularia* 103, 218, 231, 463
– *lunata* 108, 371, 403
cyclopiazonic acid 255, 258
*Cylindrobasidium laeve* 119
cyproconazole 441
cytohalasin 254
cytokine 199

cytotoxicity 249
– testing 493

**D**

*Dacrymyces* 155
– *capitatus* 121
– *punctiformis* 121
– *stillatus* 119, 122, 154
– *tortus* 121
damp 297, 299, 300, 405
dampness 277, 284, 285, 286, 287, 288, 291, 292, 293
– and mold 20
– self-reported 285
– water indicator 409
dechlorogriseofulvin 255
decontamination 427
degradation 21, 447
dehumidification 393
deoxynivalenol (DON) 245, 247, 248
*Dermatophagoides*
– *farinae* 185, 217
– *pteronyssinus* 185, 217
Deuteromycetes 101
developmental stage 60
dew point 326, 331
– principle 325
di-aldehydes 252
dichloran 18% glycerol agar 494
1,2-dichloroethane 278
dichloromethane 278
dielectric
– constant 316, 329
– permittivity 315, 316, 317, 318
diffusion
– coefficient 312
– resistance 17, 55
3,8-dihydroxy-6-methoxy-1-methylxanthone 255
dimensionless parameter 73
dioxin 294

*Diplomitoporus lindbladii* 119, 150, 151
direct microscopic methods 364
disequilibrium 27
disinfectant 407
dithiocarbamate derivative 470
DNA
– -array 167
– -based technique 158
– -chip 167
dolabellanes 252
DON – *See*: deoxynivalenol
*Donkioporia* 451
– *expansa* 117, 119, 125, 129, 149, 150, 160, 161, 162, 164, 167
*Drechslera* 371
– *hawaiiensis* 403
dry rot
– fungus, American 135
– fungus, small 134
– fungus, soft 132
– fungus, true 123
– fungus, yellow-margin 134
– mine fungus 134
durability 325
dust 363, 397
– analysis 364
– settled 363, 364

**E**

EC Biocidal Products Directive (BPD) 439, 440
eco-friendy 335
economic cost 20
efficacy 470
effused tramete 146
electrical
– impedance principle 328
– property 315
– resistance 315
electromagnetic radiation 359

electrostatic
- disturbances 358
- effect 359
EMC – *See*: equilibrium moisture content
*Emericella* 111, 402
- *nidulans* 108
- *rugulosa* 401, 402, 403
encapsulation 479, 480
*Encephalitozoon cuniculi* 166
endotoxin 288
energy 62
- conservation 15, 17
- efficient 15
- metabolism 446
*Engyodontium album* 418
entrapment 480
enumeration of spore clumps 366
environmental
- sampling 354
- tobacco smoke 183
*Epicoccum* 103, 107, 218
- *nigrum* 108, 110
- *purpurascens* 93, 110
epidemiological studies 286
equilibrium 24, 25, 36, 37, 305
- moisture content (EMC) 454
- relative humidity 308
- water content 25
ergosterol 233, 371
*Escherichia coli* 196, 478
European Committee for Standardization
  (CEN) 337
*Eurotium* 43, 102, 103, 106, 107, 111,
           191, 218, 246, 288, 339
- *amstelodami* 48, 108, 401, 402
- *chevalieri* 108
- *echinulatum* 43
- *herbariorum* 108, 339, 392
- *repens* 48, 217, 259
- *rubrum* 108

evidence
- clinical 291
- inadequate or insufficient 290
- limited or suggestive 290
- scientific 286, 290
*Exophiala* 103
- *dermatitidis* 108
exposure 288
- assessment 194
- response 184
*Exserohilum* 371
exudate formation 92

**F**
*Fellomyces* 113
ferrirubin 261
fiberboard 386
*Fibroporia* 141
*Fibulomyces mutabilis* 119
*Filobasidiella neoformans* 165
filter membranes 362
filtration samples 362
finishing materials 63, 495
fluoride 279
fluorine based preservatives 438
*Folsomia candida* 217
*Fomitopsis*
- *pinicola* 119
- *rosea* 119
formaldehyde 278, 294, 295
frequency 68, 72, 77, 80
- dependency 76
fruiting bodies 102
fugacity 37
fumigaclavine B 250
fumitremorgins 258
fumonisins 257
functional coatings 478, 479
fungal 405
- autolysis 215
- bioaerosols 357, 370

– colony    87
– fragmentation    214, 215, 217
– fragments    214, 224, 491
– glucan    193
– growth    305, 339, 341, 464
– hyphae    212
– indoor growth    18
– micro-particles    213
– particulates    212, 231
– products    109
– propagules    357
– resistance    18, 335, 336, 337, 338, 339, 348, 349
– response    42, 70
– subcellular fragments    222
– survival    87
fungicide    443, 469, 470, 472
– content    471
Fungi Imperfecti    101
fungus    284, 338, 422, 435, 470
– anamorphic    101
– arid cellar    139
– brown cellar    137
– cellar    135
– dermatophytic    356
– marmoreus cellar    139
– olive cellar    139
– phylloplane    60, 246, 355
– wood destroying    436
*Fusarium*    48, 103, 106, 111, 112, 245, 264, 355, 409, 463
– *culmorum*    108
– *graminearum*    195
– *oxysporum*    88, 89, 91, 264
– *solani*    93, 108
– *verticillioides*    108
fusigen    261

**G**

gamma ray    321
– transmission    314

*Geomyces*    103
– *pannorum*    108
geosmin    255
*Geotrichum*    103
– *candidum*    108
germination    43
*Gigartinaceae*    457
gill polypore    146
– fir    147
– timber    147
– yellow-red    147
*Gliocladium*    355
– *roseum*    93
– *virens*    471
*Gliomastix*    103
gliotoxin    109, 249, 250, 258
*Gloeophyllum*    119, 146
– *abietinum*    119, 125, 129, 146, 147, 148, 164
– *sepiarium*    117, 119, 125, 129, 146, 147, 148, 160, 161, 164, 167
– *trabeum*    117, 119, 125, 129, 146, 147, 148, 158, 160, 164, 444
glucan    197
– β    198, 288
– β-1,3-D    197, 198, 199, 222, 231, 233, 356, 372
– β-1,6-D    197, 198
glycoamylase    90
*Glycyphagus*    217
*Grandinia*    119
gravimetric determination    314
gravimetry    324
gravity settle plates    363
*Grifola frondosa*    121, 122
griseofulvin    255
growth
– assessment    340, 341, 342
– curve    346
– linear    43
– model    68

– optimum conditions 46
– pattern modeling 342
– response 68
– stages 45, 57
– successive 46, 50
growth rate 79
– effective 72
– relative 74
guidelines 277, 280, 296, 383, 386, 395, 396, 397, 404, 406, 409, 413, 417, 419, 420, 421, 422, 424, 429
– *See also*: WHO Guidelines
– targets 292
gyroso-reticulate hymenophore 126

**H**

haemoglobin-like protein 90
hair hygrometer 327, 328, 331
Halogen Immunoassay (HIA) 224, 225, 226, 229, 230, 371
hardwood 436, 445
hay fever 183
health 299, 405, 407
– effects 405, 407
– hazards 289
– impact assessment 296
– impact management 296
– implications 425
– outcomes 290
– problems 286
– related impacts 416, 424, 425
– risks 277, 414
– variables for adults 190
– variables for children 190
heat
– capacity 319
– conductivity 324
heating 18
– adequate 298
heavy metals 294
*Helminthosporium* 231

helvolic acid 258
hemicellulosis 436, 437, 438
*Heterobasidion annosum* 121, 457
HIA – *See*: Halogen Immunoassay (HIA)
house dust
– components 284
– mites 185, 186
housing conditions 190
HPLC-MS/MS 247
human skin 355
*Humicola grisea* 93
humidity 69, 298, 305, 311, 335, 340, 341, 350
– air 17
– cycles 73
– fluctuations 68, 78
– high 293
– reduction 299
humidity conditions 308
– ambiguous 342
– critical 465
– steady-state 42
HVAC system 391, 392
hydrogen sulfide 279
hydrophiles 50, 52
hydrophobic 454, 479
hydrophobins 90
hydroxyemodin 259
hydroxy-roridin 249
hygrothermal modeling 31
*Hylotrupus baljulus* 451
hypersensitivity 20
– pneumonitis 231, 291
hyphal
– fragments 212, 221
– growth 84
*Hyphoderma* 119
– *praetermissum* 121
– *puberum* 119
*Hyphodontia* 119
– *alutaria* 121

– *arguta* 121
– *aspera* 121
– *breviseta* 121
– *floccosa* 121
– *juniperi* 121
– *microspora* 119
– *nespori* 121
– *pruniacea* 121
– *radula* 121
– *spathulata* 117
Hyphomycetes 101
*Hypochniciellum molle* 119
*Hypochnicium geogenium* 121
*Hypocrea* 263
*Hypoxylon* 121
hysteresis 26, 309, 310

**I**

identification 110
– of indoor wood-decay fungi 156
idiopathic pulmonary hemosiderosis 249, 252
immunological methods 158
impedance based
– capacitive 331
– resistive 331
imperfect stage 102
impregnate 442
indole-3-acetic acid 261
indoor 413
– climate dynamics 17, 63, 491
– combustion of fuels 284
– environments 219
– environments, highly transient 311
– polypores 139
indoor air 105
– healthy 281
– pollutants 295
– quality 284
– spora 106
inertia effects 41

inertial samplers 360
inflammatory mediators 201
infrared radiation 319
ink bottle effect 310
inoculation 339
inorganic borates 441
insulation 16, 389, 392
– levels 62
intercalation of biocides in nano particles 458
interior 335, 337, 339, 349, 384
– surfaces 292
interleukin-8 199
internal dosemetry 492
International Organization for Standardisation (ISO) 337
intervention 291, 297, 298, 426
– recommendations 299
– successful 300
intra-cellular 55
ISO – *See*: International Organization for Standardisation
isopleth 48, 49, 50, 51, 67
isothermal
– conditions 313
– moisture diffusivity 312
isothiazolines 470

**J**

jet-to-plate distance 367

**K**

kebony 454
kojic acid 258
kotanin 257

**L**

*Laccaria bicolor* 165
lacquers 338
lactones 263
*Laetiporus sulphureus* 121

laminating materials 339

lead 279

*Lecanicillium* 213

  – *lecanii* 110

*Leccinum* 121

*Lentinus*

  – *lepideus* 119, 158

  – *tigrinis* 471

*Leucogyrophana* 119, 128, 132, 157, 165

  – *mollusca* 119, 123, 129, 132, 133, 163

  – *pinastri* 119, 123, 125, 129, 133, 134, 163, 166, 444

  – *pulverulenta* 123, 133, 134

  – *romellii* 121

liability 20

lignin 436, 475

  – content 455

*Limnoria* 450

*Limulus polyphemus* 198

liposcelis 217

liquid 305

  – impingement 358

  – impingers 360, 361

  – moisture content 312

  – phase 27

  – water 28, 468

  – water penetration 29

logistic

  – growth curve 344

  – model 343, 348

long term

  – climate variations 81

  – variations 41

LOSP vacuum treatments 441

*Lycoperdon pyriforme* 119

*Lyctus* 451

**M**

macrocyclic trichothecenes 109, 233, 248, 249, 250, 252, 253

macromolecule synthesis 446

*Malassezia* 356

MALDI-TOF mass spectrometry 167, 168

malformin A, B and C 257

2% malt extract agar 494

management of health effects 408

manganese 279

measuring 305, 311

  – non-destructive 314

  – techniques 305

mechanical

  – collection efficiency 361

  – ventilation 289

media for detection and isolation 494

mediator release 198

*Melanogaster broomeanus* 121

meleagrin 259, 260

membrane samples 362

*Memnoniella* 103, 110, 372

  – *echinata* 110, 113, 195

mercury 279

*Meruliporia* 122

  – *incrassata* 123, 128, 131, 135, 163, 165, 166, 447

*Merulius* 126, 128

  – *tremellosus* 121

mesophiles 47

metabolite 90

  – production 247

metal type formulations 442

microarrays 167

microbial

  – activity 463

  – burden 421

  – growth 292, 383, 405

  – volatile organic compound (MVOC) 109, 231, 288, 294, 370, 419, 425, 494

micro-climate 54, 67, 69, 331

micro-electrode 332

micro-nuclear magnetic resonance 333

micro-organisms 288, 443, 456, 457

microwave 317

minimum inhibitory concentration    471
moisture  286, 287, 288, 298, 300, 331, 358,
              396, 404, 408, 428, 452, 468
 – availability                           289
 – control                                427
 – diffusivity                            311
 – excess                            287, 293
 – problem                           285, 287
 – retention                          54, 496
 – sorption                               474
 – transport                         287, 311
moisture content    305, 306, 311, 314, 315,
                        321, 324, 448
 – by mass                                306
 – initial                                414
 – measurement                            305
 – measuring                              313
mold        277, 284, 286, 287, 291, 297, 299,
           300, 384, 388, 392, 399, 400, 404,
                             406, 408
 – claims                                  20
 – concentration in the air               494
 – dogs                                   419
 – -free environment                       17
 – in a wine cellar                       186
 – occurrence rates                       286
 – phylloplane                            406
 – -produced arsine gas                   184
 – remediation     383, 397, 399, 400, 402,
            403, 404, 405, 407, 409, 413
 – remediation strategies                 386
 – resistant                               63
 – sensitization                          193
 – storage                                246
 – toxic black                             22
 – water damaged                          246
 – xerophilic          42, 48, 49, 53, 356
mold growth   298, 385, 389, 390, 391, 415,
                        418, 472, 491
 – resistance of coatings                 338
 – susceptibility                         414

mold infestation
 – causes                                 414
 – sanitation                             416
*Mucor*                             106, 463
 – *circinnolides*                        108
 – *plumbeus*                             108
 – *racemosus*                            108
MVOC – *See*: microbial volatile organic
   compound
mycobiota                            101, 105
 – indoor                                 491
 – in food                                107
mycophenolic acid      200, 201, 250, 260
mycotoxin           231, 245, 246, 288, 423,
                             424, 425
 – antibodies in serum                    248
 – production                             265
 – testing                                493
*Myrothecium*                             355

**N**

nano-containers                      479, 480
nanoparticles                             478
nanotechnology                            473
naphtalene                                295
naphtha-g-pyrones                         257
naturally infested materials              247
*Neocallimastix*                          448
neoechinulin A and B                      201
*Neosartorya*                             111
neutron                                   314
 – method                                 324
 – scattering                             321
nickel                                    279
nigragillin                               257
*Nigrospora*                              463
nitrogen dioxide            279, 280, 295
3-nitropropionic acid                     258
NMR – *See*: Nuclear Magnetic Resonance
*Nocardiopsis*                            425
Nomarski differential interference        366

nomenclature 110
non arsenic system 441
non-gonomorphic particles 212, 213, 214,
   215, 216, 218, 219, 222, 224, 225, 227,
      229, 230, 231, 232, 233, 234
non-viable spores 107
Nuclear Magnetic Resonance (NMR) 314,
   322, 323, 324, 331
   – imaging 323
   – mouse 323
nutrients 90
nutrition 447

**O**

objective bronchial reactivity 188
occupational environments 219
ochratoxin A 255, 257, 258, 259
*Oidiodendron* 103
   – *griseum* 108
   – *rhodogenum* 108
*Oligoporus* 119
   – *placenta* 119, 122, 129, 130, 139, 140,
      141, 145, 146, 158, 160, 161, 164
   – *rennyi* 119
*Oniscus asellus* 217
operational taxonomic units 107
ophiobolins 256
*Ophiostoma*
   – *picea* 457
   – *piliferum* 457
   – *pluriannulatum* 457
orange jelly 154
organic
   – additives 44
   – metal compounds 470
*Oscillatoria lutella* 464
osmolarity 55
outdoor air 105
   – pollens 284
outdoor environment 219
oxygen 447

oyster rollrim 152
ozone 280
   – and photochemical oxidants 279

**P**

*Paecilomyces* 103, 107, 260, 365, 463
   – *lilacinus* 108, 260
   – *variotii* 108, 246, 260, 363, 463
PAH 295
paint 44, 338, 339, 467
   – acrylate 467
   – alkyd emulsion 467
   – on wood 43, 468
   – properties 466
partial β-tubulin sequences 113
particle 423
   – counts 418
particulate matter 279
passive 70, 73
   – air sampling 363
   – biomass 70, 72
*Paxillus panuoides* 152
PCB – *See*: polychlorinated biphenyl
PCDD – *See*: polychlorinated
   dibenzodioxin
PCDF – *See*: polychlorinated dibenzofuran
PCR – *See*: polymerase chain reaction
penicillic acid 258
*Penicillium* 21, 28, 43, 46, 47, 60, 84, 88,
   95, 97, 102, 103, 106, 107, 110, 112, 113,
      191, 192, 193, 197, 211, 218, 224, 231,
         263, 355, 363, 365, 366, 372, 398, 399,
            409, 450, 463
   – *bialowiezense* 260
   – *brevicompactum* 93, 108, 112, 200,
      246, 260, 288, 401, 402, 403, 463
   – *canescens* 93

– *chrysogenum*        44, 45, 56, 57, 58,
    59, 60, 61, 75, 76, 77, 85, 92, 93, 108,
    110, 112, 195, 196, 200, 213, 217, 229,
    246, 252, 255, 259, 260, 262, 342, 346,
                347, 398, 400, 402, 448
– *citreonigrum*                    108, 403
– *citrinum*                        108, 402
– *commune*                    108, 109, 246
– *corylophilum*                    108, 246
– *crustosum*                            108
– *decumbens*                            108
– *expansum*                         44, 108
– *frequentans*                          110
– *funiculosum*                     108, 471
– *glabrum*                    108, 110, 112
– *glaucum*                         110, 471
– *herquei*                               93
– *implicatum*                           403
– *martensii*                             49
– *melinii*                         216, 232
– *nigricans*                             93
– *notatum*                              110
– *olsonii*                         108, 112
– *oxalicum*                             259
– *palitans*                             108
– *purpurogenum*                         402
– *rugulosum*                            108
– *sclerotiorum*                         403
– *simplicissimum*                       108
– *spinulosum*                           108
– subgenera *Aspergilloides*             355
– subgenera *Biverticillium*             355
– subgenera *Furcatum*                   355
– subgenus *Penicillium*            113, 363
– *variabile*                            108
– *verrucosum*                       46, 110
*Peniophora*
– *gigantea*                             457
– *pithya*                               121
pentachlorphenol                         441
peptiabiotics                            263

*Perenniporia*
– *medulla-panis*                        121
– *tenuis*                               119
perfluorochemicals                       294
performance tests                        482
pergillin                                256
permeability                             468
*Phanerochaete*                          119
*Phellinus*
– *contiguus*                  119, 155, 156
– *pini*                                 121
*Phialides*                              103
*Phialophora*                       103, 418
– *fastigiata*                           108
– *verrucosa*                            108
*Phlebiopsis gigantea*                   119
*Phoma*       103, 106, 246, 355, 365, 463
– *glomerata*                            108
– *herbarum*                             347
– *macrostoma*                           108
photocatalytic TiO2                      478
phthalates                               294
phylloplane species                      191
*Picea abies*                            442
pigment volume concentration              44
*Pinus sylvestris*                       453
plasters                                  44
plastics                                 338
platinium                                279
*Pleurotus*
– *cornucopiae*                          121
– *dryinus*                              119
– *ostreatus*                            119
– *pulmonarius*                          121
plumbing leaks                           390
*Pluteus cervinus*                  121, 122
PM2.5                                    279
PM10                                     279
policy                                   296
– making                            296, 498
– recommendations                        417

– targets 296
pollutant 277, 284
polychlorinated biphenyl (PCB) 278
polychlorinated dibenzodioxin (PCDD) 278
polychlorinated dibenzofuran (PCDF) 278
polymerase chain reaction (PCR) 493
polynuclear aromatic hydrocarbons 278
*Polyporaceae* 366
polypore
– broad-spored white 143
– mine 143
– oak 149
– reddish sap 146
– row 146
– small-spored white 143
– white 143
– yellow 143
*Polyporus* 141
– *vaporarius* 141
polysaccharides 56
polyvinylacetate 44
pore radius 449
*Poria* 141
– *monticola* 141, 444
– *placenta* 141
– *vaporaria* 122, 141
*Poroconidia* 103
porous material 306
*Postia placenta* 141
preconditioning of materials 341
preservative agents 441
primary colonizers 46, 246, 288
protein-based techniques 157
proteins 445
protozoa 284
PR toxin 259
*Pseudomonas* 464
psychrometers 326
psychrophiles 47
*Ptychogaster rubescens* 122

*Pycnoporellus fulgens* 121
*Pycnoporus cinnabarinus* 121
pyridine-N-oxide derivatives 470
*Pyronema domesticum* 108
α-pyrones 263
2-pyrovoylaminobenzamide 260
pyrovoylaminobenzamides 259

**Q**
quantitative PCR 373

**R**
radon 279, 295
*Radulomyces confluens* 119, 122
Raistrick phenols 260
RAPD analysis 159
RCS Standard and High-flow devices 360
rDNA
– RFLP analysis 160
– sequences for identification 162
– sequences for phylogenetics 165
– sequencing 162
– use 160
relative humidity 22, 28, 36, 38, 47, 67, 75, 306, 308, 312, 325, 330, 342, 391, 464
– ambient 491
– capacitive sensor 328
– resistive sensors 328
– threshold value 54
remediation 396, 405, 406, 408
reservoir 54, 63
*Resinicium bicolor* 119
resistance 324, 481
– moisture meter 316
respiratory
– infections 20
– symptoms 189
rhinitis 20, 492
– seasonal 231
*Rhizoctonia solani* 85, 87
*Rhizomucor pusillus* 47

*Rhizopus*   106, 463
- *stolonifer*   108
*Rhodotorula*   113
- *glutinis*   347
- *mucilaginosa*   108
- *rubra*   471
risk
- assessment   415
- factors for respiratory symptoms   187
- management   297
roquefortine   259, 260
- C   200
roridin   249, 250

## S

*Saccharomyces cerevisiae*   471
sanitation   417, 426, 428
satratoxin   250, 253
- F   249
- G   109, 233, 249
- G-albumin   253, 264
- H   109, 233, 249
saturated salt solutions   308
scanning   321
- electron microscopy   332
*Schizophyllum*   102
- *commune*   89, 91, 108, 121
*Schizopora paradoxa*   119
*Schizosaccharomyces*   113
*Schizotrix*   464
*Scopulariopsis*   103, 113
- *brevicaulis*   108, 113, 418
- *candida*   108
- *fusca*   108, 418
*Scytonema hofmanii*   464
SDS-PAGE – *See*: sodium dodecyl sulfate polyacrylamide gel electrophoresis
secalonic acid   259
secondary
- colonizers   46, 246
- metabolites   247

second priority   295
- compounds   294
secretion   89
- extra-cellular   56
*Serpula*   122
- *himantioides*   119, 123, 125, 127, 128, 129, 132, 157, 159, 160, 161, 162, 163, 165, 166, 167
- *lacrymans*   55, 91, 93, 94, 108, 114, 117, 119, 122, 123, 124, 125, 126, 127, 128, 129, 130, 131, 134, 135, 136, 141, 157, 158, 159, 160, 161, 162, 163, 164, 165, 166, 167, 168, 449
- *lacrymans* var. *lacrymans*   126, 166
- *lacrymans* var. *shastensis*   126, 166
serum albumin   248
settle plates   363
simple sequence repeats (SSR) analysis   166
*Sistotrema*   102, 114
- *brinkmannii*   43, 108, 121
slimy heads   105
sodium dodecyl sulfate polyacrylamide gel electrophoresis (SDS-PAGE)   157
soft rot   437
- fungi   438
softwood   436, 444
sol-gel   476
- coatings   474
- deposition   475
- technology   473
sorption
- curve   309
- isotherm   26, 308, 310
spatial resolution   313
species-specific priming PCR (SSPP)   161
spirocyclic drimanes   248, 250, 252
sporangiospores   95
spore trap   365
- samplers   364

*Sporobolomyces* 102, 113, 402
– *roseus* 108, 471
*Sporodesmium ehrenbergii* 217
SSPP – *See*: species-specific priming PCR (SSPP)
SSR analysis – *See*: simple sequence repeats analysis
stachybotryamide 252, 253
stachybotrylactams 252, 253
*Stachybotrys* 22, 48, 103, 105, 107, 109, 110, 113, 218, 248, 249, 252, 262, 363, 395, 396, 398, 399, 409, 493
– *atra* – *See*: *Stachybotrys chartarum*
– *chartarum* 46, 108, 113, 184, 195, 227, 232, 246, 249, 250, 253, 254, 265, 372, 395, 398, 400, 418, 424
– *chartarum sensu latto* 200
– *chlorohalonata* 113, 195, 196, 249, 250
– (*Memnoniella*) *echinata* 108, 110, 195, 246, 253
stachylysin 252
stalkless paxillus 152
steady-state conditions 67, 69, 70
Stellac 454
*Stemphylium* 371
*Stereum rugosum* 121
sterigmatocystin 109, 201, 248, 254, 255, 256
– -guanine 264
*Sterigmatomyces* 113
*Streptomyces* 425, 457
styrene 278
suction curve 308
sulfenic acid derivatives 470
sulfone derivatives 470
sulphur 280
– dioxide 279
surface
– condensation 31, 32
– conditions 28, 311
– growth 21

– relative humidity 51
– sampling 493
– temperature 50, 51, 54, 62
– tension 309
survival 92
susceptible 396
– materials 496
sustainability 42
*Sydowia polyspora* 468
*Syncephalastrum racemosum* 108
syncytium 84
synthetic polymeric materials 338

**T**
T-2 tetraol 247
T-2 toxin 245, 247
*Talaromyces macrosporus* 86
tandem mass spectrometry 247
*Tapinella* 122
– *panuoides* 119, 152, 153
Taqman-based qPCR methodology 372
TDR – *See*: Time-Domain Reflectometry
tebuconazole 441
temperature 67, 300, 308, 325, 449
– dry-bulb 327
– incubation 340
– ratio 51, 52
– ratio criterion 50, 53
– steady-state 47
– wet-bulb 326, 327, 331
tentoxin 261
tenuazonic acids 261
*Teredo navalis* 450
tertiary colonizers 246
tetrachloroethylene 278, 295
Thallic 103
thermal
– bridge 31, 51, 293
– conductivity 319, 330, 331
– insulation 17, 18, 51, 52, 62, 300

– performance of the building envelope 495
– properties 319
thermophiles 47
Thermowood 454
*Thiophtalimide derivatives* 470
*Thuja plicata* 449
thujaplicine 450, 456
Time-Domain Reflectometry (TDR) 317, 324
time-of-wetness (TOW) 42, 56, 58, 59, 63, 67, 74, 75, 76, 77, 80, 83, 451
time to decay 450, 451
TMC-120A 201, 250, 256
tobacco smoking 279
toluene 278
*Tomentella* 119
– *crinalis* 121
total suspended particles (TSP) 279
*Trametes* 119
– *hirsuta* 119
– *ochracea* 121
– *pubescens* 121
– *versicolor* 121
transient conditions 27
transport 312
– of water 94, 332
*Trechispora* 119, 154
– *farinacea* 119, 155
– *invisitata* 121
– *microspora* 121
– *mollusca* 119
*Trentepohlia odorata* 464
triazine derivatives 470
*Trichaptum abietinum* 121
trichloroethylene 278, 295
*Trichoderma* 103, 111, 246, 263, 355, 372, 418, 437, 457, 463
– *arundinaceum* 263
– *atroviride* 263
– *brevicompactum* 263

– *citrinoviride* 263
– *harmatum* 263
– *harzianum* 93, 108, 213, 215, 216, 263, 264
– *longibrachiatum* 108, 263
– *protrudens* 263
– *turrialbense* 263
– *viride* 46, 108, 263, 471
trichodermin 252, 263
trichodermol 263
trichothecenes 249, 264, 493
trichoverrins 253
*Tritirachium* 103
– *oryzae* 108
TSP – *See*: total suspended particles
*Turdus merula* 128
turgor 55
TVOC concentrations 192
*Tyromyces placenta* 141
*Tyrophagus putrescentiae* 217

**U**
*Ulocladium* 103, 213, 218, 246, 261, 262, 371, 372, 463
– *alternariae* 108
– *atrum* 108, 262, 471
– *chartarum* 108, 262
urea derivatives 470
US Environmental Protection Agency 439

**V**
Vaillantii group 141
vanadium 279
vapor 27, 305
– phase 23
variotin 261
varnishes 338, 339
ventilation 16, 17, 18, 54, 62, 287, 293, 298, 300, 350, 452, 495
– systems 299
verrucarin 249, 253

verruculogen 258
versicolorins 255
*Verticillium* 103
– *biguttatum* 87
– *lateritium* 93
– *lecanii* 110, 213
Vesicle Supply Center 86
vinyl chloride 278
viomellein 259
vioxanthin 259
viriditoxin 260
VOCs – *See*: volatile organic compounds
volatile organic compounds (VOCs) 109, 294
volumetric air sampling 353, 360
*Volvariella bombycina* 121

**W**
wallboard 46
*Wallemia* 102, 103, 107, 218, 246, 356
– *sebi* 108, 191, 288, 401, 402
wallpaper 46
water 298, 305, 307, 308
– availability 42, 449
– contact angles 477
– content 25, 27
– damage 287, 390, 394
– disasters 393
– in materials 24
– leaks 388, 390
– potential 449
– relations 41
– retention 58
– temperature interactions 47
– tightness 29
– tolerance 42
water activity 24, 25, 28, 36, 307, 394, 449
– lower limit 43
– minimum 42
– optimum 42
waterlogging 393

water-repellent surface treatments 452
water vapor 29, 30, 307
– partial pressure 306
– pressure 55
– production 62
wet/dry cycles 71
wet-rot fungi 135
white rot 437
– fungi 450
WHO Guidelines 296
– *See also*: guidelines
– for Air Quality 278, 280, 281
– for Indoor Air Quality 283
wild merulius 132
wood 466, 467
– coated 27, 481
– coatings 464
– components 436
– degradation 436
– efficacy of preservatives 481
– modification 453
– preservatives 438, 442
– protection 435, 438, 446
– rot 435
wood materials
– durability 481
– resistance 481
Woronin bodies 84

**X**
xanthocillin 259
– X 251
xanthomegnin 258
*Xerocomus* 119
xerophiles 50, 52
*Xestobium* 451
X-ray
– absorption 321
– method 324

## Y
yeast                        113, 402, 419

## Z

zygomycetes                  101
*Zygomycete sporangia*       106
Zygomycota                   101